Advances in Sustainability Scie and Technology

CH00706793

Series Editors

Robert J. Howlett, Bournemouth Univ. & KES International,
Shoreham-by-sea, UK

John Littlewood, School of Art & Design, Cardiff Metropolitan University,
Cardiff, UK

Lakhmi C. Jain, University of Technology Sydney, Broadway, NSW, Australia

The book series aims at bringing together valuable and novel scientific contributions that address the critical issues of renewable energy, sustainable building, sustainable manufacturing, and other sustainability science and technology topics that have an impact in this diverse and fast-changing research community in academia and industry.

The areas to be covered are

- Climate change and mitigation, atmospheric carbon reduction, global warming
- Sustainability science, sustainability technologies
- Sustainable building technologies
- Intelligent buildings
- Sustainable energy generation
- Combined heat and power and district heating systems
- Control and optimization of renewable energy systems
- Smart grids and micro grids, local energy markets
- Smart cities, smart buildings, smart districts, smart countryside
- Energy and environmental assessment in buildings and cities
- Sustainable design, innovation and services
- Sustainable manufacturing processes and technology
- Sustainable manufacturing systems and enterprises
- Decision support for sustainability
- Micro/nanomachining, microelectromechanical machines (MEMS)
- Sustainable transport, smart vehicles and smart roads
- Information technology and artificial intelligence applied to sustainability
- Big data and data analytics applied to sustainability
- Sustainable food production, sustainable horticulture and agriculture
- Sustainability of air, water and other natural resources
- Sustainability policy, shaping the future, the triple bottom line, the circular economy

High quality content is an essential feature for all book proposals accepted for the series. It is expected that editors of all accepted volumes will ensure that contributions are subjected to an appropriate level of reviewing process and adhere to KES quality principles.

The series will include monographs, edited volumes, and selected proceedings.

More information about this series at http://www.springer.com/series/16477

Arnab Hazra · Rupam Goswami
Editors

Carbon Nanomaterial Electronics: Devices and Applications

 Springer

Editors
Arnab Hazra 🄳
Department of Electrical
and Electronics Engineering
Birla Institute of Technology
and Science
Pilani, Rajasthan, India

Rupam Goswami
Department of Electronics
and Communication Engineering
School of Engineering
Tezpur University
Napaam, Assam, India

ISSN 2662-6829 ISSN 2662-6837 (electronic)
Advances in Sustainability Science and Technology
ISBN 978-981-16-1054-7 ISBN 978-981-16-1052-3 (eBook)
https://doi.org/10.1007/978-981-16-1052-3

This Springer imprint is published by the registered company Springer Nature Singapore Pte Ltd.
The registered company address is: 152 Beach Road, #21-01/04 Gateway East, Singapore 189721,
Singapore

Foreword

The pleasure and comfort of living in a technologically advanced world is owed to the research behind it. In an era of sustainable technology, carbon nanomaterials hold a key position under the umbrella of electronics and applications. Carbon nanomaterials with unique properties and in various forms offer some of the best solutions for applications in electronic devices. Right from sensors to possible alternatives to silicon, carbon nanomaterials have come a long way into modernity, getting matured as a research form as it moved through every step of the chronological clock.

I believe that an area of research forms a trinity with two other footprints—knowledge network and commercialization. The interdependence among the troika has the perfect potential to steer the research zone into efficient implementation and extension into other allied areas. Hence, the throughput of expertise into the literature is important from time to time in order to cause a progress in the area. Compilation of a monograph on *carbon nanomaterials* is to be applauded because apart from bringing researchers working on similar areas together to showcase their expertise, it can serve as a motivational reference as well for young scholars.

This monograph has been propelled by a Symposium on Carbon Nanomaterial Electronics held in 2019, at Birla Institute of Technology and Science, Pilani (BITS Pilani), India, and I am immensely pleased with how an event of like-minded intellectuals has given rise to a delightful outcome like this. In the contemporary era of carbon nanomaterials and their applications in electronics, I believe that this erudite volume shall go a long way in inspiring and enriching young minds in this area of study and work.

Souvik Bhattacharyya
Vice Chancellor
Birla Institute of Technology and Science
Pilani, India

Preface

Carbon nanomaterials have aligned themselves as one of the most demanding areas of interest in nanoscience and nanotechnology since the beginning of the twenty-first century. In the past twenty years, there have been tectonic shifts in synthesis, characterization, and applications of this class of nanomaterials. Excellent properties coupled with their ability to get modified or conjugated have enriched their scope of applications ranging from sensors, waste management, optoelectronics, drug science, energy, and transistors. Graphene, fullerenes, carbon nanotubes, and other various nanostructures of carbon have been popularly explored to commercialize them in diverse domains in electronics and increase the benchmarks of their utility in this era of sustainable development and intelligent materials.

This book aims to bring together a thematic collection of technical treatise on carbon nanomaterial electronics, with emphasis on synthesis, characterization, and applications. The chapters of the book have, therefore, been divided into three major thematic areas, i.e., (i) synthesis and characterizations of carbon nanomaterials and nanocomposites, (ii) theoretical and computational study of carbon nanomaterials, and (iii) devices and applications of carbon nanomaterials with the aim of providing the overview and advancement in the domain of carbon nanomaterial electronics. Categorically divided into three sections, a total of sixteen chapters is expected to orient the reader towards getting an overall idea of the scope and contemporary ideas on the subject. On the other hand, from the perspectives of a thematic community of researchers, this compilation attempts to offer the developments to carry the motivation forward to contribute on a larger scale.

The motivation behind the idea of a thematic concept of the book, however, stems from the successful outcome resulting out of a gathering of researchers and scholars at the Symposium on Carbon Nanomaterial Electronics which was held during November 8–9, 2019, at Birla Institute of Technology and Science Pilani, Rajasthan, India. The event, one of the first of its kind in India on carbon nanomaterials, saw the exchange of ideas and innovation among the participants belonging to similar areas of interest. The event was hosted under the Scheme for Promotion of Academic and Research Collaboration (SPARC), an initiative of Ministry of Higher Education, Government of India, in collaboration with Tel Aviv University,

Israel. The success of the event showed the need for initiatives to build a knowledge network on carbon nanomaterials, which can serve to strengthen the development of the 'carbon' community.

Pilani, India   Arnab Hazra
Napaam, India   Rupam Goswami

Contents

About the Editors

Arnab Hazra received his M.Tech. and Ph.D. from Indian Institute of Engineering Science and Technology (IIEST), Shibpur, India, in 2011 and 2015, respectively. Presently, he is an Assistant Professor in the Department of Electrical and Electronics Engineering, Birla Institute of Technology and Science (BITS), Pilani, Rajasthan, India. He worked as a visiting scientist at Tel Aviv University, Israel and Sensor Laboratory, University of Brescia, Italy in 2018. His current research interest includes nanoscale materials based devices for gas sensing and memristive applications and 2D materials based field effect devices. He has published about seventy research articles in the journals and conferences of international repute, two book chapters and received two Indian patents. He received best Ph.D. thesis award by Indian National Academy of Engineering (INAE) and ISSS, India, in 2015 and 2016, respectively. He also received IEI young engineer award by the Institute of Engineers, India, in 2016.

Rupam Goswami obtained his M.Tech. in 2014, and Ph.D. in 2018 from National Institute of Technology Silchar. India. Currently he is an Assistant Professor at the Department of Electronics and Communication Engineering, School of Engineering, Tezpur University, India. Before joining Tezpur University, he worked as an Assistant Professor at Birla Institute of Technology and Science Pilani, Rajasthan, India. His research interests include simulation and modeling of TFETs, HEMTs, FinFETs, and memristors. He has published his work in 2 books, 16 international peer-reviewed journals, and 13 international peer-reviewed conferences.

Introduction

Arnab Hazra⊙, Yossi Rosenwaks, and Rupam Goswami

Carbon is one of the most exciting elements with the ability to show a wide range of chemical and physical properties. Depending on the structural geometry of the atoms, carbon has been traditionally divided into three popular categories, i.e., amorphous carbon, "hard" diamond, and "soft" graphite. These carbon allotropes have been used in the production of a wide range of consumer goods in numerous spheres of human activities [1–3].

Carbon has the electron configuration of $1s^2 2s^2 2p^2$, where four outer shell valence electrons may hybridize into sp, sp^2, and sp^3, corresponding to a carbon atom bound to 2, 3, and 4 neighboring atoms, respectively. Diamond is configured entirely with sp^3 bonded carbon exhibiting ultra-hardness, wide bandgap, thermal conductivity, and chemical inertness. On the other hand, graphite is configured entirely with sp^2 bonded carbon exhibiting softness, electrical conductivity, thermal conductivity, and high chemical reactivity [4, 5].

The number of carbon allotropes has extensively increased during the last two decades of the twentieth century after the successful invention of new low-dimensional carbon forms. The era of nanoscale carbon began with the discovery of zero-dimensional fullerenes in 1985, one-dimensional carbon nanotubes (CNTs)

A. Hazra (✉)
Department of Electrical and Electronics Engineering, Birla Institute of Technology and Science (BITS), Pilani 333031, Rajasthan, India
e-mail: arnab.hazra@pilani.bits-pilani.ac.in

Y. Rosenwaks
Faculty of Engineering, School of Electrical Engineering, Tel-Aviv University, 69978 Tel-Aviv, Israel
e-mail: yossir@tauex.tau.ac.il

R. Goswami
Department of Electronics and Communication Engineering, School of Engineering, Tezpur University, Napaam 784028, Assam, India
e-mail: rupam21@tezu.ernet.in

© The Author(s), under exclusive license to Springer Nature Singapore Pte Ltd. 2021
A. Hazra and R. Goswami (eds.), *Carbon Nanomaterial Electronics: Devices and Applications*, Advances in Sustainability Science and Technology,
https://doi.org/10.1007/978-981-16-1052-3_1

1

in 1991, and two-dimensional graphene in 2004 [1, 2]. Nanocarbons are mostly composed of sp^2 bonded graphitic carbon where fullerenes exhibit some sp^3 character. Rolling up a single-layer graphene sheet into the lowest possible tube forms a single-walled carbon nanotube, and bending it up into the lowest possible sphere makes a fullerene. However, the fullerene and CNT are not actually synthesized from graphene. All these nanocarbons have no bulk equivalents, whereas graphite is made up of large stacks of graphene [3–6].

All the carbon allotropes with reduced-dimensionality are engineered or synthesized artificially. Therefore, the availability of these carbon nanoforms does not depend on natural reserves (like diamonds). Despite this, large scale adoption of these nanocarbons has been hindered by expensive techniques of synthesis, which is principally dominated by a few fundamental issues like the scale of the production, infrastructure, purity of the materials, etc. [1, 2, 7].

Zero-dimensional fullerenes are often called a molecular form of carbon that shows various forms and sizes by clustering 30–3000 carbon atoms on a spherical surface. Buckminsterfullerene or C$_{60}$ is the most commonly investigated fullerene, which is highly symmetric and contains 60 carbon atoms, situated at the vertices forming twelve pentagons and twenty hexagons [5, 8]. The diameter of C$_{60}$ is 0.7 nm. Except for C$_{60}$, other fullerene equivalents such as C$_{70}$ and C$_{80}$ have been formed and investigated widely. Fullerenes have been synthesized by using a variety of techniques, which include arc discharge, sputtering, and electron beam ablation. In most of these processes, graphite target is used as the source of carbon [2, 9].

One-dimensional carbon nanotubes (CNTs), due to their exceptional properties, are widely accepted for a variety of technical applications. The properties of carbon nanotubes may vary by their diameter, length, chirality, and the number of layers. Structurally, CNTs may be classified as single-walled nanotubes (SWCNTs) and multi-walled nanotubes (MWCNTs). SWCNTs show a diameter of around 1–3 nm and a length of a few micrometers, whereas MWCNTs show a diameter of 5–40 nm and a length of around 10 μm. CNTs exhibit extraordinary mechanical and electrical properties in the form of rigidity, elasticity, strength, and extremely high thermal and electrical conductivity. CNTs also offer a very high aspect ratio (i.e., length to diameter ratio) within the range of 10^2–10^7. SWCNTs are classified as armchair CNTs (electrical conductivity higher than copper), zig-zag CNTs (semiconducting in nature), and chiral CNTs (also having semiconducting properties). Three basic methods are used for the production of carbon nanotubes, i.e., arc discharge (i.e., vaporizing graphitic electrodes), laser ablation, and chemical vapor (CVD). In recent days, CVD has emerged as the most commonly used technique for CNT synthesis as it requires simpler process steps as compared to the other two methods [1–4, 10–12].

In graphene, carbon atoms show sp^2-hybridization along with σ- and π-bonds in a two-dimensional hexagonal lattice having a distance of 0.142 nm between two neighboring atoms. Though graphene was investigated theoretically long back, the material was obtained only recently. So, comprehensive studies on graphene are still continuing. Graphene exhibited several unique physical properties like high mechanical rigidity, high thermal stability, extremely high electrical mobility as well as conductivity. Also, graphene is highly prevalent in its oxide forms like graphene oxide (GO) and reduced graphene oxide (rGO), which are considered for a variety of applications. Two-dimensional graphene, although often mentioned as the mother of all graphitic carbon, was discovered as the final nanoform of carbon by the mechanical exfoliation of graphite using the "scotch tape" method. Meanwhile, various methods were explored for the production of graphene, including chemical, electrochemical, laser ablation, and CVD methods. High-quality graphene for electronic applications is produced mostly by CVD synthesis, which is potential and cost-effective [1–3, 13–15].

Low-dimensional materials having length scales of a few nanometers exhibit completely different properties as compared to bulk materials. The size-related quantum confinement effect in the nanoscale materials regulates their electrical, optical, and magnetic properties. In this nanoscale research domain, carbon plays a significant role as it offers nanoforms of all three-dimensionalities, i.e., zero, one, and two in the forms of fullerenes, carbon nanotubes, and graphene, respectively. Also, these carbon nanomaterials exhibit high crystallinity, superior electronic, optical, mechanical, and thermal properties along with low densities and high effective surface areas [4, 5]. Owing to all these superlative properties, all the carbon nanostructures became a part of the diverse family of applications, including multifunctional composites, catalysts, nanoelectronics, sensors, energy storage devices, drug delivery, biomedical, environmental, and agricultural applications [1, 2, 4]. Besides the above list, carbon-based nanomaterials also have numerous potential applications in different domains. Therefore, the huge acceptability and popularity of carbon nanomaterials in multipurpose applications was the main motivation behind this monograph. Three major thematic areas, i.e., (i) synthesis and characterizations of carbon nanomaterials and nanocomposites, (ii) theoretical and computational study of carbon nanomaterials, and (iii) versatile applications of carbon nanomaterials, were considered for the current monograph.

In the first chapter of theme one, numerous properties of carbon allotropes and the various methodology adopted by the researchers to synthesize those allotropes are discussed in detail. Beyond the popular low-dimensional carbon allotropes like fullerene, CNT, and graphene, the discussion is also extended for comparatively less popular nanoforms of carbon like graphene quantum dot, carbon nanofibers, carbon nanohorns, graphene nanoribbons, etc. Furthermore, allotropes of carbon are explored for the possible functionalization with metal-based nanoparticles, biomolecules, etc., to generate smart materials in order to obtain high-performance devices. For a better understanding of the formation mechanism along with their direct visualization down to nanometer scale structural analysis of those carbon nanotropes, scanning tunneling microscopy (STM) would be a useful tool with extremely high spatial resolution. The second chapter is mainly focused on STM imaging of some of the recent carbon nanotropes such as C_{60}, CNTs, and graphene to bring together the atomic scale structure and their related material properties. In the third and fourth chapters of the current theme, a brief overview of the state-of-the-art research, along with detailed discussions on some recently developed conventional carbon-based nanocomposites (mostly oxides and nitrides) are discussed to portray their structures, properties, and synthesis techniques. In the fifth chapter, advanced research is also added to the area of metal-fullerene nanocomposite to study its tunable optical and structural properties. Emphasis is given on the applicability of that carbon-based nanocomposite for fundamental and industrial applications and related significant challenges.

In the second theme, theoretical and computational modeling of various carbon nanostructures such as fullerene, carbon nanotubes, graphene, carbon quantum dots, etc., is reviewed critically. The impact of theoretical and computational approaches in understanding the physics of these carbon nanostructures is also highlighted in the first chapter of the current theme. Therefore, the ability of computational and theoretical techniques to predict and provide insights into the structure and properties of systems play a crucial part in substantiating experimental findings. After a widespread discussion on computational modeling of carbon nanoforms, a relatively focused area, i.e., theoretical aspects of nanoscale edge effects on the electronic, magnetic, and transport properties of nano-graphene systems, is reviewed in the second study under this theme. The fundamental aspects of the impact of carrier doping and the mechanism of electric-field and chemical modification induced half-metallicity in graphene nanoribbons and their essence in spintronic device applications are deliberated. The motivation behind the study was to explore the role of edge states, which enable graphene nanoribbons to acquire such diverse and exciting physical and chemical properties and thus can be utilized in future carbon-based nanomaterial device applications. Carbon nanotube field-effect transistors (CNTFET), another highly emerging area in carbon nanomaterial electronics, is covered in the third study under the same theme. Almost all the aspect of CNFET, including the device structures, fabrication steps, various conduction models, different parameters that affects the performance of CNFET, and the possible applications of CNFET are described in this part.

In the last theme, versatile applications of carbon nanomaterials in multipurpose sensors, flexible electronics, optoelectronic devices, energy storage, medicine, etc., are considered. Even though tremendous advancement in carbon nanomaterials-based electronic devices and sensors has already been achieved, a few challenges need to be addressed before the commercialization of carbon nanomaterials-based devices. Strategies such as chemically tuning and enhancing the properties of carbon nanomaterials are important for further improvement of carbon nanomaterial-based device performance. The first chapter under the current theme focuses on understanding the basic electronic properties of graphene, CNT, and carbon quantum dots/fullerenes and their applications in electronic devices (field-effect transistors, diodes, etc.), optoelectronics, and various chemical and physical sensors. The second chapter of this theme focuses on three critical and distinct applications of carbon nanomaterials, especially fullerenes, CNTs, and graphene for biosensing, medicine, and wastewater treatment. The enormous demand for advanced touchscreen gadgets has drawn significant scientific attention in the last few years due to the substantial developments in the field of flexible and portable electronics. Graphene is an emerging material in this aspect due to good electrical conductivity, high optical transparency, and mechanical stretchability that makes graphene a better choice for flexible electronics and display devices. Also, graphene possesses sufficient robustness to be used in a harsh environment. Therefore, in the third chapter of this theme, a brief review of the basic understanding of touchscreen technology and the importance of graphene in flexible touchscreens, as well as challenges for the commercialization of graphene-based touchscreens, are discussed. Recent progress shows that the synthesis and scalable manufacturing of carbon nanotubes (CNTs) remain critical to exploit various commercial applications. In this context, the fourth chapter of the current theme focuses on the growth of CNTs either by in-situ or ex-situ synthesis techniques followed by its alignment during growth or post-growth processing. This chapter deals with the various mechanism of CNTs alignment, its process parameters, and the critical challenges associated with the individual technique. Numerous novel applications utilizing the characteristics of aligned CNTs are also discussed in this study. Infra-red (IR) radiation is the thermal radiation that is characterized by the temperature of the emitting source. Hence, IR photodetectors can be used for a number of applications such as surveillance in defense, non-contact thermometry, and non-contact human access control, bolometers and terahertz, etc. Graphene and its derivatives, such as graphene oxide, graphene nanoribbons, graphene quantum dots, etc., have revealed a wide range of novel physical properties and led to a spectrum of functional devices. Due to its small yet tunable bandgap, through controlled reduction, graphene oxide is a potential choice for IR detection devices. Under this framework, the fifth chapter of the current theme focuses on the physical attributes of graphene and its derivatives for potential Infra-red detection applications. The gas sensing behavior of metal loaded reduced graphene oxide (rGO) is discussed in the sixth chapter of the current theme. Synthesis and characterizations of palladium nanoparticle doped rGO and its utility towards low concentration hydrogen detection are described in this chapter. Solar energy holds the best potential for meeting the planet's long-term energy needs; however, as of now, more than 70% of the global energy demand is being fulfilled by non-renewable sources. Recently, perovskite solar cells (PSCs) have attracted enormous interest because they can combine the benefits of low-cost and high-efficiency with the ease of processing. However, there are still some hurdles to commercialize the perovskite solar cells. Among potential candidates, carbon-based materials provide a good alternative because of their suitable work function, high-carrier mobility, electrical conductivity, stability, and flexibility. Under this context, the seventh chapter of the current theme discusses detailed information about different carbon-based materials and their properties, which make them a front-runner in future generation perovskite solar cells. The study also covers the different advantages like flexibility, photostability, thermal stability, and scalability, which will lead to a pathway towards the commercialization of PSCs using carbon-based electrodes. The eighth chapter of this theme addresses the emerging carbon nanomaterials for organic and perovskite-based optoelectronic device applications. In the first phase of this chapter, the role of multidimensional carbon nanomaterials in organic optoelectronic devices such as organic solar cells (OSCs) and organic light-emitting diodes (OLEDs) is discussed. In the second phase, the role of different carbon nanomaterials in perovskite-based optoelectronic devices like a solar cell, photodetector, light-emitting diode, etc., is covered in detail.

The current monograph covers almost all aspects of carbon nanomaterials-based electronic devices for versatile applications, which is highly relevant in the contemporary research context.

References

1. Zaytseva O, Neumann G (2016) Carbon nanomaterials: production, impact on plant development, agricultural and environmental applications. Chem Biol Technol Agric 3(17), 26 pp
2. Clancy AJ, Bayazit MK, Hodge SA, Skipper NT, Howard CA, Shaffer MSP (2018) Charged carbon nanomaterials: redox chemistries of fullerenes, carbon nanotubes, and graphene. Chem Rev 118:7363–7408
3. Power AC, Gorey B, Chandra S, Chapman J (2018) Carbon nanomaterials and their application to electrochemical sensors: a review. Nanotechnol Rev 7:19–41
4. Scarselli M, Castrucci P, Crescenzi MD (2012) Electronic and optoelectronic nano-devices based on carbon nanotubes. J Phys: Condens Matter 24:313202, 36 pp
5. Jariwala D, Sangwan VK, Lauhon LJ, Marks TJ, Hersam MC (2013) Carbon nanomaterials for electronics, optoelectronics, photovoltaics, and sensing. Chem Soc Rev 42:2824–2860
6. Kour R, Arya S, Young S-J, Gupta V, Bandhoria P, Khosla A (2020) Review—recent advances in carbon nanomaterials as electrochemical biosensors. J Electrochem Soc 167:037555
7. Rauti R, Musto M, Bosi S, Prato M, Ballerini L (2019) Properties and behavior of carbon nanomaterials when interfacing neuronal cells: How far have we come? Carbon 143:430–446
8. Patel KD, Singh RK, Kim H-W (2019) Carbon-based nanomaterials as an emerging platform for theranostics. Mater. Horiz 6:434–469
9. Mondal K, Balasubramaniam B, Gupta A, Lahcen AA, Kwiatkowski M (2019) Carbon nanostructures for energy and sensing applications. J Nanotechnol 2019:1454327, 3 pp
10. Schroeder V, Savagatrup S, He M, Lin S, Swager TM (2019) Carbon nanotube chemical sensors. Chem Rev 119:599–663
11. Pan M, Yin Z, Liu K, Du X, Liu H, Wang S (2019) Carbon-based nanomaterials in sensors for food safety. Nanomaterials 9:1330, 23 pp
12. Baptista FR, Belhout SA, Giordani S, Quinn SJ (2015) Recent developments in carbon nanomaterial sensors. Chem Soc Rev 44:4433–4453
13. Hazra A (2020) Appropriate gate electrostatic for highest gas sensitivity in reduced graphene oxide FET. IEEE Trans Electron Devices 67:5111–5118
14. Hazra A, Basu S (2018) Graphene nanoribbon as potential on-chip interconnect material—a review, C J Carbon Res 3:49, 27 pp
15. e Dutta D, Hazra A, Hazra SK, Das J, Bhattacharyya S, Sarkar CK, Basu S (2015) Performance of a CVD grown graphene based planar device for hydrogen gas sensor. Measur Sci Technol 26:115104, 11 pp

Synthesis and Characterizations

Synthesis of Carbon Allotropes in Nanoscale Regime

Abhyavartin SelvamⓘⒹ, **Rahul Sharma**ⓘⒹ, **Soumyaditya Sutradhar**ⓘⒹ, and **Sandip Chakrabarti**ⓘⒹ

Abstract Since the last 30 years, incredible amount of research has been performed toward finding novel, smart, and cost-effective materials for device applications. Carbon among other materials is one of the most versatile elements present in nature that can produce different allotropes due to the existence of its variable hybridizations. Moreover, graphene is being considered as the mother of other carbon allotropes as they are the structurally derived allotropes of different dimensionalities such as fullerene, graphene quantum dots (0-D), carbon nanotubes, nanohorns, nanofibers, graphene nanoribbon (1-D), graphene (2-D), graphite and diamond (3-D) and are being implemented for various device applications. The synthesis methodologies of these allotropes including arc discharge, laser ablation, and chemical vapor deposition (CVD) techniques are discussed in this chapter to produce 0-D, 1-D, and 2-D carbon allotropes. CVD is considered as the most reliable technique for bulk production of highly crystalline graphene and its derivatives, single-crystalline diamonds, CNTs, and aligned CNTs on certain pre-treated substrates which are beneficial for device applications. Further, solid-state synthesis approaches such as ball milling and annealing have been adopted to generate CNTs, while graphene and offshoots have been synthesized by employing wet milling, top-down, and bottom-up processes. Also, it is noteworthy to mention that the bottom-up processes have been proven to be more effective compared to the top-down approaches for device fabrications. Furthermore, allotropes of carbon are known to be functionalized with metal-based nanoparticles, biomolecules, etc. to generate smart materials in order to obtain high-performance devices.

Keywords Carbon · Allotropes · Synthesis routes · Functionalization · Composites

A. Selvam · R. Sharma · S. Chakrabarti (✉)
Amity Institute of Nanotechnology, Amity University, Noida 201303, UP, India
e-mail: schakrabarti@amity.edu

S. Sutradhar
Department of Physics, Amity University Kolkata, Kolkata 700135, India

9

1 Introduction

Carbon is the sixth element of the periodic table with many allotropes having nanodimensionalities. Several carbon-based allotropes with reduced dimensionalities have been reported till date. Many of them have paved the path for numerous device applications such as sensors, supercapacitors, transistors, memory devices, photovoltaic devices, etc. [1]. In order to match the demand for technological developments in recent years, large-scale production of cost-effective and high-performance devices is essential. This can be achieved by the applications of different carbon allotropes due to their extraordinary electrical conductivity, thermal and mechanical properties, and significant conformation making it reliable for wide range of functionalities. The allotropes of carbon that are classified by their dimensionalities and each possessing a significant characteristic can be synthesized using numerous methodologies such as top-down, bottom-up, wet chemical methods, solid-state methods, vaporization, etc. These synthesis techniques are evident to assist in large-scale production with high crystallinity which is very essential for device applications.

Initially, 3-D materials such as diamond and graphite were used for device applications. The research on carbon-based nanomaterial has caught the attention of scientists and technologists after the discovery of fullerene, a 0-D carbon allotrope, by Smalley and co-workers in 1985 [2]. Furthermore, the discovery of carbon nanotubes (CNT), a 1-D carbon allotrope, by Iijima [3] in 1991 provided a major boost in the field of carbon-based nanomaterials for device applications. Several forms of CNTs have been studied extensively and their synthesis methodologies were also standardized. Scientists over the last decade have done immense research on other 1-D allotrope such as nanohorns, nanofibers, nanoribbons, etc. to explore their possibilities for device applications. Finally, the discovery of graphene, a 2-D carbon allotrope, by Andrew Geim and co-workers in 2004 has revolutionized the application of carbon nanomaterials in devices [4]. Graphene is a single layer of sp^2 hybridized carbon nanostructure comprising of honeycomb lattice. It demonstrates extraordinary electronic, thermal, and mechanical properties which are essential for device applications. Derivatives of graphene have captured the attention of material scientists leading to the development of quantum dots, a 0-D derivative of graphene having remarkable photoelectronic properties along with graphene nanoribbons which is a 1-D derivative known to enhance the performance of many devices.

Surface functionalization implies the modification of the chemistry of material surfaces with certain chemical species to modify or generate specific properties of the allotropes for a precise application. Functionalization of carbon-based nanomaterials is mainly carried out by two methods, namely, covalent and non-covalent functionalizations. Covalent functionalization deals with the attachment of hydrogen and halogens thereby protecting electrical properties of the material whereas non-covalent functionalization is favored as it does not disrupt the extended π-delocalization in the allotrope [5]. Further modification on functionalized carbon-based nanomaterials can be obtained on forming nanocomposites with polymers which has shown to improve device performance when compared to undoped materials in terms of mechanical, electrical, and opto-electrical properties. Nanocomposites are regarded as solid structures with dimensions in nanometer range with repeating lengths among different phases comprising the structure. Such materials conventionally are made of an inorganic (host) solid along with an organic constituent or vice versa. Otherwise, they are found to comprise two or more inorganic/organic phases in some blended form with the limitation that at least one of the phases or features is at nanoscale in size [6]. Moreover, apart from the features of the constituents, interfaces are highly responsible for enhancing or restricting the overall functionality of the system. Due to the large surface area of nanostructures, nanocomposites exhibit numerous interfaces between the amalgamated phases of components. Enhanced performance in devices having nanocomposite materials usually result from the interaction of different phases at the interfaces.

This chapter discusses certain properties of carbon allotropes with elucidation on various synthesis methodologies taken up by researchers for the generation of different allotropes in bulk amounts and having appreciable quality, which on functionalization have shown to be fabricated on to devices, thereby enhancing its performance and functionality.

2 Structure and Properties of Carbon Allotropes Related to Devices

Carbon-based nanomaterials are a subject of interest for several researchers and industrialist for its remarkably unprecedented and versatile functionalities. This class of materials is extremely reliable in several applications owing to its spectacular electronic, optoelectronic, mechanical, and chemical properties. These materials are particularly targeted for its properties, structure, shape, bonding, charge transport in confinement, and so on which have been widely explored and discussed for device applications. Moreover, demonstration of graphene as the basal structure for all the allotropes [7] can be understood from Fig. 1.

2.1 Zero-Dimensional Allotropes

2.1.1 Fullerene

Buckyball or Fullerene is named after the American architect Richard Buckminster Fuller for his inspiring design of geodesic domes, which is a closed caged structure made by the periodic arrangement of graphene layers having the chemical formula C_{60}. It is made up of truncated icosahedral symmetry with 32 faces, 60 corners, and 90 edges composed of 20 hexagons and 12 pentagons where the 6:6 ring bonds (between two hexagons) are regarded to be double bonds with shorter bond lengths

Graphene - Mother of all Allotropes (2-D)

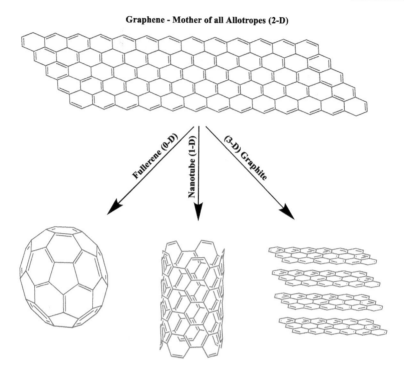

Fig. 1 Graphene represented as the origin of all carbon allotropes. Materials when enveloped, rolled, and stacked form 0-D buckyball, 1-D nanotube, and 3-D graphite respectively (Redrawn)

than the 6:5 bonds (between a hexagon and a pentagon) with an average bond length of 1.4 Å [8]. It was first discovered in 1985 by Kroto et al. at Rice University [2] who were awarded with Nobel Prize in Chemistry in 1996. Fullerenes are also known to have other isomers, C_{70}, C_{76}, C_{78}, and C_{84} [9]. The electronic structure of C_{60} is represented by a fivefold degenerate HOMO (hu) and a threefold degenerate LUMO (t1u), having a band gap of 1.8 eV (or 2 eV in certain literature) [10] thus showing semiconductor behavior. Although, C_{60} and C_{70} are observed to be active at different wavelengths, in the sense that C_{60} is photochemically reactive in UV region, whereas C_{70} is reactive at visible spectrum of light, making the latter a visible light photocatalytic material. Open-shell structures of fullerenes are energetically less favorable. Structures with isolated pentagons are favored over those composed of neighboring five-membered rings causing resonance destabilization. The size of the cage further determines thermodynamic properties such as melting point and bond energy per carbon atom [10]. Additionally, the non-homogeneous distribution of the components that cause the arch over the whole surface increases the strain of the cage. The unnatural bond edges of the carbon atoms impressively destabilize the atom which is established by the isolated pentagon rule (IPR). An additional outcome of the IPR is that the structures best resembling a circular shape are generally stable.

The quantity of twofold bonds arranged in five-membered rings is limited. Thus, five-membered rings associated through a six-membered having one placed in "Meta" position are chosen over the "Para" variation as it avoids adjacent pentagons and double bonds with five-membered rings. "Ortho"-arranged pentagons are as of now disfavored by the IPR [10].

2.1.2 Graphene Quantum Dots (GQDs)

Carbon quantum dots (CQDs), widely regarded as carbon dots (CDs), are a 0-D allotrope composed of carbon having size <10 nm [11]. Its attractive photolumines-cent and photophysical properties such as exceptional photostability and photocat-alytic activity, along with appreciable conductivity and extensive optical absorption [12], make it an appropriate material for viable photoelectrical device. However, most of the luminescent CQDs exhibit fluorescence only on excitation arising from surface defects by trapping excitons, thus posing a hindrance in the performance of carrier injection for optoelectronic applications [13, 14]. This can be overcome with graphene quantum dots (GQDs) owing to its non-zero bandgap property [15]. GQDs have significance inherited from both graphene and CDs, which could be idealized as smithereens of graphene. The dimensions and heights of graphene sheets range from 1.5–100 nm to 0.5–5 nm, respectively [16]. Overall, GQDs are believed to be in the range from 3 to 20 nm constituting of about five layers of graphene sheets (ca. 2.5 nm) [17]. Frequently, GQDs are found to be circular and elliptical shapes, nevertheless triangular, quadrate, and hexagonal dots also known to be possible [18, 19]. The modification of 2-D graphene sheets into 0-D GQDs with a radius smaller than exciton Bohr's radius [20] generates modification in electron distribution thus demonstrating new phenomena by means of quantum confinement and edge effects, as observed in CDs [21] leading to fluorescent properties along with different chem-ical and physical functionalities in contrast to other allotropes. Furthermore, GQDs are superior against to organic dyes and semiconductive QDs, in terms of their excel-lent photostability, low toxicity, and biocompatibility [22–24], along with suitable surface grafting.

2.2 One-Dimensional Allotropes

2.2.1 Carbon Nanotubes (CNTs)

In 1991, a Japanese physicist Iijima [3] revolutionized the field of carbon-based nanomaterials with his discovery of the carbon nanotubes (CNTs). It received great recognition by researchers all over the world for its unique set of characteristics. The CNTs are further classified into single-walled carbon nanotubes (SWCNTs) by Iijima and Ichihashi [25] and Bethune et al. [26] which is just a layer of graphene rolled up into hollow cylindrical product and multi-walled carbon nanotubes (MWCNTs) can be thought of several layers of graphene rolled up forming a similar structure. These elongated tubes comprise of carbon hexagons orchestrated in a concentric fashion with the two closures typically topped by fullerene-like structures containing pentagons [20]. Further, CNTs can be structurally classified as zig-zag, armchair, and chiral.

- Zig-Zag: The graphene layer(s) is/are rolled up in a manner to make the perfect parts of the open tube be a zig-zagged edge [10].
- Armchair: In contrast with the zig-zag hollow cylinders, the graphene sheet is turned by 30° previously rolling up [10].
- Chiral: In the event that the angle of turning the graphene layer before rolling up is $0° \leq \theta \geq 30°$, chiral nanotubes are gotten [10].

Furthermore, branchings such as T-junction and Y-junctions are multi-terminal (3, 4 T) structure with evidences showing to have promising electronic potential [27, 28].

CNTs are quite popular for certain characteristic features, for instance, large thermal and electrical conductivities along with remarkable tensile strength and flexibility as a result of its low coefficient of thermal expansion, at times showing ballistic transports, but are also known for its thermal insulation properties. CNTs are much more rigid compared to steel but are exceptionally flexible. Scanning probe microscopy studies elucidate its remarkable elasticity as it regains its original shape on removing stress. Furthermore, ellipsometry experiments of aligned MWCNTs show that the dielectric function of CNTs to be analogous to that of graphite to illustrate the dielectric function aligned parallel to CNTs is similar to that of graphite's parallel alignment. A similar fashion is observed even for perpendicular alignments [29]. Moreover, SWCNTs have shown to be dominant over MWCNTs in device application owing to its better conduction properties and their ability to behave as metallic or semiconducting materials based on their diameter and chirality. The existence of the degenerate point between the valence band (VB) and conduction band (CB) where the two bands meet under normal energy dispersion spectrum of a SWCNT [30, 31] implies that the distance between the two bands is approximately zero, thus expressing high conductivity. Therefore, an electron can easily be excited form VB to CB using an extremely low excitation energy. Thus, all armchair SWCNTs are theoretically expected to exhibit metallic character.

2.2.2 Carbon Nanofibers (CNFs)

Carbon nanofibers are cylindrical fibrous structures with various stacking conformations of graphene sheets, like stacked platelet, ribbon, or herringbone [32]. CNFs have lengths measured up to few micrometers, while its diameter ranging from a couple tens of nanometres to several hundreds of nanometres. Although CNFs have controllable size and graphite arrangement, the mechanical strength and electric properties are analogous to that of CNTs [33]. Moreover, CNFs are distinguished according to the conformation of graphene sheets with different shapes, giving rise to more edge sites on the exterior wall of CNFs than CNTs [34]. Furthermore, CNFs are known to possess larger functionalized surface area in contrast to CNTs, which implies that the surface-active groups to volume ratio of these materials is far greater than that of the glassy-like surface of CNTs [35]. Additionally, CNFs exhibit unique set of physical and chemical properties for instance, excellent electrical conductivity, large surface area, biocompatibility, characteristic chemical functionalities, and convenience in fabrication [35] thus making it an ideal candidate for device applications.

2.2.3 Carbon Nanohorns (CNHs)

Carbon nanohorns (otherwise called nanocones) are conical carbon nanostructures developed from a sp^2 carbon sheet. They give a practical and valuable option in contrast to CNTs, in a wide scope of utilizations. They additionally have their own exceptional characteristics due to their explicit cone-shaped morphology [36]. CNHs are known to comprise of a funnel-shaped upper area and subsequently a short round and hollow nanotube area. It is understood from literature that tips of CNHs usually consist of five pentagons, and a sixth pentagon away from the tip which is a necessity for the nanohorn walls to proceed equidistant to the nanohorn "axis" in the exact way as a CNT [37]. The property of inversion is attributable to elastic energy per carbon which occurs only if the number of pentagons is less enough to bring the conical section from the tip to the base [38, 39]. Nevertheless, during synthesis CNH tends to aggregate into spherical clusters having diameter of ~100 nm, thus restricting further functionalization [36]. Recently, it is observed that creating dahlia-like clusters enables CNH to overcome such limitations [36]. CNH is also well known for its porosity enabling it reliable for adsorption and storage activities. In contrast to SWCNTs, CNH has one-of-a-kind symmetries that employ chemical reactions to take place in the cavity of CNHs influenced via free molecular movement and clubbing of captured charge carriers, leading to various energy applicative prospects [36]. Pristine CNHs exhibit semiconducting features, majorly induced by oxygen and carbon dioxide adsorption [40]. Further investigation shows the porosity of CNH can be classified into two types, namely, interstitial pores with size varying according to temperature and internal pores [41, 42]; moreover, the micro-porosity of CNHs increases on compression at elevated pressure [43]. Further, these nanohorn dahlias displayed sequestration of supercritical CH_4 [44].

2.2.4 Graphene Nanoribbons (GNRs)

Graphene nanoribbons, a quasi-1-D derivative of graphene, are shown to have two possible edge geometries, namely, armchair and zig-zag having nomenclatures AGNRs and ZGNRs, respectively, with absolutely contrasting electronic properties as a result to their dissimilar boundary conditions [45]. AGNRs can be determined into three categories, in the order of $N_a = \{3p\}$, $\{3p + 1\}$, and $\{3p + 2\}$ where "p" is a positive integer with widely varying electronic properties and N_a is the number of dimer lines along the ribbon. The periodic AGNRs with $N_a = \{3p + 2\}$ show metallic behavior, whereas the other two show semiconducting behavior within the tight binding (TB) formalism [46]. Relating to its novel edge symmetry, ZGNRs express a wide variety of electronic properties by means of its width and edge shape, tunability in terms of electronic configuration and bandgap through structural or chemical adjustments as well as doping and external perturbations [47]. The periodic ZGNRs with AFM ground state are believed to show significant property, i.e., half-metallicity, unlike conventional metals or semiconductors, it has zero bandgap for electrons with one spin orientation; however, the other spin channel remains semiconducting or insulating, leading to anti-ferromagnetism [48, 49]. The magnetism in ZGNRs is known to rise from π-orbitals of carbon confined at the edge [47]. These properties, along with the ballistic electronic transport and quantum Hall effect [50, 51], make GNRs to be regarded to be a versatile material for forthcoming nanoelectronic and spintronic devices [52, 53].

2.3   Two-Dimensional Carbon Allotrope

2.3.1   Graphene (GR)

In 2004, Geim and Novoselov extracted single-atom-thick crystallites having a honeycomb conformation [4]. Although its late discovery, GR is widely regarded as the mother of all carbon allotropes. It is considered as the base structure of graphite, large fullerenes, nanotubes, and others. It has a phenomenal intrinsic carrier mobility of 200,000 cm^2/V^{-1} s^{-1} [54] and exhibits excellent thermal as well as electrical conductivity. GR has a wide surface area of 2630 m^{2}g^{-1} [55] which is greater than SWCNTs, while possessing remarkable mechanical, elastic properties, and so on. From the four valence states of GR (s, p_x, p_y, p_z), three atomic orbitals (s, p_x, p_y) form strong covalent bonds with three neighboring carbon atoms forming sp^2 hybridization thus conforming into chicken-wire-like structure, while one "p_z" orbital is found to overlap with three neighboring atoms of its own to relate p and $p*$ orbitals resulting in HOMO and LUMO states, respectively. The three valence electrons form the single σ bonds, whereas the fourth electron forms 1/3rd of a π bond with each surrounding atom [56]. Furthermore, the pure GR resorts to a semi-metallic property as p and $p*$ bands of GR come in contact at the Fermi level (E_f) at the corners of the Brillouin zone (BZ) [57]. GR is found to exhibit rapid electronic transfer, owing to its high specific surface area and remarkable chemical stability, and has thus become a viable material for energy storage devices along with its robust conductivity, and GR layers display enhanced electrical properties, such as room temperature Hall effect [58], ballistic transport [59], large electron–hole density, and tunable bandgap [60]. Moreover, monolayer GR is found to express massless Dirac fermions [61] and Landau level of quantization [62], which could open doors to wide range of applicative possibilities. Conventionally, defects in GR are classified into two categories of, viz. point defects and 1-D line of defects. Point defects are further sub-categorized as Stone–Wales defect, single vacancies (SV), and multiple vacancies. In Stone–Wales defect, the four hexagons are modified into two pentagons and two heptagons by tilting one of the C–C bonds by an angle of 90° [63]. SV defects in GR have a missing atom, which involves a Jahn–Teller distortion to reduce the overall energy. Two of the three dangling bonds are bonded with each other and directed toward the vacant atom thus forming a five-membered and a nine-membered ring [63]. In the case of multiple vacancies, GR is composed of double vacancies (DV) created through the alliance of two SV or the removal two neighboring atoms [64], hence appearing as two pentagons and one octagon instead of four hexagons in perfect GR. Although the formation energy of DV is similar to that of SV (ca. 8 eV), but since there is vacancy of two atoms, the energy permission atom (4 eV/atom) of the former is lesser than the latter which makes DV more thermodynamically stable as compared to SV [65, 66]. 1-D line of defects and edge defects are other kinds of defects one can come across. Line defects can be thought of as irregular attachment of two crystals of GR by a chain of pentagons, heptagons, and hexagons such that the grain boundary is not uniform, and the defects along the boundary are not uniform [67]. Edge defects involve the formation of zig-zag and armchair arrangements [68]. Such point defects and SV defects affect the GR's electrical conductivity which is why the actual conductivity varies from the ideal value. Nevertheless, these defects are inevitable in current experimental conditions, but lowering the defects does improve its conductivity [69]. Furthermore, in contrast to such intrinsic defects, integration of foreign atoms to create an extrinsic defect by means of nitrogen or boron has shown to improve the conductivity of GR by causing resonance scattering [70]. Additionally, extrinsic defects decrease thermal conductivity by 40% [71], while point defects and SV are the cause for the decrease in Young's modulus [72]. Apart from surface defects, surface corrugation of GR also impacts its properties elucidated in Slater–Koster tight binding model [73]. These corrugations include wrinkles, ripples, and crumples which are categorized according to their aspect ratio, physical dimensions, topology, and order [74]. Wrinkles have large aspect ratio of > 10, while ripples are isotropic with an aspect ratio of ~1, although both occur in 2-D plane [75]. Crumples are composed of folds and wrinkles occurring isotropically in either 2-D or 3-D plane which may or may not be ordered [76]. These corrugations cause suppressed electron transport [75], mobility, weak localization (on fluctuating the position of Dirac point) [77], extensive corrosion, and oxidation as well as quantum corrections [78] which occurs due to the quantum interference of electrons moving in undefined trajectories [79]. Thus, it is imperative to acquire wrinkle-free GR to utilize its absolute potential in devices. GR synthesized via CVD and transferred over a substrate can reduce wrinkles by (1) increasing the thickness of Ni substrate during its growth, (2) immersing PMMA/GR film in DI water [80] before transfer, and (3) transferring GR to hydrophobic substrates [81].

2.4 Three-Dimensional Carbon Allotropes

2.4.1 Graphite

Graphite is a 3-D carbon allotrope of sp^2 hybridization having each carbon atom bonded with three other atoms, thus forming a planar array of fused hexagons. It is further pigeonholed as a progression of GR layers along the XY-plane, stacked in Z-direction [82]. Within a layer of GR, the carbon atoms are located at the corners of hexagons. Further each atom of carbon display three σ bonds along its edges. Graphite with only three out of four valence electrons contributing implies p$_z$ orbital does not have a role in the hybridization. Hence, they create a π-cloud which is delocalized across the layer of GR [82]. Graphite, owing to its layered structure and van der Waals force acting between the layers hence allows the layers to glide past each other [82] which can be exploited for its appreciable thermal as well as electrical conductivity besides its well-known lubricative property. Klett et al. [83] reported graphitic foam having thermal conductivity of ca. 182 W/(m • K), and low density between 0.2 and 0.6 g/cm^3. Graphitic form has been implemented as lucrative heat exchangers and heat sinks [84]. However, it shows highest thermal conductivity only in the Z-direction as a result of the rigid nature of covalent bonding which is its limitation [85]. Moreover, there are two forms of graphites, viz., alpha (hexagonal) and beta (rhombohedral), showing comparable physical properties, apart from the circumstance that the stacking of GR layers is in contrast with each other [83]. The electrical desirability of graphite is due to the vast electron delocalization within the carbon layers enabling valence electrons to move without hindrance [82], while exhibiting resistance to high temperatures and thermal shock, along with its high melting point giving rise to graphite crucibles [86]. Also, electrodes using graphite powder have shown to be potential in many experimental techniques due to its high conductivity and resistance to corrosion [87].

2.4.2 Diamond

Diamond is another 3-D allotrope of carbon in which the atoms are organized in a detailed type of cubic lattice called diamond cubic. Diamond is the hardest natural material as a result of its conformation, in which individual carbon atoms have four neighboring atoms joined to it by covalent bonds, exhibiting sp^3 hybridization state [10]. Diamond as a semiconductor has a wide bandgap of 5.45 eV which is a factor for its larger breakdown field strength in contrast to narrow bandgap semiconductors [88]. Further, it influences luminescent activities owing to its various luminescent centers such as donor–acceptor pairs and structural defects, which can be verified through cathodoluminescence, pre-breakdown, injection, and double injection luminescence along with X-ray luminescence [89]. Moreover, in electronics, diamonds possess high mobility of charge carriers, making remarkable delta-doped FETs and Schottky diodes [90, 91]. Diamonds are regarded to be an appreciable device at high temperatures as it tends to get oxidized at higher temperatures, i.e., 500–700 °C in air and 1400–1700 °C in inert environment [92] as a result of the robust covalent bonding and low phonon scattering. Nanodiamonds are an artificial arrangement of carbon atoms having structure and properties analogous to that of a natural diamond [93]. They are found to remain stable due to surface terminations with oxygen-containing functional group [94] and nitrogen-containing functional groups hence showing a complex behavior, which can be exploited for various device applications. A serendipitous creation is the diamond-like carbon (DLC), consisting of a combination of sp^2 and sp^3 hybridization energy states [95, 96]. DLC exhibits remarkable properties such as great hardness, low coefficient of friction, excellent wear and corrosion resistance, good chemical stability and biocompatibility, large electrical resistivity, infrared transparency, high refractive index, and excellent smoothness [96, 97].

3 Synthesis Routes

Synthesis routes can be classified into top-down and bottom-up approaches which are further categorized as physical, chemical, and biological methods for the synthesis of nanomaterials. Zero-dimensional allotrope, i.e., fullerene is known to be synthesized via partial combustion of hydrocarbons, whereas GQDs are prepared via hydrothermal, solvothermal, and microwave-assisted techniques. In addition to these, few more techniques produce fullerenes such as chemical vapor deposition (CVD), which can also be seen to produce 1-D allotropes, for example, CNTs, CNFs, CNHs, and GNRs. Furthermore, the 2-D allotropes such as GR and its derivatives are synthesized through techniques like epitaxial growth, liquid phase exfoliation, and sonication. Finally, 3-D allotropes such as nanodiamonds are produced by microwave-assisted CVD and ultrasonication methods, whereas pyrolysis of carbon compounds results in artificial graphite and pyrolytic graphite that are analogous to the natural material in terms of thermal and electrical conductivities.

3.1 Partial Combustion of Hydrocarbons

Although conventional combustion methods are apparently continuous and scalable process, nevertheless when hydrocarbons are incompletely burnt, carbon black is obtained [98]. Hence, under appropriate conditions this technique gives rise to a large-scale production of fullerenes. The most essential component necessary to improve the efficiency of production is the use of porous refractory plate burner [98]. Nominal amount of fullerene has been detected in lamp black using mass spectrometry [99]. On further fine-tuning of the experimental conditions, the yield of fullerene was increased substantially in smoking flames. Benzene is traditionally used as carbon precursor used to produce fullerenes. It is being blended in with oxygen and argon and consumed in a laminar flame [100]. The subsequent blend contains soot, polycyclic aromatic compounds, and a specific division of fullerenes which make up 0.003–9.0% of the soot's complete mass [100]. Further, the combustion was carried out in an atomic C/O ratio of 0.99/1.28, a combustion chamber pressure of 5.33 kPa (40 Torr), and cold gas velocity at the burner surface of 0.7–1.67 m/s was maintained at a temperature of 298 K [98]. The laminar flame was sustained stable for a minimum of 6 h before collecting the fullerene from the soot-collecting tank after each hour of experiment [98]. Moreover, hydrocarbons like toluene or methane also provided wonderful result. The selection of appropriate experimental parameters leads to an increased C_{70} concentration in the obtained soot [100]. Additionally, the effect of gas pressure is pertinent to control the C_{70} /C_{60} ratio, as the amount of C_{70} is found to improve on raising the pressure [98, 100]. Temperature and concentration of oxygen in the experimental settings also have a pivotal role in the effectiveness of fullerene production.

3.2 Chemical Vapor Deposition (CVD)

This method is most commonly exploited in synthesizing low-dimensional carbon-based nanomaterials which is also known to assist in bulk synthesis of desired compounds. A conventional CVD strategy includes heating of catalysts to suitable temperatures ranging from 750 to 1200 °C in a quartz tube furnace under the constant flow of gaseous carbon sources such as hydrocarbon gases. These gases are mixed with process gases such as, H_2, Ar, or He with varied ratios. The presence of catalysts inside the quartz tube specifically produces CNTs. Catalytic nanoparticles of Fe, Co, or Ni or their oxides are regularly utilized in CVD method to grow CNTs [101]. Usually, the synthetic route of CNTs in a CVD technique includes the breakdown of hydrocarbons, dilution and immersion of carbon particles over the catalysts. Similarly, CNFs are also produced on decomposing hydrocarbon on a metal catalyst between 500 and 1000 °C [102, 103]. Moreover, GNRs are also being produced using the same method where solid monomers are heated at 200–300 °C to form prepolymers initially followed by the dehydrogenation of these polymers at an elevated temperature between 400 and 450 °C giving rise to the formation of GNRs [104]. The yield can be improved on cleaning the quartz tube with conc. HNO_3 at 1000 °C initially [105]. CVD approach is also known to produce polycrystalline diamonds. This method assisted via microwave irradiation (MWCVD) has shown to give rise to nanocrystalline diamonds or nanodiamonds. The reaction is carried out at 800 °C under the exposure of H_2 and CH_4 with the gas flow controlled using throttle valves while maintaining the entire system at a pressure of 20–30 torr along with a microwave generator operated at a frequency of 2.45 GHz [106]. Another technique with the same principles of CVD is hot filament CVD (HFCVD) which have been employed to deposit the microcrystalline diamonds grown from 1:99% mixture of CH_4 and H2 and the temperature ranging between 500 and 1000 °C at a pressure of 2–20 mbar [107, 108].

3.3 Hydrothermal/Solvothermal

Hydrothermal synthesis is an approach that involves chemical reactions to be carried out in an aqueous solution at temperatures greater than the boiling point of water while solvothermal synthesis occurs in a non-aqueous solution at temperatures higher than that boiling point of that particular solvent [109, 110]. Both hydrothermal and solvothermal reactions are carried out in a specially sealed container or high-pressure autoclave at subcritical or supercritical conditions of solvent. Hydrothermal process is used to produce GQDs under autogenous pressure. The hydrothermal treatment of the oxidized GR sheets at 200 °C results in deoxidization, and the strength of vibrational absorption band of C–O/COOH tends to decline. This leads to the vibration band of epoxy groups to disappear, such that eventually the size of GR sheets declines rapidly obtaining ultrafine GQDs on dialysis [111]. Similarly, through solvothermal process GQDs are synthesized by blending hydrogen peroxide with graphite which is stirred for 5 min to form homogeneous mixture. This solution is taken in an autoclave for further thermal treatment at 170 °C for 5 h. The resultant product under vacuum filtration forms raw GQDs. Further, on evaporation the residual solvent is dissolved in DI water thus producing pure GQDs [112].

3.4 Microwave Irradiation

Microwave irradiation technique is usually employed in one-pot synthesis approaches as well as in the assistance of other techniques. It is an electromagnetic irradiation process which introduces mobile electric charges having high-frequency waves to rise the temperature in a short time thus improving the rate of crystallization [113]. Microwave irradiation has explicitly proven beneficial as it enables spontaneous fabrication of robust GQDs into porous networks [114, 115] to further exploit it for its diverse applications. For the preparation of GQDs, a solution of graphene oxide (GO) is taken as the precursor. A homogeneous solution of GO, concentrated HNO_3, and H_2SO_4 is heated and refluxed in a microwave oven for 1–5 h operated at a power of 240 W. The resultant transparent brown suspension with black precipitate was brought to room temperature and exposed to mild ultrasonication for few minutes. The obtained product is found to be marginally basic [116]. The suspension was filtered to remove the large tracts of GO and further dialyzed in a dialysis bag to achieve pure GQDs.

3.5 Epitaxial Growth

Epitaxy or epitaxial growth is defined as the oriented overgrowth of film material and is known for its growth of single crystal films. Conventionally, GR sheets

(GS) are epitaxially grown on silicon carbide (SiC). Epitaxial growth is a well-established method for the large-scale production of GR as it enables large-size and single-domain GR production in a systematic manner [117]. In this process, GS are grown on SiC wafers, via the sublimation of silicon atoms from the surface, followed by graphitization of the residual carbon atoms at an elevated temperature (>1000 °C) under ultrahigh vacuum [118]. Further, the resultant GS demonstrate crystalline domains having an extensive diameter of several hundred millimeters. Moreover, the surface of the substrate used to grow GS complements the mobility of charge carriers, for instance, GS prepared at the silicon end shows a range of 500–2000 cm^2V^{-1} s^{-1}, whereas the GS grown on the carbon side displays an improved mobility range of 10,000–30,000 cm^2 V^{-1} s^{-1} [118]. This method uninterruptedly yields single-crystalline monolayer GR (MLG) or bilayer GR (BLG) films with thickness based on the size of the substrate(s). Nevertheless, this method has certain drawbacks, such as limited film sizes and heavy production cost which allows this technique to be employed only for studying at research scale and niche applications. Moreover, this approach can be implemented in the synthesis of other allotropes such as GNRs, CNTs, and CNFs. Teeter et al. [119], Vo et al. [120] epitaxially grew chevron GNRs on Cu (111) and Au (111) substrates using 6,11-Dibromo-1,2,3,4-tetraphenyltriphenylene, ($C_{42}Br_2H_{26}$) as the precursor. This precursor molecule is deposited on the substrates by subliming it at 160 °C with the help of Knudsen cell evaporator. On Cu(111) substrate, the allotrope forms at room temperature, whereas on Au(111) substrate, polymerization occurs only on raising the temperature. CNTs are also prepared through this approach using fullerendione as the precursor to initially prepare hemispherical fullerenes, which is further engineered to produce SWCNTs. Thermal oxidation of fullerendione is carried out by varying the temperature to rupture the carbon cage. Additional amorphous carbon is etched away along with oxygen-containing functional group via an open-end growth process, with EtOH as the carbon source [121]. Incidentally, SWCNTs grown from fullerene caps display a step-like diameter distribution. Further engineering of the fullerene caps by altering the thermal oxidation temperatures shows that the temperature is indirectly proportional to the diameter of SWCNTs [121]. However, this approach of synthesis makes it difficult to obtain uniformly sized carbon caps to prepare SWCNTs, and thus the growth efficiency is found to be extremely less. Interestingly, Lin's team [122] presented an uninterrupted epitaxial fabrication of CNFs by spinning CNT films carried out through gas phase pyrolytic carbon deposition. Super aligned CNT (SACNT) arrays are taken as CNT source and the entire process is effectuated in a quartz tube. Further, the obtained CNFs undergo graphitization in a furnace filled with argon for an hour maintained at a temperature of ~2900 °C, resulting in G-CNF.

3.6 Sonochemical

Sonochemical approach refers to the process of using sound energy to agitate particles or discrete fibers in a liquid through a frequency >20 kHz, which is also referred to as ultrasonication techniques [123]. Furthermore, sonication can be conducted using either an ultrasonic bath or an ultrasonic probe sonicator. This approach has been widely used in the production of GR nanoribbons (GNR) where graphite powder is dispersed in a solution of 1,2-dichloroethane (DCE) and poly (m-phenylenevinylene-co 2,5-dioctoxy-p-phenylenevinylene) (PmPV) and the reaction is carried out in a bath sonication for an hour after which a homogeneous suspension of GS and a plethora of macroscopic aggregates is obtained. The dispersion is then centrifuged for 5 min at 15,000 rpm to remove the aggregates and larger GS, and eventually a supernatant containing thin GS and GNRs are obtained [124]. Furthermore, GQDs are also synthesized sonochemically via oxidative cutting of GO in the presence of KMnO$_4$ as the oxidizing agent [125]. This technique is further assisted with intermittent microwave heating for 30 min at a fixed temperature of 90 °C while varying the power usage by 100, 200, 300, and 400 W giving rise to a range of yield from 75 to 81%. Similarly, SWCNT bundles are prepared on dispersing silica powder in ferrocene-dissolved p-xylene solution which is treated under sonication at room temperature for 20 min [126]. The change in color of the powder to gray indicates the deposition of carbon on its surface. On ultrasound irradiation, silica powder fragmented to sizes less than several hundreds of micrometers, which is attributable to inter-particle collisions of silica [126]. Subsequent filtration using HF solution gives rise to fibrous structures, which on purification displays morphological resemblance to SWCNT bundles produced via laser ablation. Moreover, ultrasonication of PAH such as benzene leads to the cavitation of benzene which further produces organic polymers and other solid carbon particles consisting of nanodiamonds which can be retrieved through filtration [127].

3.7 Liquid Phase Exfoliation (LPE)

Liquid phase exfoliation involves the assistance of sonication to exfoliate GR from graphite in the presence of ideal solvents. This strategy of GR synthesis can be summarized in three approaches: (1) dispersal of graphite into solvent, (2) exfoliation, and (3) purification [128]. GR flakes produced from LPE involves surfactant-free exfoliation of graphite via wet chemical dispersion, followed by ultrasonication in organic solvents. During ultrasonication, the growth and collapse of the micrometer-sized bubbles are as a result of pressure fluctuations impacting the bulk material thus inducing exfoliation. After exfoliation, the solvent–GR interaction balances the inter-sheet attractive forces. The ideal solvents used to disperse GR are those which have minimized the interfacial tension [128]. It is eventually purified using EtOH. This approach further gives rise on GQDs and GNRs. Sarkar et al. [129] synthesized GQDs via sonochemical-assisted LPE technique wherein a solution of ethyl acetoacetate and NaOH is kept under constant stirring for 30 min followed by filtration. Further, a mixture consisting of the filtrate and graphite powder is sonicated using a probe sonicator for 2 h. On cooling, the dark brown mixture is centrifuged to further eradicate unexfoliated graphite and heavier GQDs, which is then filtered once again using a 0.22 μm syringe filter. The subsequent filtrate on dialysis after 24 h results in a yellow solution of GQDs which is further dried using a rotary evaporator system forming powdered GQD. In another report, Xiaolin Li's group of researchers [52] developed a simple method based on LPE to synthesize GNRs. They demonstrated the exfoliation of commercially available graphite by heating to 1000 °C for a short duration in the presence of argon. The exfoliated graphite along with 1,2-dichloroethane (DCE) solution of PmPV undergoes sonication for 30 min to make a homogenous suspension. Further centrifuging eliminates large chunks of materials, leaving behind GNRs of varying topographies which were mostly composed of smooth edges.

3.8 Pyrolysis

This method involves pyrolysis of carbon compounds such as petrol coke and hydrocarbons for the production of artificial graphite and pyrolytic graphite, respectively. In order to synthesize artificial graphite, the petrol coke is initially heated at 1400 °C to eliminate impure volatile components associated with petrol coke, which is further consumed between 800 and 1300 °C [10]. This undergoes further graphitization via resistive heating. It is noteworthy to mention that the entire process is enclosed by sand since silicon exhibits useful catalytic action as it develops into SiC which gives rise to artificial graphite on subsequent decomposition of silicon carbide [10]. Similarly, pyrolytic graphite is prepared through the thermolysis of hydrocarbons at 2000 °C, followed by graphitization at 3000 °C [10].

4 Functionalization and Composites

Adjustments of parameter in the synthesis techniques of carbon materials although improve the quality and quantity of resultant yield; nonetheless, further modification is necessary to improve the functionalities of the materials to achieve the highest possible performance on fabrication of devices. This can be obtained on functionalizing these allotropes with several organic compounds such as polymers and biomolecules as well as many inorganic materials, viz., metal, metal oxides, chalcogenides, perovskite, and so on thereby imparting modifications in its structural configurations. Functionalization is the post-synthesis modifications which alters the structural conformation of the material through the addition or elimination of atoms in the pristine material which in turn brings about novel characteristics to the desired allotrope. Such modifications can be further exploited by creating heterostructure and nanocomposites resulting in devices with impressive efficiencies. The creation of heterostructure imparts functional enhancement as the resulting improvement rises from the development of interfacial sites in the compound.

4.1 Fullerene

Fullerene functionalized with conjugated oligomers leads to rapid photo-induced electron transfer thereby giving rise to efficient photovoltaic devices. With the help of azomethine ylide, several linear blends of naphthalene/thiophene-conjugated polymers [130], linear and branched oligophenylenevinylene (OPPV), and linear oligophenyleneethynylene (OPE) form heterostructures with C_{60} [131, 132]. Moreover, several other structures have been prepared by adopting quadrupole hydrogen bond strategy. Fullerenes conjugated with oligo(p-phenylenevinylene)s (OPVs), thereby results in functionalized materials with enhanced photochemical as well as electrochemical properties [133]. These materials are well known for plastic solar cell applications. Recently, mono- and dimetallofullerenes (for instance, La@C_{82} and LaR$_2$R@C_{80}) have been magnificently synthesized, along with other complexes of fullerenes such as metal nitrides (e.g., Sc$_3$N@C_{2n}). Furthermore, fullerene-based complex species with trapped metal carbides (M_nC_2, $n = 2 \pm 4$), hydrogenated metal carbides (Sc$_3$CH), metal nitrogen carbides (M_xNC, $x = 1, 3$), metal oxides (Sc$_y$O$_z$, $y = 2, 4$; $z = 1 \pm 3$), and metal sulfides (M_2S) [134] have also been reported by several other groups. Additionally, fullerenes having methano bridges between two six-membered rings forming methanofullerenes, specifically 6,6-phenyl-C_{61}-butyric acid methyl ester (PCBM), 6,6-phenylC_{71}-butyric acid methyl ester (PC$_{70}$BM), and 6,6-thienyl-C_{61}-butyric acid methyl ester (ThCBM) [135], are largely employed as electron acceptors in applications like photovoltaic devices and solar cells. Furthermore, polymer/fullerene nanocomposites are decorated from functionalized fullerene as well as its derivatives to bring about remarkable modification in its properties which can be further exploited to improve the efficacy of devices. Since the performance of conventional bilayer fullerene-based devices is hindered due to the reduced surface area of heterojunctions, it is overcome by employing fullerene derivatives [136]. Functionalized fullerenes such as PCBM allow for the development of bulk heterojunction (BHJ) solar cells, in which the polymer acts as donor layer while the fullerene derivative as the acceptor layer. The creation of nanocomposites develops an improved interfacial area in contrast to the original structure. Several such polymers have been investigated demonstrating great BHJ efficiencies. For instance, a solar cell based on the polymer composite MEH-PPV/PCBM displayed an increase in PCE by 5.5% from 2.9% [137]. Another report demonstrated an enhancement in performance of P3HT/PCBM solar cells by up to 4.5 ± 5%. The semi-crystalline nature of P3HT in contrast to other amorphous polymers is significant in terms of providing fitting morphology for BHJ organic photovoltaic devices, along with its satisfactory charge transport functionality. A precise control over the growth and morphology of the composite's active layer was achieved, which in turn ameliorated the interfaces in contact with electrodes [138, 139]. Furthermore, Li et al. [140] developed fullerenated polycarbonate (PC) through facile process involving C_{60} derivative and PC in the presence of AlCl$_3$ as the catalyst. It was reported to improve the electron accepting ability of the nanocomposite, thereby enhancing the optical hinderance of the heterostructure. Moreover, it demonstrated an appreciable thermal stability; however, its glass transition temperature (T_g) is nearly 30 °C, which is lower than PC (148.5 °C).

4.2   GQDs

GQDs are functionalized using wet chemical methods in combination with 1,2-ethylenediamine as well as $SOCl_2$, thus exhibiting visible photoluminescence property [141]. Furthermore, GQDs functionalized with valine are utilized for the detection of mercury ions owing to its powerful fluorescence and appreciable photostability. Moreover, it is observed to be sensitive as well as selective in terms of fluorescent response to Hg^{2+} ions [142]. Fluorescence wavelengths of GQDs can be conveniently tuned (blue to orange) on controlling the quantity of the primary-amine functionalization. Similarly, GQDs on interaction with acetic acid enables effective electrochemical detection of cardiac Troponin I (cTnI) as a result of hydrogen bonding interactions employed by the carboxylic group of GQDs [143]. Further modifications include engineering GQD nanocomposites using polymers, such as PANI, PPy, PTh, MIP, etc. The most commonly used polymer for applications involving charge-transfer doping, electrical properties is PANI. Integration of GQD with PANI effectively improves the electrical, mechanical, and optical properties. Dinari and co-workers [144] have reported PANI-GQD to be Pt-free electrode for dye-sensitized solar cell (DSSC) application producing an energy conversion competence of 1.6%. Similarly, Liu et al. [145] showed GQD/PANI to be a viable micro-supercapacitor with a rate capability of up to 1000 V/s and maintaining stability even after 1500 cycles. Moreover, a GQD/PANI composite sensor was employed for the detection of NH_3 by Gavgani et al. [146]. Further modifications were implemented to this sensor by co-doping it with S and N for a remarkable response toward 300 ppm of NH_3 with 15 kΩ resistance. This material has also been adjusted to sense volatile organic compounds (VOCs) like toluene, methanol, acetone, ethanol, chlorobenzene, and propanol with responses 42.3%, 0.5%, 0.45%, 0.5%, 0.48%, 0.51%, and 0.48%, respectively. QDs are generally toxic, which was then replaced with GQDs; however, GQDs were further improved with a conjugated polymer, polypyrrole (PPy) owing to its outstanding electronic, optical, and thermal properties [147, 148]. PPy/GQD composites are reliable materials in applications such as supercapacitors, electrodes, and sensors. For instance, Zhou et al. [149] designed a fluorescent-based sensor for the detection of DA with splendid sensitivity and selectivity toward the target. The nanocomposite is made with PPy microsphere, thereby creating a core/shell configuration, displaying a stronger fluorescence emission compared to pristine GQD. This sensor exhibits a linearity in range of 5–8000 nM, having a 10 pM detection limit. Moreover, the composite possessed a fluorescence emission peak at 465 nm with an excitation peak at 365 nm. Another class of conjugated polymer investigated at large is polythiophene (PTh) for its redox states, optoelectronic, and charge transport abilities, as well as environmental stability. Incidentally, PTh has been widely blended with GQDs to develop high-performance photovoltaic, light-emitting diode, and field-effect transistors [150–152]. To illustrate, a dye-sensitized solar cell made using ITO/modified PTh/graphite in the presence of N719 dye demonstrated a PCE of 1.76%, whereas GQD/PPy displayed an improvement of 2.09% efficiency [153]. Interestingly, Zhou et al. [154] developed a molecularly imprinted polymer (MIP) complex with GQD for the detection of paranitrophenol (4-NP), involving GO modification with 3-aminopropyltriethoxysilane (APTS) as the precursor to form GQDs via hydrothermal treatment. MIP is further obtained through sol–gel polymerization with silica-coated GQD using 4-NP as the template, APTS monomer, and tetraethoxysilane crosslinker. Subsequent removal of template creates MIP-coated GQD with imprinted which is implemented for 4-NP sensing. Likewise, other polymers have also been studied for device applications. For instance, Yang et al. [155] reported a solution-processable inverted bulk heterojunction (BHJ) polymer solar cell (PSC) through the creation of GQD/Cs$_2$CO$_3$. They displayed a 56% enhancement in PCE and 200% improvement in the stability of PSC as well as external quantum efficiency (EQE) of 60%.

4.3 CNTs

CNTs are functionalized non-covalently as it not only preserves the conjugation of CNT sidewalls, but also does not affect the main structure. This type of functionalization involves the use of aromatic compounds, surfactants, and polymers, under the influence of π–π stacking or hydrophobic action. In case of aromatic compounds, pyrene, porphyrin, and its derivatives are interacted on the sidewalls of CNTs via π–π stacking [156]. Hecht et al. [157] have demonstrated the fabrication of CNTs/FET via non-covalent functionalization by means of zinc porphyrin derivative resulting in direct photo-induced electron transfer within the system. Furthermore, the application of pyrene functionalized CdSe on CNT by Hu et al. [158] via self-assembly was also investigated on light-induced charge transfer. Additionally, Kymakis et al. [159] studied the hole transport layer (HTL) of SWCNT with P3HT and PCBM in bulk heterojunction (BHT) photovoltaic devices by varying the thickness of SWCNT showed an improvement in PCE of 3%. Similarly, Bergeret et al. [160] recorded SWCNT/P3HT and PCBM having a PCE of 3.6% with partial crystallization of RR-P3HT. June and co-workers [161] projected a PCE of nearly 4.4% with SWCNTs homogeneously blended with similar polymer as Kymakis et al. Other reports showed SWCNTs introduced with copper-phthalocyanine derivative (TSCuPc). MWCNT was investigated to have a PCE of 3.8% with P3HT and PCBM [162]. Recently, Reddy and Divya [163] reported small amounts SWCNT induced with poly(9,9-di-n-octylfluorene-alt-benzothiadiazole) (F8BT) and poly(9,9-dioctylfluorene) (F8) to improve electron–hole injection resulting in an effective organic light-emitting diode (OLED). Reddy and Divya [163] implanted a diketone ligand, i.e., 4,4,5,5,5-pentafluoro-3-hydroxy-1- (phenanthren-3-yl)pent-2-en-1-one (Hpfppd) onto MWCNT recording an overall quantum yield of 27% in the wavelength range of roughly 330 nm to 460 nm bringing its activity from UV to visible spectrum. Additionally, Fanchini et al. [164] showed improved conductivity of SWCNT films on treating it with nitric acid and thionyl chloride while Kim and Geng [165] attained a fourfold sheet conductance by spraying nitric acid onto the SWCNT films.

4.4 CNFs

The chemical stability of CNFs enables its surface modification with ease and in a controllable fashion. CNFs are functionalized in five ways: wet etch method, plasma treatment, photochemical functionalization, thermal treatment, and through the addition of linkers and polymers. In wet etch technique, CNFs are exposed to harsh acidic conditions in order to attach oxygen-containing groups to defect sites at the surface of CNFs [166]. Similarly, accumulation of other groups, such as –COOH, is carried out via plasma treatment of CNFs. This method is well known and is utilized frequently for the alteration of hydrophobicity of any materials [167]. Another approach involves the illumination of UV to influence a chemical reaction, thus enabling functionalization of —H groups on CNF's sidewalls. Baker et al. [168] functionalized CNFs to have COO^- as well as NH_3^+ groups on treating CNFs with protected acid and amine terminating molecules, i.e., methyl ester protected undecylenic acid and tBOC-10-aminodec-1-ene (tBOC-tertbutyloxycarbamate). Furthermore, exposing CNFs to elevated temperatures is conventionally relied on for the removal of oxygen complexes and defects, thus inducing graphitization of CNFs [169, 170]. Nevertheless, it has also been reported to eliminate amorphous material and further oxidation of graphitic components [171]. Moreover, addition of linkers and polymeric components to CNFs is another prominent method of functionalization giving rise to enhanced functionalities. In addition, polymers can also form heterostructures with pristine CNFs as well as functionalized CNFs for it to be utilized to its maximum potential. Composites of CNFs are widely used for several device applications. The conductivity of CNFs can further be enhanced on creating CNF/epoxy composites with appropriately high CNF concentration [172]. On increasing CNF content from 1 to 2%, the resistance and reactance of the composite decrease, whereas on further increase of CNF, the resistance and reactance increase. However, in the case of capacitance, composites with 2% CNFs exhibit least value [172]. Further engineering of electrical properties can be done on altering the concentration of CNF content. For instance, Lee et al. demonstrated the electrical properties of nonwoven [173] and silk fabric [174] decorated with CNF/PVDF-HFP heterostructures. They reported the composite to display an increase in electrical properties of silk fabric with 4% increase in CNF.

4.5 CNHs

Chemical functionalization of CNH undergoes via two pathways: (1) covalent bonding of organic molecules and (2) non-covalent conglomeration of several functional groups. CNHs are first exposed to temperatures higher than 550 °C in oxygen environment that leads to the generation of holes on both tips and sidewalls of CNHs. This process improves the specific surface area and electrochemical capacitive property, showing suitable adsorption of large molecules such as iodine [175, 176]. However, oxidation occurring at lower temperatures affects those CNHs having lower thickness. Subsequent controlling of its porosity converts it into an ideal substrate for adsorption of gases as well as water [177–179]. The functionalization of oxidized CNHs with photoactive molecules achieved great accolades, particularly as a proficient electron donor–acceptor hybrid material for energy conversion-related applications. Furthermore, acyl chloride modified CNHs treated with the α-5-(2-aminophenyl)-α-15-(2- nitrophenyl)-10,20-bis(2,4,6-trimethyl-phenyl)-porphyrin (H_2P) resulted in CNH–H_2P hybrid material, thus demonstrating amazing photophysical properties [180]. Similarly, photoelectrochemical electrodes constructed from CNH−H2P hybrid material exhibited an incident photon-to-photocurrent efficiency (IPCE) of 5.8% [181]. Many researchers have also focused CNHs in fuel cell applications arena, for instance, Kubo and teams [182] loaded 60% wt Pt on oxidized SWCNH obtaining a powerful direct methanol fuel cell (DMFC) at 40 °C having a power density of 76 mW $\cdot$ cm^{-2} at 0.4 V. Similarly, unprotected Pt nanoclusters were introduced on N_2-doped SWCNH thereby creating an efficient cathode catalyst [183]. It was observed that no loss in catalysis after 15,000 cycles at 30 °C from 0.6 to 1.1 V versus RHE, thus maintaining its chemical and structural stabilities. Furthermore, SWCNH loaded with PtRu nanoparticles had a 60% greater catalytic activity than commercially available carbon black when used as an electrocatalyst supported by H_2-fed proton exchange membrane fuel cell (PEMFC) and DMFC [184]. Additionally, Brandão et al. have reported a Pt-free electrocatalysts such as RuSe deposited on SWCNHs along with carbon black as an oxygen reduction reaction to support the electrocatalyst [185]. SWCNH has also been doped with heteroatom, for instance, on heating SWCNH with urea at 800 °C results in N-doped SWCNH [186] with an improved surface area of 1836 m^{2}g^{-1}, and later a surface modification was studied on continuous doping of SWCNH with Fe and N at 900 °C led to 30 mV, which in overall had a maximum power density of 35mW cm^{-1} enhancement along with the oxygen reduction reaction to increase even after 1000 cycles. Furthermore, photoelectrochemical studies were also conducted on this material for which Hasobe and colleagues [181] studied the effect of dyes on porphyrin functionalized SWCNH solar cells which was applied to an optically transparent electrode (OTE) cast with SnO_2 (OTE/SnO_2) nanofilms through electrophoretic deposition. The SWCNHs-H2P/SnO2/OTE cell showed an IPCE of 5.8% at 0.2 V *versus* SCE. The same group had also investigated SWCNHs-Zn porphyrin supramolecular assembly for photon-induced electron transfer applications [187]. Attaching crown ether to functionalized ZnP along with ammonium cation on SWCNHs via spacer forming Crown-ZnP and SWCNH-sp-NH^{3+}, respectively, and further fabrication on OTE/SnO$_2$ (OTE/SnO$_2$/SWCNH-sp-NH^{3+} -Crown-ZnP) brought an IPCE of 9% [190]. Moreover, on doping SWCNH with dimeric porphyrin {(H2P)$_2$} gave an IPCE value of 9.6% at 430 nm [188]. Lodermeyer et al. [189] recently introduced solid-state dye-sensitized solar cells (DSSCs) with SWCNH and further optimized it with guanidinium thiocyanate (GuSCN), 1-butyl-3-methylimidazolium tetrafluoroborate ([BMIM][BF$_4$]), and 4-tert-butylpyridine (TBP) obtaining an efficiency of 7.84%.

4.6   GNRs

Although GNRs are known to possess exceptional characteristics on its own, it can be further enhanced through functionalization enabling its role in larger domains of devices. Reduced GR nanoribbons (RGNRs) undergo non-covalent modification on immersing it into a solution of iron (III) meso-tetrakis(N-methylpyridinum4-yl) porphyrin (FeTMPyP) exhibiting improved electrocatalytic behavior toward O$_2$ reduction, as a result of the beneficial interaction between FeTMPyP and RGNRs [190]. The functionalized nanoribbons saturate a specific edge with a molecule or a functional molecule while leaving the supplementary edge to undergo saturation by hydrogen thus notably affecting the electronic structure near to Fermi level [191]. Similarly, modifying an edge of ZGNR with two hydrogen atoms while the other remaining with one hydrogen atom converts the anti-ferromagnetic nanoribbon into ferromagnetic [192]. Similarly, single-edge decoration of ZGNRs creates half-semiconductors with varying band gaps for each spin further exhibiting spin-polarized half-semiconductor or a semiconductor–metal transition [193]. Moreover, functionalization of either or both the edges of ZGNRs with functional groups, viz., –O, –F, –OH, –NH$_2$, –CH, –BH, and –B was studied [191, 194–196]. Substitution of carbon atoms in the nanoribbon by introducing impurities such as B or N to the arrangement is a unique method to tune the electronic characteristics of GNRs thus inducing semiconductive, half-metallic, or metallic functionalities [197]. Further, half-metallicity was attained on substituting the middle zig-zag chains of a ZGNR by B-N chains [198].

4.7   Graphene

The functionalization of GR is done by two methods: covalent and non-covalent. Nucleophilic substitution reaction is one such covalent method in which various primary amines and amino acids are attached with surface of graphite oxide in order to improve solubility [199]. Electrophilic substitution reaction is another such method to functionalize GR which involves the elimination of a hydrogen atom under the influence of an electrophile [200–206]. Electrophilic substitution of aryl diazonium salt on the surface of surfactant-wrapped GR causes the functionalization of GR resulting into functionalized GR being highly dispersible in DMF [201, 202]. Furthermore, a non-covalent approach can be carried out through the wrapping of polymers, adsorption of surfactants, or minor aromatic compounds, and functionalizing GR with porphyrins or biomolecules such as deoxyribonucleic acid (DNA) and peptides [207]. Stable dispersal of reduced GO in several organic solvents can be done via non-covalent functionalization with ammonia–hydrazine mixture [208]. Hence, functionalized GR is appreciably been dispersed in many solvents and exhibits satisfactory electrical conductivity. A novel non-covalent functionalization approach to exfoliate and stabilize chemically converted GR along with low-temperature exfoliated GR (LTEG) in aqueous solution is performed using thionine thereby improving electrical conductivity [209]. Wang et al. [210] have reported 3-Hexylthiophene (P3HT)/fullerene/GR ternary structure as an effective organic polymer for solar cell applications. For the similar application, Hsu's team [211] reported a sandwiched structure composed of GR/tetracyanoquinodimethane (TCNQ)/GR stacked films and showed P3HT/PCBM on multi-layered GR/TCNQ as a bulk heterojunction polymer solar cell with remarkable conductivity and transparency. Similarly, Wang and his co-workers [212] demonstrated the photovoltaic applications of a solution-processable functionalized GR (SPF Graphene). This is achieved on introducing P3HT hybrid thin films onto the functionalization groups of SPFGraphene. Also, Liu' group [213] developed a flexible film from GO along with polyethylene terephthalate (PET) for electronic applications. GR has also been investigated for energy storage applications, and certain composites like GR/CNT have been applied into lithium–ion batteries [214]. Another form of energy storage, i.e., supercapacitors or ultracapacitors involves the incorporation of highly conductive PANI with GO or reduced GO (rGO); moreover, rGO sheets processed with polyethylemimine (PEI) along with MWCNT are shown to give an average capacitance of 120 F/g at a scan rate of 1 V/s [215]. Further, Xu et al. [216] studied hydroxypropyl-b-cyclodextrin (HPCD) in composite with GR and tetra-phenyl-porphyrin (TPP) forming HPCD/GO/TPP to exhibit great electrocatalytic performance in the reduction and oxidation of hemoglobin (Hb) with high sensitivity having a detection limit of 5×10^9 M. In another report, Wua et al. [217] developed a nanocomposite film composing of (GOD)/Pt/GR/chitosan being rapid and sensitive in glucose quantification process. Several other composites of GR are being investigated for many such device applications.

4.8   Graphite

Graphite is found to be functionalized via surface adsorption techniques: physisorption and chemisorption. Physisorption functionalization is derived from physical interactions, viz., hydrogen bonding, hydrophobic, and π–π stacking interactions, as well as Van der Waals and electrostatic forces. This approach of modification only negligibly affects the chemistry of pure graphite [218]. Graphite functionalized via this method involves the interaction with 4-n-octyl-4'-cyanobiphenyl [219], trimesic acid [220], pyridine [221, 222], pentafluoropyridine [221], 9-anthracene carboxylic acid [223] deoxyribonucleic acid [224], quinolone [225], 1-pyrenesulfonic acid sodium salt [226], 1,3,6,8-pyrenetetrasulfonic acid [227], pyrene [228], fluorinated benzene derivatives [221], and 1-pyrenecarboxylic acid [229]. Chemisorption method occurs via covalent bonding of functionalizing molecules on the surface of graphite. However, it is impractical to only functionalize the base-plane surface without concurrent modification of edge-plane surface [230]. For instance, Hummers method was employed to develop a non-conductive insulating material from graphite. This modification arises from the base-plane functionalization which causes hybridization changes from sp^2 to sp^3 [231, 232].

4.9 Diamond and Nanodiamond

Diamonds are found to have three different terminating atoms, namely, hydrogen, oxygen, and nitrogen which is caused during its synthesis. Functionalization of H-terminating diamonds is done through photochemical approach and diazonium grafting. Photochemical approach has been used to functionalize diamonds with several short and long alkenes bearing amines [233], carboxylic acids [234, 235], thiols [236, 237], ethers [238, 239], alcohols [240–242], perfluorinated alkanes [243], pyridine [244], ferrocene [245], and phosphoric acids [246]. Grafting of diazonium salts enables the creation of C–C bonds among diamond's surface and aromatic organic components. Grafting is done electrochemically by immersing a boron-doped polycrystalline diamond (BDD) electrode in a solution of diazonium salt [247]. Moreover, O-terminating diamonds are functionalized via silanization, esterification, and thermal alkylation. Silanization was performed on OH- terminating diamond surfaces by Notsu et al. [248] by submerging BDD in an EtOH solution containing 2% APTES for 2 h. This approach of splicing improved water contact angle enhanced the charge-transfer kinetics with $[Fe(CN)_6]^{3-/4-}$ redox pair as a result of substantial electrostatic interactions with anion amine. Esterification in OH– terminating BDDs is reported to bind pyrene alkyl carboxylic acid [249], biotin [250], and benzophenone [251] covalently. It further demonstrated that biotin grafting utilizes its strong biotin attraction to fluorescently detect FITC-labeled DNA, while pyrene on BDD's surface exhibits photoelectrochemical nature. Benzophenone demonstrates attractive photochemical immobilization of DNA, proteins, and peptides owing to its effective photoactive group which is stable under ambient conditions and protic solvent. In the case of N– terminating diamonds, it is essentially analogous to OH– terminating diamonds. The primary-amine atoms form amid bonds with molecules consisting of activated RCOOH moieties. Further, bifunctional linkers, viz., succinyl chloride and activated esters of adipic, trimesic as well as terephthalic acid can achieve a layer of COOH functional group on NH_2 terminated surface [252, 253]. High energy ions are currently being used to implant defects in nanodiamond lattices

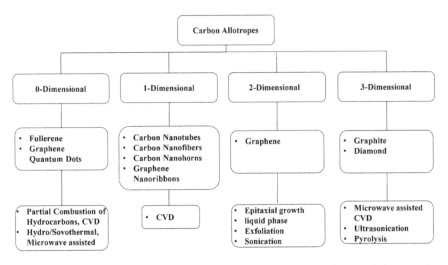

Fig. 2 Flowchart categorizing different allotropes of carbon according to their respective dimensionalities and their synthesis routes

[254]. Such defects incorporated into the sp^3 structure create color centers, thus introducing new atoms or vacancies into the lattice hence forming fluorescence nanodiamonds. Nanodiamonds make up for a variety of color centers such as N-vacancy centers, Ni–N complex centers, and Si vacancy centers [255], which upon photo-excitation result in the emission of bright and non-bleaching fluorescence with emission wavelengths extending from visible to NIR, henceforth making it viable for fluorescence tracking and imaging [256]. Furthermore, Bollina et al. [257] investigated the performance of diamond composites with Ag, Al, and Cu as heat spreader. In addition, further blending with boron, silicon, and chromium improved mechanical and thermal conductions of the composites. However, nanodiamonds in contrast to diamonds are better explored to form composites with polymers such as thermosetting, thermoplastic, and elastomers. These composites have wide applications in electronics, biomedical, packaging, membranes, and so on [258]. Moreover, the synergistic activity of nanodiamonds with polymers and other fillers enhances its mechanical properties, bioimaging, and drug deliver functionalities [259]. To summarize, the flowchart in Fig. 2 highlights the contents covered in this chapter.

5 Conclusion

This chapter covers topics like structure and properties of several carbon allotropes, elucidating its remarkable characteristics which enables its effective role in various device applications. Furthermore, it reviews the recent updates in carbon materials for various device applications by comparing different literatures to get a wide scope of understanding from scratch. Overall, carbon materials have shown to put the field of nanotechnology on the map for researchers and scientists to easily access its unprecedented functionalities and versatility. Carbon-based nanomaterials have achieved a plethora of attention for applications like super or ultracapacitors, organic photovoltaic devices, sensors, various FETs and diodes, memory devices, and so on. Moreover, emphasis on graphene being the mother of all carbon-based materials and the base for the development of different nanostructures of carbon has been put forward. Additionally, structural descriptions also include sp^2 and sp^3 hybridization states giving rise to several differences in electronic properties among the allotropes. In order to achieve such extraordinary properties, it is imperative to understand its method of production. Many synthesis techniques have been concisely discussed to comprehend the bulk productions of allotropes with pristine qualities in accordance to achieving high efficiency for large-scale device fabrication. Similar methodologies are being adopted to synthesize various carbon nanomaterials with minor modification in the parameters during the experimental conditions. However, these materials in its pure form are incapable of being obtained to its absolute potential in many instances. In such cases, its performance is enhanced through post-synthesis functionalization and composite creation with other organic and inorganic molecules. Moreover, through post-synthesis modifications, it has been observed to reduce defects such as wrinkles, crumples, and ripples in graphene which hinders its brilliance in devices. Eventually, the highly efficient carbon-based nanomaterial can be fabricated to devices at a large scale, thereby conquering markets for its best possible performance in conventional electronic components. Additionally, these materials are cost-effective, easy to manufacture, and convenience in storage and transport which make it extremely attractive in industrialization.

References

1. Tiwari SK, Kumar V, Huczko A et al (2016) Magical allotropes of carbon: prospects and applications, critical reviews in solid state and materials sciences. Crit Rev Solid State Mater Sci 41:257–317
2. Kroto H, Heath J, O'Brien S et al (1985) C60: buckminsterfullerene. Nature 318:162–163
3. Iijima S (1991) Helical microtubules of graphitic carbon. Nature 354:56–58
4. Novoselov KS, Geim AK, Morozov SV, Jiang D, Zhang Y, Dubonos SV, Grigorieva IV, Firsov AA (2004) Electric field effect in atomically thin carbon films. Science 306(5696):666–669
5. Speranza G (2019) The role of functionalization in the applications of carbon materials: an overview. C. https://doi.org/10.3390/c5040084
6. Camargo PHC, Satyanarayana KG, Wypych F (2009) Nanocomposites: synthesis, structure, properties and new application opportunities. Mater Res 12(1):1–39
7. Geim AK, Novoselov KS (2007) The rise of graphene. Nat Mater 6(3):183–191
8. Haddon RC (1997) C60: sphere or polyhedron? J Am Chem Soc 119:1797–1798
9. Diederich F, Ettl R, Rubin Y, Whetten RL, Beck R, Alvarez M et al (1991) The higher fullerenes: isolation and characterization of C76, C84, C90, C94, and C70O, an oxide of D5h–C70. Science 252:548–451
10. Krueger A (2010) Carbon materials and nanotechnology. Wiley-VCH, Weinheim, pp 33–122

11. Yulong Y, Xinsheng P (2016) Recent advances in carbon-based dots for electroanalysis. Analyst 141:2619–2628
12. Namdari P, Negahdari B, Eatemadi A (2017) Synthesis, properties and biomedical applications of carbon-based quantum dots: an updated review. Biomed Pharmacother 87:209–222
13. Li D, Jing PT, Sun LH, An Y, Shan XY et al (2018) Near-infrared excitation/emission and multiphoton-induced fluorescence of carbon dots. Adv Mater. https://doi.org/10.1002/adma.201705913
14. Li H, He X, Kang Z, Liu Y (2010) Water-soluble fluorescent carbon quantum dots and photocatalyst design. Angew Chem Int Ed 49:4430–4434
15. Campuzano S, Sedeño PY, Pingarrón JM (2019) Carbon dots and graphene quantum dots in electrochemical biosensing. Nanomaterials. https://doi.org/10.3390/nano9040634
16. Sun H, Wu L, Wei W, Qu X (2013) Recent advances in graphene quantum dots for sensing. Mater Today 16:433–442
17. Li L, Wu G, Yang G, Peng J, Zhao J, Zhu JJ (2013) Focusing on luminescent graphene quantum dots: current status and future perspectives. Nanoscale 5:4015–4039
18. Kim S, Hwang SW, Kim MK et al (2012) Anomalous behaviors of visible luminescence from graphene quantum dots: interplay between size and shape. ACS Nano 6(9):8203–8208
19. Peng J, Gao W, Gupta BK et al (2012) Graphene quantum dots derived from carbon fibers. Nano Lett 12:844–849
20. Tian P, Tang L, Teng KS, Lau SP (2018) Graphene quantum dots from chemistry to applications. Mater Today Chem 10:221–258
21. Zhu S, Song Y, Wang J, Wan H, Zhang Y, Ning Y, Yang B (2017) Photoluminescence mechanism in graphene quantum dots: quantum confinement effect and surface/edge state. Nano Today 13:10–14
22. Yao X, Niu X, Ma K, Huang P, Grothe J, Kaskel S, Zhu Y (2017) Graphene quantum dots-capped magnetic mesoporous silica nanoparticles as a multifunctional platform for controlled drug delivery, magnetic hyperthermia, and photothermal therapy. Small. https://doi.org/10.1002/smll.201602225
23. Zhang D, Wen L, Huang R, Wang H, Hu X, Xing D (2018) Mitochondrial specific photodynamic therapy by rare-earth nanoparticles mediated near-infrared graphene quantum dots. Biomaterials 153:14–26
24. Sahub C, Tuntulani T, Nhujak T, Tomapatanaget B (2018) Effective biosensor based on graphene quantum dots via enzymatic reaction for directly photoluminescence detection of organophosphate pesticide. Sens Actuators B Chem 258:88–97
25. Ijima S, Ichihashi T (1993) Single-shell carbon nanotubes of 1-nm diameter. Nature 363(6430):603–605
26. Bethune DS, Kiang CH, De Vries MS, Gorman G, Savoy R, Vazquez J, Beyers R (1993) Cobalt-catalyzed growth of carbon nanotubes with single-atomic-layer walls. Nature 363(6430):605–607
27. Chernozatonskii L (2003) Three-terminal junctions of carbon nanotubes: synthesis structures, properties and applications. J Nanopart Res 5:473–484
28. Satishkumar BC, Thomas PJ, Govindaraj A, Rao CNR (2000) Y-junction carbon nanotubes. Appl Phys Lett Doi 10(1063/1):1319185
29. de Heer WA, Bacsa WS, Châtelain A, Gerfin T, Baker RH, Forro L, Ugarte D (1995) Aligned carbon nanotube films: production and optical and electronic properties. Science 268(5212):845–847
30. Ouyang M, Huang JL, Lieber CM (2002) Fundamental electronic properties and applications of single-walled carbon nanotubes. Acc Chem Res 35:1018–1025
31. Ouyang M, Huang JL, Lieber CM (2002) Scanning tunneling microscopy studies of the one-dimensional electronic properties of single-walled carbon nanotubes. Annu Rev Phys Chem 53:201–220
32. Serp P, Corrias M, Kalck P (2003) Carbon nanotubes and nanofibers in catalysis. Appl Catal A-Gen 253(2):337–358

33. Vamvakaki V, Tsagaraki K, Chaniotakis N (2006) Carbon nanofiber-based glucose biosensor. Anal Chem 78(15):5538–5542
34. Kim SU, Lee KH (2004) Carbon nanofiber composites for the electrodes of electrochemical capacitors. Chem Phys Lett 400:253–257
35. Huang J, Liu Y, You T (2009) Carbon nanofiber based electrochemical biosensors: a review. Anal Methods. https://doi.org/10.1039/b9ay00312f
36. Karousis N, Martinez IS, Ewels CP, Tagmatarchis N (2016) Structure, properties, functionalization, and applications of carbon nanohorns. Chem Rev 116:4850–4883
37. Ajima K, Yudasaka M, Murakami T, Maigne´ A, Shiba K, Iijima S (2005) Carbon nanohorns as anticancer drug carriers. Mol Pharm 2(6):475–480
38. Shenderova OA, Lawson BL, Areshkin D, Brenner DW (2001) Predicted structure and electronic properties of individual carbon nanocones and nanostructures assembled from nanocones. Nanotechnology 12:191–197
39. Zhang S, Yao Z, Zhao S, Zhang E (2006) buckling and competition of energy and entropy lead conformation of single-walled carbon Nanocones. Appl Phys Lett . https://doi.org/10.1063/1.2358109
40. Urita K, Seki S, Utsumi S, Noguchi D, Kanoh H, Tanaka H, Hattori Y, Ochiai Y, Aoki N, Yudasaka M et al (2006) Effects of gas adsorption on the electrical conductivity of single-wall carbon nanohorns. Nano Lett 6:1325–1328
41. Murata K, Kaneko K, Kokai F, Takahashi K, Yudasaka M, Iijima S (2000) Pore structure of single-wall carbon nanohorn aggregates. Chem Phys Lett 331:14–20
42. Ohba T, Omori T, Kanoh H, Yudasaka M, Iijima S, Kaneko K (2005) Interstitial nanopore change of single wall carbon nanohorn assemblies with high temperature treatment. Chem Phys Lett 389:332–336
43. Bekyarova E, Hanzawa Y, Kaneko K, Albero JS et al (2002) Cluster-mediated filling of water vapor in intratube and interstitial nanospaces of single-wall carbon nanohorns. Chem Phys Lett 366:463–468
44. Bekyarova E, Murata K, Yudasaka M, Kasuya D, Iijima S, Tanaka H, Kahoh H, Kaneko K (2003) Single-wall nanostructured carbon for methane storage. J Phys Chem B 107:4681–4684
45. Nakada K, Fujita M, Dresselhaus G, Dresselhaus MS (1996) Edge state in graphene ribbons: nanometer size effect and edge shape dependence. Phys. Rev. B 54:17954–17961
46. Son YW, Cohen ML, Louie SG (2006) Energy gaps in graphene nanoribbons. Phys Rev Lett. https://doi.org/10.1103/PhysRevLett.97.216803
47. Shen H, Shi Y, Wang X (2015) Synthesis, charge transport and device applications of graphene nanoribbons. Synth Met. https://doi.org/10.1016/j.synthmet.2015.07.010
48. Son YW, Cohen ML, Louie SG (2006) Half-metallic graphene nanoribbons. Nature 444:347–349
49. Duttaa S, Pati SK (2010) Novel properties of graphene nanoribbons: a review. J Mater Chem 20:8207–8223
50. Guimarães MHD, Shevtsov O, Waintal X, van Wees BJ (2012) From quantum confinement to quantum Hall effect in graphene nanostructures. Phys Rev B. https://doi.org/10.1103/PhysRevB.85.075424
51. Baringhaus J, Ruan M, Edler F (2014) Exceptional ballistic transport in epitaxial graphene nanoribbons. Nature 506(7488):349–354
52. Li X, Wang X, Zhang L, Lee S, Dai H (2008) Chemically derived, ultrasmooth graphene nanoribbon semiconductors. Science 319(5867):1229–1232
53. Daniela E, Nemnes GA, Ioan U (2014) Spintronic devices based on graphene nanoribbons with transition metal impurities. Towards Space Appl INCAS BULLETIN 6(1):45–56
54. Chen J, Jang C, Xiao S et al (2008) Intrinsic and extrinsic performance limits of graphene devices on SiO_2. Nature Nanotech 3:206–209
55. Bonaccorso F, Colombo L, Yu G (2015) Graphene, related two-dimensional crystals, and hybrid systems for energy conversion and storage. Science. https://doi.org/10.1126/science.1246501

56. Allen MJ, Tung VC, Kaner RB (2010) Honeycomb carbon: a review of graphene. Chem Rev 110:132–145
57. Williams JR, DiCarlo L, Marcus CM (2007) Quantum hall effect in a gate-controlled p-n junction of graphene. Science 317(5838):638–641
58. Novoselov KS, Jiang Z, Zhang Y, Morozov SV, Stormer HL, Zeitler U, Maan JC, Boebinger GS, Kim P, Geim AK (2007) Room-temperature quantum hall effect in graphene. Science 315(5817):1379
59. Areshkin DA, Gunlycke D, White CT (2007) Ballistic transport in graphene nanostrips in the presence of disorder: importance of edge effects. Nano Lett 7(1):204–210
60. Zeng Z, Tang K et al (2013) Tunable band gap in few-layer graphene by surface adsorption. Sci Rep. https://doi.org/10.1038/srep01794
61. Novoselov KS, Geim AK, Morozov SV et al (2005) Two-dimensional gas of massless dirac fermions in graphene. Nature 438:197–200
62. Yin LJ, Bai KK, Wang WX, Li SY, Zhang Y, He L (2017) Landau quantization of dirac fermions in graphene and its multilayers. Front Phys. https://doi.org/10.1007/s11467-017-0655-0
63. Meyer JC, Kisielowski C, Erni R, Rossell MD, Crommie MF, Zettl A (2008) Direct imaging of lattice atoms and popological defects in graphene membranes. Nano Lett 8(11):3582–3586
64. Liu L, Qing M, Wang Y, Chen S (2015) Defects in graphene: generation, healing, and their effects on the properties of graphene: a review. J Mater Sci Technol 31(6):599–606
65. Krasheninnikov AV, Lehtinen PO, Foster AS, Nieminen RM (2006) Bending the rules: contrasting vacancy energetics and migration in graphite and carbon nanotubes. Chem Phys Lett 418(1–3):312–136
66. Liang Z, Xu Z, Yan T, Ding F (2013) Atomistic simulation and the mechanism of graphene amorphization under electron irradiation. Nanoscale 6:2082–2086
67. Malola S, Häkkinen H, Koskinen P (2010) Structural, chemical, and dynamical trends in graphene grain boundaries. Phys Rev B. https://doi.org/10.1103/PhysRevB.81.1
68. Ö. Girit C, Meyer JC, Erni R, et al (2009) Graphene at the edge: stability and dynamics. Science 323(5922):1705–170865447
69. Tian W, Li W, Yu W, Liu X (2017) A review on lattice defects in graphene: types, generation, effects and regulation. Micromachines. https://doi.org/10.3390/mi8050163
70. Biel B, Blase X, Triozon F, Roche S (2009) Anomalous doping effects on charge transport in graphene nanoribbons. Phys Rev Lett. https://doi.org/10.1103/PhysRevLett.102.096803
71. Chien SK, Yang YT, Chen CK (2011) Influence of hydrogen functionalization on thermal conductivity of graphene: Nonequilibrium molecular dynamics simulations. Appl Phys Lett . https://doi.org/10.1063/1.3543622
72. Hao F, Fang D, Xu Z (2011) Mechanical and thermal transport properties of graphene with defects. Appl Phys Lett. https://doi.org/10.1063/1.3615290
73. Kim EA, Neto AHC (2008) Graphene as an electronic membrane. EPL. https://doi.org/10.1209/0295-5075/84/57007
74. Deng S, Berry V (2016) Wrinkled, rippled and crumpled graphene: an overview of formation mechanism, electronic properties, and applications. Mater. Today 19(4):197–212
75. Xu K, Cao P, Heath JR (2009) Scanning tunneling microscopy characterization of the electrical properties of wrinkles in exfoliated graphene monolayers. Nano Lett 9(12):4446–4451
76. Cranford SW, Buehler MJ (2011) Packing efficiency and accessible surface area of crumpled graphene. Phys Rev B. https://doi.org/10.1103/PhysRevB.84.205451
77. Jung S, Rutter GM, Klimov NN et al (2011) Evolution of microscopic localization in graphene in a magnetic field from scattering resonances to quantum dots. Nat Phys 7:245–251
78. de Parga ALV, Calleja F, Borca B, Passeggi MCG Jr, Hinarejos JJ, Guinea F, Miranda R (2008) Periodically rippled graphene: growth and spatially resolved electronic structure. Phys Rev Lett. https://doi.org/10.1103/PhysRevLett.100.056807
79. Schneider M, Brouwer PW (2014) Quantum corrections to transport in graphene: a trajectory-based semiclassical analysis. New J Phys. https://doi.org/10.1088/1367-2630/16/7/073015

80. Liu N, Pan Z, Fu L, Zhang C, Dai B, Liu Z (2011) The origin of wrinkles on transferred graphene. Nano Res. https://doi.org/10.1007/s12274-011-0156-3

81. Calado VE, Schneider GF, Theulings AMMG, Dekker C, Vandersypen LMK (2012) Formation and control of wrinkles in graphene by the wedging transfer method. Appl Phys Lett. https://doi.org/10.1063/1.4751982

82. Chung DDL (2002) Review graphite. J Mater 37(8):1475–1489

83. Klett J, Hardy R, Romine E, Walls C, Burchell T (2000) High-thermal-conductivity, mesophase-pitch-derived carbon foams: effect of precursor on structure and properties. Carbon 38:953–973

84. Lin W, Yuan J, Sundén B (2011) Review on graphite foam as thermal material for heat exchangers. Paper presented at the World Renewable Energy Congress, Sweden, 8–13 May 2011

85. Klett JW, Mcmillan AD, Gallego NC, Walls CA (2004) The role of structure on the thermal properties of graphite foams. J Mater 39:3659–3676

86. Chernyavets AN (2008) Production of high-quality graphite crucibles for metallurgy. Solid Fuel Chem 42(2):98–102

87. Deprez N, McLachlan DS (1988) The analysis of the electrical conductivity of graphite conductivity of graphite powders during compaction. J Phys D Appl Phys 21(1):101–107

88. Champion FC (1963) Electronic properties of diamond. Butterworths, London

89. Bull C, Garlick GFJ (1950) The luminescence of diamonds. Proc Phys Soc A 63(11):1283–1291

90. Wort CJH, Balmer RS (2008) Diamond as an electronic material. Mater Today 11(1–2):22–28

91. El-Hajj H, Denisenko A, Kaiser A, Balmer RS, Kohn E (2008) Diamond MISFET based on boron delta-doped channel. Diam Relat Mater 17(7–10):1259–1263

92. Field JE (1979) The properties of diamond. Academic Press, New York

93. Mochalin VN, Shenderova O, Ho D, Gogotsi Y (2011) The properties and applications of nanodiamonds. Nat Nanotechnol 7(1):11–23

94. Nafisi S, Maibach HI (2017) Nanotechnology in cosmetics. Cosmetic Sci Technol. https://doi.org/10.1016/B978-0-12-802005-0.00022-7

95. Moriguchi H, Ohara H, Tsujioka M (2016) History and applications of diamond-like carbon manufacturing processes. Sci Tech Rev 82:52–58

96. Ito H, Yamamoto K (2017) Mechanical and tribological properties of DLC films for sliding parts. Kobelco Technol Rev 35:55–60

97. Roy RK, Lee KR (2007) Biomedical applications of diamond-like carbon coatings: a review. J Biomed Mater Res B Appl Biomater 83(1):72–84

98. Takehara H, Fujiwara M, Arikawa M, Diener MD, Alford JM (2005) Experimental study of industrial scale fullerene production by combustion synthesis. Carbon 43(2):311–319

99. Hepp H, Siegmann K, Sattler K (1995) Multiphoton ionization mass spectroscopy of fullerenes in methane diffusion flames. Proc Mater Res Soc Symp 359:517–522

100. Howard JB, McKinnon JT, Johnson ME, Makarovsky Y, Lafleur AL (1992) Production of C60 and C70 fullerenes in benzene-oxygen flames. J Phys Chem 96(16):6657–6662

101. Terranova ML, Sessa V, Rossi M (2006) The world of carbon nanotubes: an overview of CVD growth methodologies. Chem. Vap. Deposition 12:315–325

102. Feng L, Xie N, Zhong J (2014) Carbon nanofibers and their composites: a review of synthesizing, properties and applications. Materials 7(5):3919–3945

103. Zahid MU, Pervaiz E, Hussain A, Shahzad MI, Niazi MBK (2018) Synthesis of carbon nanomaterials from different pyrolysis techniques: a review. Mater Res Express. https://doi.org/10.1088/2053-1591/aac05b

104. Chen Z, Zhang W, Palma CA, Rizzini AL et al (2016) Synthesis of graphene nanoribbons by ambient-pressure chemical vapor deposition and device integration. J Am Chem Soc 138(47):15488–15496

105. Sakaguchi H, Kawagoe Y, Hirano Y, Iruka T, Yano M, Nakae T (2014) Width-controlled sub-nanometer graphene nanoribbon films synthesized by radical- polymerized chemical vapor deposition. Adv Mater. https://doi.org/10.1002/adma.201305034

106. May PW (2000) Diamond thin films: a 21st-century material. Philos Trans R Soc a 358(1766):473–495
107. Matsumoto S, Sato Y, Tsutsumi M, Setaka N (1982) Growth of diamond particles from methane-hydrogen gas. J Mater Sci 17:3106–3112
108. Matsumoto S, Sato Y, Kamo M, Setaka N (1982) Vapor deposition of diamond particles from methane. Jpn J Appl Phys 21(Part 2, no 4):183–185
109. Xu M, Li Z, Zhu X, Hu N, Wei H, Yang Z, Zhang Y (2013) Hydrothermal/solvothermal synthesis of graphene quantum dots and their biological applications. Nano Biomed Eng 5(2):65–71
110. Sangam S, Gupta A, Shakeel A, Bhattacharya R, Sharma AK, Suhag D, Chakrabarti S et al (2018) Sustainable synthesis of single crystalline sulphur-doped graphene quantum dots for bioimaging and beyond. Green Chem. https://doi.org/10.1039/C8GC01638K
111. Pan D, Zhang J, Li Z, Wu M (2010) Hydrothermal route for cutting graphene sheets into blue-luminescent graphene quantum dots. Adv Mater. https://doi.org/10.1002/adma.200902825
112. Tian R, Zhong S, Wu J, Jiang W, Shen Y, Jiang W, Wang T (2016) Solvothermal method to prepare graphene quantum dots by hydrogen peroxide. Opt Mater 60:204–208
113. Grewal AS, Kumar K, Redhu S, Bhardwaj S (2013) Microwave assisted synthesis: a green chemistry approach. Int Res J Pharm App Sci 3(5):278–285
114. Nguyen HY, Le XH, Dao NT et al (2019) Microwave-assisted synthesis of graphene quantum dots and nitrogen-doped graphene quantum dots: Raman characterization and their optical properties. Adv Nat Sci Nanosci Nanotechnol. https://doi.org/10.1088/2043-6254/ab1b73
115. Cala BF, Soriano ML, Sciortino A et al (2018) One-pot synthesis of graphene quantum dots and simultaneous nanostructured self-assembly via a novel microwave-assisted method: impact on triazine removal and efficiency monitoring. RSC Adv. https://doi.org/10.1039/c8ra04286a
116. Li LL, Ji J, Fei R et al (2012) A facile microwave avenue to electrochemiluminescent two-color graphene quantum dots. Adv Funct Mater. https://doi.org/10.1002/adfm.201200166
117. Mishra N, Boeckl J, Motta N, Iacopi F (2016) Graphene growth on silicon carbide: a review. Phys Status Solidi a. https://doi.org/10.1002/pssa.201600091
118. Yang W, Chen G, Shi Z et al (2013) Epitaxial growth of single-domain graphene on hexagonal boron nitride. Nat Mater 12(9):792–797
119. Teeter JD, Costa PS, Pour MM (2017) Epitaxial growth of aligned atomically precise chevron graphene nanoribbons on Cu(111). ChemComm 53(60):8463–8466
120. Vo TH, Shekhirev M, Kunkel DA (2013) Large-scale solution synthesis of narrow graphene nanoribbons. Nat Commun. https://doi.org/10.1038/ncomms4189
121. Zhang F, Hou PX, Liu C, Cheng HM (2016) Epitaxial growth of single-wall carbon nanotubes. Carbon 102:181–197
122. Lin X, Zhao W, Zhou W et al (2017) Epitaxial growth of aligned and continuous carbon nanofibers from carbon nanotubes. ACS Nano 11:1257–1263
123. Xu H, Zeigera BW, Suslick KS (2013) Sonochemical synthesis of nanomaterials. Chem Soc Rev 42:2555–2567
124. Ling C, Setzler G, Lin MW et al (2011) Electrical transport properties of graphene nanoribbons produced from sonicating graphite in solution. Nanotechnology. https://doi.org/10.1088/0957-4484/22/32/325201
125. Nair RV, Thomas RT, Sankar V et al (2017) Rapid, acid-free synthesis of high-quality graphene quantum dots for aggregation induced sensing of metal ions and bioimaging. ACS Omega 2:8051–8061
126. Jeong SH, Ko JH, Park JB, Park W (2004) A sonochemical route to single-walled carbon nanotubes under ambient conditions. J Am Chem Soc 126(49):15982–15983
127. Galimov ÉM, Kudin AM, Skorobogatskii VN et al (2004) Experimental corroboration of the synthesis of diamond in the cavitation process. Dokl Phys 49:150–153
128. Ciesielski A, Samorı̀ P (2014) Graphene via sonication assisted liquid-phase exfoliation. Chem Soc Rev. https://doi.org/10.1039/c3cs60217f
129. Sarkar S, Gandla D, Venkatesh Y et al (2016) Graphene quantum dots from graphite by liquid exfoliation showing excitation-independent emission, fluorescence upconversion and delayed fluorescence. Phys Chem Chem Phys. https://doi.org/10.1039/C6CP01528J

130. Guldi DM, Luo C, Swartz A et al (2002) Molecular engineering of C60-based conjugated oligomer ensembles: modulating the competition between photoinduced energy and electron transfer processes. J Org Chem 67(4):1141–1152
131. Nierengarten JF (2004) Chemical modification of C60 for materials science applications. New J Chem 28:1177–1191
132. Campidelli S, Deschenaux R, Eckert JF, Guillon D, Nierengarten JF (2002) Liquid-crystalline fullerene-oligophenylenevinylene conjugates. Chem Commun (Camb) 6:656–657
133. Rispens MT, Sánchez L, Beckers EHA et al (2003) Supramolecular fullerene architectures by quadruple hydrogen bonding. Synth Met 135–136:801–803
134. Zhaoa J, Huang X, Jin P, Chen Z (2015) Magnetic properties of atomic clusters and endohedral metallofullerenes. Coord Chem Rev 289–290:315–340
135. Penkova AV, Acquah SFA, Piotrovskiy LB (2017) Fullerene derivatives as nano-additives in polymer composites. Russ Chem Rev 86(6):530–566
136. Kausar A (2016) Advances in polymer/fullerene nanocomposite: a review on essential features and applications. Polym Plast Technol Eng 56(6):594–605
137. Yu G, Gao J, Hummelen JC, Wudl F, Heeger AJ (1995) Polymer photovoltaic cells: enhanced efficiencies via a network of internal donor-acceptor heterojunctions. Science 270(5243):1789–1791
138. Ma W, Yang C, Gong X, Lee K, Heeger AJ (2005) Thermally stable, efficient polymer solar cells with nanoscale control of the interpenetrating network morphology. Adv Funct Mater. https://doi.org/10.1002/adfm.200500211
139. Li G, Shrotriya V, Huang J (2005) High-efficiency solution processable polymer photovoltaic cells by self-organization of polymer blends. Nat Mater 4(11):864–868
140. Li F, Li Y, Ge Z, Zhu D, Song Y, Fang G (2000) Synthesis and optical limiting properties of polycarbonates containing fullerene derivative. J Phys Chem Sol 61:1101–1103
141. Qian Z, Ma J, Shan X et al (2013) Surface functionalization of graphene quantum dots with small organic molecules from photoluminescence modulation to bioimaging applications: an experimental and theoretical investigation. RSC Adv 3:14571–14579
142. Xiaoyan Z, Zhangyi L, Zaijun L (2017) Fabrication of valine-functionalized graphene quantum dots and its use as a novel optical probe for sensitive and selective detection of Hg^{2+}. Spectrochim Acta A 171:415–424
143. Lakshmanakumar M, Nesakumar N, Sethuraman S (2019) Functionalized graphene quantum dot interfaced electrochemical detection of cardiac troponin I: an antibody free approach. Sci Rep. https://doi.org/10.1038/s41598-019-53979-5
144. Dinari M, Momeni MM, Goudarzirad M (2016) Dye- sensitized solar cells based on nanocomposite of polyaniline/graphene quantum dots. J Mater Sci 51:2964–2971
145. Liu W, Yan X, Chen J, Feng Y, Xue Q (2013) Novel and high-performance asymmetric micro-supercapacitors based on graphene quantum dots and polyaniline nanofibers. Nanoscale 5:6053–6062
146. Gavgani JN, Hasani A, Nouri M, Mahyari M, Salehi A (2016) Highly sensitive and flexible ammonia sensor based on S and N co-doped graphene quantum dots/polyaniline hybrid at room temperature. Sens Actuat B Chem 229:239–248
147. Bruchez M, Moronne M, Gin P, Weiss S, Alivisatos AP (1998) Semiconductor nanocrystals as fluorescent biological labels. Science 281:2013–2016
148. Jeon SS, Kim C, Ko J, Im SS (2011) Spherical polypyrrole nanoparticles as a highly efficient counter electrode for dye-sensitized solar cells. J Mater Chem 21:8146–8151
149. Zhou X, Ma P, Wang A, Yu C, Qian T, Wu S, Shen J (2015) Dopamine fluorescent sensors based on polypyrrole/graphene quantum dots core/shell hybrids. Biosens. Bioelectron 64:404–410
150. Woo CH, Holcombe TW, Unruh DA, Sellinger A, Frechet JM (2010) Phenyl Vs alkyl poly-thiophene: a solar cell comparison using a vinazene derivative as acceptor. J Chem Mater 22:1673–1679
151. Puniredd SR, Kiersnowski A, Battagliarin G, Zajaczkowski W, Wong WWH, Kirby N, Mullen K, Pisula W (2013) Polythiophene–perylene diimide heterojunction field-effect transistors. J Mater Chem C 1:2433–2440

152. Das S, Samanta S, Chatterjee DP, Nandi AK (2013) Thermosensitive water-soluble poly (ethylene glycol)-based polythiophene graft copolymers. J Polym Sci Part A: Polym Chem 51:1417–1427

153. Routh P, Das S, Shit A, Bairi P, Das P, Nandi AK (2013) Graphene quantum dots from a facile sono-fenton reaction and its hybrid with a polythiophene graft copolymer toward photovoltaic application. ACS Appl Mater Interf 5:12672–12680

154. Zhou Y, Qu ZB, Zeng Y, Zhou T, Shi G (2014) A novel composite of graphene quantum dots and molecularly imprinted polymer for fluorescent detection of paranitrophenol. Biosens Bioelectron 52:317–323

155. Yang HB, Dong YQ, Wang X, Khoo SY, Liu B (2014) Cesium carbonate functionalized graphene quantum dots as stable electron-selective layer for improvement of inverted polymer solar cells. ACS Appl Mater Interf 6:1092–1099

156. Jeon IY, Chang DW, Kumar NA, Baek JB (2011) Functionalization of carbon nanotubes. Carbon Nanotubes—Polymer Nanocompos. https://doi.org/10.5772/18396

157. Hecht DS, Ramirez RJA, Briman M, Artukovic E, Chichak KS, Stoddart JF, Gruner G (2006) Bioinspired detection of light using a porphyrin-sensitized singlewall nanotube field effect transistor. Nano Lett 6:2031–2036

158. Hu L, Zhao Y-L, Ryu K, Zhou C, Stoddart JF, Gruner G (2008) Light-induced charge transfer in pyrene/CdSe-SWNT hybrids. Adv Mater 20:939–946

159. Kymakis E, Servati P, Tzanetakis P et al (2007) Effective mobility and photocurrent in carbon nanotube polymer composite photovoltaic cells. Nanotechnology. https://doi.org/10.1088/0957-4484/18/43/435702

160. Bergeret C, Cousseau J, Nunzi JM, Habak DH (2011) Improving the current density JSC of organic solar cells P3HT:PCBM by structuring the photoactive layer with functionalized SWCNTs. Sol Energy Mater Sol Cells 95:53–56

161. Jin SH, Park SH, Jeon S, Hong SH, Jun GH (2012) Highly dispersed carbon nanotubes in organic media for polymer: fullerene photovoltaic devices. Carbon 50:40–46

162. Raïssi M, Vignau L, Cloutet E, Ratier B (2015) Soluble carbon nanotubes/phthalocyanines transparent electrode and interconnection layers for flexible inverted polymer tandem solar cells. Org Electron 21:86–91

163. Reddy MLP, Divya V (2013) Visible-light excited red emitting luminescent nanocomposites derived from Eu^{3+}-phenanthrene-based fluorinated b-diketonate complexes and multi-walled carbon nanotubes. J Mater Chem C 1:160–170

164. Fanchini G, Eda G, Chhowalla M, Parekh BB (2007) Improved conductivity of transparent single-wall carbon nanotube thin films via stable postdeposition functionalization. Appl Phys Lett 10(1063/1):2715027

165. Kim KK, Geng HZ (2007) Effect of acid treatment on carbon nanotube-based flexible transparent conducting films. J Am Chem Soc 129:7758–7759

166. Ros TG, van Dillen AJ, Geus JW, Koningsberger DC (2002) Surface oxidation of carbon nanofibers. Chem Eur J. https://doi.org/10.1023/A:1024744131630

167. Hea P, Dai L (2004) Aligned carbon nanotube-DNA electrochemical sensors. Chem Commun. https://doi.org/10.1039/B313030B

168. Baker SE, Tse KY, Hindin E, Nichols BM, Clare TL, Hamers RJ (2005) Covalent functionalization for biomolecular recognition on vertically aligned carbon nanofibers. Chem Mater 17(20):4971–4978

169. Zhou H, Sui ZJ, Zhu J et al (2007) Characterization of surface oxygen complexes on carbon nanofibers by TPD. XPS and FT-IR. Carbon 45(4):785–796

170. Lim S, Yoon SH, Mochida I (2004) Surface modification of carbon nanofiber with high degree of graphitization. J Phys Chem B 108(5):1533–1536

171. Klein AV, TE Melechko McKnight, Retterer ST, Rack PD et al (2008) Surface characterization and functionalization of carbon nanofibers. J Appl Phys 10(1063/1):2840049

172. Poveda RL, Gupta N (2016) Electrical properties of CNF/Polymer composites. In: Carbon nanofiber reinforced polymer composites. Springer briefs in materials. Springer, Cham, p 71

173. Lee SH (2011) Mechanical and electrical properties of nonwoven coated with CNFs/PVDF-HFP composite. J Korean Soc Cloth Text 13:279–284
174. Lee SH (2012) Physical properties of silk fabrics coated by carbon nanofibers/poly(vinylidenefluoride-hexafloropropylene) composites. J Textile Sci Eng 49:119–125
175. Yang CM, Kim YJ, Endo M, Kanoh H, Yudasaka M, Iijima S, Kaneko K (2007) Nanowindow-regulated specific capacitance of supercapacitor electrodes of single-wall carbon nanohorns. J Am Chem Soc 129:20–21
176. Yang C-M, Kim Y-J, Miyawaki J, Kim YA, Yudasaka M, Iijima S, Kaneko K (2015) Effect of the size and position of ion-accessible nanoholes on the specific capacitance of single-walled carbon nanohorns for supercapacitor applications. J Phys Chem C 119:2935–2940
177. Comisso N, Berlouis LEA, Morrow J, Pagura C (2010) Changes in hydrogen storage properties of carbon nano-horns submitted to thermal oxidation. Int J Hydrogen Energy 35:9070–9081
178. Ohba T, Kanoh H, Kaneko K (2011) Superuniform molecular nanogate fabrication on graphene sheets of single wall carbon nanohorns for selective molecular separation of CO_2 and CH_4. Chem Lett 40:1089–1091
179. Ohba T, Kanoh H, Kaneko K (2012) Facilitation of water penetration through zero-dimensional gates on rolled-up graphene by cluster-chain-cluster transformations. J Phys Chem C 116:12339–12345
180. Pagona G, Sandanayaka ASD, Araki Y, Fan J, Tagmatarchis N, Charalambidis G, Coutsolelos AG, Boitrel B, Yudasaka M, Iijima S et al (2007) Covalent functionalization of carbon nanohorns with porphyrins: nanohybrid formation and photoinduced electron and energy transfer. Adv Funct Mater 17:1705–1711
181. Pagona G, Sandanayaka ASD, Hasobe T, Charalambidis G, Coutsolelos AG, Yudasaka M, Iijima S, Tagmatarchis N (2008) Characterization and photoelectrochemical properties of nanostructured thin film composed of carbon nanohorns covalently functionalized with porphyrins. J Phys Chem C 112:15735–15741
182. Kosaka M, Kuroshima S, Kobayashi K, Sekino S, Ichihashi T, Nakamura S, Yoshitake T, Kubo Y (2009) Single-wall carbon nanohorns supporting Pt catalyst in Direct methanol fuel cells. J Phys Chem C 113:8660–8667
183. Zhang L, Zheng N, Gao A, Zhu C, Wang Z, Wang Y, Shi Z, Liu Y (2012) A robust fuel cell cathode catalyst assembled with nitrogen-doped carbon nanohorn and platinum nanoclusters. J Power Sources 220:449–454
184. Brandao L, Passeira C, Gattia DM, Mendes A (2011) Use of single wall carbon nanohorns in polymeric electrolyte fuel cells. J Mater Sci 46:7198–7205
185. Eblagon KM, Brandao L (2015) RuSe electrocatalysts and single wall carbon nanohorns supports for the oxygen reduction reaction. J Fuel Cell Sci Technol 10(1115/1):4029422
186. Unni SM, Bhange SN, Illathvalappil R, Mutneja N, Patil KR, Kurungot S (2015) Nitrogen-induced surface area and conductivity modulation of carbon nanohorn and its function as an efficient metal-free oxygen reduction electrocatalyst for anion-exchange membrane fuel cells. Small 11:352–360
187. Vizuete M, Escalonilla JGM, Fierro JLG, Sandanayaka ASD, Hasobe T, Yudasaka M, Iijima S, Ito O, Langa FA (2010) Carbon nanohorn-porphyrin supramolecular assembly for photoinduced electron-transfer processes. Chem A 16:10752–10763
188. Pagona G, Zervaki GE, Sandanayaka ASD, Ito O, Charalambidis G, Hasobe T, Coutsoleos AG, Tagmatarchis N (2012) Carbon nanohorn-porphyrin dimer hybrid material for enhancing light-energy conversion. J Phys Chem C 116:9439–9449
189. Lodermeyer F, Costa RD, Casillas R, Kohler FTU, Wasserscheid P, Prato M, Guldi DM (2015) Carbon nanohorn-based electrolyte for dye-sensitized solar cells. Energy Environ Sci 8:241–246
190. Zhang S, Tang S, Lei J, Dong H, Ju H (2011) Functionalization of graphene nanoribbons with porphyrin for electrocatalysis and amperometric biosensing. J Electroanal Chem 656(1–2):285–288

191. Gunlycke D, Li J, Mintmire JW, White CT (2007) Altering low-bias transport in zigzag-edge graphene nanostrips with edge chemistry. Appl Phys Lett 10(1063/1):2783196
192. Xu B, Yin J, Xia YD, Wan XG, Jiang K, Liu ZG (2010) Electronic and magnetic properties of zigzag graphene nanoribbon with one edge saturated. Appl Phys Lett 10(1063/1):3402762
193. Sodi CF, Csányi G, Piscanec S, Ferrari AC (2008) Edge-functionalized and substitutionally doped graphene nanoribbons: electronic and spin properties. Phys Rev B. https://doi.org/10.1103/PhysRevB.77.165427
194. Maruyama M, Kusakabe K, Tsuneyuki S, Akagi K, Yoshimoto Y, Yamauchi J (2004) Magnetic properties of nanographite with modified zigzag edges. J Phys Chem Solids 65(2–3):119–122
195. Wu M, Pei Y, Zeng XC (2010) Planar tetracoordinate carbon strips in edge decorated graphene nanoribbon. J Am Chem Soc 132(16):5554–5555
196. Hod O, Barone V, Peralta JE, Scuseria GE (2007) Enhanced half-metallicity in edge-oxidized zigzag graphene nanoribbons. Nano Lett 7(8):2295–2299
197. Zheng XH, Wang XL, Abtew TA, Zeng Z (2010) Building half-metallicity in graphene nanoribbons by direct control over edge states occupation. J Phys Chem C 114(9):4190–4193
198. Dutta S, Manna AK, Pati SK (2009) Intrinsic half-metallicity in modified graphene nanoribbons. Phys Rev Lett. https://doi.org/10.1103/PhysRevLett.102.096601
199. Wang A, Yu W, Huang Z et al (2016) Covalent functionalization of reduced graphene oxide with porphyrin by means of diazonium chemistry for nonlinear optical performance. Sci Rep. https://doi.org/10.1038/srep23325
200. Bekyarova E, Itkis ME, Ramesh P, Berger C, Sprinkle M, de Heer WA, Haddon RC (2009) Chemical modification of epitaxial graphene: spontaneous grafting of aryl groups. J Am Chem Soc 131(4):1336–1337
201. Lomeda JR, Doyle CD, Kosynkin DV, Hwang W-F, Tour JM (2008) Diazonium functionalization of surfactant-wrapped chemically converted graphene sheets. J Am Chem Soc 130(48):16201–16206
202. Zhu Y, Higginbotham AL, Tour JM (2009) Covalent functionalization of surfactant-wrapped graphene nanoribbons. Chem Mater 21(21):5284–5291
203. Sun Z, Kohama S, Zhang Z, Lomeda JR, Tour JM (2010) Soluble graphene through edge-selective functionalization. Nano Res 3(2):117–125
204. Avinash MB, Subrahmanyam KS, Sundarayya Y, Govindaraju T (2010) Covalent modification and exfoliation of graphene oxide using ferrocene. Nanoscale. https://doi.org/10.1039/c0nr00024h
205. Pham VH, Cuong TV, Hur SH, Oh E, Kim EJ, Shin EW, Chung JS (2011) Chemical functionalization of graphene sheets by solvothermal reduction of a graphene oxide suspension in N-methyl-2-pyrrolidone. J Mater Chem 21(10):3371–3377
206. Fang M, Wang K, Lu H, Yang Y, Nutt S (2009) Covalent polymer functionalization of graphene nanosheets and mechanical properties of composites. J Mater Chem 19:7098–7105
207. Bagherzadeh M, Farahbakhsh A (2015) Surface functionalization of graphene. Fundamentals and emerging applications, Graphene Materials. https://doi.org/10.1002/9781119131816.ch2
208. Choi E-Y, Han TH, Hong J, Kim JE, Lee SH, Kim HW, Kim SO (2010) Noncovalent functionalization of graphene with end-functional polymers. J Mater Chem 20(10):1907–1912
209. Chen C, Zhai W, Lu D, Zhang H, Zheng W (2011) A facile method to prepare stable noncovalent functionalized graphene solution by using thionine. Mater Res Bull 46(4):583–587
210. Wan J, Wang Y, He D, Wu H, Wang H, Zhou P, Fu M (2012) Influence of polymer/fullerene-graphene structure on organic polymer solar devices. Integr Ferroelect 137(1):1–9
211. Hsu CL, Lin CT, Huang JH, Chu CW, Wei KH, Li LJ (2012) Layer-by-layer graphene = TCNQ stacked films as conducting anodes for organic solar cells. ACS Nano 6(6):5031–5039
212. Wang J, Wang Y, He D, Wu H, Wang H, Zhou P, Fu M, Jiang K, Chen W (2011) Organic photovoltaic devices based on an acceptor of solution-processable functionalized graphene. J Nanosci Nanotechnol 11(11):9432–9438
213. Liu J, Xie L, Huang W (2011) Transparent, conductive, and flexible graphene films from large-size graphene oxide. Integrat Ferroelect 128(1):105–109

214. Yoo E, Kim J, Hosono E, Zhou HS, Kudo T, Honma I (2008) Large reversible Li storage of graphene nanosheet families for use in rechargeable Lithium ion batteries. Nano Lett 8:2277–2282
215. Yu D, Dai L (2009) Self assembled graphene = carbon nanotubes hybrid films for super capacitor. J Phys Chem Lett 1:467–470
216. Xu C, Wang X, Wang J, Hu H, Wan L (2010) Synthesis and photoelectrical properties of b-Cyclodextrin functionalized graphene materials with high bio-recognition capability. Chem Phys Lett 498:162–167
217. Wua H, Wanga J, Kanga X, Wanga C, Wanga W, Liua J, Aksayb IA, Lina Y (2009) Glucose biosensor based on immobilization of glucose oxidase in platinum nanoparticles/graphene/chitosan nanocomposite film. Talanta 80:403–406
218. Georgakilas V, Otyepka M, Bourlinos AB, Chand V (2012) Functionalization of graphene: covalent and non-covalent approaches, derivatives and applications. Chem Rev 112(11):6156–6214
219. Foster JS, Frommer JE (1988) Imaging of liquid crystals using a tunnelling microscope. Nature 333:542–545
220. Griessl S, Lackinger M, Edelwirth M, Hietschold M, Heckl WM (2002) Self-assembled two-dimensional Molecular host-guest architectures from trimesic acid. Single Mol 3(1):25–31
221. Bourlinos AB, Georgakilas V, Zboril R, Steriotis TA, Stubos AK (2009) Liquid-phase exfoliation of graphite towards solubilized graphenes. Small 5(16):1841–1845
222. Chen IWP, Huang CY, Jhou SHS, Zhang YW (2014) Exfoliation and performance properties of non-oxidized graphene in water. Sci Rep. https://doi.org/10.1038/srep03928
223. Bose S, Kuila T, Mishra AK, Kim NH, Lee JH (2011) Preparation of non-covalently functionalized graphene using 9-anthracene carboxylic acid. Nanotechnology. https://doi.org/10.1088/0957-4484/22/40/405603
224. Manohar S, Mantz AR, Bancroft KE, Hui CY, Jagota A, Vezenov DV (2008) Peeling single-stranded DNA from graphite surface to determine oligonucleotide binding energy by force spectroscopy. Nano Lett 8(12):4365–4372
225. Kim J, Song SH, Im H-G, Yoon G, Lee D, Choi C, Kim J, Bae B-S, Kang K, Jeon S (2015) Moisture barrier composites made of non-oxidized graphene flakes. Small 11(26):3124–3129
226. Yang H, Hernandez Y, Schlierf A, Felten A, Eckmann A, Johal S, Louette P, Pireaux JJ, Feng X, Muellen K (2013) A simple method for graphene production based on exfoliation of graphite in water using 1-pyrenesulfonic acid sodium salt. Carbon 53:357–365
227. Dong XC, Shi YM, Zhao Y, Chen DM, Ye J, Yao YG, Gao F, Ni ZH, Yu T, Shen ZX et al (2009) Symmetry breaking of graphene monolayers by molecular decoration. Phys Rev Lett. https://doi.org/10.1103/PhysRevLett.102.135501
228. Zhang YH, Liu CJ, Shi WQ, Wang ZQ, Dai LM, Zhang X (2007) Direct measurements of the interaction between pyrene and graphite in aqueous media by single molecule force spectroscopy: understanding the π-π interactions. Langmuir 23(15):7911–7915
229. An X, Simmons T, Shah R, Wolfe C, Lewis KM, Washington M, Nayak SK, Talapatra S, Kar S (2010) Stable aqueous dispersions of noncovalently functionalized graphene from graphite and their multifunctional high-performance applications. Nano Lett 10(11):4295–4301
230. Aliyeva S, Alosmanov R, Buniyatzadeh I, Azizov A, Maharramov A (2019) Recent developments in edge-selective functionalization of surface of graphite and derivatives—a review. Soft Mater. https://doi.org/10.1080/1539445X.2019.1600549
231. Hummers WS, Offeman RE (1958) Preparation of graphitic oxide. J Am Chem Soc. https://doi.org/10.1021/ja01539a017
232. Dreyer DR, Park S, Bielawski CW, Ruoff RS (2010) The chemistry of graphene oxide. Chem Soc Rev 39(1):228–240
233. Yang W, Auciello O, Butler JE, Cai W et al (2002) DNA-modified nanocrystalline diamond thin-films as stable, biologically active substrates. Nat Mater 1(4):253–257
234. Chong KF, Loh KP, Vedula SRK, Lim CT et al (2007) Cell adhesion properties on photochemically functionalized diamond. Langmuir 23(10):5615–5621

235. Zhong YL, Chong KF, May PW, Chen Z-K, Loh KP (2007) Optimizing biosensing properties on undecylenic acid-functionalized diamond. Langmuir 23(10):5824–5830
236. Young SL, Kellon JE, Hutchison JE (2016) Small gold nanoparticles interfaced to electrodes through molecular linkers: a platform to enhance electron transfer and increase electrochemically active surface area. J Am Chem Soc 138(42):13975–13984
237. Young SL, Hutchison JE (2019) Selective deposition of metals onto molecularly tethered gold nanoparticles: the influence of silver deposition on oxygen electroreduction. Chem Mater 31:2750–2761
238. Lasseter TL, Clare BH, Abbott NL, Hamers RJ (2004) Covalently modified silicon and diamond surfaces: resistance to nonspecific protein adsorption and optimization for biosensing. J Am Chem Soc 126(33):10220–10221
239. Hamers RJ, Butler JE, Lasseter T, Nichols BM, Russell JN, Tse K-Y, Yang W (2005) Molecular and biomolecular monolayers on diamond as an interface to biology. Diam Relat Mater 14(3–7):661–668
240. Ruther RE, Rigsby ML, Gerken JB et al (2011) Highly stable redox-active molecular layers by covalent grafting to conductive diamond. J Am Chem Soc 133(15):5692–5694
241. Yao SA, Ruther RE, Zhang L, Franking RA, Hamers RJ, Berry JF (2012) Covalent attachment of catalyst molecules to conductive diamond: CO_2 reduction using "smart" electrodes. J Am Chem Soc 134(38):15632–15635
242. Nie B, Yang M, Fu W, Liang Z (2015) Surface invasive cleavage assay on a maskless light-directed diamond DNA microarray for genome-wide human SNP mapping. Analyst 140(13):4549–4557
243. Henke AH, Saunders TP, Pedersen JA, Hamers RJ, (2019) Enhancing electrochemical efficiency of hydroxyl radical formation on diamond electrodes by functionalization with hydrophobic monolayers. Langmuir 35(6):2153–2163
244. Kondo T, Tamura A, Kawai T (2009) Cobalt phthalocyaninemodified boron-doped diamond electrode for highly sensitive detection of hydrogen peroxide. J Electrochem Soc 156(11):F145–F150
245. Kondo T, Hoshi H, Honda K, Einaga Y, Fujishima A, Kawai T (2008) Photochemical modification of a boron-doped diamond electrode surface with vinylferrocene. J Phys Chem C 112(31):11887–11892
246. Kondo T, Taniguchi Y, Yuasa M, Kawai T (2012) Polyoxometalate-modified boron-doped diamond electrodes. Jpn J Appl Phys. https://doi.org/10.1143/jjap.51.090121
247. Wang J, Firestone MA, Auciello O, Carlisle JA (2004) Surface functionalization of ultra-nanocrystalline diamond films by electrochemical reduction of aryldiazonium salts. Langmuir 20(26):11450–11456
248. Notsu H, Fukazawa T, Tatsuma T, Tryk DA, Fujishima A (2001) Hydroxyl groups on boron-doped diamond electrodes and their modification with a silane coupling agent. Electrochem Solid-State Lett 4(3):H1–H3
249. Bouvier P, Delabouglise D, Denoyelle A, Marcus B, Mermoux M, Petit J-P (2005) Erratum: photosensitization of boron-doped diamond by surface grafting of pyrene groups. Electrochem Solid-State Lett 8:E57
250. Delabouglise D, Marcus B, Mermoux M, Bouvier P et al (2003) Biotin grafting on boron-doped diamond. ChemComm. https://doi.org/10.1039/b308185k
251. Delamarche E, Sundarababu G, Biebuyck H, Michel B et al (1996) Immobilization of antibodies on a photoactive self-assembled monolayer on gold. Langmuir 12(8):1997–2006
252. Takahashi K, Tanga M, Takai O, Okamura H (2003) DNA preservation using diamond chips. Diamond Relat Mater 12(3–7):572–576
253. Yang J-H, Song K-S, Zhang G-J, Degawa M, Sasaki Y, Ohdomari I, Kawarada H (2006) Characterization of DNA hybridization on partially aminated diamond by aromatic compounds. Langmuir 22(26):11245–11250
254. Nagl A, Hemelaar SR, Schirhagl R (2015) Improving surface and defect center chemistry of fluorescent nanodiamonds for imaging purposes—a review. Anal Bioanal Chem 407(25):7521–7536

255. Alkahtani MH, Alghannam F, Jiang L, Almethen A et al (2018) Fluorescent nanodiamonds: past, present, and future. Nanophotonics 7(8):1423–1453
256. Fu C-C, Lee H-Y, Chen K, Lim T-S et al (2007) Characterization and application of single fluorescent nanodiamonds as cellular biomarkers. Proc Natl Acad Sci USA 104(3):727–732
257. Bollina R, Landgraf J, Wagner H, Wilhelm R, Knippscheer S, Tabernig B (2006) Performance, production, and applications of advanced metal diamond composite heat spreader. Paper presented at IMAPS 39th international symposium on microelectronics, San Diego, 8–12 October 2006
258. Mochalin VN, Gogotsi Y (2015) Nanodiamond–polymer composites. Diam Relat Mater 58:161–171
259. Karami P, Khasraghi SS Hashemi M, Rabiei S, Shojaei A (2019) Polymer/nanodiamond composites—a comprehensive review from synthesis and fabrication to properties and applications. Adv Colloid Interf Sci. https://doi.org/10.1016/j.cis.2019.04.006

Scanning Tunneling Microscopy (STM) Imaging of Carbon Nanotropes: C_{60}, CNT and Graphene

Subhashis Gangopadhyay and Sushil

Abstract The discovery of fullerenes and other nanosized carbon allotropes has opened a vast new field of possibilities in nanotechnology and has become one of the most promising research areas. Carbon nanomaterials have a wide-scale scientific as well as technological importance because of their distinctly different physical, chemical and electronic properties. 'Carbon Nanotropes', the nanoscale carbon allotropes such as *Buckminsterfullerenes* (C_{60}), *Carbon nanotubes* (CNTs) and *Graphene*, show a huge potential toward various devices, sensors and catalytic applications and therefore draw a wide-scale industrial attentions. A better understanding of their formation mechanism along with their direct visualization down to nanometer-scale structural analysis is of high technological demand. Recent advancements in nanoscience and nanotechnology make it possible to study the growth/synthesis along with structure and bonding by analyzing the atomic-scale imaging of these carbon nanotropes. In this aspect, scanning tunneling microscopy (STM) would be a useful tool with extremely high spatial resolution. This chapter is mainly focused on STM imaging of some of the recent carbon nanotropes such as C_{60}, CNTs and graphene to bring together the atomic-scale structure and their related material properties.

1 Introduction

Carbon is one of the most fascinating and useful materials with its single entity and various forms for technology as well as mankind. It is an indispensable and fourth most common element in our solar system [1]. At the same time, it is the second most common element of the human body building up about 18% of our body mass [2]. The cantenation property of carbon as well as its ability to form bonds with other light elements makes it an essential material for nature which eventually leads to the wonders of the formation of life. Since a very early stage of civilization, charcoal and carbon black have been used as materials for drawing

S. Gangopadhyay (✉) · Sushil
Birla Institute of Technology and Science (BITS), Pilani, Rajasthan, India
e-mail: subha@pilani.bits-pilani.ac.in

© The Author(s), under exclusive license to Springer Nature Singapore Pte Ltd. 2021   47
A. Hazra and R. Goswami (eds.), *Carbon Nanomaterial Electronics: Devices and Applications*, Advances in Sustainability Science and Technology,
https://doi.org/10.1007/978-981-16-1052-3_3

and/or writing whereas in later stages, graphite does the same. However, in the last few decades, carbon-based nanomaterials have attracted enormous scientific interest and technological importance. Due to its exclusively new/superior mechanical, electrical, chemical, thermal and other materialistic properties, carbon in nanoscale became one of the most promising candidates for the field of nanoscience and nanotechnology. Modern experimental techniques also make it possible to grow/form various forms of carbon-based nanomaterials and their characterization down to atomic scale. All these scientific and technological progresses resultantly manifold the applications of carbon-based nanostructures in various aspects of electronics, sensors, energy science and even in the field of medical science such as drug delivery.

Carbon: A Unique Material with Many Allotropes

In general, carbon possesses a wide range of kinetically as well as thermodynamically stable phases toward the ambient condition which eventually results in the formation of various allotropes of it. Among these, diamond and graphite are the only known allotropes of carbon which appear naturally in the earth. Both show some distinctly different material properties with respect to each other such as hardness, thermal conductivity, lubrication behavior and electrical conductivity [3–9]. In the last few decades, many more allotropes of carbon have been discovered which include ball, tube, sheet, ribbon or even dot-shaped nanostructures and their combinations, such as Bucky ball, Carbon nanotube, Graphene, Nano-ribbons and Carbon dots [10–13]. Depending on these definite shape and size effects, these C-based nanostructures have various levels of quantum-mechanical confinement effect which can significantly influence their electronic properties. According to Samara Carbon Allotrope Database (SACADA), it possible to have more than 500 hypothetical allotropes of carbon with three-dimensional periodicities. However, a few established allotropes of carbon such as (i) Diamond, (ii) Graphite, (iii) Lonsdaleite, (iv) C_{60} (Bucky ball), (v) C_{540}, (vi) C_{70} (Fullerite), (vii) Carbon Nanotubes (CNTs) and (viii) Amorphous Carbon are briefly introduced here [14, 15] along with the recent scientific sensation of the two-dimensional (2D) nanomaterials (ix) '*Graphene*' [16–18]. A schematic representation of the structural bonding between the C atoms for different carbon allotropes is depicted in Fig. 1 [15].

Diamond: Diamond is a solid form of elemental carbon with a crystal structure called diamond cubic. It is a metastable allotrope of carbon which consists of extended networks of sp^3 hybridized carbon atoms with an interatomic spacing of about 0.154 nm (Fig. 1a). These C atoms form a face-centered cubic (FCC) crystal structure. Diamonds are widely used as jewelry because of their luster and durability. The color of a diamond is generally determined by the trace elements such as nitrogen, sulphur and boron present within it [19]. Due to their extraordinary hardness, diamonds are vastly used in the industry to cut, grind, or drill other materials. It has been reported that synthetic (laboratory-made) diamonds are usually formed at a relatively high pressure (5–6 Gpa) and temperature ranging 900–1400 °C [20].

Graphite: Another solid form of elemental carbon is known as graphite. It is chemically more stable than the diamond, however, the conversion of diamond to graphite

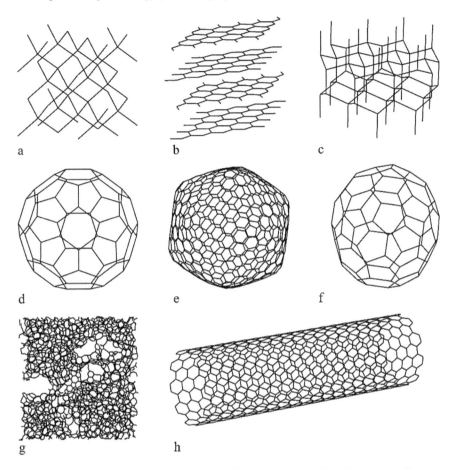

Fig. 1 Schematic illustration of atomic bonds for some major carbon allotropes: **a** diamond, **b** graphite, **c** lonsdaliete, **d** C_{60} (Bucky ball), **e** C_{540}, **f** C_{70}, **g** amorphous carbon and **f** CNT. Reprinted with permission from Reprinted from Wikipedia [15]

in standard conditions is not observed. Graphite has a planar structure of densely packed honeycomb lattice of carbon atoms stacked in layers. Here, C atoms appear in an sp^2 hybridized planer structure, with an interatomic separation of 0.142 nm, which are finally stacked in layers (Fig. 1b). Each of these layers is held together by a weakly unsaturated π-bond, whereas the inlayer C atoms have strong covalent σ-bonds. Each of these single layers of carbon atoms in a two-dimensional honeycomb lattice is known as 'graphene' which is expected to be the strongest material, 100 times harder than steel [21]. More details of graphene will be discussed later.

Lonsdaleite: Lonsdaleite, also called hexagonal diamond [22, 23], is another allotrope of carbon which was found in meteorites containing graphite. It is reported to be much harder than diamond (58%) on the (100) crystal face, and has more resistance to indentation pressure (152 Gpa) as compared to diamond (97 Gpa)

Orthogonal projections					
Centered by	Vertex	Edge 5–6	Edge 6-6	Face Hexagon	Face Pentagon
Image	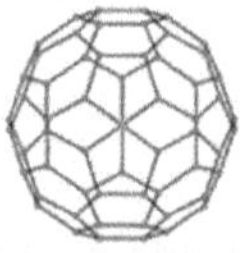	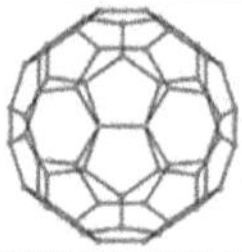	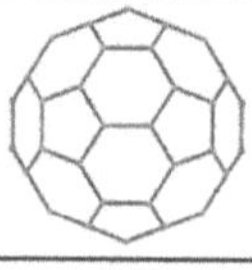	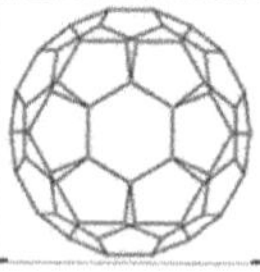	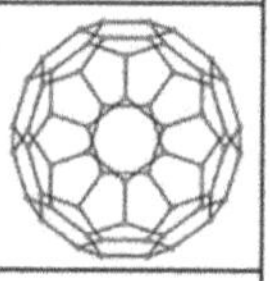
Projective symmetry	[2]	[2]	[2]	[6]	[10]

Fig. 2 Schematic illustration of a C_{60} molecule with different orthogonal projections along with their structural symmetries (60 vertices, 90 edges and 32 faces)

[24]. Within the lonsdaleite crystal, the interlayer C-bonds formed an eclipse-like conformation (Fig. 1c).

C_{60} or Bucky Ball: It is another allotrope of carbon, formed as a molecule made out of 60 carbon atoms in a shape of a hollow sphere that resembles a soccer ball. It was first reported/discovered by Harry Kroto and Richard Smalley in 1985 [25]. This new form of carbon exhibits a truncated icosahedron structure formed by 20 hexagons and 12 pentagons, where every carbon atom is bonded to three other carbon atoms (Fig. 1d). This truncated icosahedron structure appears with 60 vertices (C atoms) forming a total of 90 edges among which 60 edges are between pentagon-hexagon and 30 are between hexagon-hexagon. A schematic representation of various orthogonal projections of a C_{60} molecule and their related structural symmetries are summarized in Fig. 2. Although each C atom of this molecule is bonded with another 3 C atoms, due to its curved morphology the hybridization of C_{60} falls under an intermediate stage between the graphite (sp^2) and the diamond (sp^3), and finally appears to be about $sp^{2.28}$ hybridizations. Two different types of C–C bonds are found within a C_{60} molecule: (a) double bond between two hexagonal rings is slightly shorter as compared to (b) single bond between a hexagon and a pentagon. The average bond length is found to be about 0.14 nm. In the laboratory, C_{60} molecule, which is also known as 'buckminsterfullerene', is formed by vaporizing graphite under an electrical arc in low-pressure helium or argon ambient condition. The cage structure of fullerene is predicted to be used for various encapsulation processes down to nanometer scale. A very useful application of this would be the controlled drug delivery. They can also act as hollow cages to trap other dangerous molecules/substances within our body.

Fullerenes: After the discovery of C_{60} in 1985, the traditional concept of carbon-based allotropes has changed quite rapidly. This unanticipated discovery of C_{60} can also be considered as the starting of the new era of synthetic carbon allotropes. Afterwards, many new allotropes of carbon have been discovered [14]. Among them, different groups of Fullerene are found to be of high scientific as well as techno-logical significance which attracted huge attention to researchers all over the world. In principle, the fullerene group is made up of many carbon atoms connected by single and double bonds to form a closed or partially closed mesh such as sphere, ellipsoid, tube, and even a single atomic layer sheet, with fused rings of five to seven atoms. Fullerenes with a closed mesh topology can informally be denoted by the total number of carbon atoms (n) such as C_n (for example, C_{540} or C_{70} as shown in Fig. 1e and f). Another form of closed mesh fullerene structure is carbon nano-onions. Concentric fullerenes are enclosed one inside the other in an onion-like layered structures. In addition, an open or partially closed mesh such as tubes, sheet and other configurations is also available within the fullerene group. Some of these special types are discussed in the following. Depending on their structural geometries, fullerenes have very high potential usages in the field of drug delivery, lubricants and catalyst applications.

Amorphous Carbon: Unlike all the above-mentioned crystalline allotropes of carbon, it does not have any long-range crystalline symmetry (Fig. 1g). Similar to any other glassy materials, some short-range symmetry can be observed within it. During the large-scale production of amorphous carbon, however, some micro- or nano-crystals of graphite or diamond can also be found. Although coal and carbon black are informally known as amorphous carbon, however, they do not represent true amorphous carbon in normal condition.

Carbon Nanotube (CNT): It is one of the most important members of the fullerene family and has a very high impact on electrical conductivity. It was first discovered by S. Iijima in the year 1991 [26]. The name is basically representative of its cylin-drical shape. A CNT is a rolled-up graphene sheet forming a cylindrical shape. The aspect ratio (length to diameter ratio) of CNTs could be extremely high (up to a million), which is significantly higher than any other material. Depending on their layered thickness, they can further be classified into different categories such as (a) single-walled carbon nanotubes (SWCNTs), (b) double-walled carbon nanotubes (DWCNTs) and (c) multi-walled carbon nanotubes (MWCNTs). SWCNTs are formed when a single sheet of the graphene layer is rolled up to form a nanotube struc-ture, whereas, in the case of DWCNTs, two graphene layers are rolled up concentri-cally. However, MWCNTs have two different rolling possibilities with multiple layer thickness. On the one hand, multiple SWCNTs of different diameters are arranged coaxially, whereas on the other hand single graphene sheet is rolled up on itself like a paper roll where the layer spacing between two consecutive layers is also maintained to about 3.4 Å. In principle, the properties of DWCNTs are very similar to those of SWCNTs, but have much better stability against any exposed chemicals [27].

Depending upon their rolling up directions, single-walled CNTs can further be categorized into three different types such as (i) zig-zag, (ii) arm-chair and (iii) chiral. These classifications can be described in terms of the chiral vector:

$$\mathbf{R} = n\mathbf{a} + m\mathbf{b}(n, m) \tag{1}$$

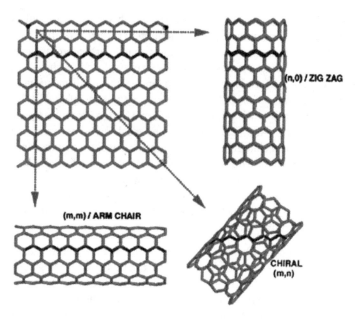

Fig. 3 Schematic illustration of a graphene sheet with different chiral vectors and corresponding SWCNTs with different chirality: arm-chair, zig-zag and chiral

where n and m are integers, and **a** and **b** represent the unit vectors of graphene lattice. A schematic representation of various chiral vectors is depicted in Fig. 3, where the arrow represents the rolling directions of the graphene sheet which resultantly forms SWCNTs of different chirality (Fig. 3) [28, 29]. Apart from these, there are many more hybrid structures of nanotubes available which are known as graphenated (g-CNTs) [30, 31], shortest (extreme CNT) [32], torus (doughnut-shaped) CNTs [33], etc. (Fig. 4).

Graphene: The recently discovered new allotrope of carbon, *Graphene*, a single-layer graphite, is considered to be the first 2D crystal, where all carbon atoms are arranged within a plane in a densely packed honeycomb lattice. Each carbon atom within a graphene layer is connected to three nearest C atoms by a σ-bond (sp^2 hybridization) and contributes the fourth electron which extends over the sheet to the conduction band. A similar kind of bonding is also (partially) observed in other fullerenes. The conduction band electronics of graphene exhibit a very unusual property where the charge carriers show a linear energy dependency on momentum instead of the usual quadratic nature. As a result, field-effect transistors with bipolar conduction or ballistic charge transport over a long distance can be possible with graphene-based electronics. In addition to many exciting electronic properties, graphene is a very good conductor of heat and electricity within its plane.

Since its discovery in 2004 by Andre Geim and Konstantin Novoselov, graphene has drawn huge attraction to the scientific community as well as device industry,

Fig. 4 Schematic diagram of single-layer graphene; sp^2 hybridized carbon atoms within a planer structure

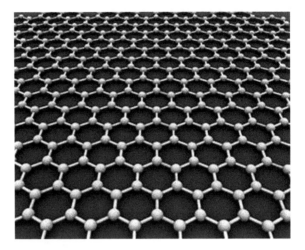

both because of its many superior physical, chemical and electronic properties [17, 18, 34–36] and vast possibilities of wide-range potential applications [37–39]. It is predicted to be the high-tech material for the next generation. Its amazing electronic properties, excellent mechanical stability and flexibility make graphene an ideal material for future technologies. It is described as being much lighter than a feather, far stronger than a diamond, much more conductive than copper and as flexible as rubber. Therefore, it is now at the heart of a worldwide scientific contest to exploit its properties and develop techniques to commercialize it. Due to its extraordinary material properties, graphene could unlock a new era of super energy-efficient gadgets, cheap, quick-charge batteries, wafer-thin flexible touch-screen computing, and a sturdier light-weight automobile chassis. Graphene has the potential to completely revolutionize entire semiconductor industries, creating bendable phones and touch screens, tiny self-powered oil and gas sensors and even synthetic blood. Strategic metal experts are already calling graphene the most important substance created since synthetic plastic a century ago. However, plastics are simply a conservative comparison whereas graphene has far more high-tech uses and potentials to be much more lucrative.

Future Possibilities: Conceptually, many other ways to construct new carbon allotropes are very much possible by controlled alteration of the hybridized carbon atoms' periodic network. A number of scientists are currently investigating various possible structural modifications of these synthetic carbon-based allotropes. All these efforts also indicate that this challenge is not yet over and many more are to come.

2 Experimental Techniques

Fundamental building blocks of any material are the atoms/molecules and their bonding configurations, which have dimensions in the order of nanometer length scale. Visible bulk materials are simply the assembled of these atoms and molecules into a very large entity. Therefore, nanoscale domains of any material may act as a key element and have a very significant impact, which can finally determine its material properties. Hence, a detailed understanding of the formation of these nanoscale building blocks and their atomic-scale structural characterization would be of high practical importance for both science and technology. Recent experimental advancements in nanoscience and nanotechnology have made it possible to grow/synthesis various nanoscale carbon allotropes (nanotropes) in a controlled manner and characterize their surface structures down to atomic scale. In this section, the formation mechanism of some selective synthetic carbon nanotropes is discussed. In addition, some basic features of scanning tunneling microscopy toward the atomic-scale surface analysis are also explained.

2.1 Growth/Synthesis of Carbon Nanotropes

The recent additions to the carbon allotrope family are the caged fullerenes (C_{60} and other), CNTs, graphene and its derivatives with intermediate sp-sp^2 hybridizations. At the same time, many other synthetic nanoscale-sized allotropes are also in this queue. Altogether, these nanometer-scale allotropes are called '*Nanotropes*' [14]. Various experimental methods have been reported for the successful growth/synthesis of these different forms of synthetic carbon allotropes. Among these, chemical vapor deposition (CVD), arc discharge process under a controlled ambient, use of plasma discharge, laser ablation mechanism, flame synthesis, etc., are mostly used.

Bucky Ball (C_{60}) Synthesis: In the laboratory, C_{60} molecules are usually made by blasting together pure carbon sources, such as graphite. However, the presence of hydrogen can drastically reduce the possibility of this fullerene formation. Direct vaporization of carbon source also does not result in an effective formation of C_{60}. Irradiation of graphite source using a laser in a low-pressure helium ambient condition was found to be useful for C_{60} formation [40]. This method also results in a small fraction of other fullerene derivatives such as C_{70} or C_{84}. Therefore, it also requires several follow-up techniques to purify the C_{60} molecules such as vacuum evaporation at low temperature (400 °C) or stepped chemical filtration. In general, the production of a single type of fullerene is always difficult as it requires a very harsh environment. For large-scale production of C_{60}, the radio-frequency-plasma technique or the arc discharge vaporization techniques are usually employed. The phase purity of the fullerene is also found to be a real problem here, as these combustion-based mechanisms result in about 98% pure fullerene or fullerene mixtures. However, the growth conditions for a specific target material can be modulated depending on its requirements [41].

CNT Growth: In general, arc discharge, laser ablation and CVD methods are largely used to grow high-quality CNTs on a solid support with the assistance of metal catalysts. However, these methods are largely affected by the material purities of the as grown CNTs. Residual metal (catalyst) or supports as well as amorphous carbon may appear within the final products which also demands a post-growth purification treatment. It has to be taken into account that an extensive purification treatment may also cause a degradation of the as-grown CNTs [42]. Therefore, an optimization between the growth methods and the purification process must be achieved. The longest CNT ever produced was achieved using the CVD growth method and found to be about 86 mm long. Some recent techniques for CNT synthesis using different carbon sources are listed below.

Thermal decomposition (950 °C) of CH_4 in Ar ambient using a bimetallic (Fe-Mo) catalyst has found to be successful to grow high purity CNTs of 20–45 nm, having 70% of SWCNTs and rest of DWCNT [43]. CNTs have also been synthesized using pyrolysis of mustard oil (650 °C), which results in the formation of water-soluble (ws)-CNTs of diameter about 9–30 nm [44]. DC arc discharge of coal was performed in a He atmosphere using a graphite anode and a mixture of coal and CuO micro-granule cathode, which produced CNTs with many branches (BCNTs) [45]. Operating parameters for the arc discharge were 70–80 A and 25–30 V with a pressure of 50 kPa. Arc vaporization of Chinese coal has also been reported to successfully synthesize CNTs [46]. In this process, initially, coal was mixed with binder (coal tar) and pressed at a pressure of 10–20 MPa before having any heat treatment at 900 °C. Afterwards, this was used as anode whereas cathode was the high purity graphite. The arc chamber was operated under He ambient, at a pressure of about 30–70 kPa within a range of 30–40 V and 0–70 A. After arcing, CNTs were formed and deposited onto the cathode. These CNTs were found to be 4–70 μm long and having a diameter of 2–15 nm.

Graphene Formation: The graphene layer was first experimentally produced by the repeated splitting of a mechanically exfoliated graphite, using scotch tape [47]. Several attempts have been made using this but a major drawback of it is the lack of reproducibility. However, CVD-based decomposition of hydrocarbons on transition metal (TM) surfaces at elevated temperature is found to be a practical solution to this challenge, with a controlled formation of a large-scale graphene layer [48–55]. Transition metals are well known for their catalytic activity which generally arises due to their ability to adopt multiple oxidation states. For instance, they serve as vital catalysts in C–C coupling reactions such as the Heck, Sonogashira, Suzuki, Stille and Ullmann reactions [56, 57]. Controlled vacuum or reactive annealing of a TM surface at low pressure, followed by slow cooling, would be a useful technique for high-quality graphene growth. Exploiting these mechanisms, recent approaches of epitaxial graphene growth using TM thin films as a substrate have become very much popular.

Thin metal films or foils are used as catalytic substrates for the decomposition of hydrocarbon at an elevated temperature at relatively lower pressure. As soon as graphene layers are formed, they strongly stick to the metal surface, protecting from any further sticking/decomposition of newly arrived hydrocarbon molecules. Formation of the graphene layer is also reported by sacrificial growth or segregated carbon, using Ni, Cu or Co films. Depending on the TM property, two types of segregation processes of C atoms toward the film surface are found: (a) surface segregation (Cu) or bulk segregation (Ni). The solubility of carbon within a metal is the driving force which strongly depends on the temperature. Single- or poly-crystalline substrates of group *3d–5d* metals have been reported as a catalytic surface for graphene growth. Graphene layers prepared on such surfaces, i.e., Ru (0001) [58–60], Rh (111) [61–63], Ir (111) [64–68] and Pt (111) [69, 70], usually appears with so-called *Moire´* structures in STM imaging. This is mainly due to the relatively large lattice mismatch between the graphene layer and metal surface.

2.2 STM: A Powerful Microscopic Tool

Scanning tunneling microscopy (STM) is one of the most useful techniques which offered a real-space visualization of the surface atomic structure for the first time. It shows a direct visualization of the surface atomic structure down to a sub-nanometer scale resolution [71, 72]. In addition, it can also provide very local information about the surface electronic states. Moreover, it can also be used for surface nano-lithography. Atomic or molecular level surface manipulations toward the fabrication of surface nano-patterns or modification of bond alignments have also been performed using the STM tip. Investigations of surface reaction dynamics can also be studied using this technique.

The working principle of STM is based on one-dimensional (1D) quantum mechanical tunneling of electrons from the sample surface atom to the atom of the tip apex (or vice versa) through the vacuum gap. As the tunneling probability of electrons depends exponentially on the gap distance, the tunneling current is extremely sensitive which provides a unique atomic-scale spatial resolution of STM. The first atomic-resolution STM imaging showed resolutions between 1 nm and 10 pm for the lateral and the vertical scales, respectively. With technological advancement, modern STMs are capable of vertical and lateral resolution down to the pico-meter length scale. However, the spatial resolution limit of the STM and the image formation mechanism are still an open question related to the overlapping of the electronic orbitals of tip and surface atoms.

Since its invention in 1981 by G. Binnig and H. Rohrer at IBM Zurich, STM has become one of the most useful techniques to investigate the surface atomic reconstructions, surface diffusion kinetics, overlayer growth mechanism, as well as local electronic density of states (DOS) down to the atomic scale. It has widely been used in the field of physics, chemistry and biology to study the surface of organic and inorganic nano-objects, and even its limitation toward the insulating material surface due to the charging effect.

3 Results: STM Imaging of Carbon-Based Nanotropes

Within this section, we will discuss the surface atomic structures of some selective carbon allotropes with a special focus toward their atomic-scale surface structures, bonding configuration and formation mechanism. More precisely, high-resolution STM imaging of a few reported results of C_{60} nucleation, CNT growth and graphene formation on various crystalline substrates and their related structural and electronic properties will be discussed [73–81].

3.1 C_{60} Nucleation, Growth and Self-assembly: Atomic-Scale Studies

The adsorption of a large organic molecule on a solid surface, its interface interaction and bonding mechanism are of high practical importance in order to understand the molecular functionality in electronic devices. There exists a significant amount of literature reporting on the initial adsorption process of C_{60} molecules, its self-assembled island formation and finally the closed-pack layered growth morphology on various crystalline surfaces, using scanning tunneling microscopy (STM) as the main characterizing tool [73 and references therein]. An overall observation suggests that C_{60} molecules bond to the semiconducting surfaces in a different manner as compared to that observed on metal surfaces. Among various semiconducting substrates, in particular, silicon attracts more attention due to its technological compatibilities. Usually, bare Si surfaces appear with many dangling bonds whereas metallic surface has free electrons. An electronic transfer from the semiconducting (Si) surfaces into the lowest unoccupied molecular orbital (LUMO) of C_{60} molecules is reported for the initial nucleation layer of C_{60}. This finding eventually results in strong covalent bond formation between C_{60} molecules, and Si surface atoms and molecules are chemisorbed. On the other hand, weak Van der Walls interaction between the metal surface and C_{60} molecules with a negligible LUMO occupation results in a physisorption process. Hence, the initial nucleation of C_{60} molecules can strongly depend on the choice of the substrate surface. However, for C_{60} coverage above 1 ML, only Van der Walls interaction between themselves are responsible, and the impact of the substrate surface almost disappears. In the following, we will discuss some reported STM studies of initial absorption of C_{60} molecules on two different types of crystalline surface, along with real-space imaging of their sub-molecular orbitals and self-assembled layered growth mechanism.

(i) C_{60} on Si (111): Adsorption Coverage and Bonding

STM images of C_{60}/Si (111)-7 × 7 surface with an increasing C_{60} coverage are shown in Fig. 5 [73]. Clean Si (111) surface appears in 7 × 7 reconstruction with a clear contrast of faulted and unfaulted half unit cells (Fig. 5a). Initially, C_{60} molecules start to nucleate on the Si surface at room temperature in a scattered way throughout the surface. C_{60} molecules are randomly distributed without having any kind of surface segregation toward the closed-pack island formation (Fig. 5b−d). In the case of C_{60} coverage of about 0.9ML, the surface structure looks slightly different with a little surface ordering as can be seen in (Fig. 5e). The surface appears with a

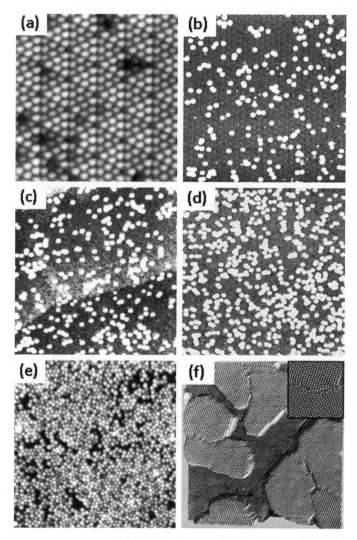

Fig. 5 STM images of C_{60} adsorbed Si (111)-7 × 7 surface with increasing coverages. Reprinted with permission [73]

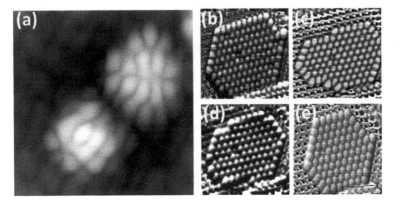

Fig. 6 **a** Sub-molecular features of an isolated C_{60} molecules adsorbed on Si (111)-7 × 7 surface. **b**−**e** Rapid rearrangements of closed-pack molecular islands at room temperature

few vacancies, but there is no sign of any C_{60} nucleation for the second layer. This finding was explained in terms of a strong molecule–substrate interaction. As a result, the nucleation of the second molecular layer of C_{60} only starts after the successful completion of the first monolayer. Figure 5f shows the closed-pack layered islands structure of C_{60} molecules formed at a coverage of 3 ML. For multilayer C_{60} films, surface morphology also differs drastically and appears with the ordered structure of hexagonal closed-pack islands, having two rotational domains (shown within the inset). This is due to a weak Van der Waals interaction between the C_{60} molecules themselves.

A closer look at STM imaging of C_{60} molecules covalently bonded with the surface Si atoms usually appears with many inter-molecular superstructures, even at room temperature (Fig. 6a). These superstructures basically represent the surface electronic density of states of different molecular orbitals at a specific bias voltage condition. It may also have a very strong influence from the tip apex physical geometry or any kind of tip functionalization which can significantly modify the image quality. However, for physisorbed C_{60} molecules (multilayer films), this type of superstructures is not observed for room temperature STM imaging. This is explained in terms of thermally excited continuous and rapid rotation of the molecule at a particular adsorption site. Moreover, rapid rearrangements of the closed-pack molecular islands are also observed during room temperature STM imaging (Fig. 6b−e). This finding also confirms a very weak Van der Waals bonding between the C_{60} molecules, and during room temperature imaging this potential barrier can easily be overcome by the thermal energy or tip-induced forces to the molecule.

C_{60} on Si (111): Nucleation Sites and Adsorption Geometry

Large-scale statistical analysis of STM images of covalently bonded C_{60} molecules on Si (111)-7 × 7 surface also reveals some specific nucleation sites for C_{60} molecules. Although no ordered structure of C_{60} is formed within the first nucleation layer, five preferential adsorption sites on Si (111)-7 × 7 unit cell have been reported [74].

These specific sites are marked as (i) corner hole (*CH*), (ii) middle of unfaulted half (*Mu*), (iii) middle of faulted half (*Mf*), (iv) top of the rest-atoms in unfaulted half (*Cu*) and (v) above the rest-atoms in unfaulted half (*Cf*). The top two rows of Fig. 7 represent the STM images (top row) and schematic representations (second row) of various adsorption sites, respectively. Among these, *Cf* and *Cu* adsorption sites are found to be most preferable for C_{60} nucleation, with an adsorption probability of about 43% and 34%, respectively [74].

High-resolution STM imaging of these nucleation site-specific C_{60} molecules along with density function theory studies have further suggested the orientation geometry of differently adsorbed molecules [74]. The bottom three rows of Fig. 7 clearly depict the HR-STM images (middle row), simulated DFT images (fourth row) and schematic orientation geometries (bottom row) of different C_{60} molecules. Individual STM images of C_{60} molecules adsorbed at each nucleation site appear with very distinct sub-molecular fine structures. By careful comparison between the experimental STM images and the corresponding simulated images, orientations of

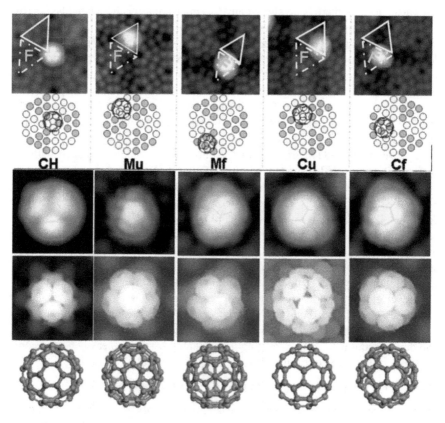

Fig. 7 HR-STM images of C_{60} molecules adsorbed at different nucleation sites on Si (111)-7 × 7 surface, with specific molecular orientation geometries. Reprinted with permission [74]

C_{60} molecules on different adsorption sites have also been proposed. STM images of C_{60} molecules adsorbed on the *CH* site have a hexagonal face attached to the Si surface whereas molecules on the *Mu* site have a pentagon facing the Si surface. Similarly, C_{60} molecules absorbed on the *Mf* site have a vertex (apex) atom facing downward. However, C_{60} molecules on the *Cu* and on the *Cf* sites are found to have one edge of the 6–6 and 6–5 bonds, respectively, facing the substrate surface. All these findings clearly confirm that C_{60} adsorptions on five available nucleation sites offer stable adsorption geometries due to strong covalent interactions between the adsorbed molecules and the Si surface. As a result, distinctly different and very unique sub-molecular fine structures of individual molecular orbitals are nicely observed during the STM imaging at room temperature.

C_{60} on Si(111):Rotational Symmetry of Molecule and Substrate

STM images of chemisorbed fullerene molecule on Si (111)-7 × 7 at room temperature also support the rotational symmetry of C_{60}. In this part, various possible rotations of the molecule with respect to the substrate symmetry are discussed (Fig. 8) [75]. C_{60} molecule marked within the yellow box is found to be adsorbed on the top of the rest-atoms in the faulted part, i.e., *Cf* site facing 5–6 bonds edge toward the Si surface (Fig. 8a). Inset shows the sub-molecular superstructures of the C_{60} molecule with a mirror-symmetry axis (marked as a red arrow) of a regular pentagon. As discussed earlier, depending on the projection plane, the C_{60} molecule can have various rotational symmetries. In the case of 5–6 bond edges, the symmetry is found to be twofold (Fig. 2). At the same time, if we ignore the effect of stacking fault of the Si (111) substrate, the crystal plane also offers a sixfold rotational symmetry. Therefore, for this particular *Cf* adsorption site on Si (111)-7 × 7 surface, the C_{60} molecule may have 12 different rotational symmetries with respect to the Si (111) crystal axis. Various possible rotational orderings of C_{60} molecules are nicely summarized here with the help of high-resolution STM imaging (bottom two rows of Fig. 8) [75]. Altogether, 12 symmetry directions with respect to the [0 1–1] crystal direction of Si (111) surface are recorded, having an offset of $(15 \pm 30n)°$ rotations where n is an integer whose values range from 1 to 6. These results suggest that a 5–6 bond-faced C_{60} molecules have 12 energy-equivalent states particularly labeled by in-plane molecular orientations. These in-plane molecular orientations do not change spontaneously at room temperature due to the strong covalent bonding between the C_{60} molecule and the Si surface. However, tip-induced rotational and translational rearrangements of C_{60} molecules at room and low temperatures have also been performed and reported for both Si (111) and Si (100) substrate surfaces. This type of molecular manipulation may offer a new dimension to device physics in the field of quantum information storage and molecular switching.

(ii) C_{60} on Cu (100) at Low Temperature

In general, high-resolution STM imaging of C_{60} molecules on metal or metal-terminated silicon surfaces do not appear with intra-molecular superstructures at room temperature. The basic reason behind this type of finding is mainly related to

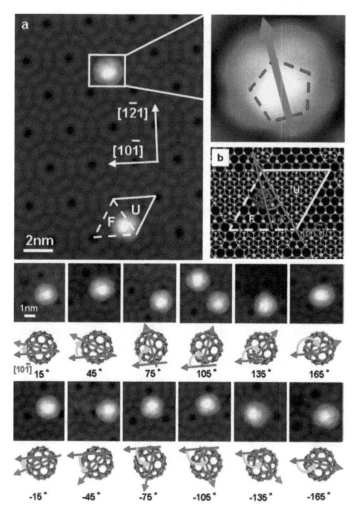

Fig. 8 High-resolution STM images of C_{60} molecules adsorbed on Mu site of Si (111)-7 × 7 surface, represented with tenfold rotational symmetry of the pentagonal face. Reprinted with permission [75]

the weak interaction between the molecule and substrate atoms. At room temperature, the thermal energy of the molecule may easily overcome the interaction potential which promotes rapid molecular rotation. Therefore, to observe any orientational ordering of the C_{60} molecule on a metal surface, thermal energy has to be drastically reduced. In the following, we will discuss the reported low-temperature STM imaging of C_{60} molecules adsorbed on an atomically clean Cu (100) surface (Fig. 9) [76].

C_{60} molecules were deposited on a clean Cu (100) surface at room temperature under an ultra-high vacuum (UHV) condition (10^{-9} Pa) followed by subsequent annealing at 500 K for better ordering. STM imaging was performed at a cryogenic

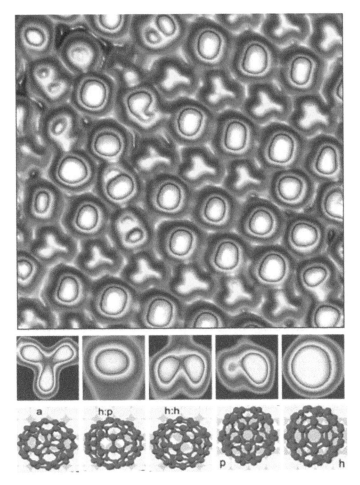

Fig. 9 Low-temperature STM imaging of C_{60} molecules adsorbed on Cu (100). High-resolution images reveal various possible orientations of the C_{60} molecules on the surface. Reprinted with permission [76]

temperature (8 K) which confirms the orientations of the absorbed C_{60} molecules are frozen on the metal surface, as there is no sufficient thermal energy available for rapid molecular rotations. C_{60} molecules appeared with a nearly hexagonal array, having alternately arranged bright and dim stripes that originated from the single and double missing row reconstructions of Cu (100) surface, respectively (Fig. 9a). Similar to earlier studies of C_{60} adsorption on Si (111), high-resolution STM imaging also suggests five distinct/specific orientational ordering of the molecule on the Cu (100) surface. These molecular orientations are denoted here as (i) hexagon ring (*h*), (ii) hexagon-pentagon bond (*h:p*), (iii) hexagon-hexagon bond (*h:h*), (iv) apex/vertex atom (*a*) and (v) pentagon ring (*p*). The middle row of Fig. 9 represents the high-resolution STM images, whereas the bottom row depicts the schematic representations of corresponding molecular orientations. Statistical analysis of the orientational ordering suggests that the hexagon ring (*h*) and hexagon-pentagon edge (*h:p*) are the most probable orientations with an occupancy of about 31% and 56%, respectively. Cu (100) surface reconstruction was also found to have a significant impact on molecular orientation. The hexagon and pentagon rings are mainly observed on the double missing rows of the Cu (100) surface, whereas *h:p* and *h:h* edges as well as vertex (*a*) orientations are mostly found on single missing rows of the Cu (100) surface. Apart from the sub-molecular imaging, tip-induced manipulation of these molecular orientations has also been reported. Below a certain threshold tip-surface separation, orientational reordering may occur due to the electro-mechanical forces exerted by the tip [76].

3.2   CNT Atomic-Scale Imaging: Structure and Electronics

Due to the recent trend of miniaturization of electronic devices, carbon nanotubes (CNTs) are considered as one of the most prominent candidates to address this issue. Its superior electronic properties may govern as an interplay between a molecule and a bulk material. CNTs can possess ballistic electrical conductivity, a quantum mechanical behavior, at a relatively high operating temperature. It is known that the quantum confinement of electronic states is very much dependent on the real-space size distribution of the material and its electronic wave functions. A real example of this type of finding is the strange electronic behavior of CNTs. Depending upon the chirality (atomic-scale wrapping, which is already discussed earlier), CNTs can either be metallic or semiconducting. Such a drastic change in electrical conductivity originated from its chirality making it is of high technological interest to determine. However, it is not a simple task without having a clear idea about the atomistic structure of CNTs. In this aspect, STM can play an important role as diffraction techniques such as LEED or RHEED which are not useful due to the curved morphology of CNTs. Apart from the atomic-scale visualization, STM can also provide very local information about the surface electronic density of states (DOS).

CNT is nothing but a wrapped graphene sheet of hexagonal lattice forming a hollow cylinder. Hence, basic knowledge of graphene imaging is necessary for a proper understanding of the STM images of curved CNT morphology. A general discussion on STM images of the graphene layer is provided later. In the following, we will discuss some reported STM studies on CNTs with a special focus toward the basic understanding of complex patterns observed during STM imaging and their underlying physics for precisely determining the chirality [77, 78].

Direct Visualization of Chirality and Electronic States

Theoretically, to describe the properties of any CNT completely, tube diameter and its chiral angle have to be specified. Usually, the chiral vectors are represented by a

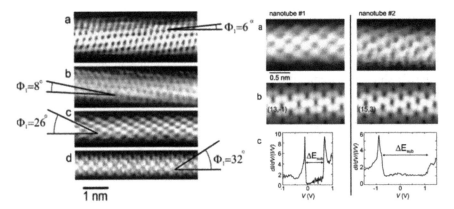

Fig. 10 (Left) STM images of atomically resolved carbon nanotubes, presented with apparent angle (Φ_1) between honeycomb lattice and the tube axis. (Right) STM and STS comparison between semiconducting and metallic CNTs. Reprinted with permission [77]

pair of lattice indices (*n*, *m*), which eventually control the electronic and mechanical properties of the nanotube. To determine the chirality of any CNT, the best way is to directly visualize the atomic arrangements with the tube. In this aspect, STM would be the most useful technique for real-space imaging down to atomic scale, along with local electronic information (DOS) of the tube surface. It has already experimentally confirmed that CNTs can either be metallic or semiconducting in nature. Theoretically, a CNT is considered to be metallic when the chiral vector (n, m) satisfies the condition that (n-m)/3 is an integer. In this part, we will discuss the reported STM studies of the determination of chirality of CNTs using atomically resolved STM images as well as its electronic properties using STS [77].

Figure 10 (left) represents atomically resolved STM images of various chiral CNTs. In principle, STM images of CNTs should appear with a hexagonal atomic structure similar to graphene. However, a distorted hexagon or even triangular patterns of dark and bright dots can also be observed (Fig. 10c). Various parameters can significantly affect image distortion. Among them, the curved surface morphology of CNTs and the tip apex geometry are considered to be most important. Anyway, these distortions may have an impact on the lattice constant but do not have any significant impact on chiral structure determination. A slight overestimation of the chiral angle is reported for the cylindrical shape of the CNTs. In addition, a height scan line profile across the nanotube can provide the diameter of the tube. Apparently, chiral angle Φ_1 is measured as the offset angle between the tube axis and the honeycomb lattice direction. For arm-chair CNTs, the chiral angle Φ remains close to 0° and none of the C–C bonds are found to be parallel to the tube axis. Hence, in STM imaging it appears with a hexagonal structure (Fig. 10 (left) a and b). In contrast, triangular patterns are generally observed for nanotubes with a chiral angle Φ close to 30° (zig-zag), having bonds parallel to the tube axis

(Fig. 10 (left) c and d). Figure 10 (right) shows a direct comparison between semi-conducting and metallic nanotubes using STM, DFT (middle row) and STS analysis. STM appearance (top row) of the metallic tube (Tube 2) is found more hexagonal as compared to the semiconducting one (Tube 1). These findings are nicely reproduced with DFT simulation studies (middle row). Normalized, dI/dV spectra (bottom row) are also very much complementary with the microscopic findings and clearly depict the position of HOMO and LUMO for Tube 1 (semiconducting) from where the bandgap can easily be estimated [77].

Interference Pattern-Dependent Chirality: Interference patterns of the surface electronic orbitals of near-end cap SWCNTs of reported studies using high-resolution STM imaging have been discussed [78]. Apart from the above-mentioned direct imaging discussed above, by systematically studying these interference patterns, with the help of tight-binding simulations (which reproduce the observed patterns), it is also possible to understand the electronic properties of the CNTs. All STM imaging was performed at liquid N_2 temperature and images are presented in the derivative mode for better visibility. A detailed comparison of STM images between end cap arm-chair and zig-zag CNTs are depicted in Fig. 10. It has been found that these interference patterns are highly sensitive to the chirality of the corresponding nanotubes. Simulated interference patterns produced with the tight-binding model of Bloch states (k and –k) can nicely reproduce the observed interference patterns. These findings, in principle, can be categorized by the positions of conduction band minima (k_{min}) within the Brillouin zone, which further provides information about the chirality of the nanotubes.

For both CNTs, periodic patterns are observed in the STM image on the right-hand side of nanotubes, which represent the continuous cylindrical nature of the so-called 'normal' tube structure. Hexagonal rings around the dark areas represent the honeycomb lattice of CNTs. From these periodic patterns, the chirality of both CNTs can be measured as discussed in the earlier section. The arm-chair CNT was found to have a chirality $(28, -13)$ with a diameter of 1.9 nm (Fig. 10a), whereas the zig-zag CNT showed to have a chiral vector $(15, -1)$. However, the hexagonal patterns presented here are slightly distorted due the curved surface morphology of CNTs. However, a very different surface pattern is observed toward the left-hand side of the STM images, near end-cap of the tube. In the case of arm-chair CNT, a combination of bright spots and wavy lines are found, aligned with rotations of 60° with the tube axis (Fig. 10a). The average separation between these bright dots along the tube axis is found to be 0.72 nm (~3a, where a represents the lattice constant of graphene). This finding clearly indicates the metallic nature of the CNT as the energy level crosses the Fermi level at $k = 2\pi/3a$. However, in the case of zig-zag CNT, bright spots are aligned perpendicular to the tube axis and appear with a larger periodicity of 4a which indicates its semiconducting nature (Fig. 10a).

3.3 Graphene Formation on TM Surface: Mechanism and Surface Imaging

Epitaxial growth of single-layer graphene is of great technological importance because of its wide range of potential applications, especially for future generation electronic devices. Transition metals are well known for their catalytic activity which can provide a suitable platform for the CVD of graphene growth via thermal decomposition of the hydrocarbons on the TM surfaces. Sacrificial growth of graphene layer on TM surface during vacuum annealing would also be a useful approach. In this part, we will discuss some reported STM studies of the graphene formation on the TM surface with a special focus toward its formation mechanisms [79–81].

CVD Growth Via Hydrocarbon Decomposition: Growth Mechanism

Real-time microscopic observation of graphene layer formation would be one of the most exciting experiences for any material scientist as it can undoubtedly help in many ways to understand the growth kinetic and thermodynamic stability of the graphene films. In addition, it can also provide the gateway for better/improved growth methodology. Due to its high CVD growth temperature and atomically small dimensions (thickness), it is technologically very challenging to realize such tiny happenings. In this aspect, elevated temperature STM imaging can be one of the very few available technical methods.

Different transition metals have their unique catalytic properties which reflect during the graphene formation at elevated temperatures. In addition, the quality of the graphene can significantly depend on the ability of the transition metal to facilitate the dissociation process of hydrocarbon and carbon diffusion toward the segregation process. Two different groups of TMs have been reported to promote the CVD growth of graphene via two separate growth mechanisms such as (i) bulk mediated (Ni) or surface mediated (Cu) segregation of carbon atoms [79]. It can strongly be determined by the solubility of C atoms at the elevated growth temperature. In the case of bulk segregation, initially, C atoms dissolve into the bulk metals at a high temperature to reach carbon saturation. Afterwards, they segregate to the top surface to form the graphene layer which expands in a carpet growth mode. Ru, Ir, Ni, or Co are found to have such bulk segregation. However, graphene growth on the Cu surface is somehow different. The growth model for graphene formation shows a surface diffusion here. Carbon aggregates at the surface of Cu without diffusing much into the bulk, which further forms the graphene layer via surface diffusion. Apart from this, orientations of crystalline facets can also have a significant impact on the graphene growth mechanism. Moreover, the as-grown graphene layer may have different interaction energies with the substrate atoms, as depicted in Fig. 11b [80]. Yellow color represents the elements having weak interaction with the as-grown graphene layer, whereas red and blue represent elements with strong graphene interaction.

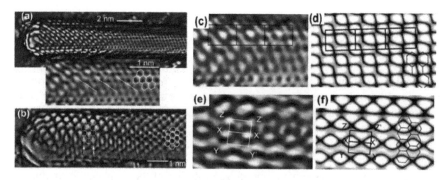

Fig. 11 STM derivative images of end cap CNTs: (**a**) arm-chair and (**b**) zig-zag. Comparison of STM images and simulation results: (**c**) and (**d**) arm-chair CNT, and (**e**) and (**f**) zig-zag CNT, respectively. Reprinted with permission [78]

Surface Diffusion of C Atoms: Graphene Formation on Cu (111) Surface

The CVD growth of graphene layer on a single crystalline Cu (111) surface has been widely achieved by thermal decomposition of hydrocarbons (ethylene, methane, etc.) under a UHV ambient condition. Here, we discuss a reported STM study of graphene formation on a Cu (111) surface at 1000 °C, with special attention to the structural and electronic properties of the as-grown graphene layer [81]. As discussed in the earlier CVD growth process, the graphene layer is generally formed by carbon segregation to the surface. Thermal dissociation of ethylene molecules on the catalytic metal surface acts as a carbon source which finally results in multiple islands of graphene formation, with a variable size distribution. As the carbon solubility of Cu is extremely low, graphene growth is mainly dominated by surface diffusion. Therefore, the initial surface morphology of the Cu substrate can also play a very significant role to control the growth kinetics, as can be seen in Fig. 12. Graphene islands of monolayer height scale initially nucleate with two predominant domain orientations. With increasing coverages, these islands meet each other to form a layered growth morphology. As a result, large numbers of domain boundaries are formed within the graphene layer. This might be a real technological challenge for high-quality graphene formation as these domain boundaries can act as a scattering center for electrons and hence lower the carrier mobility.

For larger view analysis of graphene island formation on Cu (111) surface, STM topography (Fig. 12a) along with differential conductance (dI/dV) imaging (Fig. 12b) have been employed to differentiate between the as-grown graphene islands and the bare Cu (111) metal surface. Differential conductance dI/dV image of graphene layer recorded here (highlighted with dotted lines) appears with a darker contrast as compared to the bare Cu (111) surface (Fig. 12c). Various graphene domains are formed which eventually meet at the domain boundaries (Fig. 12c). A closer view of the STM imaging of graphene domains shows that a honeycomb structure appears with *Moire´* patterns related to subsurface electronic information that originated from the substrate–film lattice mismatch (Fig. 12d).

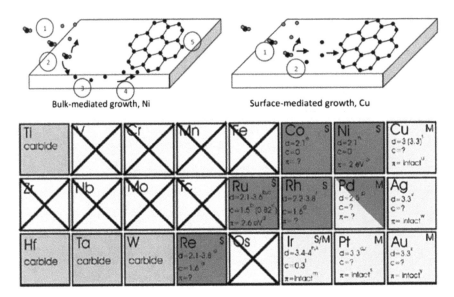

Fig. 12 (Top) Bulk (left) and surface (right) diffusion mediated graphene growth mechanism (bottom) Interaction energies of TMs with graphene layer. Reprinted with permission [79]

Bulk Diffusion of C Atoms: Graphene Formation on Ru (0001)

The formation of millimeter length-scale graphene of monolayer thickness has been reported on a Ru (0001) crystal using the CVD growth method [80]. Single-crystalline graphene monolayer formation on a Ru (0001) surface by thermal annealing in the presence of carbon precursors is shown in Fig. 13. In contrast to graphene growth on Cu (111) (Fig. 12), here the formation mechanism is very different. Instead of surface diffusion of the added carbon atoms, initially they diffuse into the bulk of the Ru crystal. As the solubility reaches the saturation limit, surface segregation of C atoms starts to occur which finally leads to the graphene formation. In addition to the initial surface morphology of Ru (0001), the hexagonal crystal lattice of it may also influence the growth process and hence finally the structural quality of the graphene layer. Figure 13 represents the STM images of the graphene layer without having any domain boundaries (Fig. 12b). High-resolution STM images confirm that the as-grown graphene layer is highly ordered, continuous, having perfect crystallinity [80] (Fig. 14).

4 Conclusion

In summary, we have briefly discussed various allotropes of carbon, their synthesis and structures, and other properties with atomic-scale precision. STM analysis of some of these selective allotropes of nanoscale size (nanotropes) has been discussed

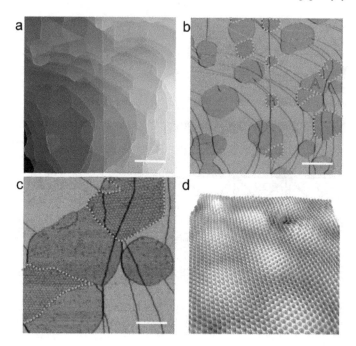

Fig. 13 STM imaging of epitaxial graphene formation Cu (111) surface at elevated temperature, using ethylene as hydrocarbon source. Graphene coverage 0.35 ML. Reprinted with permission [81]

in detail with a special focus toward their structure, bonding, formation mechanism as well as correlated electronic properties. Overall, these can be concluded in three parts. (i) The adsorption process of C_{60} molecules on Si (111)-7 × 7 surface at room temperature has been found to be an interplay between the substrate surface lattice structure and the molecular symmetry of the C_{60} itself. (ii) STM imaging of SWCNTs has successfully determined the chirality of the nanotube which further reflects on their electronic properties. (iii) Graphene growth on catalytic TM surface using CVD has been found to occur in two ways such as bulk and surface diffusion process of C atoms, which finally lead toward the surface segregation for successive graphene layer formation.

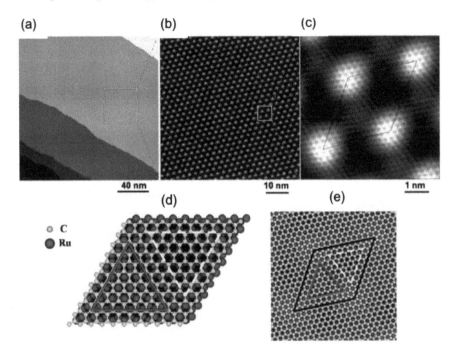

Fig. 14 STM imaging of epitaxial graphene formation on Ru (0001) surface. Reprinted with permission [79]

Acknowledgements The author (SG) gratefully acknowledges the instrumental support from Prof. Philip Moriarty of University of Nottingham, UK, for some of the STM images of C_{60} molecules.

References

1. Zhang Y, Yin Q-Z (2012) Carbon and other light element contents in the Earth's core based on first-principles molecular dynamics. Proc Natl Acad Sci USA 109:19579–19583. https://doi.org/10.1073/pnas.1203826109
2. Pace NR (2001) The universal nature of biochemistry. Proc Natl Acad Sci USA 98:805–808. https://doi.org/10.1073/pnas.98.3.805
3. Ferrari AC, Robertson J (2000) Interpretation of Raman spectra of disordered and amorphous carbon. Phys Rev B 61:14095–14107. https://doi.org/10.1103/PhysRevB.61.14095
4. Wei L, Kuo PK, Thomas RL, Anthony TR, Banholzer WF (1993) Thermal conductivity of isotopically modified single crystal diamond. Phys Rev Lett 70:3764–3767. https://doi.org/10.1103/PhysRevLett.70.3764
5. Hodkiewicz J, Scientific TF (2005) Characterizing carbon materials with Raman spectroscopy. Prog Mater Sci 50:929–961
6. Titirici M. (2013) Sustainable Carbon Materials from Hydrothermal Processes; John Wiley & Sons Ltd.: Chichester UK.
7. Dai L, Chang DW, Baek J-B, Lu W (2012) Carbon nanomaterials for advanced energy conversion and storage. Small 8:1130–1166. https://doi.org/10.1002/smll.201101594

8. Viswanathan B, Neel P, Varadarajan T (2009) Methods of Activation and Specific Applications of Carbon Materials. In: Viswanathan B (ed). NCCR IIT Madras: Chennai India.

9. Kaneko K, Ishii C, Ruike M, kuwabara H, (1992) Origin of superhigh surface area and micro-crystalline graphiticstructures of activated carbons. Carbon 30:1075–1088. https://doi.org/10.1016/0008-6223(92)90139-N

10. Titirici M-M, White RJ, Brun N, Budarin VL, Su DS, Del Monte F, Clark JH, MacLachlan MJ (2015) Sustainable carbon materials.Chem Soc Rev 44:250–290. https://doi.org/https://doi.org/10.1039/C4CS00232F

11. Loos M. Allotropes of Carbon and Carbon Nanotubes. In Carbon Nanotube Reinforced Composites; Elsevier:Amsterdam The Netherlands 2015; pp. 73–101.

12. Deng J, You Y, Sahajwalla V, Joshi RK (2016) Transforming waste into carbon-based nanomaterials. Carbon 96:105–115. https://doi.org/10.1016/j.carbon.2015.09.033

13. Rodríguez-Reinoso F (1998) The role of carbon materials in heterogeneous catalysis.Carbon 36:159–175. https://doi.org/https://doi.org/10.1016/S0008-6223(97)00173-5

14. Hirsch A (2010) The era of carbon allotropes. Nat Mater 9:868. https://doi.org/10.1038/nmat2885

15. Wikipedia. Allotropes of carbon [Internet]. Available from: https://en.wikipedia.org/wiki/Allotropes_of_carbon#mediaviewer/File:Eight_Allotropes_of_Carbon.png.

16. Allen MJ, Tung VC, Kaner RB (2010) Honeycomb carbon: A review of graphene. Chem Rev 110:132–145. https://doi.org/10.1021/cr900070d

17. Geim AK, Novoselov KS (2007) The rise of graphene. Nat Mater 6:183–191. https://doi.org/10.1142/9789814287005_0002

18. Novoselov KS, Fal VI, Colombo L, Gellert PR, Schwab MG, Kim K (2012) A roadmap for graphene. Nature 490:192–200. https://doi.org/10.1038/nature11458

19. Kaiser W, Bond WL (1959) Nitrogen a major impurity in common type I diamond. Phys Rev 115:857. https://doi.org/10.1103/PhysRev.115.857

20. Pal'yanov YN, Sokol AG, Borzdov YM, Khokhryakov AF, Sobolev NV, (1999) Diamond formation from mantle carbonate fluids. Nature 400:417. https://doi.org/10.1038/22678

21. Lee C, Wei X, Kysar JW, Hone J (2008) Measurement of the elastic properties and intrinsic strength of monolayer graphene. Science 321:385. https://doi.org/10.1126/science.1157996

22. Frondel C, Marvin UB (1967) Lonsdaleite a hexagonal polymorph of diamond.Nature 214:587. http://dx.doi.org/https://doi.org/10.1038/214587a0.

23. Bundy FP Kasper JS (1967) Hexagonal diamond: a new form of carbon. J Chem Phys 46:3437. http://dx.doi.org/https://doi.org/10.1063/1.1841236.

24. Lonsdaleite [Internet]. Available from: https://www.answers.com/topic/lonsdaleite.

25. Kroto HW, Heath JR, O'Brien SC, Curl RF, Smalley RE (1985) C60: Buckminsterfullerene Nature 318:162–163

26. Iijima S (1991) Helical microtubules of graphitic carbon. Nature 354:56. https://dxdoiorg/101038/354056a0

27. Pfeiffer R, Pichler T, Kim Y, Kuzmany H (2008) Double-wall carbon nanotubes In: Jorio A Dresselhaus G Dresselhaus M (eds) Carbon Nanotubes Vol 111 Springer Berlin Heidelberg 495. https://dxdoiorg/101007/978-3-540-72865-8_16

28. Ando T (2009) The electronic properties of graphene and carbon nanotubes. NPG Asia Mater 1:17. https://dxdoiorg/101038/ asiamat20091

29. Baughman RH, Zakhidov AA, Heer AW (2002) Carbon Nanotubes—the Route Toward Applications Science 297:787–792. https://doi.org/https://doi.org/10.1126/science.1060928

30. Parker CB, Raut AS, Brown B, Stoner BR, Glass JT (2012) Three-dimensional arrays of graphenated carbon nanotubes. J Mater Res 27:1046.https://dxdoiorg/101557/jmr201243

31. Yu K, Lu G, Bo Z, Mao S, Chen J (2011) Carbon nanotube with chemically bonded graphene leaves for electronic and optoelectronic applications. J Phys Chem Lett 2:1556. https://doi.org/10.1021/jz200641c

32. Zhao X, Liu Y, Inoue S, Suzuki T, Jones RO, Ando Y (2004) Smallest carbon nanotube is 3Å in diameter. Phys Rev Lett 92:125502. https://dxdoiorg/101103/PhysRevLett92125502

33. Martel R, Shea HR, Avouris P (1999) Rings of single-walled carbon nanotubes. Nature 398:299. https://dxdoiorg/101038/18589
34. Novoselov KS, Jiang Z, Zhang Y, Morozov SV, Stormer HL, Zeitler U, Maan JC, Boebinger GS, Kim P, Geim AK (2007) Room-Temperature Quantum Hall Effect in Graphene. Science 315:1379. https://doi.org/10.1126/science.1137201
35. Castro Neto AH, Guinea F, Peres NMR, Novoselov KS, Geim AK (2009) The electronic properties of grapheme. Rev Mod Phys 81:109–162. https://doi.org/10.1103/RevModPhys.81.109
36. Du X, Skachko I, Duerr F, Luican A, Andrei EY (2009) Fractional quantum Hall effect and insulating phase of Dirac electrons in grapheme. Nature 462:192–195. https://doi.org/10.1038/nature08522
37. Westervelt R M (2008) Graphene nanoelectronics. Science 324–325
38. Britnell L, Gorbachev RV, Jalil R, Belle BD, Schedin F et al (2012) Field-Effect Tunneling Transistor Based on Vertical Graphene Heterostructures. Science 947-950. https://doi.org/https://doi.org/10.1126/science.1218461
39. Lin Y, Dimitrakopoulos C, Jenkins KA, Farmer DB, Chiu HY et al (2010) 100-GHz Transistors from Wafer-Scale Epitaxial Graphene. Science 327:662. https://doi.org/https://doi.org/10.1126/science.1184289
40. Kratschmer W, Lamb LD, Fostiropoulos K, Huffman DR (1990) Solid C60: a new form of carbon. Nature 347:354. https://dxdoiorg/101038/347354a0
41. Nano-C Fullerene Technology [Internet] Available from: http:// www.nano-ccom/fullerene tech.html
42. Cheng J, Meziani MJ, Sun YP, Cheng SH (2011) Poly(ethylene glycol)-conjugated multi-walledcarbon nanotubes as an efficient drug carrier for overcoming multidrug resistance. Toxicol Appl Pharmacol 250:184–193. https://doi.org/10.1016/j.taap.2010.10.012
43. Liu BC, Lyu SC, Lee TJ, Choi SK, Eum SJ, Yang CW, Park CY, Lee CJ (2003) Synthesis of single- and double-walled carbon nanotubes by catalytic decomposition of methane. Chem Phys Lett 373:475. https://dxdoiorg/101016/S0009-2614(03)00636-5
44. Dubey P, Muthukumaran D, Dash S, Mukhopadhyay R, Sarkar S (2005) Synthesis and characterization of water-soluble carbon nanotubes from mustard soot. Pramana 65:681–697. https://dxdoiorg/101007/BF03010456
45. Wang Z, Zhao Z, Qiu J (2006) Synthesis of branched carbon nanotubes from coal. Carbon 44:1298–1352. https://dxdoiorg/101016/jcarbon200512030
46. Qiu J, Li Y, Wang Y, Li W (2004) Production of carbon nanotubes from coal. Fuel Process Technol 85:1663 https://dxdoiorg/101016/jfuproc200312010
47. Novoselov KS, Geim AK, Morozov SV, Jiang D et al (2004) Electric Field Effect in Atomically Thin Carbon Films. Science 306:666–669. https://doi.org/10.1126/science.1102896
48. Loginova E, Bartelt NC, Feibelman PJ, McCarty KF (2009) Factors influencing graphene growth on metal surfaces. New J Phys 11:063046. https://doi.org/10.1088/1367-2630/11/6/063046
49. Martoccia D, Willmott PR, Brugger T, Björck M et al (2008) Graphene on Ru(0001): A 25x25 supercell. Phys Rev Lett 101:126102. https://doi.org/10.1103/PhysRevLett.101.126102
50. Oshima C, Tanaka N, Itoh A, Rokuta E, Yamashita K, Sakurai T (2000) A heteroepitaxial multi-atomic-layer system of graphene and h-BN. Surf Rev Lett 7:521–525. https://doi.org/10.1142/S0218625X00000683
51. Mccarty KF, Feibelman PJ, Loginova E, Bartelt NC (2009) Kinetics and thermodynamics of carbon segregation and grapheme growth on Ru(0001). Carbon 47:1806–1813. https://doi.org/10.1016/j.carbon.2009.03.004
52. Coraux J, N'Diaye AT, Busse C, Michely T, (2008) Structural coherency of graphene on Ir(111). Nano Lett 8:565–570. https://doi.org/10.1021/nl0728874
53. Kim KS, Zhao Y, Jang H, Lee SY et al (2009) Large-scale pattern growth of graphene films for stretchable transparent electrodes. Nature 457:706–710. https://doi.org/10.1038/nature07719
54. Sutter PW, Flege JI, Sutter EA (2008) Epitaxial Graphene on Ruthenium Nature Materials 7:406–411. https://doi.org/10.1038/nmat2166

55. Wintterlin J, Bocquet ML (2009) Graphene on metal surfaces. Surf Sci 603:1841–1852. https://doi.org/10.1016/j.susc.2008.08.037
56. Astruc D (2007) Palladium Nanoparticles as Efficient Green Homogeneous and Heterogeneous Carbon—Carbon Coupling Precatalysts: A Unifying View. Inorg Chem 46:1884. https://doi.org/10.1021/ic062183h
57. Piao Y, Jang Y, Shokouhimehr M, Lee IS, Hyeon T (2007) Facile Aqueous-Phase Synthesis of Uniform Palladium Nanoparticles of Various Shapes and Sizes. Small 3:255. https://doi.org/10.1002/smll.200600402
58. Marchini S, Guenther S, Wintterlin J (2007) Scanning tunneling microscopy of graphene on Ru(0001). Phys Rev B 76:075429. https://doi.org/10.1103/PhysRevB.76.075429
59. Wang B, Guenther S, Wintterlin J, Bocquet ML (2010) Periodicity, work function and reactivity of graphene on Ru(0001) from first principles. New J Phys 12:043041. https://doi.org/10.1088/1367-2630/12/4/043041
60. Altenburg S, Kröger J, Wang B, Bocquet M-L, Lorente N, Berndt R (2010) Graphene on Ru(0001): Contact Formation and Chemical Reactivity on the Atomic Scale. Phys Rev Lett 105:236101. https://doi.org/10.1103/PhysRevLett.105.236101
61. Sicot M, Leicht P, Zusan A, Bouvron S, Zander O et al (2012) ACS Nano 6:151. https://doi.org/10.1021/nn203169j
62. Wang B, Caffio M, Bromley C, Fruechtl H, Schaub R (2010) Coupling Epitaxy, Chemical Bonding, and Work Function at the Local Scale in Transition Metal-Supported Graphene. ACS Nano 4:5773. https://doi.org/10.1021/nn101520k
63. Voloshina EN, Dedkov YS, Torbruegge S, Thissen A, Fonin M (2012) Graphene on Rh(111): Scanning tunneling and atomic force microscopies studies. Appl Phys Lett 100:241606. https://doi.org/10.1063/1.4729549
64. Coraux J, N'Diaye AT, Engler M, Busse C, Wall D, Buckanie N, Meyer, Heringdorf FJM, Gastel R, Poelsema B and Michely T (2009) Growth of graphene on Ir(111). New J Phys 11:023006 (22pp).
65. Busse C, Lazic P, Djemour R, Coraux J et al (2011) Graphene on Ir(111): Physisorption with Chemical Modulation. Phys Rev Lett 107:036101. https://doi.org/10.1103/PhysRevLett.107.036101
66. Starodub E, Bostwick A, Moreschini L et al (2011) In-plane orientation effects on the electronic structure, stability, and Raman scattering of monolayer graphene on Ir(111) Phys Rev B 83:125428. https://doi.org/https://doi.org/10.1103/PhysRevB.83.125428
67. Kralj M, Pletikosic I, Petrovic M, Pervan P et al (2011) Graphene on Ir(111) characterized by angle-resolved photoemission. Phys Rev B 84:075427. https://doi.org/10.1103/PhysRevB.84.075427
68. Pletikosic I, Kralj M, Milun M, Pervan P (2011) Finding the bare band: Electron coupling to two phonon modes in potassium-doped graphene on Ir(111). Phys Rev B 85:155447. https://doi.org/10.1103/PhysRevB.85.155447
69. Land T, Michely T, Behm R, Hemminger J, Comsa G (1992) STM investigation of single layer graphite structures produced on Pt(111) by hydrocarbon decomposition. Surf Sci 264:261. https://doi.org/10.1016/0039-6028(92)90183-7
70. Sutter P, Sadowski J, Sutter E (2009) Graphene on Pt(111): Growth and substrate interaction. Phys Rev B 80:245411. https://doi.org/10.1103/PhysRevB.80.245411
71. Binnig G, Rohrer H, Gerber Ch, Weibel E (1982) Surface studies by Scanning Tunneling Microscopy. Phys Rev Lett 49:57
72. Binnig G, Rohrer, (1986) Scanning Tunneling Microscopy. IBM J Res Develop 30:4
73. Gangopadhyay S, Woolley RAJ, Danza R, Phillips MA, Schulte M, Wang L, Dhanak VR, Moriarty PJ (2009) C_{60} submonolayers on the Si (111)-(7× 7) surface: Does a mixture of physisorbed and chemisorbed states exist? Surf Sci 603:2896–2901
74. Du X, Chen F, Chen X, Wu X, Cai Y, Liu X, Wang L (2010) Adsorption geometry of individual fullerene on Si surface at room-temperature. Appl Phys Lett 97:253106. https://doi.org/10.1063/1.3529446

75. Liu L, Liu S, Chen X, Li C, Ling J, Liu X, Cai Y, Wang L (2013) Switching Molecular Orientation of Individual Fullerene at Room Temperature. SCIENTIFIC REPORTS 3:3062. https://doi.org/10.1038/srep03062
76. Néel N, Limot L, Krögerand J, Berndt R (2008) Rotation of C60 in a single-molecule contact. PHYSICAL REVIEW B 77:125431. https://doi.org/10.1103/PhysRevB.77.125431
77. Venema LC, Meunier V, Lambin Ph, Dekker C (2000) Atomic structure of carbon nanotubes from scanning tunneling microscopy. PHYSICAL REVIEW B 61:2991–2996. https://doi.org/10.1103/PhysRevB.61.2991
78. Furuhashi1 M, Komeda T (2008) Chiral vector determination of carbon nanotubes by observation of interference patterns near the end cap. Phys Rev Lett 101:185503. https://doi.org/10.1103/PhysRevLett.101.185503
79. Seah C-M, Chai SP, Mohamed AR (2014) Mechanisms of graphene growth by chemical vapour deposition on transition metals. Carbon 70:1–21. https://doi.org/10.1016/j.carbon.2013.12.073Get
80. Batzill M (2012) The surface science of graphene: Metal interfaces CVD synthesis nanoribbons chemical modifications and defects. Surf Sci Rep 67:83–115. https://doi.org/10.1016/j.surfrep.2011.12.001
81. Gao L, Guest JR, Guisinger NP (2010) Epitaxial Graphene on Cu (111). Nano Lett 10:3512–3516. https://doi.org/10.1021/nl1016706

Carbon: A Phantom for Nanocomposite-Driven Applications

Sakaray Madhuri, Chidurala Shilpa Chakra, Thida Rakesh Kumar, Konda Shireesha, Sai Kumar Pavar, and Velpula Divya

Abstract Clever combinations of elements store energy in chemical form like a battery and then release energy pulses whenever and wherever it is needed. Every chemical element in the periodic table is special, but some elements are more special than others. An essential element of life has to multitask. Carbon, the sixth element, is unwonted in its impact on our lives. Carbon lies at the heart of progression intriguing the emergence of planets, life, and us. And, more than any other entirety, carbon has greased the rapid emergence of new technologies. If we discover to replenish our rhapsodically beautiful carbon-rich world, then we may hope to leave a peerless, high-end legacy for all the generations to come. Fullerenes, graphene, carbon nanotubes, fluorescent carbon quantum dots, activated carbon, and carbon black belong to the carbon family with tremendous optical, physical, mechanical, and thermal properties. Among them, carbon nanocomposites can be synthesized with the amalgamation of different elements. Carbon nitride with covalent network compound are unlinked into beta carbon nitride and graphitic carbon nitride (g-C_3N_4) that are relatively new type of carbon based material retaining high photoresponsiveness, high intrinsic photoabsorption, semiconductive properties, high stability under physiological conditions and good biocompatibility. Use of sunlight as a sustainable source for energy generation, environmental medicament photocatalysts for heavy metal pollutant control, and water splitting by use of polymeric materials with incorporation of carbon can be achieved. An oxocarbon consists of a single carbon and single oxygen which has the ability to polymerize at the atomic level, thus forming very long carbon chains. The smart material can be obtained by homogenizing with carbon nanocomposites synthesized in an inexpensive process like printing and roll to roll which are ideal for flexible energy generation and storage. To overcome the extremely high volume change by alloying reaction with lithium for commercialization, carbonaceous materials are induced to improve the structural stability of the electrodes. Lithium-ion batteries (LIBs) are considered as efficacious and practical technology

S. Madhuri · C. Shilpa Chakra (✉) · T. Rakesh Kumar · K. Shireesha · S. K. Pavar · V. Divya
Center for Nano Science and Technology, Institute of Science and Technology, JNTU, 500085 Hyderabad, Telangana, India
e-mail: shilpachakra.nano@jntuh.ac.in

for electrochemical energy storage. Due to high theoretical capacities, electrochemically active metal oxides materialize as promising candidate for the anodes in LIBs. Carbon coating can productively improve the surface chemistry of active material and electrode conductivity and protect the electrode from interacting with electrolyte, enhancing the shelf life of batteries, etc. The fascinating properties of these materials are observed in the emerging strategies for tailoring carbon-based nanocomposites in catalytic organic transformation properties, energy storage, absorbents, biomedical, textile, sensor, molecular imaging, bioimaging, drug, and gene delivery.

Keywords Sixth element · Energy · Coating · Flexible · Nanocomposites · Nitrides · Oxides

1 Introduction

Carbon

Carbon known as the King of the elements forms a diverse range of nanomaterial compounds with the utmost ten million compounds that belong to Group 14 of the periodic table sitting at position 6 with the 15th most abundant nonmetallic as well as tetravalent that makes four electrons free for all to form covalent chemical bonds. Pure carbon has less toxicity. Carbon that comes from Latin was discovered in the prehistoric period of mankind in the forms of charcoal and soot. Earth has around 4360 million gigatonnes of carbon, 2000 ppm in core and 120 ppm in the mantle as well as crust. The non-identical allotropes such as graphite is a good electrical conductor and soft, and diamond is a low electrical conductor and hard. Fullerene includes buckyballs, carbon nanotubes, carbon nanobuds, and nanofibers shown in Fig. 1.

Multiple exotic allotropes have been scrutinized, such as glassy carbon (GC), nano-foam, lonsdaleite, and carbyne. Dubbed Q-carbon has been created in 2015 from North Carolina state university which exhibits fluorescence, hardness, and ferromagnetism. Atomic carbon allotropes [1] are solids under normal conditions with chemical resistivity which requires elevated temperatures to react even with oxygen. Due to triple point 10.8 ± 0.2 MPa and $4,600 \pm 300$ K, it has the highest sublimation point. Carbon is thermodynamically prone to repel oxidation non conjecturally compared to copper and iron with electronegativity -2.5. Hydrocarbons have their primary use of carbon in the fossil fuel and crude oils. Plastics are artificially produced from synthetic carbon polymers with regular intervals of nitrogen and oxygen atoms and pyrolysis of fibers of synthetic polyester that forms the main polymer chain. Carbon black can be used in carbon paper, printing ink, oil paint, watercolors, laser printer toner, and fillers in tires. The nanocomposites can be mainly synthesized by using sol–gel process, in-situ intercalative polymerization, and in-situ polymerization.

Carbon and its derivatives carbon nanotubes (CNTs), carbon nanoparticles, graphene, and graphene oxide have captivated attention due to electrical conductivity,

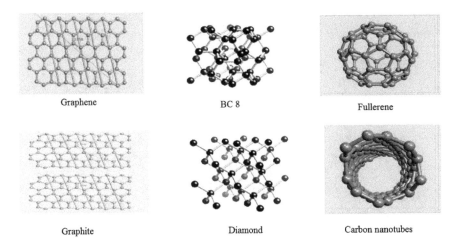

Fig. 1 Allotropes of carbon

biocompatibility, high surface area, thermal and mechanical stability, regular structure, and chemical inertness. Their potential applications are found in energy storage, catalysts, absorbents, biomedical, sensors, textiles, and many more. The characteristics of carbon with electronic properties of semiconductors are mentioned in Table 1. The incorporation of carbon nanocomposites into metals or metal oxides is found to have a broad range of applications. A footloose atom-thick sheet, sp^2 hybridized carbon is graphene that can be stacked to form 1-dimensional CNTs. Transfiguration of graphene and CNTs is explored due to the long-range π-conjugation and luminous physiochemical properties. The solid-state structures of carbon are shown in Fig. 2.

Preparation

Noncovalent functionalization is a physical absorption involving weak π-interactions that improve hydrophobicity with outshine metal–carbon interaction. This process can be done by wet and dry synthesis.

Metal nanoparticles are incorporated with carbon, Nobel and non-Nobel metal supported carbon nanocomposites and bimetallic carbon nanocomposites.

Table 1 Characteristics of carbon allotropes

Isomer	Graphene	Nanotube	Fullerene
Hybridization	sp^2	sp^2	sp^2
Bond length(Å)	1.40(CQC)	1.44(CQC)	1.42(CQC)
Dimension	2D	1D	0D
Density (g/cm^3)	1.72	1.22	2.26
Electronic properties	Zero-gap semiconductor	Metal/semiconductor	Semiconductor

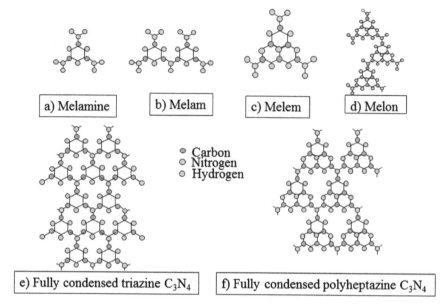

a) Melamine   b) Melam   c) Melem   d) Melon

○ Carbon
○ Nitrogen
○ Hydrogen

e) Fully condensed triazine C_3N_4 f) Fully condensed polyheptazine C_3N_4

Fig. 2 Structural content for carbon solid-state structures: **a** melamine, **b** melam, **c** melem, **d** melon, **e** fully condensed triazine C_3N_4, **f** fully condensed polyheptazine C_3N_4

Table 2 Phase and bandgap of seven graphitic carbon nitrides

Phase	Bandgap(eV)
a$-C_3N_4$	5.5
b$-C_3N_4$	4.85
cubic$-C_3N_4$	4.3
g$-$h$-$triazine	2.97
g$-$otriazine	0.93
g$-$h$-$heptazine	2.88
pseudocubic C_3N_4	4.13

2 Carbon Derivatives and Its Applications

The derivatives of carbon can be described as follows:

Graphite that arefound in India, Mexico, Russia, Greenland, and US exists in amorphous, vein, and flakes which occurs in plates crystallized in the metamorphic rock. At normal pressures, carbon takes forms: graphite trigonally bonded in plane fused with hexagonal rings and delocalization in the outer electrons of the atom to form a π-cloud. This makes graphite conducting and also account for its vital stability when compared to diamond at room temperature which is thermodynamically stable and acts as a lubricant, electrical conductor and thermal insulation. Graphene, a 2-dimensional magnitude sheet of carbon in which the atoms are choreographed in a hexagonal lattice. Graphite are fundamentally different from 3-dimensional ones as they can be used as neutron moderators and annealing to 250 °C can release the energy steadily.

Applications: pencil leads, sticks of compressed charcoal, textiles, carbide end-mills.

Diamonds are applied in the semiconductor business for their heat conducting property and in drill tips for grinding as well as polishing applications. The hardest material is the synthetic nanodiamond which is abrasive with good electrical insulator crystallizes in the cubic system which is three times greater than the density of graphite. Dust of diamond can be used as abrasive and it is harmful if inhaled. These results descend in oodles of electrical conductivity for carbon than for most of the metals. The delocalization also accounts for the energetic stability of graphite over diamond at room temperature.

Fullerenes are fused by pentagons, hexagons, and heptagons of carbon atoms that can be wrapped into cylinders, ellipses, and spheres. Carbon nanotubes are structurally resemblant to buckyballs; singly atom is trigonally bonded into a curve forming a hollow cylinder. Carbon nanotubes, a nanocarbon form, have tremendous mechanical and chemical properties contributing from the last two decades and their applications in diverse fields.

Carbon forms compounds that are of organic, inorganic, and organometallic.

Catenation is the property by which carbon forms very strong and stable long chains which form a countless number of compounds. Carbon fixation causes photo-synthesis that makes it possible to form organic compounds which are used by plants and animals. The structural formula for methane and carbon cycles is referred to in Fig. 3. The enormous amount of carbon can be used by plants by photosynthesis as their intake is carbon dioxide.

Hydrocarbon in its simplest form of organic family and its backbone can be substi-tuted by heteroatoms that include nitrogen, sulfur, oxygen, phosphorous, lithium, and magnesium which form many groups in biological compounds. Categorization and study of these can be termed by functional groups. Lubricants, plastics, refrigerators, fossil fuels, and petrochemicals use different forms of hydrocarbon.

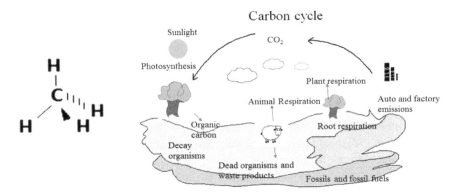

Fig. 3 Methane structural formula and carbon cycle

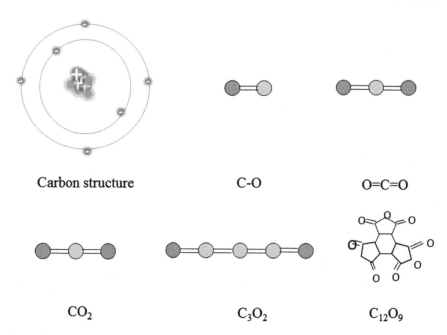

| Carbon structure | C–O | O=C=O |

| CO_2 | C_3O_2 | $C_{12}O_9$ |

Fig. 4 Figures of oxides of carbon

Prevailingly carbon-containing compounds when hooked up with minerals or which do not contain bonds to the other carbon atoms, halogens, or hydrogen are reckoned solely from classical organic compounds. The definition is not rigid as the rubric of some compounds can vary from author to author. Simplest oxides which are unstable form inorganic compounds such as carbon dioxide, carbon monoxide, cyanide, cyaphide, carbon suboxide, dicarbon monoxide, and carbon trioxide.

Carbon metal covalently bonded forms organometallic compounds fencing alkyl metals, metallocenes, and transition metal carbine. Grouping of hetero atoms occurs in a large number of organic compounds.

2.1 Energy Storage

Energy generation techniques to store energy take a pivotal step in the present generation [2]. The anodes of battery constitute carbon allotropes such as graphitic carbon palliating electronic conductivity which has a higher capacity of ~ 372 mA h g^{-1} intercalated with lithium insinuated compounds which may have flung up to ~ 524 mA h g^{-1} [3] residing with intralayer voids between them. As graphite shares the same structure as that of graphitic carbon nitride, it shows good cyclability, voltage range, and high conductivity. Many reports have been examined, and due to cycling, irreversible capacity losses were significant with the limiting factor of electrical resistance. To trim buffer additives that are electronically conductive are interposed as intermittently the additives does not boost theoretical capacity [4]. The incorporation of carbon nitride may lead to practicable anodes for batteries. The coulombic efficiency can further be increased by doping N with the open structure for the anodes with high discharge capacities with histrionic electrochemical performance furnishing extra sites for the lithium exchange and storage by replacing C atoms with N atoms. Irreversible reactions can be reduced by pyridinic-N in the graphitic carbon nitride [5] supporting the lithium storage by improving its capacity with superior conductivity. This N type describes the electrodes based on carbon in the lithium-ion batteries.

2.2 Photo Catalysis

Photocatalysis has been a distributary for thriving applications such as water splitting, energy conversion and storage, and solar technology and has its impact on conversion of carbon dioxide into bearing fuels, light-driven water splitting, enervate waste pollutants, selectivity regarding toxicity, ease in the synthesis, band structure, and electronic nature. Enhanced photocatalytic property and water splitting have drawn considerable interest during the decade and even until now [6]. Photocatalyst has to felicitously capture as much as light transpiring from the sun in the visible range at the surface of the earth to split into its elemental constituents at the molecular level. By engineering these hybrids, tuning the already existing graphitic carbon nitride is done by revamping the electron–hole pairs mobility and preventing its recombination. This leads to the improved redox photocatalytic activity driving toward energy generation producing hydrogen fuel [7]. The face-to-face interface explicates the 2D heterojunction nano-architectured carbons. This exhibits various limitations; complexities, queries, chemistry and physics involved, and environmental impact, which has drawn significant response to photocatalysis [8]. Due to the wide semiconducting nature of carbon nitrides, they have turned out to play a reminiscent role in the modern interest in photocatalytic water splitting by doping a co-catalyst. Future development is done by understanding the composition, structure, and different properties of the compounds. Graphitic carbon nitrides have $2.5-2.8$ eV of bandgap and its tuning plays a major role by overbearing the visible range that can be determined by p–p* transitions. Coupling with metal oxides by hydrothermal and deposition heating increases the photocatalytic performance of the composites [9]. In factuality, this photocatalytic activity was amended by incorporating graphene nanostructures reduced Graphene Oxide (rGO) which increases the efficiency of the reaction at a low cost.

2.3 Photoluminescence (PL)

Semiconducting materials having wide band gaps exhibit PL property; graphitic carbon nitride with 2.7 eV emits blue PL when it is dissolved in solvents ranging from 400 to 650 nm with tremendous stability compared with carbon dots. In addition to ferreting metal ions and bio-molecules, $g-C_3N_4$ can also respond to temperature. The $g-C_3N_4$ nanodots are accounted as nitrogen-rich carbon with the luminescent property with sumptuous stability than carbon dots. Photoluminescence gives way to $g-C_3N_4$ nanosheets for a potent counter response to the copper ions for chemical sensors.

3 Introduction to Carbon Nitride and Carbon Oxide

Carbon exists in two main forms as oxocarbon or honest form of carbon monoxide (CO) and carbon dioxide (CO_2). In accession to suboxide (C_3O_2), a form of carbon also forms. These forms are shown in Fig. 5. Carbon when reacts with oxygen forms carbon monoxide and carbon dioxide which are parts of the carbon cycle. Carbon monoxide is weighed as a senseless killer and poison. Since CO_2 is naturally produced through metabolism and consumed by photosynthesis. As of now to help appraise environmental issues, Carbon Dioxide Information Analysis Center (CADIAC) provides data regarding the gases and climate change.

Some of the oxides are stable at room temperature and decompose to oxocarbons when they get warmed. Intermediate chemical reactions are observed momentarily which exist in the gas phase or detected by matrix isolation. Oxocarbon's molecular orbital is similar to that of nitrogen (iso-electronic). The electronegativity of oxygen is greater than carbon. They are highly stabilized with low energy with three filled bonding orbitals and two nonbonding orbitals. One of them is closer to carbon atomic orbitals and another is closer to the oxygen atomic orbitals.

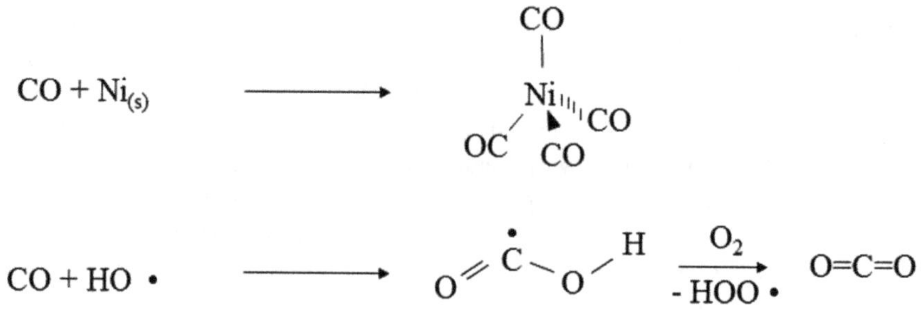

Fig. 5 Reactions of carbon oxide with Nickel and hydroxyl compounds

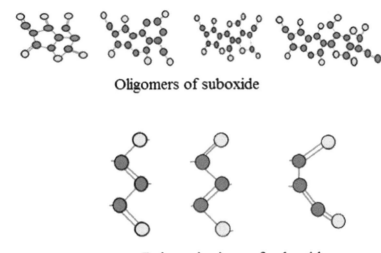

Oligomers of suboxide

Polymeric rings of suboxide

Fig. 6 Polymeric rings and Oligomers of Suboxide

Nuclephilic

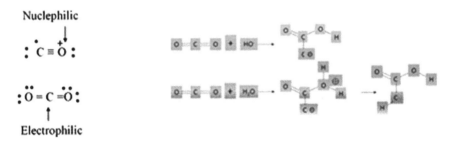

Electrophilic

Fig. 7 Nucleophiles, Reaction of CO_2 with nucleophiles

3.1 Carbon Oxide and Its Applications

3.1.1 Carbon Oxide

Carbon oxide is a stable diamagnetic molecule. It is formed when graphite gets combusted in presence of heat in the limited amount of oxygen. Having a molecular weight 28.01 g/mol with a boiling point $-191.6\,°C$, it is electron-rich as carbon being odorless gets dissolved in water (0.0352vol/vol) yet flammable with dangerous toxic gases which do not give any heads-up also reacts with electrophiles. It strongly binds with the oxygen-carrying protein and prevents oxygen binding which does not break down by processes of the body stopping the circulation of hemoglobin that leads to suffocation.

Structure of Carbon Oxide: Using valence electrons, CO can have a Lewis structure and CO_2 can have three resonance structures and signify the strong chemical bonding between the two. By auditing the electronic configurations of oxygen: $1s^2 2s^2 2p^4$ and carbon: $1s^2 2s^2 2p^2$. Carbon has four valence electrons and oxygen has six valence electrons so the S and P orbital are employable for the bonding. This valence bonding approach bonds with the over lapping of two P orbitals and one σ bond in the formation of $C\equiv O$.

Reactivity of CO: As an example, the reaction is followed, where nickel is tangled in the center with CO and also reacts with hydroxyl radical which is very poor.

CO is evolved when steam reacts with red-hot coke with H_2 as a byproduct which can be used as industrial fuel also called as water gas. As for lab-scale purpose, CO is prepared as sulfuric acid removes the substitutes of water because CO readily gets oxidized to form carbon dioxide, heating formic acid (HCOOH) or oxalic acid ($H_2C_2O_4$) in presence of sulfuric acid (H2SO4). CO is useful as gaseous fuel and used as a metallurgical reducing agent for metal oxides at higher temperatures (Eg: CuO and Fe_2O_3).

3.1.2 Carbon Suboxide (C_3O_2)

Carbon suboxide (acid anhydride of malonic acid) is a linear molecule with a turbulent smell; Dehydration of malonic acid produces C_3O_2 with phosphorous pentaoxide at $140-150\ °C$. In presence of light, it decomposes to ketene (C_2O) and reacts with water to produce the acid. The polymers and oligomers of carbon suboxide are shown in Figs. 8 and 9. It polymerizes to a linear chain with lactone six-membered ring

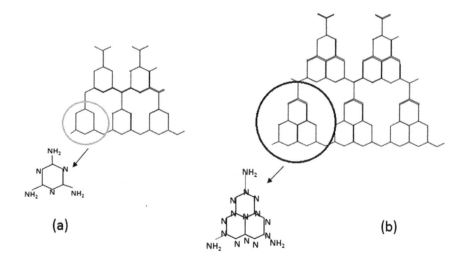

Fig. 8 Tectonic structures of g–C_3N_4 (**a**) triazine, (**b**) tri-s-triazine heptazine structures can be viewed as graphite whose lattice is partially substituted with nitrogen atoms [10]

Fig. 9 Structure of Beta carbon nitride

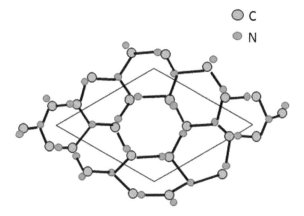

instinctively into carbon and oxygen polymer with 3:2 ratio with carbon backbone having alternate single and double bonds.

3.1.3 Carbon Dioxide

The carbon dioxide used as a building block in organic synthesis having a molecular weight 44.04 g/mol freezes to form dry ice at −78.5 °C. It is produced when any geometry of carbon gets combusted in overage of oxygen with odor and colorless as it is heavier than air. It is toxic if a large amount of concentration is inhaled. The Earth's atmosphere contains 0.04 percent volume of carbon dioxide and has increased excessively by burning fossil fuels causing greenhouse effect allowing UV to penetrate the earth's surface and reradiated as infrared radiation. When it is absorbed by water, it cannot abscond into space which is a reason to boost the temperature day by day having its impact on agriculture, climate, and environment. As the end product of fermentation, combustion, and metabolism, the human body produces ~1.04 kg of CO_2 per day via cellular respiration through the energy obtained by breaking down the fats, sugars, and amino acids. CO_2 tends to fall during summer, as plants consume most of it, and increase during winter. It is covalently bonded centrosymmetric with no net dipole and has a triple bond charge. It serves as a catalyst for the plants to carry out photosynthesis. Two sigma antibonding and two sigma bonding are made by combining carbon's 2 s and 2p orbitals. Carbon's 2p(x) combines with oxygen's group orbital (y) and carbon's 2p(y) combines with oxygen's group orbital (x) thus making 2 pi bonding and 2 pi antibonding orbitals.

CO_2 which is electron-poor acts as an electrophile through the Lewis structure. When carbonates get heated up, they produce CO_2 as a byproduct, and when CO_2 reacts with nucleophiles, they produce calcium oxide.

Calcium carbonate ($CaCo_3$) when heated (burned in excess of oxygen) yields carbon dioxide and calcium oxide (CaO).

$$CaCo_3 + HeatCo_2 + Cao.$$

CO_2 is useful in.

- Beverage industry
- Fertilizers
- Growth of plants in the green house
- Dry ice
- Fire extinguishers
- Inert environment for raw materials
- Solvent extraction.

3.2 Carbon Nitride and Its Applications

A makeshift for design and optimization to reach current applications in energy harvesting, photocatalysts, fuel cell, and electrolyzer catalysts [11] is momentous which reckon on the idiomatic set of chemical, optical, and electronic properties. They are produced from substituting carbon atoms with nitrogen in the structure. Like carbon derivatives, carbon nitrides gained a long history from 1834, Berzelius coined it "melon". Because of its uncataloged structure, insolubility, and chemical inertness, basic solvents were not fully reputed. Another string of planar carbon nitride layers formed by polytriazine imide linked units furnish scrum for Li^+ Br, Cl, and H^+ ions. Carbon nitride has two covalent compounds: Graphitic carbon nitride and Beta carbon nitride. It is an acute time to understand the structural, physical, chemical, and mechanical properties of these materials in terms of their function. Carbon nitride developed with intralayer chemical reactivity led to the catalytic applications either intrinsically or incorporated with metal/metal oxides. There are three different classes of carbon nitride: C_xN_y materials classified into triazine(TGCN)-based and heptazine(HGCN)-based graphitic carbon nitride; $C_xN_yH_z$ compounds divided into monomer/dimer and polytriazine imide; N-doped carbons that are polymeric, cross-linked polymeric, and heptazine-based thermolysis products with graphitic domains.

The instinct to capture light from the sun in the visible range and splitting them into elemental components en route to energy harvesting through exothermic recombination leading to the application in photocatalysts [12, 9].

In solid-state lithiation with the incorporation of carbon nitride with reduced crystallinity triazine-based compound, the electrochemical storage capacity of the battery in anode boosts the conductivity of electrons [13].

3.2.1 Graphitic Carbon Nitride

Graphitic carbon nitride (g−C_3N_4) with the general formula $(C_3N_3H)_n$ is a stable allotrope with 2D conjugated polymer that belongs to the family of carbon nitride with substantial substructures currish on heptazine and poly(triazine imide) units with different conditions of reaction, condensation, and reactivities. Notably, most of the materials synthesized to date boast not only C and N but also tectonic quantities of H as an imperative component of their structures.

Carbon's owed porosity property makes it ideal for the electrodes of batteries and supercapacitors, carbon dioxide storage, solid-state hydrogen, and hydrogen storage [14]. Surface area, active edges, and accessible channels can be increased by the incarnation of porosity to $g-C_3N_4$ which promotes light-harvesting, reactions at the surface, mass transfer, and charge separation with semimetals [15] 2-Dimensional materials have perceived Herculean absorption in the past decade to have been cited for over 40,000 times till now. Similar to graphene, $g-C_3N_4$ has weak van der Waals forces between the layers which could be due to liquid exfoliation from bulk $g-C_3N_4$ in water which shows extreme density of states near to the conduction band by density functional calculations with tunable surface area [16].

It can be synthesized from the polymerization of cyanamide, melamine, and dicyandiamide using in-situ or ex-situ synthesis [16]. A recent report suggests that a new chemical vapor deposition (CVD) [17] method favors the formation of graphitic carbon nitride layers by heating the fusion of uric acid and melamine with alumina and also with silver nanocomposites and titanium dioxide [18]. The principle calculations prognosticated seven angles of graphitic carbon nitride as shown in the table.

Using triazine ring, this formation can be identified using characterization tools such as Fourier transform infrared spectroscopy, photoluminescence spectroscopy, and X-ray photoelectron spectroscopy.

$g-C_3N_4$ is electron-rich with high physiochemical stabilities and groovy electronic band structures, earth abundant, and can be used for energy storage [19], sensing, imaging, catalytic applications, insulators, inert coatings, medical coatings, and tribological coatings in its micron-sized graphitic carbon form. The main application of this in the area of a fuel cell with the polymer electrolyte or with proton exchange membrane is owing to their wide bandgap that reduces the agglomeration of carbonaceous materials caused by etching. To improve the performance and selectivity, chemical modifications, doping, and protonation make it amenable to have metal-free heterogeneous catalysts [20]. The basic applications of $g-C_3N_4$ are works as metal-free catalyst in photocatalysis and no decomposition, for differentiating oxygen activation in oxidation reactions.

Applications: $g-C_3N_4$ finds its applications in wide bandgap semiconductors, photocatalysts, heterogeneous catalysts, and energy storage materials.

3.2.2 Beta Carbon Nitride

Beta carbon nitride ($\beta-C_3N_4$) [10] is a 3D molecule in one layer with a carbon atom having four separate bonds, each nitrogen atom having three. It has a strong bond in the stable crystal lattice which is a super hard material and it can be synthesized by mechanochemical processing such as ball milling. The milling process reduces the grain size to nano-size sub-grain forms. Dissociation of NH_3 molecules into monoatomic nitrogen that adhered to the surface of carbon results in the nanoscale amorphous–flake structure of $\beta-C_3N_4$ that can be found purging under argon atmosphere. Nano-sized carbon reacts with free nitrogen atoms sequel with high surface area and particle dimension forming $\beta-C_3N_4$ [21]. Nanorods with no defects of $\beta-C_3N_4$ can also be formed by thermal annealing which can change flakes into rod or sphere-like structures of compound purging with NH_3 gas. These have hemispherical ends with faster growth along axis direction having prismatic morphology with high yield and low-cost method for the synthesis.

Diode sputtering, solvothermal process, ion implantation pulsed laser ablation shock wave compression have sequels in thin amorphous films of $\beta-C_3N_4$ other than nanorods and powder [22]. $\beta-C_3N_4$ was confirmed by characterization techniques: transmission electron microscopy (TEM), X-Ray Diffraction (XRD), Fourier transform infrared spectroscopy (FTIR), and selected area electron diffraction (SAED) with a lattice constant a = 6.36A° and c = 4.64A°. It exhibits hardness more or less compared to a diamond and has a bulk modulus of 4.27Mbar; a diamond with the bulk modulus 4.43Mbar even utilized modifying iodine in the visible light range [23].

Applications: optical and electronic engineering, wear resistance, tribological factors, substitute for the uses in place of a diamond, and carbon nitride with TiN as seeding layers.

4 Characterization Techniques

Characterization of Carbon Materials: The study of the basic chemistry involved and structure of the solid-state carbon compounds to till is under development. The flash heating from 900 °C to 1000 °C CHN (O) analyzers in an inert atmosphere following oxidation and reduction reactions related compounds are determined [24]. Characterization of carbon and its derivatives is one of the vital parts of carbon study and research. It includes the investigation of carbon and its allotropes structure and its properties: defects and the number of layers in the graphene based on spectroscopic and microscopic studies. Under this section, the characterization techniques that are used for carbon and its allotropes to explore include XRD, ultraviolet-visible spectroscopy, scanning tunneling microscopy (SEM), TEM, and Raman spectroscopy which will be propounded specifically.

4.1 X-Ray Diffraction

X-Ray diffraction is used to arbitrate the structure of the crystalline compounds and polymers. Most of the graphitic carbon nanomaterials exhibit peaks with broad features amorphous to crystalline nature. The main peak is observed with inter-planar spacing. The phase validation of compounds is based on data from unit cell magnitude. Example of graphitic carbon nitride performed on Phillips Xpert SUPER powder X-ray diffractometer with Cu Ka radiation and a strong sharp peak at 27.5° matched with the graphitic carbon nitride with (0 0 2) matched with the previous record. Graphitic carbon nitride prepared by condensation of cyanamide thermally in nitrogen environment gives two peaks of desorption at 145° and 297°. The temperature difference of the two desorption peaks alludes to changing of surface basic intensity.

4.2 Transmission Electron Microscopy (TEM)

TEM is used to study the quality of the structure and these images are formed when the electron beam gets intersected and transmitted through the specimen [10]. This is a direct stronger coulombic interaction method between the electrons, not a field-to-field interaction as X-rays. Electrons are scattered better in thicker areas than in thinner areas which results in a thickness contrast. Phase contrast arises from the interference of the primary beam and the diffracted beams and this is responsible for the appearance of fringes in TEM images [25].

4.3 X-Ray Photoelectron Spectroscopy (XPS)

X-Ray photoelectron spectroscopy [10] is used for surface and chemical analysis of materials. Surface analytical technique delving 1–10 nm samples represents bulk composition. Powder samples are accumulated at the top of carbon tape, when irradiation of X-rays spectrum with the known wavelength having binding energies (BE) is passed through the sample, it delivers the local coordination, oxidation state, bonding fitted with Gaussian–Lorentzian to referee the concentrations of relative various species [25]. The contamination by fretting the surface of the sample can be analyzed with the beam of argon atoms.

4.4 Raman Spectroscopy

A qualitative measure of CNT alignment and structure is done by comparing the G-band parallel Vs perpendicularly aligned sample directions. The structure quality can be accessed when the monochromatic laser interacts with the phonons which leads to shifts in laser energy due to scattering. Three main peaks exhibits in carbon nanostructures—D, G, and 2D. D at 1350 cm^{-1} depicts disorders in the bonding of carbon due to backscattering of phonon to conserve momentum. G peak at 1580 cm^{-1}

attributed to vibrational modes. 2D at 2700 cm^{-1} refers to second-order Raman scattering near Dirac point. Raman spectrum of mono, bi, three, and four layers of graphene on Ni electrodes deposited on alumina substrate can be obtained.

4.5 Ultraviolet-Visible Spectroscopy (UV–Vis)

UV–Visible is an analytical technique that readily determines the concentration of substances and rates of reaction that takes place. It is used to measure the absorbance of ultraviolet or visible light spectrum of material, quantitatively. UV-Visible technique estimates the chemical compounds in a complex mixture, energy bandgap, functional group detection, identifying the unknown compound and measuring the purity of the substance.

The light source emits the ultraviolet and visible region radiated energy sources, moved to monochromator having prism. Rotating slits are used for the dispersion to increase the wavelength, and transfer to slit for selection of beam energy is monochromatic, further divided to double beam spectrometer. The double beam source is passed through sample and reference cell which are made from quartz. The detector receives the two photocells from the sample and reference beams. To get clear and recordable signals, the alternate current generated in the detector amplified a number of times to improve intensity signals. These recorded signals produce a spectrum of the desired compound. The electron transition occurs at π bonds for pristine graphene and single-layer graphene oxide around 250, 270, and 230 nm [26]. The 550 nm wavelength shows transmittance of 97% for the single-layer graphene.

4.6 Fourier Transform Infrared Spectroscopy

Infrared spectra are obtained by powder transmission or attenuated total reflectance (ATR) that belong to a powerful family for chemical analysis of polymers, solid-state/molecular compounds, and structure demystification. The IR spectroscopy is applied to $C_xN_yH_z$ compounds—melem, melam, melon—to understand the structures of these riveting materials. Crystalline melamine gives two sharp peaks of N–H stretching at 3469 and 3419 cm^{-1} that are affected by H-bonding that consists of broader peaks at 3334 and 3132 cm^{-1} within molecular solid. The IR spectrum for melamine, melem, melon, and melam can be elucidated.

4.7 Scanning Electron Microscopy

SEM is a technique employed to study the morphology, topography, crystallography, chemical composition, orientation of grain, and structure in the nano range with a clear image with <2 nm special resolution. Electrons emitted from the tip applied by cathode fall on the material surface after surpassing all the lenses. Then the signal captured by the detector gives the image of active material.

Using silica template the structure of carbon nitride can be etched. The small black spots of about 60 nm that are the connections between spherical pores that originated from the neighboring silica spheres can be seen.

5 Future Perspectives

Carbon plays its application in almost every industry either new or improved technologies for production and application in renewable energy. The family of carbon is graphene, fullerene, nanotubes, and its derivatives; carbon quantum dots, nanohorn, nanoribbons, capsules, and cages. With doping and surface functionalization of these derivatives' architecture, it has applications in energy storage, catalysis, hydrogen storage, and many more. The oxides of carbon have their own applications in these categories either organic or inorganic species. Issues need to be resolved taking environmental pollution personally. The chemical production renders on fossil fuels. The acceleration of oxides and nitrides of carbon to recyclable active material precede growth in the development of chemical industry. So the harmless transformations of carbon oxides and carbon nitrides would be highly reposed. The use of carbon sources in energy storage, photocatalysis, and solar with the amalgamation of metal oxides and chalcogenides with hybrid heterojunctions is to be increased and it will be the "crown jewel" in further future applications.

Acknowledgements I thank Technical Education Quality Improvement Programme (TEQIP-III) and DST File No: SP/YO/2019/1599(G) for supporting me to carry out my work at the Center for Nanoscience and Technology, Institute of Science and Technology, Jawaharlal Nehru Technological University, Hyderabad.

References

1. Shijun Yu, Xu Han, Dawei Yu, Yanming Chen and Xiaoli Wang (2013) Synthesis and application of carbon-based nanocomposites, Applied mechanics and materials 345.172. Vol. 345 (2013) pp 172–175.
2. Maya Krishnan gopiraman, Ick soo kim.(2018) Carbon nanocomposites: preparation and its application in catalytic organic transformations, nanocomposites- recent evolutions, 81109.
3. Notarianni M, Liu J, Vernon K, Motta N (2016) Synthesis and applications of carbon nanomaterials for energy generation and storage. Journal of Nanotechnology 7:149–196
4. Tehrani, M., & Khanbolouki, P. (2017). Carbon Nanotubes: Synthesis, Characterization, and Applications. Advances in Nanomaterials, 3–35.
5. Luo Y, Yan Y, Zheng S (2019) Huaiguo Xue and Huan Pang (2019) Graphitic carbon nitride based materials for electrochemical energy storage. J. Mater. Chem. A 7:901–924

6. Feng Y, Shen J, Cai Q, Yang H, Shen Q (2015) The preparation and properties of a gC$_3$N$_4$/AgBr nanocomposite photocatalyst based on protonation pretreatment. New J. Chem. 39(2):1132–1138

7. Yang Q (2015) Holeygraphitic carbon nitride nanosheets with carbon vacancies for highly improved photocatalytic hydrogen production. Adv. Funct. Mater. 25(44):6885–6892

8. Guo Y, Kong F, Wang C, Chu S, Yang J, Wang Y, Zou Z (2013) Molecule-induced gradient electronic potential distribution on a polymeric photocatalyst surface and improved photocatalytic performance. J. Mater. Chem. A 1(16):5142–5147

9. Junjiang Zhu, Ping Xiao, Hailong Li and Sonia A. C. Carabineiró (2014) Graphitic Carbon Nitride: Synthesis, Properties, and Applications in Catalysis, 2014, 6, 16449–16465, Appl. Mater. Interfaces.

10. T.S miller, A. Belen Jorge, T. M. Suter, A. Sella, F. Cora and P. F. McMillan, (2017) Carbon Nitrides: synthesis and characterization of a new class of functional materials. Phys. Chem. Chem. Phys. 2017(19):15613–15638

11. Wang K, Li Q, Liu B, Cheng B, Ho W, Yu J (2015) Sulfur-dopedg-C$_3$N$_4$ with enhanced photocatalytic CO$_2$-reduction performance. Appl. Catal. B 176:44–52

12. Kumar S, Kumar A, Bahuguna A, Sharma V, Krishnan V (2017) Two-dimensional carbon-based nanocomposites for photocatalytic energy generation and environmental remediation applications. Journal of Nanotechnology 8:1571–1600

13. Xiao J, Xie Y, Nawaz F, Wang Y, Du P, Cao H (2016) Dramatic coupling of visible light with ozone on honeycomb-like porous g-C$_3$N$_4$ towards superior oxidation of water pollutants. Appl. Catal. B 183:417–425

14. Fang JW, Fan HQ, Li MM, Long CB (2015) Nitrogen self-dopedgraphitic carbon nitride as efficient visible light photocatalyst for hydrogen evolution. J. Mater. Chem. A 3(26):13819–13826

15. Dong F, Zhao Z, Sun Y, Zhang Y, Yan S, Wu Z (2015) An advanced semimetal–organic bi spheres–g-C$_3$N$_4$ nanohybrid with SPR-enhanced visible-light photocatalytic performance for NO purification. Environ. Sci. Technol. 49(20):12432–12440

16. Xu J, Wu HT, Wang X, Xue B, Li YX, Cao Y (2013) A new and environmentally benign precursor for the synthesis of mesoporous g-C$_3$N$_4$ with tunable surface area. Phys. Chem. Chem. Phys. 15(13):4510–4517

17. Zhu Y-P, Ren T-Z, Yuan Z-Y (2015) Mesoporous phosphorus-doped g-C$_3$N$_4$ nanostructured flowers with superior photocatalytic hydrogen evolution performance. ACS Appl. Mater. Interfaces 7(30):16850–16856

18. Y. Chen, W. Huang, D. He, Y. Situ, H. Huang (2014) Construction of heterostructured g-C$_3$N$_4$/Ag/TiO$_2$ microspheres with enhanced photocatalysis performance under visible-light irradiation. ACS Appl. Mater. Interfaces 6(16), 14405–14414 (2014).

19. Sevilla M, Mokaya R (2014) Energy storage applications of activatedcarbons: supercapacitors and hydrogen storage. Energy Environ. Sci. 7(4):1250–1280

20. Liu X, Dai L (2016) Carbon-Based Metal-Free Catalysts. Nat. Rev. Mater. 1:16064

21. Ma TY, Tang Y, Dai S, Qiao SZ (2014) Proton-functionalized twodimensional graphitic carbon nitride nanosheet: an excellent metal-/label-free biosensing platform. Small 10(12):2382–2389

22. Liu G, Wang T, Zhang H, Meng X, Hao D, Chang K, Li P, Kako T, Ye J (2015) Nature-inspired environmental phosphorylation boosts photocatalytic H$_2$ production over carbon nitride nanosheets under visible-light irradiation. Angew. Chem. Int. Ed. 127(46):13765–13769

23. Zhang G, Zhang M, Ye X, Qiu X, Lin S, Wang X (2014) Iodinemodified carbon nitride semiconductors as visible light photocatalysts for hydrogen evolution. Adv. Mater. 26(5):805–809

24. Sun XD, Li YY, Zhou J, Hai Ma C, Wang Y, Zhu JH, Facile, (2015) synthesis of high photocatalytic active porous g-C$_3$N$_4$ with ZnCl$_2$ template. J. Colloid Interface Sci. 451:108–116

25. Bykkam S, Narsingam S, Mohsen Ahmadipour T, Dayakar KV, Rao Ch, Chakra S, Kalakotla S (2015) Few layered graphene sheet decorated by ZnO nanoparticles for anti-bacterial application. Superlattices Microstruct S0749–6036(15):00233–00235

26. Shiraishi Y, Kofuji Y, Sakamoto H, Tanaka S, Ichikawa S, Hirai T (2015) Effects of surface defects on photocatalytic H_2O_2 production by mesoporous graphitic carbon nitride under visible light irradiation. ACS Catal. 5(5):3058–3066

Carbon-Based Nanocomposites: Processing, Electronic Properties and Applications

Manab Mallik and Mainak Saha

Abstract The last two decades have witnessed a large volume of research revolving around structure–property correlation in carbon-based nanocomposites, synthesized by several methods. The electronic properties of carbon-based nanocomposites vary mainly as a function of the kind of reinforcement, method of synthesis, and structure-dependent parameter. The structure-dependent parameter is highly influenced by the reinforcement and method of synthesis and plays a vital role in determining the ionic and electronic transport phenomenon in these materials. In other words, the interaction between electrons and the equilibrium 0-D (point) defects, along with different types of 2-D interfaces, plays an imperative function in the understanding of electronic properties, apart from the physical and chemical properties of these materials. The present chapter offers a concise overview of the state of the art on research and detailed discussions on some recent developments in understanding the electronic properties of some conventional carbon-based nanocomposites (synthesized by different techniques) based on the structure–property correlation in these materials. Finally, some of the significant challenges in this field have been addressed from industrial and fundamental viewpoints.

Keywords Carbon · Nanocomposites · Oxides · Nitrides · Electronics · Sensors

1 Introduction

Since the last two decades, the structure–property correlation of a "hybrid" material with integrated properties has gained much more attention to the design of a number of "complex" materials [1]. Nanocomposites are one such class of materials that are considered as a potential candidate for various advanced applications ranging from nanoelectronics to energy-storage industries due to their exceptional electronic, mechanical, and chemical properties.

M. Mallik (✉) · M. Saha
Department of Metallurgical and Materials Engineering, National Institute of Technology Durgapur, Durgapur, India
e-mail: manab.mallik@mme.nitdgp.ac.in

© The Author(s), under exclusive license to Springer Nature Singapore Pte Ltd. 2021 97
A. Hazra and R. Goswami (eds.), *Carbon Nanomaterial Electronics: Devices and Applications*, Advances in Sustainability Science and Technology,
https://doi.org/10.1007/978-981-16-1052-3_5

Nanocomposites are defined as composite or multiphase materials in which, as a minimum, one phase has at least one dimension in the nanoscale (10^{-9} m) range [2]. Nanocomposites have extraordinary physical, mechanical, and chemical properties, which make these materials appropriate alternatives to conventional materials and composites. Among different nanocomposites, carbon-based nanocomposites have witnessed a large volume of attention in the last two decades due to their structure-dependent electronic properties, low density, and large specific surface area [1, 3]. Carbon-based nanocomposites contain different carbonaceous materials (graphite/nanotube/C_{60}) as matrix with various reinforced phases such as metal oxides/sulfides/nitrides, etc.

In recent times, carbon nanotube(CNT)-based nanostructured composites have proven to be promising materials for application in electrochemical supercapacitors, gas and biological sensors, electromagnetic absorbers, and photovoltaic cells due to the high charge transfer capacity of adsorbates [3]. This has arisen due to limited charge sensitive conductance in single-walled (SW) CNTs, like in graphite [4, 5]. Hence, the novel strategies of combining CNTs with transition metals like Ru, Pt, etc. [6, 7] and metallic oxides or conducting polymers have been attempted [8], and a higher specific value of capacitance than CNTs in supercapacitor applications has been obtained. However, the exceptionally high price of these materials limits their application in commercial supercapacitors [8]. Besides, the need to cope up with the ever-increasing requirements for greater energy density and power output of rechargeable Li-ion batteries (LIBs) [9] has necessitated the usage of CNT-based nanocomposites both as cathode and anode materials for such application. Recent work has also highlighted the use of graphite (with a suitable binder) as anode material for potassium-ion batteries (PIBs) to improve energy efficiency than that of LIBs and sodium-ion batteries (SIBs) [10].

Moreover, in the recent decade, the "correlative microscopy" methodology [11–16] has been used extensively for structural characterization. One of the main reasons as to why the "correlative microscopy" methodology has not been used widely to study the structure–property correlation in C-based nanocomposites may be attributed to the challenges involved in sample preparation. In addition, other challenges include the characterization of a light element such as C and understanding the complex atomic structures in nanocomposites. This has also been the reason behind a limited understanding of the structure-dependent parameter. Thus, most of the literature published in this field can address the structure–property correlation as a function of only process parameters (and not structure-dependent parameter) in C-based nanocomposites. The present chapter focuses on the structure-dependent electronic properties of nanocomposites from the state of the art in order to create a new paradigm in the field of research on functional materials.

2 Why Is Carbon so Interesting?

Among all the elements in the periodic table, carbon turns out to be extremely interesting, not only due to its ability to form several versatile isotopes but also because of the ability to give rise to several completely different properties, on account of being bonded differently with other atoms and molecules. In addition to the costly diamond and graphite, various other forms of carbon include nanotubes, fullerene molecules, graphene, etc. (Fig. 1) [17–22]. Different forms of carbon allotropes are shown in Fig. 1 [21–24]. Table 1 also lists a few of the common allotropes of carbon.

In the well-known diamond cubic crystal structure with every carbon atom bonded to four other carbon atoms by strong sp^3 covalent bonds, the absence of free (mobile) electrons makes a diamond a hard yet a highly electrically insulating material. On the other hand, graphite is formed by the stacking of 2-D hexagonal flat sheets of sp^2-bonded carbon atoms with the sheets weakly bonded by Van der Waals forces. This

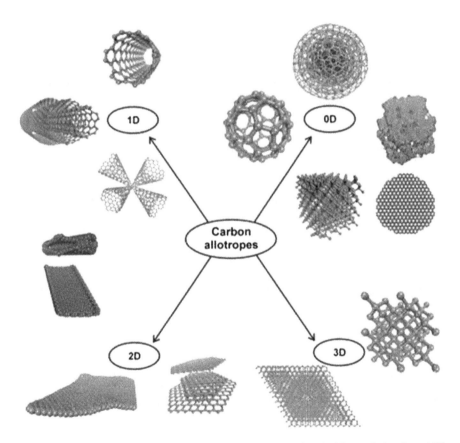

Fig. 1 Structures of different forms of carbon allotropes (Reproduced with permission from ACS Publications [21])

Table 1 Properties of different carbon allotropes (*p: in the plane; c: along c axis*)

Property	Graphite [21]	Diamond [22]	C_{60} [17, 18]	CNT		Graphene [20]
				SWCNT [20]	MWCNT [20]	
Crystal structure		Hexagonal	Diamond cubic	Buckminsterfullerene	Hexagonal	Hexagonal
Density (g/cm3)	2.09–2.23	3.5	1.5	0.8	1.8	2.27
Electrical conductivity (S/cm)	$p: 2.5 \times 10^4$ $c: 3.3$	10^{-2}	10^{-5}	10^2–10^6	10^3–10^5	10^7–10^8
Thermal conductivity (W/mK)	$p: 3000$ $c: 6$	900–2320	0.4	~6000	~2000	~5000
Coefficient of thermal expansion (K^{-1})	$p: -1 \times 10^{-6}$ $c: 29 \times 10^{-6}$	1–3×10^{-6}	6.2×10^{-5}	Insignificant	Insignificant	$p: -1 \times 10^{-6}$ $c: 29 \times 10^{-6}$
Thermal stability in air (°C)	450–650	777	>600	>600	>600	450–650
Surface area (m^2/g)	Variable	10–50	>100	>100	>100	2630
Modulus (GPa)	$p: 1000$ $c: 36.5$	500–1000	14	1200	1000	1000

bond interestingly leads to the high mobility of C between the sheets and primarily accounts for the high electronic conductivity of graphite. The narrow bandgap between conduction and valence bands in graphite also accounts for excellent electrical conductivity in graphite. Before the onset of carbon nanotubes (CNTs) [25, 26] and then graphene, the most investigated artificially synthesized allotrope of C was fullerene C_{60} molecule with an icosahedral structure (having twenty hexagons and twelve pentagons) in a truncated manner. Charge doping may also be employed as a method to induce superconductivity in the semiconducting C_{60} molecule [17, 18, 25].

For defining CNTs, a chiral vector ($c = na_1 + ma_2$, with a_1 and a_2 being the unit vectors of the 2-D hexagonal lattice of graphene and n and m being integers) is used. Based on the number of graphene layers present, CNTs may be broadly classified into two types: single-walled CNTs (SWCNTs) and multi-walled CNTs (MWCNT) developed from one or limited single layers of graphene, with an interlayer distance of nearly 0.34 nm. Graphene, on the other hand, exhibits extraordinary electronic properties with electrons behaving as Dirac fermions and showing extremely high mobility together with holes [9, 20]. Besides, the band structure of graphene with the fully occupied valence band with the empty conduction band (both touching at six points) renders a zero-gap semiconductive nature to graphene [20].

2.1 Graphitic Carbon Nitride-Based Nanocomposites

Graphitic carbon nitride (designated as g-C_3N_4) are semiconductor photocatalysts with a medium bandgap. Moreover, they have been reported to possess non-toxicity and excellent thermal and chemical stability. In recent times, the photocatalytic nature of these nanocomposites is being investigated for wastewater treatment applications. Based on dopants (doped with g-C_3N_4), these nanocomposites have been classified as (i) metal-free g-C_3N_4, (ii) noble metals/g-C_3N_4 heterojunction, (iii) non-metal doped g-C_3N_4, transition, and (iv) post-transition metal-based g-C_3N_4 nanocomposite [21, 25]. Moreover, these nanocomposites have unique electronic configuration enabling them to be coupled with other functional materials for enhanced performance [25]. However, investigations in these materials are in the preliminary stage, and only limited information (till date) is known about these nanocomposites.

2.2 Carbon Nanotube(CNT)-Based Nanocomposites

Apart from the wide usage of CNTs owing to their properties [22–25], certain exceptional features render CNTs a perfect supplement for a large group of materials, comprising metals, ceramics and polymers. The present state of research in CNTs is mainly aimed at synthesizing CNT-based composites as an alternative to conventional materials [25]. However, compared to CNTs, graphene-based materials exhibit numerous advantages, which include: a unique combination of mass production coupled with a low cost of manufacturing [25] and excellent electronic conductivity [26]. In the context of electronic properties, it is highly interesting to discuss the intensively studied RuO_2-reinforced CNT used to design supercapacitor devices, having a much higher cost and possessing toxic characteristics among inorganic supercapacitor electrode materials [3]. The use of CNT-based metal oxides/conducting polymers as electrochemical supercapacitor electrode materials has not only resulted in an extremely high specific capacitance (~180 F/g [4, 5]) but has also led to higher electronic conductivity when compared to that inactivated carbon [3–6]. CNTs may be synthesized by several methods ranging from arc discharge to different chemical vapor deposition (CVD) techniques, and even industrial waste (especially from coke ovens in integrated steel plants) such as coal tar may also be used as the starting material for the same [27]. The radius of a CNT (typically varying between 100 nm to 20 cm) strongly depends on synthesis parameters [3], and it strongly influences the electronic properties of the material. CNTs are reported to be covered by half of a fullerene-like molecule [7].

The present state of research on CNT-based nanocomposites mostly revolves around the detailed understanding of supercapacitor behavior and the stability of CNT-based metal oxides and CNT-based polymers through the engineering of the structure-based parameter which involves composition, grain size, etc. [27], in addition to enhancing the electronic, mechanical, and thermal properties of these materials [27–32]. Kavita et al. [28] have studied the influence of poly (vinyl butyral) and structural modification (based on acid-functionalized MWCNT treatment) on the thermomechanical properties of the novolac epoxy nanocomposites and reported **a rise** of nearly ~15 °C at the peak degradation temperature in comparison to the unmodified novolac epoxy. In recent times, a double-layer capacitor with fullerene-activated carbon composite electrodes has been reported by Okajima et al. [32] to possess an extremely high capacitance even on a 1 wt% ultrasonically treated electrode (loaded with C_{60}). The first attempt to synthesize a novel Pt-modified $[PW_{11}NiO_{39}]^{5-}$@reduced graphene oxide (rGO) and Pt-modified $[PW_{11}NiO_{39}]^{5-}$@multi-walled carbon nanotube (MWCNT) composites has been reported by Ensafi et al. [31]. Pyrrole-treated functionalized SWCNT exhibits excellent electrochemical performance, which includes a high capacitance and power density [33]. The plasma surface treatment of MWCNTs with NH_3, leading to an enhancement of total surface area and wettability of MWCNTs and an enhancement in capacitance, has been reported by Yoon et al. [34]. CNTs/conducting polymer composites, on the other hand, have been synthesized by several techniques, among which the most common are the in situ polymerization (both chemical and electrochemical polymerization) of monomers [35, 36].

2.2.1 Polymer/CNT-Based Composites

Owing to an excellent combination of thermal, electrical, and optical properties [25], as already discussed in the earlier section, CNTs turn out to be potential candidates in various newly emerged fields, which commonly include nanotechnology and biosensors. Polymer-reinforced CNT nanocomposites possess an excellent combination of mechanical properties and electrical conductivity for biosensor applications [35]. In the context of supercapacitor devices, the performance is largely determined by MWCNT content in the composite electrode [37]. Consequently, polypyrrole (PPy)/MWCNT composites can be employed as potential candidates for supercapacitors with high capacitance and a long life per cycle [8, 37].

The composite electrodes composed of conducting polymers reinforced with CNTs result in an enhancement of mechanical strength along with thermal and electronic conductivities [38]. PANI/MWCNT composite has been reported to possess a specific capacitance combined with large retention of capacitance even after a cosiderable number of cycles [39]. Besides, several conducting polymers, mainly polyaniline, polythiophenes, and polypyrrole, have also been reported to possess high capacitance [40]. This has rendered CNT-based conducting polymer nanocomposites to be promising materials for energy-storage application in supercapacitors [41–44] and many amperometric biosensors [41, 42]. Raman spectra of i-PANI/MWCNTs and r-PANI/MWCNTs (PANI—Polyaniline) composites are represented in Fig. 2a, wherein the Raman peaks (characteristic of PANI) were observed to undergo a significant change with temperature. Figure 2b shows the XRD patterns of PANI/MWCNT composites, and Fig. 2c illustrates PANI molecular conformations as a function of temperature [43]. The electronic properties of PANI, poly (vinylidene fluoride)

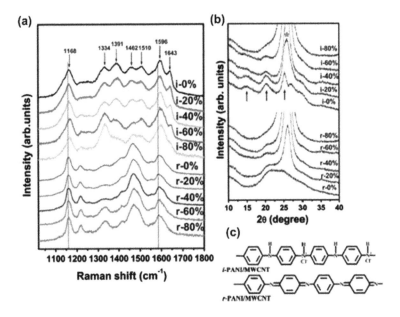

Fig. 2 Representation of **a** Raman spectra and **b** XRD patterns of PANI-reinforced MWCNT composites with variation in weight fraction of MWCNTs (* represents CNT peaks) **c** schematic image of PANI molecular conformations at various temperatures of synthesis (Reproduced with permission from Elsevier [43])

(PVDF), and MWCNTs were studied by Blaszczyk-Lezak et al. [41] wherein, the surface functionalization of MWCNT was performed at 90 °C using concentrated mixture of sulfuric and nitric acid for 24 h.

2.2.2   Activated Carbon (AC)/CNT-Based Nanocomposites

As reported by Navarro-Flores et al. [45], activated carbon (AC)/CNTs nanocomposite electrodes on being examined in an organic electrolyte (1.5 M NEt_4BF_4 in acetonitrile) render a number of interesting results, the most significant of which is a reasonably suitable compromise between energy and power density even for a CNT content of 15 wt% [45]. Besides, a high cell series resistance of ESR (~0.6 Ωcm^{-2}) and high capacitance (~88 F/g) have also been reported from this work [45]. The probability of making AC/CNT nanocomposite-based electrodes for supercapacitor devices using the electrophoretic deposition (EPD) technique has been extensively explored by Huq et al. [46] wherein, it was observed that the as-prepared AC/CNT nanocomposite electrode exhibits excellent capacitance retaining (of nearly 85%) even after a cyclic stability test for a prolonged time (~11,000 cycles [46]). Besides, AC has also been used extensively as an electrode material in electric double-layer capacitors (EDLCs) for a prolonged period owing to a remarkable combination of high capacitance, long cycle life, and most importantly, low cost of manufacturing [47]. Qiu et al. [48] have reported that activated hollow carbon fibers (ACHFs) containing CNTs and Ni nanoparticles (CNTs-Ni-ACHFs) may be synthesized using various techniques such as thermal reduction and chemical vapor deposition (CVD). However, activated carbon (AC) remains usually used absorbent, primarily due to its high surface area [49–51].

2.2.3   Metal Oxide/CNT Nanocomposites

Most of the studies on electronic properties of CNT-based RuO_2/TiO_2 nanocomposites have considered the potential window (−0.4 to +0.4 V) [53]. Alam et al. [54] have prepared $BaMg_{0.5}Co_{0.5}TiFe_{10}O_{19}$/MWCNT nanocomposites with variation in the amounts of MWCNTs (0, 4, 8, and 12 vol.%) and demonstrated that ~8% vol. of MWCNTs in the produced MXCNT-based nanocomposite possess the best microwave absorption capacity. Besides, an innovative technique to improve the interfacial bond strength by developing a coating of magnesium oxide (MgO) nanoparticles on the CNT substrate has been reported by Yuan et al. [55]. Metal oxides often find suitable candidates for pseudocapacitive electrode materials for supercapacitor devices due to their high energy density [55], unlike carbonaceous materials. Titanium dioxide (TiO_2) is presently a vital anode material for rechargeable Li-ion batteries due to an excellent combination of high cycle life coupled with high safety and low cost. Yuan et al. [52] have reported the synthesis of $MoO_2@TiO_2@CNT$ nanocomposites (sandwich structured) from CNTs through a

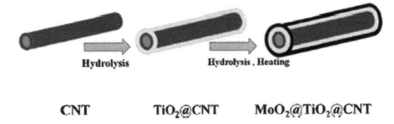

Fig. 3 Schematic of synthesis processes for MoO_2 @ TiO_2 @CNT nanocomposites (with sandwich-like structure) (Reproduced with permission from Elsevier [52])

two-stage process involving (i) Hydrolysis of CNT to form TiO_2 @CNT and finally, (ii) Hydrolysis and heating (in an inert Ar/H_2 atmosphere) of TiO_2 @CNT to form MoO_2 @ TiO_2 @CNT, as schematically illustrated in Fig. 3.

2.2.4 Carbon Fibers (CF)/CNT-Based Nanocomposites

Based on the study by Islam et al. [56], CNTs, covalently bonded with CF through an ester linkage in the absence of catalysts or coupling agents, may be employed to acquire the combined advantages of improved interfacial shear and impact strength [8]. The intention lies in the necessity for a higher level of tensile stress to detach the CNTs from CF [57]. There are two methods used to reinforce CF with CNTs using physical adsorption technique (based on weak Van der Waals interaction) to limit the reinforcing effect [58]. CF/CNT-based nanocomposites have found an extensive application in the context of lightweight automotive and aerospace components with high fuel efficiency—"The present need of the hour."

Wang et al. [58] have reported an improvement in the interfacial shear strength of CFs using graphene oxide (GO) as reinforcement with CNT. CNTs have numerous advantages as conducting wires compared to copper wires because of their size and quantum effects [59]. Tamrakar et al. [60] accumulated poly-ethylenimine (PEI) functionalized MWCNT using the EPD technique onto the glass fiber surface and reported that the EPD method provides the required thickness of CNT coating, thereby enabling control of interfacial resistivity between fiber and matrix.

2.2.5 CNT/Metal Nitride-Based Nanocomposites

Jiang et al. [61] have reported the electronic properties of CNT/metal nitride-based nanocomposites with remarkable electrochemical stability in the presence of strongly corrosive electrolytes. The presence of CNTs enhances the electronic conductivity of the CNT/TiN composites through the percolation effect (of CNTs) [61–63]. Based on the published literature in this field, the mobility of conduction electrons through the lattice to numerous scattering sites, ranging from point defects to different kinds of 2-D interfaces and numerous 3-D volume defects such as pores, etc, play a vital role in influencing the electronic properties of TiN [64]. CNTs, likewise, provide an excellent conducting path [61, 65], especially if these are located (or more likely segregated) at the grain boundaries (GBs) of the composite. The presence of CNTs at the GBs aids carrier transportability and improves the electronic conductivity of the composite [63]. This is one of the aspects of material science research where the "less well-known" structure-dependent parameter starts influencing the electronic properties of the C-Based nanocomposites. CNTs have also been reported to serve as an electrically conducting bridge connecting the domains of TiN, owing to the higher charge mobility of the CNT–TiN composite as compared to that of TiN [63]. The other reported reason is the presence of strong interfacial cohesion between CNTs and the matrix [63] due to the presence of TiN nanoparticles along the CNT wall, leading to a highly efficient transfer of electrons from the matrix to the CNT. The presence of ~12 vol.% CNTs in CNT-TiN nanocomposite has been reported to exhibit nearly ~45% increase in electrical conductivity compared to TiN [64, 65].

2.3 Vanadium Nitride (VN)/Graphene (G) Composite

Until 2017, the practical implementation of rechargeable lithium–sulfur batteries had been impeded by several issues. Among these, the shuttle effect and low Coulombic efficiency were the most significant. The first report on the development of conductive porous vanadium nitride (VN) nanoribbon/graphene (G) composite came from Sun et al. [66]. Based on both experimental and theoretical results, vanadium nitride/graphene composite has been reported to provide a strong driving force for rapid conversion to polysulfides [66]. Sun et al. [66] reported that VN exhibits lower polarization coupled with faster kinetics of redox reaction as compared to that in a reduced graphene oxide (rGO) due to its excellent electronic transport [Fig. 4]. Based on this report [66], the starting capacity reaches 1,471 mAhg^{-1}, and the capacity after 100 cycles is 1,252 mAhg^{-1} at 0.2 C, with a loss of ~15%, thereby proposing the probable use for high-energy lithium–sulfur batteries.

Based on Sun et al. [66], the VN/G nanocomposite, prepared using a two-stage Hydrothermal and Freeze-drying technique, schematically illustrated in Fig. 4, combines the benefits of both graphene and VN. Figure 5 shows the X-Ray diffraction (XRD) pattern and TG-DSC (TG: Thermogravimetry, DSC: Differential Scanning Calorimetry) curve for VN/G nanocomposite. Electron and ion transportation, along with electrolyte absorption, is facilitated by the 3-D free-standing structure of the graphene (G) network. Besides, the VN/G electrode has also been reported to possess excellent specific capacitance (with a much higher magnitude as compared to the rGO electrodes). This pathbreaking work [66] has also opened up a new avenue of research wherein even metal nitrides other than VN may be used as potential cathodes to explore energy-storage capacity for applications in Li-S batteries.

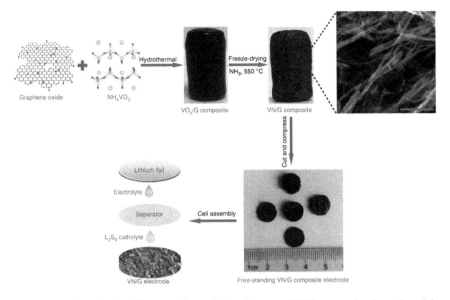

Fig. 4 The schematic view illustrates the synthesis of a porous VN/G composite electrode and the cell assemblage with consequent optical images of the fabricated material. A scale bar of 500 nm has been used (Reproduced with permission from Nature [66])

3 Nanocrystalline and Amorphous Carbon Films (CN_x)

Hydrogenated or non-hydrogenated C films, often termed as diamond-like carbon (DLC) films [3], may be either amorphous or nanocrystalline with mixed states of sp^3 and sp^2 bonding. Those with a predominantly sp^2 state of bonding are known as graphite-like amorphous carbon (GAC) films. To be extremely specific, the bonding and antibonding electronic states due to the presence of π bonds (of sp^2 carbon sites) determine the electronic properties and many physical properties of these materials [4]. In the context of electronic properties, they are highly attractive due to a wide range of DC electrical conductivity values at ambient temperature (ranging from ~10 to 101 S/m [4]). A thermally activated process dominates the temperature dependence of electronic conductivity in these materials (with $\sigma = \sigma_0 \exp(-\Delta E/kT)$; σ: intrinsic conductivity [4], σ_0: pre-exponential factor [4], ΔE: energy barrier [4], and T: absolute temperature [4]) based on hopping mechanism [4]. Among the existing theoretical models, the variable range hopping (VRH) model [8] is the most commonly used for studying the temperature dependence of electronic conductivity in these materials. For instance, the electronic conductivity of activated carbon (AC) has been reported to change from 1-dimensional (1-D) to 3-dimensional (3-D) during the graphitization process, depending on the thermal treatment temperature [4]. Besides, at temperatures below 1000 °C under high-pressure conditions, VRH is the primary

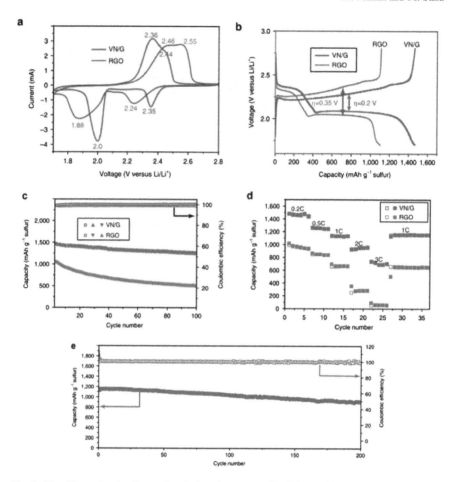

Fig. 5 Plots illustrating the electrochemical performances of VN/G and rGO cathodes. **a** CV profiles of the VN/G and rGO cathodes (scan rate of $0.1 mVs^{-1}$) between 1.7 to 2.8 V. **b** Galvanostatic charge–discharge profiles of the VN/G and rGO cathodes (at 0.2 C). **c** Cycling performance and charge (Coulombic) efficiency of the VN/G and rGO cathodes (at 0.2 C for 100 cycles). **d** Rate performance of the VN/G and rGO cathodes at different current densities. **e** Cycling stability of the VN/G cathode (at 1 C for 200 cycles) (Reproduced with permission from Nature [66])

conduction mechanism, as reported by Zhao et al. [67]. In 1-D VRH, the electrical resistivity ρ follows an exponential relationship with temperature T ($\rho = \rho_0 \exp(T_0/T)^x$; ρ: electrical resistivity at temperature T [4], ρ_0: electrical resistivity at an infinite temperature [4], and T_0: characteristic temperature [4]) with $x = 1/2$ for 1-D VRH [5]) with T_0 related to localization length ξ for the wave function [6]. Even for large 2-D (with $x = 1/3$) or 3-D ($x = 1/4$) nanoparticles, VRH conduction mechanisms have been reported, indicating the change in electronic conduction mechanism from 1-D to 2-D or 3-D [4]. The insulator–metal transition has been reported to occur between non-graphitized and the almost-graphitized regions with the material exhibiting semi-metallic behavior in the near-graphitized region and a nearly linear electrical resistivity–temperature (ρ-T) relationship. At higher sintering temperatures (1200–1600 °C), the electrical resistivity of the graphitized activated C exhibits a power-law relation with temperature ($\rho = A+BT^{3/2}$: A and B: coefficients of temperature [4]) with C behaving as a non-Fermi liquid [67].

The concentration of localized states by hydrogen in hydrogenated DLC films has been extensively reviewed by Staryga et al. [68], where the effects of nitrogen doping on the electronic transport characteristics have also been studied in amorphous CN_x films. These films owing to the localization of sp^2 hybridized state and interestingly, significant changes in the electrical properties have been observed in amorphous CN_x (a-CN_x) films as a function of Nitrogen concentration. It has been correlated to bond strength between N atoms and sp^2 and sp^3 sites of C [69]. The dynamics of hopping transport in a-CN_x have also been studied in detail using AC electrical spectroscopy measurements [70–72].

3.1 Diamond-Like Carbon (DLC) and Graphite-Like Amorphous(GAC)-Based Nanocomposites

One of the most researched topics in the context of electronic applications in DLC and GAC films is their capability to use in electron field emission devices [4] and as cold-cathodes in field emission displays [73, 74]. Incorporating boron is an effective method to enhance the oxidation resistance of various carbon-based nanocomposites, thereby avoiding the major drawback of surface oxidation in these materials. The electronic conduction of these materials is generally reported in terms of Mott VRH for localized electronic states near the Fermi level [73, 74]. Porosity, coupled with the highly inert nature, render DLC and GAC as ideal candidates for application in the form of neutral or electron donor nanoparticle species required for the preparation of hybrid materials for catalytic applications [74]. In devices such as rechargeable Li-ion batteries (LIBs) meant for energy storage with high energy density, V_2O_5/C composites have also been observed to improve high-rate performances [75].

3.2 Hard Nanocomposite Coatings with Amorphous Carbon

In the context of amorphous C(a-C)-based nanocomposite thin films and hard nanocrystalline C-based coatings, there is a lack of understanding of the influence of C atoms on the electronic properties of these materials. For instance, C segregation at GBs is generally expected to lead to the formation of localized phases at GBs, as recently reported by Meiners et al. [76], which depending on the structure

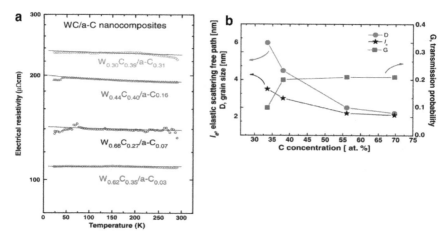

Fig. 6 a Plot illustrating DC electrical resistivity vs. temperature of a-C/WC nanocomposites. The solid lines (in the figure) represent the best fit with the GB model. **b** Elastic scattering free path (l_e), grain size (D), and electron transmission probability (G) deduced from the model as a function of atomic concentration of C (Reproduced with permission from Elsevier [4])

of GB, maybe either conducting or insulating [4]. Although, DC electrical resistivity measurements as a function of temperature may be extensively used as an experimental tool to investigate the structural evolution (involving grain size of both the matrix phase and also of the phase present at GBs) of the composite films. However, the novel "Correlative Microscopy" methodology, discussed in the introductory section of the review, has not been widely employed in these materials, till date, to understand the presence of GB phase in these materials, although an extensive review on the influence of structured interfaces on the electronic performance of composite materials has been published by Mishnaevsky [77]. Figure 6 shows the variation in electrical resistivity due to structural modifications in WC/a-C nanocomposites. The electrical resistivity vs. temperature curves suggest that the scattering of electrons against 2-D GBs and 0-D point defects turn out to be extremely significant. In such nanocomposite materials, the experimental results have been reported to be easily validated by the GB scattering model [78]. Based on calculations using this particular model (Fig. 6), the deduced grain sizes have been reported to be highly consistent with crystallite sizes estimated from different structural characterization techniques such as XRD and TEM [4]. Secondly, the calculated elastic scattering free path (l_e) values have been reported to be in line with a high density of point defects in the films [4]. Thirdly, grain size has been reported to be the main parameter controlling the electronic transport phenomenon of these films. Based on the present understanding of the vanishing of GBs in quasi-amorphous materials, the scattering of electrons is mostly dominated by the high density of 0-D (point) defects, mainly vacancies as reported by Sanjinés et al. [4].

4 Some of the Major Limitations and Future Challenges in This Field

4.1 From an Industrial Viewpoint

In the context of supercapacitors, presently, power limitation is a significant issue. That depends on the type of applied electrolyte, electrode, and constant current. A significant investigation on the effect of electrodes and their charge-storage mechanisms on supercapacitor applications is needed.

In the context of supercapacitors, presently, power limitation is a significant issue that depends on the type of applied electrolyte, electrode type, and several other parameters such as evaluated constant current, to name a few, and much further investigation on the electrodes and their charge-storage mechanisms is needed. However, the present state of research in the energy-storage capacity of supercapacitors is focused primarily on the understanding of charge-storage mechanisms (in sub-micropores) based on material design [80, 81]. Electrical energy is stored in the bulk structure of electrodes, whereas the surface area of the electrodes opposes this particular phenomenon of "charge storage inside the battery electrodes" [8]. At present, both SWCNTs and MWCNTs have been extensively used in supercapacitors owing to their high specific surface area, low electrical resistance, high cyclic stability, and low mass density [8, 79, 80]. Besides, several investigations have been carried out on exploring the charge-storage capacities of C-based materials such as activated carbon, porous carbon, CNTs, graphene, etc. [79, 80].

Davies et al. [81] reported that oxides and nitrides of transition metals, as well as conducting polymers, have the potential to improve the electrochemical performance of supercapacitors. These materials are considered as pseudocapacitive materials. Also, as reported by Ates et al. [8], EDLCs exhibit remarkably good cycling stability but lower specific capacitance, whereas, on the contrary, pseudocapacitive materials have low stability during cycling process owing to the reactions (faradaic type).

Hence, in recent times, there has been a considerable focus on utilizing a hybrid material based on the combined advantages of both EDLCs and pseudocapacitive materials. Although, at present, attempts aimed at the integration of C with redox materials (polymer, metals oxides etc.) (for the purpose of enhancing the capacitive ability of redox materials) appear advantageous; however a detailed understanding on the influence of pore size distribution and process parameters (associated with synthesis techniques) on the electrochemical performance of these materials is currently limited. It is only through a proper investigation of some fundamental aspects (in microstructural level) in these materials that supercapacitor devices may be made to render a high energy density (similar to the currently rechargeable Li-ion batteries (LIBs)) to the electrochemical capacitors (ECs), which leads to a substantial influence on the future application of high power energy-storage devices. There are numerous technological challenges to the use of CNT-based nanocomposites as electrode materials for LIBs, sodium-ion batteries (SIBs) [9], and the recently developed potassium-ion batteries (PIBs) [9]. The limiting factors mainly include poor adhesion between at CNT/nanoparticle interface and problems of uniform dispersion of nanoparticles in CNTs [8].

As discussed by Liu et al. [9], other significant challenges include the development of environment-friendly and industrial fabrication techniques for application in the battery systems. In recent times, graphene has been used as an alternative to other carbon allotropes, including CNTs, for several applications, especially with the development of various routes for synthesis [82–85]. Similar to graphene, graphene oxide (GO) also exhibits an interesting combination of different properties, including enhanced electrical conductivity and mechanical strength with a high level of nontoxicity and large electrochemical stability. These combinations of properties make both graphene and graphene oxide as suitable candidates for electrode materials for LIBs and supercapacitors [86–88]. However, graphene may be used as an electrode for rechargeable batteries such as LIBs, SIBs, and PIBs for energy-storage applications. But, the significant challenge viz. the development of an economic fabrication technique for large-scale production of these materials still needs to be overcome [87].

Besides, as reported by Guo et al. [89] and Chang et al. [90], the globally increasing demands for energy-storage applications have paved an avenue for many investigations aiming at industrial applications of Graphene-based nanocomposites, through modifications in mainly the process parameters, associated with their fabrication methods and subsequent understanding of structure–property correlation based on such modifications. It is worth mentioning that the enormous potential of graphene-based nanocomposites in energy-storage applications has been recognized by a large number of cutting-edge research [91–104] in a number of areas such as photo-electrochemical and photovoltaic devices, LIBs, supercapacitors, etc. In particular, graphene nanocomposite-based LIBs and supercapacitors presently turn out to be highly promising materials for energy-storage applications.

However, there do exist several unexplored avenues and a countless number of challenges to be addressed before one may implement graphene-based nanocomposites in real industrial applications of graphene-based nanocomposites. The most important of them all is the mass production of high and uniform quality graphene nanocomposites. Secondly, photoelectrochemical and photovoltaic efficiencies in devices based on graphene nanocomposites are still reported to be quite low even in the most prototype devices [90–94]. Hence, as the fundamental questions remain unaddressed despite numerous research from different groups getting published almost regularly, the idea of aiming for industrial applications with these materials is nearly close to a myth. However, the challenges associated, especially with the fabrication and both long- and short-term industrial applications of graphene nanocomposites, make them one of the most exciting materials for energy-storage and energy conversion-related studies and, hopefully, may be achieved in the near future.

As discussed in an extensive review on interfacial engineering of Li-ion batteries (solid-state) (LIBs) [105], the major drawbacks associated with research in the field of LIBs have been due to the fact that those have been directed only toward obtaining highly efficient ionic conductors in the form of solid-state electrodes (SSEs) rather than designing batteries for practical applications. This is further reflected in the fact that the LIBs, used today, still have a number of limitations, such as low rate capability coupled with limited energy and power density and low cycle life [105]. The limitations are mostly in terms of properties of the solid-solid interfaces (between SSEs and other cell components *such as* cathode and anode) in the LIBs. However, recently, there has been a gradual shift from seeking better SSEs to designing solid-state batteries (SSBs) with high interface stability and interfacial resistance through the design of an artificial buffer layer for the modification of buffer layer and (/or) surface of electrode. For a rational design of the artificial buffer layer, a detailed knowledge of the solid-solid interfaces viz. SSE/cathode and anode/SSE interfaces is highly essential. The challenges associated with the interface between SSE and cathode/anode are mainly due to the following:

(i)   High resistance (at the interface) and consequent uneven current distribution during the operation of the SSB due to limited physical contact at the interface between SSE and cathode/anode leading to poor rate performance and subsequent formation of Li dendrites [105].

(ii)   Miserable cycle life combined with low energy and power density associated with volume and stress may quickly deteriorate physical contact [105].

(iii)   Formation of a thick passivated layer hindering the diffusion of Li^+ across the electrode/SSE interfaces due to electrochemical or/and chemical reactions between the electrode and SSE [105].

In the context of interfacial stability with cathodes, development of a coating between the cathode materials with an inorganic artificial buffer layer has been reported to be a promising approach for improving the stability of the cathode/SSE interface, mainly for sulfide-type SSEs [105, 106]. However, volumetric changes associated with the cathode side of the cathode/SSE interface during repeated charge–discharge cycles lead to the breakage of the buffer layer and a subsequent loss of contact [105]. Design of hosting active materials with a large contact area in a porous matrix (with high ionic conductivity) has been reported to be a promising strategy to overcome the above challenge and may be developed using advanced 3-D printing techniques [105].

For the purpose of producing a high-quality artificial buffer layer on the SSE surface and/or electrode, numerous synthesis techniques have been reported to be used [105]. These fabrication techniques may be summarized as (i) top-down approach and (ii) bottom-up approach. In recent times, the magnetron sputtering procedure has gained commercial success in the synthesis of thin-film SSBs whereas atomic layer deposition has shown encouraging effects in the improvement of the cycle stability of sulfide-type SSBs [105]. On the other hand, pulsed laser deposition has been reported to be an attractive method for the preparation of high-quality buffer layers with a stoichiometry, nearly similar to that of the bulk target [105]. This has necessitated the exploration of the buffer layers through optimization of synthesis techniques. Besides, the expense for large-scale commercial production of thin-film

SSBs is also a concern that needs significant attention. Although several characterization techniques have been used to study liquid-based battery systems, a challenge still remains in understanding interfacial behavior in SSBs [105]. Synthesis of highly reliable artificial buffer layers may be employed in order to achieve mass production, in order to enable the application of SSEs and help realize the potential of SSBs in electric vehicles.

At present, the Na-ion battery (SIB) technology seems to be a replacement for LIBs due to (i) lower cost, (ii) higher abundance of Na than Li, and most interestingly, (iii) the chemistry of intercalation of SIBs, which is very similar to that of LIBs. However, a considerable volume of research needs to be done to make SIB technology keep up with the standards already set by the LIBs, as discussed in an extensive review on SIBs by Palomares et al. [106]. Several Na-based cathodic materials have been investigated, mostly oxides, fluorophosphate, and phosphates [106]. Among the materials investigated till date, oxides do not seem to be a good option owing to the complexity of insertion–extraction behavior [106]. Phosphates, and fluorophosphate, as reported by Palomares et al. [106], may be considered the right choice due to the high stability of these materials and because the inductive effect of phosphate polyanion produces enormous working potentials. However, the latter needs to be studied in detail to have a better understanding of structural characteristics and Na insertion–extraction mechanisms in these materials. This is one of the areas where C-based nanocomposites (primarily, graphite-like materials) may find application as negative electrodes in SIBs. It is primarily due to the presence of sp^2 structure that graphene sheets (of heteroatoms) produced during synthesis or post-synthesis process may be expected to benefit the insertion process of Na ions between the stacked layers (present in graphene).

As reported in a recent review on potassium-ion batteries (PIBs) [107], the cathode materials used in LIBs may not be suitable for PIBs. For instance, one of the most promising electrode materials available for LIBs is $LiFePO_4$, for which the K-analog has been reported to be practically inactive due to the demonstration of only a little activity. Besides, the larger atomic radius of K as compared to that of Li also makes it extremely hard to synthesize $KFePO_4$ with the crystal structure of pure olivine [107].

However, among the investigated cathode materials, the Prussian blue and its analogs have been reported to show excellent electrochemical properties in terms of both specific energy and capacity [107]. Its density is lower than that of layered oxide materials, and limitations include high sensitivity toward water, inadequate specific discharge capacity, etc. [107]. Hence, the significant challenge for designing new cathode materials for PIBs is to have broader and more flexible void space to accommodate high strain produced by the occupation of K ions. In the context of anode materials, the most significant discovery for PIBs is the sandwiching of K into different graphite layers, which has been reported to be advantageous over SIBs, since Na cannot be sandwiched effectively into the layers of graphite [107]. The graphite electrode used presently in PIBs has been reported to demonstrate excellent specific capacity, rate performance, and cycle life [107]. Although C-based nanocomposites have also been reported as potential candidates to store K ions [105–107], however, they are currently unable to compete with the graphite electrodes, in terms of both initial Coulombic efficiency and energy density (arising due to the lower operating potential of PIBs).

The challenge (in terms of high volume expansion) associated with graphite anode during potentiation needs to be overcome for all practical applications [107]. Alloyed anodes may be considered as promising candidates for full-cell PIBs owing to high specific capacity; however, certain limitations which include high operating potential, low initial Coulombic efficiency, high volume expansion, etc. need further investigation. The materials used for intercalating K into different layers of graphite have been reported to be potential candidates for stable anode materials for practical PIBs [107]; however, detailed investigation in this direction is essential in order to design an anode material with a combination of both high capacity and low operating potential. Apart from investigating individual electrodes in half-cell format, it is extremely important to fabricate a full-cell followed by a proper investigation of the same [107]. This is essential for the purpose of understanding the overall performance of a battery. However, at present, only a limited amount of work has been directed in this area. Hence, this turns out to be a field wherein a considerable volume of work may be done, especially in order to increase the initial Coulombic efficiency and energy density of C-based nanocomposites so as to be able to completely replace the presently used graphite electrodes in PIBs, both in terms of economy and better electronic properties. This necessitates in-depth investigations for a complete understanding of electrochemistry by finding an appropriate combination of electrodes and electrolytes.

4.2 *From the Fundamental Viewpoint*

Despite an enormous volume of published literature and reviews on electronic properties of carbon-based nanocomposites, very little is known about the role of different kinds of defects on the electronic properties of these materials, primarily due to lack of experimental information on the direct visualization of these defects. This, of course, becomes highly interesting when the carbon-based nanocomposite, of interest, turns out to be crystalline because of the onset of long-range atomic order and the pre-existing short-range order of atoms. It has three entities: (i) the lattice, (ii) 1-D dislocations [108] and, most importantly, (iii) different kinds of 2-D interfaces [109], on which the structure-dependent parameters are controlling the electronic properties, or more specifically the electron transport in the material. The simplest of the interfaces being grain boundaries (GBs).

Grain boundaries (GBs) act as electron scattering centers and, based on the GB scattering model [77] has also been proposed to study the temperature dependence of electronic conductivity in C-based hard nanocrystalline composite coatings. The segregation of CNTs to GBs in CNT/metal nitride-based nanocomposites is beneficial in the sense that it leads to the enhancement of electronic conductivity in these materials [63]. In spite of the scattering of electrons at GBs, various C-based nanocomposites have now been studied for quite some time using DC resistivity measurements [4]. One must recall some facts viz. dislocations, in any crystalline material, possess a strain field around them and hence act as regions of localized curvature in the microstructure of all crystalline materials [110].

The GBs possess five independent macroscopic degrees of freedom (DOFs) and three dependent microscopic DOFs, which help define the structure of a GB. Based on these DOFs, different types of GBs may be distinctly classified, each associated with an entity known as GB energy [14], in a polycrystalline material. Moreover, there is yet another entity associated with GB segregation known as GB excess [14]. Based on the first fact, it may be expected that there exists a small change in crystallographic orientation of regions adjacent to a dislocation from which the strain field around the same maybe calculated based on several orientations coupled with defect imaging microscopy techniques (such as electron backscatter diffraction (EBSD) (both 2-D and 3-D) for orientation imaging and either TEM and electron channeling contrast imaging (ECCI) for defect characterization) coupled with the complicated strain tensor analysis [108–110]. The reason to why the present chapter intends to address these facts is that there exist neither any experimental evidence nor theoretical calculations to account for the influence of GB energy on the electron scattering tendency from the GB. Hence, we claim that the proposed GB scattering model [77] for studying the temperature dependence of electronic conductivity in C-based nanocrystalline composite coatings also needs to incorporate an additional term accounting for GB energy, in order to incorporate the influence of GB segregation on the electron scattering tendency of GBs [14, 111]. Moreover, this will also make the GB scattering model more robust and help in understanding the temperature dependence of electronic conductivity in crystalline C-based nanocomposites. One of the main reasons why a study on electron interaction with defects has been missing in the context of C-based nanocomposites is the complexity of understanding crystal structures of these incredibly complex materials and the complexity associated with sample preparation for experimental investigations.

Besides, unlike materials for various structural applications, a very limited amount of research has actually been aimed at understanding the "structure–property correlation" for electronic applications in C-based nanocomposites, using the concept of "correlative microscopy" [11–16] which involves correlating the structural information using a number of microscopy techniques from the region of interest in a microstructure with the chemical analysis (up to the atomic level using atom probe tomography (APT) technique [12, 13], for example) from the same using techniques such as focused ion beaming (FIB) [11, 110–114]. Besides, for the sake of experimental validation, first principle calculations are also a must. As discussed in the introductory section of the present chapter, the potential reasons as to why the "correlative microscopy" methodology has not been extensively employed in understanding the atomic structure-based electronic properties of these materials are the challenges associated with sample preparation for the same and also in characterizing a light element as C, which, either in free form or bonded manner, is the most essential constituent in C-based nanocomposites. This, in particular, acts as a potential field of research wherein a lot of future investigations may be carried out in order to understand the electronic properties of C-based nanocomposites based on analysis at the atomic scale. In recent times, the emergence of artificial intelligence (AI) and machine learning (ML) guided material design [115], also seems to offer a huge potential in the design of C-based nanocomposites, which may aid material scientists to tailor the electronic properties, through microstructural modifications based on AI and ML guided design concepts. This field certainly needs numerous investigations, in the future, as presently, this avenue of research is almost completely unexplored in C-based nanocomposites.

5 Concluding Remarks

There remains absolutely no iota of doubt that C-based nanocomposites will find numerous applications in various fields shortly and that the idea of developing these C-based hybrid materials will be utilized to the maximum possible extent, owing to the presence of numerous research possibilities, especially in the field of electronics, both industrially and academically. This will be necessary to render such complex materials as C-based nanocomposites to get closer to being considered more exclusively for day-to-day electronics-related applications in daily life.

References

1. Zhao Y, Wang LP, Sougrati MT, Feng Z, Leconte Y, Fisher A, Srinivasan M, Xu Z (2017) A review on design strategies for carbon based metal oxides and sulfides nanocomposites for high performance li and na ion battery anodes. Adv Energy Mater 7(9):1–70
2. Camargo PHC, Satyanarayana KG, Wypych F (2009) Nanocomposites: synthesis, structure, properties and new application opportunities. Mater Res 12(1):1–39
3. Baibarac M, Romero PG, Cantu ML et al (2006) Electrosynthesis of the poly (N- vinylcarbazole)/carbon nanotubes composite for applications in the supercapacitor field. Eur Polymer J 42:2302–2312
4. Sanjinés R, Abad MD, Vâju Cr R, Smajda Mionić M, Magrez A (2011) Electrical properties and applications of carbon based nanocomposite materials: an overview. Surf Coat Technol 206:727–733
5. Obreja VVN (2008) On the performance of supercapacitors with electrodes based on carbon nanotubes and carbon activated material-A review. Phys E 40:2596–2605
6. Star A, Joshi V, Skarupo S, Thomas D, Gabriel JCP (2006) Gas sensor array based on metal-decorated carbon nanotubes. J Phys Chem B 110:21014
7. Lu Y, Li J, Han J, Ng HT, Binder C, Partridge C, Meyyapan M (2004) Room temperature methane detection using palladium loaded single-walled carbon nanotube sensors. Chem Phys Lett 391:344
8. Ates M, Eker AA, Eker B (2017) Carbon nanotube-based nanocomposites and their applications. J Adhesion Sci Technol 31:1977–1997
9. Liu XM, Huang ZD, Oh SW, Zhang B, Ma PC, Yuen MF, Kim JK (2012) Carbon nanotube (CNT)-based composites as electrode material for rechargeable Li-ion batteries: a review. Compos Sci Technol 72:121–144

10. Wu X, Chen Y, Xing Z, Lam CWK, Pang SS, Zhang W, Ju Z (2019) Advanced carbon-based anodes for potassium-ion batteries. Adv Energy Mater 9:1–46
11. Felfer PJ, Alam T, Ringer SP, Cairney JM (2012) A reproducible method for damage- free site-specific preparation of atom probe tips from interfaces. Microsc Res Techniq 75:484–491
12. Toji Y, Matsuda H, Herbig M, Choi PP, Raabe D (2014) Atomic-scale analysis of carbon partitioning between martensite and austenite by atom probe tomography and correlative transmission electron microscopy. Acta Mater 65:215–228
13. Gault B, Moody MP, Cairney JM, Ringer SP (2012) Atom probe crystallography. Mater Today 15:378–386
14. Herbig M, Raabe D, Li YJ, Choi P, Zaefferer S, Goto S (2014) Atomic-scale quantification of grain boundary segregation in nanocrystalline material. Phys Rev Lett 112:
15. Singh S, Wanderka N, Murty BS, Glatzel U, Banhart J (2011) Decomposition in multi-component AlCoCrCuFeNi high-entropy alloy. Acta Mater 59:182–190
16. Raabe D, Herbig M, Sandlöbes S, Li Y, Tytko D, Kuzmina M, Ponge D, Choi PP (2014) Grain boundary segregation engineering in metallic alloys: A pathway to the design of interfaces. Curr Opin Mater Sci 18:253–261
17. Hornyak GL, Tibbals HF, Dutta J, Moore JJ (2009) Introduction to nanoscience and technology. CRC Press, New York
18. Kuzmany H, Fink J, Mehring M, Roth S (ed) (2000) Electronic properties of novel materials—molecular nanostructures. AIP conference proceedings, 544
19. Dresselhaus MS, Dresselhaus G, Avouris P (2001) Carbon nanotubes: synthesis, structure, properties and applications. 80 Springer, Berlin
20. Avouris P, Chen Z, Perebeinos V (2007) Carbon-based electronics. Nat Nanotechnol 2:605
21. Georgakilas V, Perman JA, Tucek J, Zboril R (2015) Broad Family of Carbon Nanoallotropes: classification, chemistry, and applications of fullerenes, carbon dots, nanotubes, graphene, nanodiamonds, and combined superstructures. Chem Rev 115(11):4744–4822
22. Lin Y, Taylor S, Li H, Shiral Fernando KA, Qu L, Wang W (2004) Advances toward bioapplications of carbon nanotubes. J Mater Chem 14:527–541
23. Ajayan PM, Schadler LS, Braun PV (2003) Nanocomposite science and technology. Wiley-VCH, Verlag GmbH & Co, Weinheim (Germany)
24. Nalwa HS (2000) Handbook of nanostructured materials and nanotechnology. 5 Academic Press, New York
25. Modi A, Koratkar N, Lass E, Wei BQ, Ajayan PM (2003) Miniaturized gas ionization sensors using carbon nanotubes. Nature 424:171
26. Ajayan PM, Lijima S (1992) Smallest carbon nanotube. Nature 358:23
27. Qiu J, Lia Y, Wang Y, Li W (2004) Production of carbon nanotubes from coal. Fuel Process Technol 85:1663–1670
28. Kavita M, Mordina B, Tiwari RK (2016) Thermal and mechanical behaviour of poly(vinyl butyral)- modified novolac epoxy/multiwalled carbon nanotube nanocomposites. J Appl Polym Sci 133:43333–43344
29. Spitalsky Z, Tasis D, Papagelis K, Galiotis C (2010) Carbon nanotube-polymer composites: chemistry, processing, mechanical and electrical properties. Prog Polym Sci 35:357–401
30. Bouchard J, Cayla A, Devaux E, Campagne C (2013) Electrical and thermal conductivities of multiwalled carbon nanotubes-reinforced high performance polymer nanocomposites. Compos Sci Technol 86:177–184
31. Ensafi AA, Soureshjani EH, Asl MJ, Rezaei B (2016) Polyoxometalate-decorated graphene nanosheets and carbon nanotubes, powerful electrocatalysts for hydrogen evolution reaction. Carbon 99:398–406
32. Okajima K, Ikeda A, Kamoshita K, Sudoh M (2005) High rate performance of highly dispersed C_{60} on activated carbon capacitor. Electrochim Acta 51:972–977
33. Zhou C, Kumar S, Doyle CD, Tour JM (2005) Functionalized single wall carbon nanotubes treated with pyrrole for electrochemical supercapacitor membranes. Chem Mater 17:1997–2002

34. Yoon BJ, Jeong SH, Lee KH, Kim HS, Park CG, Han H (2004) Electrical properties of electrical double layer capacitors with integrated carbon nanotube electrodes. Chem Phys Lett 388:170–174
35. Koysuren O, Du C, Pan N, Bayram G (2009) Preparation and comparison of two electrodes for supercapacitors: Pani/CNT/Ni and Pani/Alizarin-treated nickel. J Appl Polym Sci 113:1070–1081
36. Wang DW, Li F, Zhao J, Ren W, Chen ZG, Tan J, Wu ZS, Gentle I, Lu GQ, Cheng HM (2009) Fabrication of graphene (polyaniline composite paper via in situ anodic electropolymerization for high performance flexible electrode. ACS Nano 3:1745–1752
37. Kim JY, Kim KH, Kim KB (2008) Fabrication and electrochemical properties of carbon nanotube/polypyrrole composite film electrodes with controlled pore size. J Power Sources 176:396–402
38. An KH, Jeong SY, Hwang HR, Lee YH (2004) Enhanced sensitivity of a gas sensor incorporating single- walled carbon nanotube–polypyrrole nanocomposites. Adv Mater 16:1005–1009
39. Cheng G, Zhao J, Tu Y, He P, Fang Y (2005) A sensitive DNA electrochemical biosensor based on magnetite with a glassy carbon electrode modified by multi-walled carbon nanotubes in polypyrrole. Anal Chim Acta 533:11–16
40. Limelette P, Schmaltz B, Brault D, Gouineau M, Autret-Lambert C, Roger S, Grimal V, Van T (2014) Conductivity scaling and thermoelectric properties of polyaniline hydrochloride. J Appl Phys 115:
41. Blaszczyk-Lezak I, Desmaret V, Mijangos C (2016) Electrically conducting polymer nanostructures confined in anodized aluminum oxide templates (AAO). Express Polym Lett 10:259–272
42. Choudhury A, Kar P (2011) Doping effect of carboxylic acid group functionalized multi-walled carbon nanotube on polyaniline. Compos Part B Eng. 42:1641–1647
43. Wang Y, Zhang S, Deng Y (2016) Semiconductor to metallic behavior transition in multi-wall carbon nanotubes/polyaniline composites with improved thermoelectric properties. Mater Lett 164:132–135
44. Taberna PL, Chevallier G, Simon P, Plée D, Aubert T (2006) Activated carbon-carbon nanotube composite porous film for supercapacitor applications. Mater Res Bull 41:478–484
45. Navarro-Flores E, Omanovic S (2005) Hydrogen evolution on nickel incorporated in three-dimensional conducting polymer layers. J Mol Catal A: Chem 242:182–194
46. Huq MM, Hsieh CT, Ho CY (2016) Preparation of carbon nanotube-activated carbon hybrid electrodes by electrophoretic deposition for supercapacitor applications. Diamond Related Mater 62:58–64
47. Khomenko V, Raymundo-Pinero E, Beguin F (2008) High-energy density graphite/AC capacitor in organic electrolyte. J Power Sources 177:643–651
48. Qiu J, Wu X, Qiu T (2016) High electromagnetic wave absorbing performance of activated hollow carbon fibers decorated with CNTs and Ni nanoparticles. Ceram Int 42:5278–5285
49. Gangupomu RH, Sattler ML, Ramirez D (2016) Comparative study of carbon nanotubes and granular activated carbon: physicochemical properties and adsorption capacities. J Hazard Mater 302:362–374
50. Rudge A, Davey J, Raistrick I, Gottesfeld S (1994) Conducting polymers as active materials in electrochemical capacitors. J Power Sources 47:89
51. Bouchard J, Cayla A, Odent S, Lutz V, Devaux E, Campagne C (2016) Processing and characterization of polyethersulfone wet-spun nanocomposite fibres containing multi walled carbon Nanotubes. Synth Met 217:304–313
52. Yuan D, Yang W, Ni J, Gao L (2015) Sandwich structured MoO2@TiO2@CNT nanocomposites with high-rate performance for lithium ion batteries. Electrochim Acta 163:57–63
53. Wang YG, Wang ZD, Xia YY (2005) An asymmetric supercapacitor using RuO2/TiO2 nanotube composite and activated carbon electrodes. Electrochim Acta 50:5641–5646
54. Alam RS, Moradi M, Nikmanesh H (2016) Influence of multi-walled carbon nanotubes (MWCNTs) volume percentage on the magnetic and microwave absorbing properties of BaMg$_{0.5}$Co$_{0.5}$TiFe$_{10}$O$_{19}$/MWCNTs nanocomposites. Mater Res Bull 73:261–267

55. Yuan QH, Zeng XS, Liu Y, Luo L, Wu J, Wang Y, Zhou G (2016) Microstructure and mechanical properties of AZ91 alloy reinforced by carbon nanotubes coated with MgO. Carbon 96:843–855

56. Islam MS, Deng Y, Tong L, Roy AK, Minett AI, Gomes VG (2016) Grafting carbon nanotubes directly onto carbon fibers for superior mechanical stability: towards next generation aerospace composites and energy storage applications. Carbon 96:701–710

57. Wu G, Ma L, Liu L, Wang Y, Xie F, Zhong Z, Zhao M, Jiang B, Huang Y (2016) Interface enhancement of carbon fiber reinforced methylphenylsilicone resin composites modified with silanized carbon nanotubes. Mater Des 89:1343–1349

58. Wang Y, Colas G, Filleter T (2016) Improvements in the mechanical properties of carbon nanotube fibers through graphene oxide interlocking. Carbon 98:291–299

59. Yang LJ, Cui JL, Wang Y et al (2016) Research progress on the interconnection of carbon nanotubes. New Carbon Mater 31:1–17

60. Tamrakar S, An Q, Thostenson ET, Rider AN, Haque BZ (Gama), Gillespie JW Jr (2016) Tailoring interfacial properties by controlling carbon nanotube coating thickness on glass fibers using electrophoretic deposition. ACS Appl Mater Interfaces. 8:1501–1510

61. Flahaut E, Peigney A, Laurent Ch, Ch. Chastel MF, Rousset A (2000) Acta Mater, 48:3803

62. Kymakis E, Alexandou I Amaratunga GAJ (2002) Single-walled carbon nanotube–polymer composites: electrical, optical and structural investigation. Synth. Met 127:59

63. Jiang L, Gao L (2005) Carbon nanotubes–metal nitride composites: a new class of nanocomposites with enhanced electrical properties. J Mater Chem 15:260–266

64. Peigney A, Laurent Ch Rousset A (1997) Key Eng. Mater. 743:132–136

65. Rao CNR, Satishkumar BC, Govindaraj A, Nath M (2001) Nanotubes. Chem Phys Chem 2:78

66. Sun Z, Zhang J, Yin L, Hu G, Fang R, Cheng HM, Li F (2017) Conductive porous vanadium nitride/graphene composite as chemical anchor of polysulfides for lithium-sulfur batteries. Nat Comm 8:14627

67. Zhao JG, Yang LX, Li FY, Yu RC, Jin CQ (2008) Electrical property evolution in the graphitization process of activated carbon by high-pressure sintering. Solid State Sci 10:1947

68. Staryga E, Bak GW (2005) Relation between physical structure and electrical properties of diamond-like carbon thin films. Diamond Relat Mater 14:23

69. Mott NF, Davis EA (1979) Electronic processes in non-crystalline materials. Clarendon Press, Oxford

70. Shimikawa K, Miyake K (1989) Hopping transport of localized π electrons in amorphous carbon films. Phys Rev B 39:7578

71. Godet C, Kleider JP, Gudovskikh AS (2007) Frequency scaling of AC hopping transport in amorphous carbon nitride. Diamond Relat Mater 16:1799

72. Vishwakarma PN, Subramanyam SV (2006) Hopping conduction in boron doped amorphous carbon films. J Appl Phys 100:

73. Zhong DH, Sano H, Uchiyama Y, Kobayashi K (2000) Effect of low-level boron doping on oxidation behavior of polyimide-derived carbon films. Carbon 38:1199

74. Sikora A, Berkesse A, Bourgeois O, Garden JL, Guerret-Piécourt C, Rouzaud JN, Loir AS, Garrelie F, Donnet C (2009) Structural and electrical characterization of boron-containing diamond-like carbon films deposited by femtosecond pulsed laser ablation. Solid State Sci 11:1738

75. Xue B, Chen P, Hong Q, Lin J, Tan KL (2001) Growth of Pd, Pt, Ag and Au nanoparticles on carbon nanotubes. J Mater Chem 11:2378

76. Meiners T, Frolov T, Rudd RE, Dehm G, Liebscher CH (2020) Observations of grain-boundary phase transformations in an elemental metal. Nature 579:375–378

77. Mishnaevsky Jr LL (2007) Computational Mesomechanics of composites. John Wiley England

78. Nigro A, Nobile G, Rubino MG, Vaglio R (1988) Electrical resistivity of polycrystalline niobium nitride films. Phys Rev B 37:3970

79. Yu Z, Tetard L, Zhai L, Thomas J (2013) Supercapacitor electrode-materials nanostructures from 0 to 3 dimensions. Energy Environ Sci 8:702–730

80. Borenstien A, Noked M, Okashy S, Aurbach D (2013) Composite carbon nanotubes (CNT)/activated carbon electrodes for non-aqueous supercapacitors using organic electrolyte solutions. J Electrochem Soc 160:A1282–A1285

81. Davies A, Yu A (2011) Material advancements in supercapacitors: from activated carbon to carbon nanotube and graphene. Can J Chem Eng 89:1342–1357

82. Stankovich S, Dikin DA, Dommett GHB, Kohlhaas KM, Zimney EJ, Stach EA, Piner RD, Nguyen ST, Ruoff RS (2006) Graphene-based composite materials. Nature 442:282–286

83. Geim AK, Novoselov KS (2007) The rise of graphene. Nat Mater 6:183–191

84. Geng Y, Wang SJ, Kim JK (2009) Preparation of graphite nanoplatelets and graphene Sheets. J Colloid Interface Sci 336:592–598

85. Zheng QB, Ip WH, Lin XY, Yousefi N, Yeung KK, Li Z, Kim JK (2011) Transparent conductive films consisting of ultralarge graphene sheets produced by Langmuir-Blodgett assembly. ACS Nano 5:6039–6051

86. Lian PC, Zhu XF, Liang SZ, Li Z, Yang WS, Wang HH (2010) Large reversible capacity of high quality graphene sheets as an anode material for lithium-ion batteries. Electrochim Acta 55:3909–3914

87. Wu ZS, Ren WC, Wen L, Gao LB, Zhao JP, Chen ZP (2010) Graphene anchored with Co3O4 nanoparticles as anode of lithium ion batteries with enhanced reversible capacity and cyclic performance. ACS Nano 4:3187–3194

88. Su FY, You CH, He YB, Lv W, Cui W, Jin FM, Li B, Yang QH, Kang F (2010) Flexible and planar graphene conductive additives for lithium-ion batteries. J Mater Chem 20:9644–9650

89. Guo Y, Wang T, Chen F, Sun X, Li X, Yu Z, Wan P, Chen X (2016) Hierarchical graphene–polyaniline nanocomposite films for high-performance flexible electronic gas sensors. Nanoscale 8(23):12073–12080

90. Chang H, Wu H (2013) Graphene-based nanocomposites: preparation, functionalization, and energy and environmental applications. Energy Environ Sci 6(12):3483

91. Geim AK (2009) Graphene: status and prospects. Science 324:1530–1534

92. Allen MJ, Tung VC, Kaner RB (2009) Honeycomb carbon: a review of graphene. Chem Rev 110:132–145

93. Rao CNR, Sood AK, Subrahmanyam KS, Govindaraj A (2009) Graphene: the new two-dimensional nanomaterial. Angew Chem Int Edit 48:7752–7777

94. Chang HX, Wu HK (2013) Graphene-based nanomaterials: synthesis, properties, and optical and optoelectronic applications. Adv Funct Mater 23:1984–1997

95. Yang K, Feng L, Shi X, Liu Z (2013) Nano-graphene in biomedicine: theranostic applications. Chem Soc Rev 42:530–547

96. Zhu Y, Murali S, Cai W, Li X, Suk WJ, Potts JR, Ruoff RS (2010) Graphene and graphene oxide: synthesis, properties, and applications. Adv Mater 22:3906–3924

97. Du X, Skachko I, Barker A, Andrei EY (2008) Approaching ballistic transport in suspended graphene. Nat Nanotechnol 3:491–495

98. Huang X, Yin ZY, Wu SX, Qi XY, He QY, Zhang QC, Yan QY, Boey F, Zhang H (2011) Graphene-based materials: synthesis, characterization, properties, and applications. Small 7:1876–1902

99. Lee C, Wei XD, Kysar JW Hone J (2008) Measurement of the elastic properties and intrinsic strength of monolayer graphene. Science 321:385–388

100. Balandin AA, Ghosh S, Bao W, Calizo I, Teweldebrhan D, Miao F, Lau CN (2008) Superior thermal conductivity of single-layer graphene. Nano Lett 8:902–907

101. Nair R, Blake P, Grigorenko A, Novoselov K, Booth T, Stauber T, Peres N, Geim A (2008) Fine structure constant defines visual transparency of graphene. Science 320:1308

102. Lin YM, Dimitrakopoulos C, Jenkins KA, Farmer DB, Chiu HY, Grill A, Avouris P (2010) 100-GHz transistors from wafer-scale epitaxial graphene. Science 327:662

103. Liao L, Lin Y-C, Bao M, Cheng R, Bai J, Liu Y, Qu Y, Wang KL, Huang Y Duan X (2010) High-speed graphene transistors with a self-aligned nanowire gate. Nature 467:305–308

104. Schwierz F (2010) Graphene transistors. Nat Nanotechnol 5:487–496
105. Du M, Liao K, Lu Q, Shao Z (2019) Recent advances in the interface engineering of solid-state Li-ion batteries with artificial buffer layers: Challenges, materials, construction, and characterization. Energy Environ Sci 12(6):1780–1804
106. Palomares V, Serras P, Villaluenga I, Hueso KB, Carretero-Gonzalez J, Rojo T (2012) Na-ion batteries, recent advances and present challenges to become low cost energy storage systems. Energy Environ Sci 5:5884
107. Rajagopalan R, Tang Y, Ji X, Jia C, Wang H (2020) Advancements and Challenges in potassium ion batteries: a comprehensive review. Adv Funct, Mater, p 1909486
108. Cottrell AH (1949) Theory of dislocations. B. Chalmers (Ed.), Progress in Metal Physics Chapter. II, pp 1–52
109. Gleiter H (1983) On the structure of grain boundaries in metals. In: Latanision RM, Pickens JR (eds) Atomistics of fracture. Springer Boston MA
110. Zaafarani N, Raabe D, Roters F, Zaefferer S (2008) On the origin of deformation-induced rotation patterns below nanoindents. Acta Mater 56(1):31–42
111. Zaafarani N, Raabe D, Singh RN, Roters F, Zaefferer S (2006) Three-dimensional investigation of the texture and microstructure below a nanoindent in a Cu single crystal using 3D EBSD and crystal plasticity finite element simulations. Acta Mater 54:1863–1876
112. Wheeler J, Mariani E, Piazolo S, Prior DJ, Trimby PJ, Drury MR (2009) The weighted Burgers vector: a new quantity for constraining dislocation densities and types using electron backscatter diffraction on 2D sections through crystalline materials. J Microscopy 233:482–494
113. Gutierrez-Urrutia I, Zaefferer S, Raabe D (2013) Coupling of Electron Channeling with EBSD: toward the quantitative characterization of deformation structures in the SEM. JOM 65(9):1229–1236
114. Stoffers A, Cojocaru-Mirédin O, Seifert W, Zaefferer S, Riepe S, Raabe D (2015) Grain boundary segregation in multicrystalline silicon: correlative characterization by EBSD, EBIC, and atom probe tomography. Prog Photovolt: Res Appl 23:1742–1753
115. Huber L, Hadian R, Grabowski B, Neugebauer J (2018) A machine learning approach to model solute grain boundary segregation. npj Comput Mater 64(1):1–8

Tuning of SPR and Structural Properties of Cu-Fullerene Nanocomposite

Rahul Singhal, Jyotsna Bhardwaj, Amena Salim, Ritu Vishnoi, and Ganesh D. Sharma

Abstract Metal-matrix nanocomposite has a multitude of applications. The local structure and optical modifications of these materials are studied using fullerene as a matrix material, incorporated by noble metal nanoparticles of Cu. These metal-fullerene nanocomposites are useful because of the amalgamation of different properties of the fullerene and metal nanoparticles. Cu being cheap and abundant in nature has an advantage over other metal nanoparticles. Cu nanoparticles being more reactive are stabilized by incorporating them in fullerene matrix and therefore can be used in various applications. The structural and optical properties (mainly SPR) of Cu-fullerene nanocomposites are tuned by different methods and synthesis procedures such as (a) Ion irradiation, (b) ion implantation, (c) thermal annealing and (d) increasing the concentration of the Cu nanoparticles in the matrix material. In this review all the factors influencing the tuning of structural and optical properties of Cu-fullerene nanocomposites are investigated in detail. Each property is studied by different characterization techniques such as, TEM, UV–visible spectroscopy and electron diffraction method.

Keywords Metal-fullerene nanocomposite · Ion irradiation · Thermal annealing · Surface plasmon resonance

1 Introduction

Metal-matrix nanocomposites have a wide range of applications in memory devices, optoelectronic devices, mechanical hardware, sensors, reinforcement materials, electromechanics, biomedical, etc. due to their bifunctional properties [1–3]. Depending on the nature of the matrix, the properties of the metal can be tuned, eg. generation of

R. Singhal (✉) · J. Bhardwaj · A. Salim · R. Vishnoi
Department of Physics, Malaviya National Institute of Technology Jaipur, JLN Marg, Jaipur 302017, India
e-mail: rsinghal.phy@mnit.ac.in

G. D. Sharma
Department of Physics, The LNM Institute of Information Technology, Jamdoli, Jaipur, India

© The Author(s), under exclusive license to Springer Nature Singapore Pte Ltd. 2021   123
A. Hazra and R. Goswami (eds.), *Carbon Nanomaterial Electronics: Devices and Applications*, Advances in Sustainability Science and Technology,
https://doi.org/10.1007/978-981-16-1052-3_6

surface plasmon resonance by using metal matrix nanocomposite thin film. Surface plasmon resonance is the collective oscillation of surface electrons in the presence of electromagnetic rays. This phenomenon is used in a number of applications such as light sensors, solar cells, bandgap limiters, etc. The formation and tuning of SPR are studied by various methods such as (a) low-energy ion irradiation, (b) high-energy ion irradiation and (c) thermal annealing and a review is presented here for Cu-fullerene nanocomposite.

Ion beam irradiation is used to alter the structural and optical behaviour of the material at atomic and sub-levels. When the accelerating ion hits the sample, it experiences a number of collisions as illustrated in Fig. 1. Depending upon the nature of collision whether an electronic collision or nuclear collision, the effect on the sample surface is studied. Nuclear loss is elastic and occurs with the ions having the energy of the order of keV $\cdot$ nucleon^{-1}. The collision results in recoil of the target

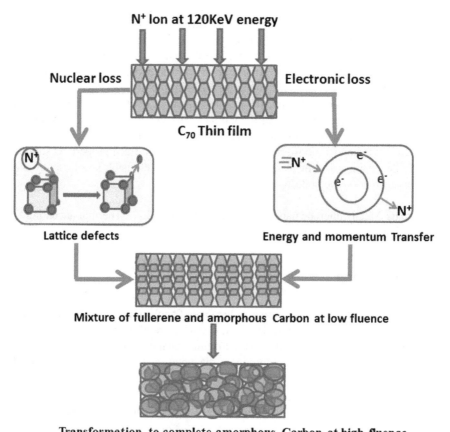

Fig. 1 Image shows electronic and nuclear losses happening after irradiating the sample by 120 meV N$^+$ ions. "Reprinted from [18], Copyright (2018), with permission from Elsevier"

Table 1 Studies done till date on CuC_{60} nanocomposite thin film to generate SPR

Nanocomposite	Irradiating energy	Metal concentration at%	SPR	References
CuC_{60}	120 MeV Ag	3%	No	[4]
CuC_{60}	120 MeV Au	1%	No	[5]
CuC_{60}	120 MeV Au	5%	No	[13]
CuC_{70}	120 MeV Au	–	No	[20]
CuC_{70}	120 MeV Ag	6.8%	No	[15]
CuC_{60}	120 MeV Ag	18%	622 nm	[9]

atom and an interstitial defect occurs. Therefore, the modifications in volume of the sample surface occurs in nuclear stopping. On the other hand, electronic stopping is inelastic in nature because the fast-moving ion transfers its energy to the neighbouring electrons in the target material which provide the ground state electrons some amount of activation energy to transit at a higher energy state. Some of the energy is used to excite the phonons and produces thermal vibrations. Electronic loss dominates at high energy (~ 1000 keV $\cdot$ nucleon^{-1}). Singhal et al. and the group have studied the formation and tuning of SPR through all the methods mentioned above and is summarized in Table 1. The method of synthesizing CuC_{60} thin film is the same in all the studies. The thermal co-deposition method is used to synthesize CuC_{60} thin films with different concentrations of metal. These films were subjected to high energy ion irradiation, low energy ion irradiation and thermal annealing. The different methods of generating and tuning the SPR are reviewed below.

1.1 Tuning of SPR by High-Energy Ion Irradiation

CuC_{60} and CuC_{70} thin films are irradiated by 120 MeV Au and Ag ions [4]. No traces of SPR can be observed when the metal concentration is low but a change in bandgap from 2.18 eV (in pristine) to 1.93 eV (at the highest fluence of 3×10^{13} ions/cm^2) is observed. This happens owing to the conversion of the fullerene molecule to the amorphous carbon by ion impact. As the concentration of metal nanoparticle is so low that the inter-particle distance is very large and therefore the growth of particles is very small. The particle size before and after ion irradiation when the $Cu(3\%)C_{60}$ nanocomposite thin film is irradiated by 120 meV Ag ions, only increases from 2.5 ± 0.05 nm to 2.6 ± 0.06 nm as calculated using transmission electron microscopy. Nevertheless, when we increase the concentration of metal nanoparticles in metal matrix nanocomposite to 18%, the pristine film itself shows SPR at ~ 622 nm and after ion irradiation at highest fluence, a red shift of 11 nm in the SPR wavelength is observed which is shown in Fig. 2. The red shift is due to the growth in the size of the metal nanoparticles with ion irradiation [9].

The growth mechanism of metal nanoparticles can be explained by Ostwald ripening and nucleation growth. In order to attain a thermodynamic stability, small

Fig. 2 Optical absorption
spectra of pristine and
120 meV Ag ion irradiated
Cu–C$_{60}$ nanocomposite thin
films. The inset shows the
SPR band variation with
irradiation. "Reprinted
by permission from
[Springer Nature] [10]
[copyright] (2019)"

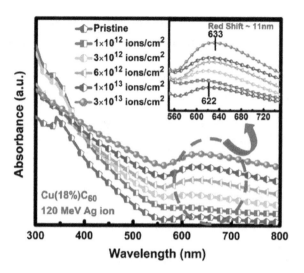

metal particles which are already present in the matrix dissolve and deposit on
colossal particles succeeded by shrinking and growth of particles. This process is
called Oswald ripening. The inset of Fig. 2 shows the variation of SPR band with
the increase in irradiation fluence, also the spread of SPR band broadens. The Drude
approximation dielectric function depends upon the frequency which is given by
the equation below [5, 8]:

$$\varepsilon_m = 1 - \frac{\omega_p^2}{(\omega^2 + i\gamma\omega)} \tag{1}$$

where ω_p is the Drude bulk plasmon frequency which is represented as $\omega_p \sqrt{\frac{ne^2}{\varepsilon_o m}}$
where, n, e, ε_o and m are free e$^-$ density, charge of electron, permittivity of vacuum,
and effective mass, respectively. γ implies the damping constant, which defines the
width of the SPR and depends on the radius of the metal nanoparticle [5, 7].

$$\gamma(r) = \gamma_0 + \frac{Av_f}{r} \tag{2}$$

where A = constant defining the scattering details, v_f = Fermi velocity, r = radius
of the nanoparticle and γ_0 = (bulk) damping constant.

If we analyze the above equation (in accordance with the particle size distribution),
the width of SPR gives two types of particle behaviour. (i) the particles smaller
than 25 nm intrinsic size effect dominate and it depends upon radius (r) of metal
nanoparticle as 1/r [6]. Hence, the width of SPR is inversely proportional to the size
of the nanoparticle. (ii) for the particles that have size >25 nm, the extrinsic size
effect dominates in which the size of the nanoparticle is directly proportional to the

bandwidth. Also, the dual behaviour can be elucidated by the surface of nanoparticles where the conduction band electrons get scattered. This broadens the SPR peak with the decrease in the size of the metal nanoparticle. Hence, we can see in Fig. 2 the SPR band broadening when metal nanoparticle size increases.

1.2 Tuning of SPR with Low-Energy Ions Irradiation

The advantage of low-energy ion over high-energy ion beam irradiation is that the nanocomposite thin films having a metal concentration less than 18% can also shows a surface plasmon resonance. CuC_{60} nanocomposite thin films having concentration 2 and 8.5% when irradiated with 120 keV N^+ ions gives a SPR at a wavelength of ~630 nm [12, 14]. Here we observe a considerable increase in particle size. When $Cu(3\%)C_{70}$ thin film is irradiated by 180 keV Ar ions, it gives tuning of SPR; firstly, SPR gives a red shift upto a moderate fluence and then with an increase in ion irradiation, a blue shift is observed [16]. We assume metal nanoparticles to be spherical in shape so that we can easily predict their absorption cross section by using Mie theory (also known as electrostatic approximation). The effective dielectric permittivity of particles, in this case, is explained by assuming particles as Drude-like particles having interband transition given earlier by Eq. (1) but with an extra expression for interband transition [5, 7, 24].

$$\epsilon(\omega) = 1 - \frac{\omega_p^2}{(\omega^2 + i\gamma\omega)} + \chi^{ib}(\omega) \tag{3}$$

When the size of metal is less than the mean free path of electron collision, we can find the damping constant for bulk but it depends on the radius of a particle given by "r", as given in Eq. (2).

In such a scenario, damping constant increases with increased collisions. The interband susceptibility is expressed by Kreibig and Vollmer [5]

$$\chi^{ib}(\omega) = \chi_1^{ib}(\omega) + i\chi_{12}^{ib}(\omega) \tag{4}$$

The cross section of light absorption is given by

$$\sigma(\omega) = \left(\frac{9V}{c}\right) \frac{\varepsilon_m^{3/2}}{1 + \chi_1^{ib} + 2\,\epsilon_m} \times \frac{\Omega^2\gamma\omega^2 + \kappa\omega^3(\omega^2 + \gamma^2)}{(\Omega^2 - \omega^2 + \kappa\omega\gamma)^2 + (\gamma\omega + \kappa\omega^2)^2} \tag{5}$$

where ω represents free space absorbing light frequency, V represents the volume of the particle, ε_m represents the dielectric constant of the surrounding medium, the resonance frequency is given by $\Omega = \dfrac{\omega_p}{\sqrt{1+\chi_1^{ib}+2\varepsilon_m}}$ and ε is the constant given by $\dfrac{\chi_2^{ib}}{1+\chi_1^{ib}+2\varepsilon_m}$.

We can deduce the dependence of resonance frequency on the surrounding medium from the above expressions and also on the shape and size of the nanoparticles. As shown in Fig. 3, the pristine film does not show the SPR band because of the tiny size of Cu nanoparticles as shown in TEM indicated in Fig. 5. When the nanocomposite film is irradiated at a fluence of 3×10^{14} ions/cm², a wide SPR peak at ~617 nm is observed in which at a higher fluence of 1×10^{15} and 3×10^{15} ions/cm², the peak is red shifted at the wavelength of 657 and 663 nm, respectively. Figure 4 shows the shift in the SPR peak with fluence.

Maxwell–Garnett (MAG) effective medium theory is generally used to calculate the peak position of SPR and also the full width half maximum in a complex system, considering not only the different shapes of particles but also the absorbing matrix. As fullerene C_{70} is an absorbing medium therefore we can effectively follow MAG theory, which affirms that the growth of metal nanoparticles gives red shift in SPR band. When both Figs. 3 and 5 are compared, we can see that with the increasing fluence, metal nanoparticle size increases and hence we get the red shift in the SPR position. But at the higher fluence of 1×10^{16} ions/cm², fullerene C_{70} completely transforms into amorphous carbon; therefore, a sharp change in the refractive index of the matrix is noticed resulting in the blue shift of the SPR band at ~645 nm [17].

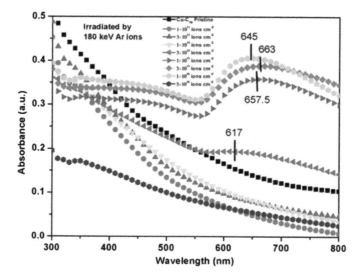

Fig. 3 UV–visible absorption spectra showing the evolution of SPR band and shift in SPR peak position with increasing fluence of 180 keV Ar ion beam. "Reprinted from [19], Copyright (2019) with permission from Elsevier"

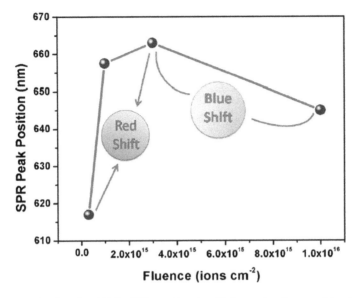

Fig. 4 Graph showing the shift in SPR wavelength with increasing fluence of ion irradiation. "Reprinted from [19], Copyright (2019) with permission from Elsevier"

1.3 Tuning of SPR by Thermal Annealing

Thermal annealing is the easiest way to generate and tune the SPR band. CuC_{60} is synthesized by the thermal co-evaporation method and is annealed at various temperatures ranging from 100 to 400 °C in the presence of a continuous flow of Ar gas [11]. The concentration of Cu in this study is 27 at%. At a temperature of 300 °C, the SPR is found at ~694 nm which is blue shifted to ~684 nm at a temperature of 400 °C. A similar study is done on CuC_{70} nanocomposite thin film having metal concentration ~4.5 at% and it shows a red shift from 585 to 621 nm at the highest annealing temperature of 400 °C [21]. This tuning of SPR from red shift to blue shift or vice versa through thermal annealing is well explained by Vishnoi et al. in their article [22]. The author has taken CuC_{70} nanocomposite thin film and annealed it at different temperatures from 100 to 350 °C. A sharp red shift is observed as shown in Fig. 6.

The effect of temperature on shifting and broadening of SPR peak can be explained by three effective components as described by Yeshchenko et al. in detail [24]: (a) relation of matrix and dielectric constant, (b) scattering of electrons and phonons in a metal matrix system and (c) thermal expansion of metal nanoparticles. The absorption coefficient $\alpha(\omega)$ of a composite thin film with non-interacting metal nanoparticles (spherical) is expressed as follows by Kreibig and Vollmer [5]:

$$\alpha(\omega) = \frac{9f\omega\varepsilon_m^{3/2}}{c} \frac{\varepsilon_2}{(\varepsilon_1 + 2\varepsilon_m)^2 + \varepsilon_2^2} \tag{6}$$

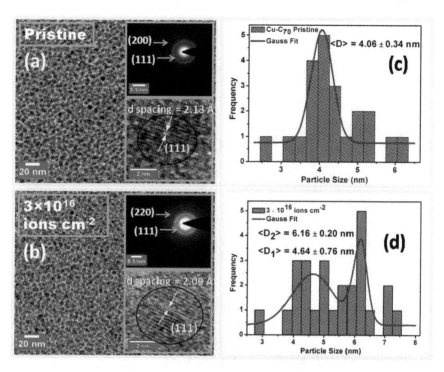

Fig. 5 TEM image and particle size distribution for pristine Cu-C$_{70}$ film (**a** and **c**) and Cu-C$_{70}$ thin films irradiated at highest fluence (**b** and **d**). Inset of **a** and **b** shows HRTEM and SAED pattern. "Reprinted from [19], Copyright (2019) with permission from Elsevier"

where f represents the filling factor of the composite and ε_m represents the dielectric constant of the host matrix. As discussed earlier that the position and width of the SPR peak of Cu nanoparticles depend on the dielectric permittivity of metal nanoparticles and the fullerene matrix, and therefore affects the absorption spectra of nanocomposite thin film. The dielectric function which depends upon frequency consists of two factors.

(i)   Intraband transition (transition which occurs in conduction band) and
(ii)   Interband transition (d bands to the conduction band).

$$\varepsilon_m = \varepsilon_f + \varepsilon_{ib} \tag{7}$$

and $\quad \varepsilon_{ib} = \varepsilon_{ib1}(\omega) + i\varepsilon_{ib2}(\omega)$

For small damping of resonance:

$$\varepsilon_1(\omega) = -2\varepsilon_m \tag{8}$$

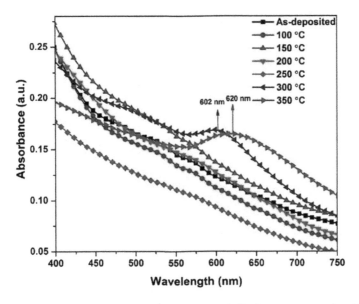

Fig. 6 Absorption spectra of as-deposited and annealed Cu-C70 nanocomposite thin films. "Reprinted from [23], Copyright (2019) with permission from Elsevier"

From Eqs. (1), (6) and (7), the SPR resonance energy for a nanoparticle is

$$\omega_{sp} = \sqrt{\frac{\omega_p^2}{1 + \varepsilon_{ib1} + 2\varepsilon_m} - \gamma^2} \tag{9}$$

Here, ε_{ib1} = the real part of interband transitions divided by the permittivity of nanoparticles. The surface plasmon damping constant is given by Eq. (2).

There are two mechanisms to explain the dependency of SPR position and width of SPR on temperature.

1. Electron–phonon scattering (effect on γ_∞) and
2. Thermal expansion of metal nanoparticles

The number of phonons increases due to an increase in temperature in the metal surface which gives the high possibility of electron–phonon scattering, affecting the damping constant γ_∞. The increase in γ_∞ will result in the broadening and red shifting of the SPR peak.

The second mechanisms concern the dependence of $\varepsilon(T)$ in the thermal expansion of nanoparticles. The size of metal nanoparticles increases with temperature which is also confirmed by TEM images as given in Fig. 7.

Due to the thermal expansion of nanoparticles, the concentration of free electrons decreases which leads to a decrease in surface plasmon frequency. This gives the red shift of the SPR wavelength with temperature. The surface plasmon damping

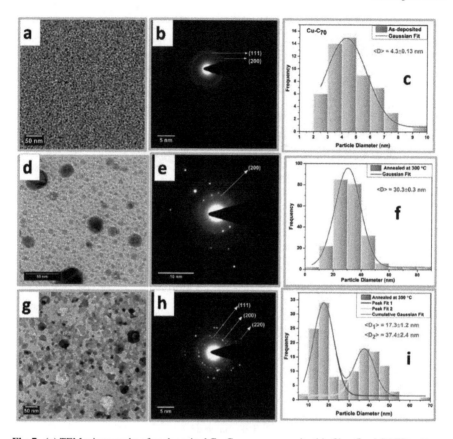

Fig. 7 (a) TEM micrographs of as-deposited Cu-C$_{70}$ nanocomposite thin film, (b–c) SAED pattern and distribution of particles for as-deposited Cu-C$_{70}$ nanocomposite thin film, (d) TEM micrographs of Cu-C$_{70}$ nanocomposite thin film annealed at 300 °C and (e–f) SAED pattern and distribution of particles for Cu-C$_{70}$ nanocomposite thin film annealed at 300 °C, (g) TEM micrographs of Cu-C$_{70}$ nanocomposite thin film annealed at 350 °C and (h–i) SAED pattern and distribution of particles for Cu-C$_{70}$ nanocomposite thin film annealed at 350 °C. "Reprinted from [23], Copyright (2019) with permission from Elsevier"

constant varies with the size of the metal nanoparticle. The change in the radius of nanoparticles through temperature is given by

$$R(T) = R_0(1 + \beta \Delta T)^{1/3} \tag{10}$$

According to the above equation, the thermal expansion will influence the frequency and volume of bulk Plasmon and also some part of the damping constant which is size dependent.

The SPR frequency which depends on temperature is given by

$$\omega_{sp} = \sqrt{\frac{\omega_{p0}^2}{(1 + \varepsilon_{ib1} + 2\varepsilon_m(T))(1 + \beta(T)\Delta T)} - \gamma^2(T)} \tag{11}$$

This means that with temperature, free electron concentration decrease and volume of nanoparticles increase. This decreases the SPR frequency which gives red shift to the SPR wavelength position. The red shifted SPR frequency decreases due to the increased damping constant which in turn is in effect due to increased electron–phonon scattering. Thermal expansion in Cu nanoparticles attributes to the red shifted SPR peak. On the contrary, the electron–phonon scattering in Cu nanoparticles gives SPR peak broadening with increasing temperature. Any changes in the refractive index of the matrix, the size of nanoparticles or their shape attribute to the shift in SPR peak.

2 Conclusion

Cu nanoparticles embedded in fullerene matrix in form of nanocomposite thin films have an advantage over other metal-fullerene nanocomposite thin films in terms of cost. Its cost-effectiveness allows this nanocomposite to be used commercially in various devices. The characteristics of Cu-fullerene nanocomposite thin films can be changed in various ways like low- and high-energy ion irradiation and thermal annealing. The main advantage of these techniques is to generate surface plasmon resonance in the nanocomposite material. Tuning of SPR in the entire visible range makes this material a competent material to be used in the active layer present in the solar cells. The generation of characteristic SPR in each nanocomposite sample can be used in the bandgap limiters. The tuning of SPR depends upon the concentration of metal in the nanocomposite material and also on the change in refractive index of the host matrix.

Acknowledgement One of the authors (Dr. Ritu Vishnoi) greatly acknowledges the financial support by DST New Delhi in terms of the Woman Scientist Project (SR/WOS-A/PM- 47/2019).

References

1. Ayesh AI, Qadri S, Baboo VJ et al (2013) Nano-floating gate organic memory devices utilizing Ag–Cu nanoparticles embedded in PVA-PAA-glycerol polymer. Synth Met 183:24–28. https://doi.org/10.1016/j.synthmet.2013.09.018
2. Barnes WL, Dereux A, Ebbesen TW (2003) Surface plasmon subwavelength optics. Nature 424:824–830
3. Bashiri S, Vessally E, Bekhradnia A et al (2017) Utility of extrinsic [60] fullerenes as work function type sensors for amphetamine drug detection: DFT studies. Vacuum 136:156–162. https://doi.org/10.1016/j.vacuum.2016.12.003

4. Inani H, Singhal R, Sharma P et al (2017) Effect of low fluence radiation on nanocomposite thin films of Cu nanoparticles embedded in fullerene C_{60}. Vacuum 142:5–12. https://doi.org/10.1016/j.vacuum.2017.04.034c

5. Kreibig U, Vollmer M (1995) Optical properties of metal clusters. Springer Ser Mater Sci 25:535. https://doi.org/10.1007/978-3-662-09109-8; U. Kreibig, C.V. Fragstein (1969) The limitation of electron mean free path in small silver particles. Zeitschrift für Physik 224(4):307–323

6. Link S, El-Sayed MA (1999) Size and temperature dependence of the plasmon absorption of colloidal gold nanoparticles. J Phys Chem B. https://doi.org/10.1021/jp9847960

7. Pinchuk A, Von Plessen G, Kreibig U (2004) Influence of interband electronic transitions on the optical absorption in metallic nanoparticles. J Phys D Appl Phys. https://doi.org/10.1088/0022-3727/37/22/012; Pinchuk A, Kreibig U, Hilger A (2004) Optical properties of metallic nanoparticles: influence of interface effects and interband transitions, Surf Sci 557(1–3):269–280

8. Rai V, Rai VN, Srivastava AK (2016) Correlation between optical and morphological properties of nanostructured gold thin film. JSM Nanotechnol Nanomed 4(1):1039

9. Sharma P, Singhal R, Vishnoi R et al (2019) Evolution of SPR in 120 MeV silver ion irradiated Cu (18%) C_{60} nanocomposites thin films. J Mater Sci Mater Electron. https://doi.org/10.1007/s10854-019-01148-9

10. Sharma P, Singhal R, Vishnoi R, Sharma GD, Kulriya P, Ojha S, Banerjee MK, Chand S (2019) Evolution of SPR in 120 MeV silver ion irradiated Cu (18%) C_{60} nanocomposites thin films. J Mater Sci Mater Electron 30(9):8301–8311

11. Singhal R, Sharma K, Vishnoi R (2018) Synthesis and characterization of Cu-C_{60} plasmonic nanocomposite. Phys B Condens Matter 550:225–234. https://doi.org/10.1016/j.physb.2018.08.027

12. Singhal R, Bhardwaj J, Vishnoi R et al (2018a) Low energy ion irradiation induced SPR of Cu-Fullerene C_{70} nanocomposite thin films. J Alloys Compd 767:733–744. https://doi.org/10.1016/j.jallcom.2018.07.108

13. Singhal R, Gupta S, Vishnoi R et al (2018b) Study on copper–fullerene nanocomposite irradiated by 120 MeV Au ions. Radiat Phys Chem 151:276–282. https://doi.org/10.1016/j.radphyschem.2018.06.048

14. Singhal R, Gupta S, Vishnoi R et al (2018c) Optical properties of Cu-C_{70} nanocomposite under low energy ion irradiation. Mater Res Express 5:035044. https://doi.org/10.1088/2053-1591/aab476

15. Singhal R, Bhardwaj J, Vishnoi R, Sharma GD (2019a) Study on Cu-fullerene C_{70} nanocomposite thin films under electronic excitations. Mater Res Express 6. https://doi.org/10.1088/2053-1591/aae440

16. Singhal R, Gupta S, Vishnoi R, Sharma GD (2019b) Synthesis and modification of Cu-C_{70} nanocomposite for plasmonic applications. Appl Surf Sci 466:615–627. https://doi.org/10.1016/j.apsusc.2018.10.029

17. Singhal R, Vishnoi R, Sharma P et al (2017) Synthesis, characterization and thermally induced structural transformation of Au_{70} nanocomposite thin films. Vacuum 142. https://doi.org/10.1016/j.vacuum.2017.05.017

18. Singhal R, Bhardwaj J, Vishnoi R, Aggarwal S, Sharma GD, Pivin JC (2018) Low energy ion irradiation studies of fullerene C_{70} thin films–An emphasis on mapping the local structure modifications. J Phys Chem Solids 117:204–214

19. Singhal R, Gupta S, Vishnoi R, Sharma GD (2019) Synthesis and modification of Cu-C_{70} nanocomposite for plasmonic applications. Appl Surf Sci 466:615–627

20. Vishnoi R, Gupta S, Dwivedi UK, Singhal R (2020) Optical and structural modifications of copper-fullerene nanocomposite thin films by 120 MeV Au ion irradiation. Radiat Phys Chem 166. https://doi.org/10.1016/j.radphyschem.2019.108442

21. Vishnoi R, Sharma K, Yogita et al (2019a) Investigation of sequential thermal annealing effect on Cu-C_{70} nanocomposite thin film. Thin Solid Films 680:75–80. https://doi.org/10.1016/j.tsf.2019.04.004

22. Vishnoi R, Singhal R, Bhardwaj J et al (2019b) Thermally induced plasmonic resonance of Cu nanoparticles in fullerene C_{70} matrix. Vacuum 159:423–429. https://doi.org/10.1016/j.vacuum.2018.10.026

23. Vishnoi R, Singhal R, Bhardwaj J, Inani H, Kumar Y, Sharma AK, Plaisier JR, Gigli L, Sharma GD (2019) Thermally induced plasmonic resonance of Cu nanoparticles in fullerene C_{70} matrix. Vacuum 159:423–429

24. Yeshchenko OA, Bondarchuk IS, Gurin VS, Dmitruk IM, Kotko AV (2013) Surf Sci 608:275–281; Yeshchenko OA, Dmitruk IM, Dmytruk AM, Alexeenko AA (2007) Influence of annealing conditions on size and optical properties of copper nanoparticles embedded in silica matrix. Mater Sci Eng B 137(1–3):247–254

Theoretical and Computational Study

Theoretical and Computational Investigations of Carbon Nanostructures

Basant Roondhe, Vaishali Sharma, and Sumit Saxena

Abstract Carbon is one of the most versatile elements in the periodic table and is known to occur in various allotropic forms. It has been widely explored since the eighteenth century and its investigation in various forms has witnessed continuous growth thereafter. The effect of these advancements has guided numerous discoveries which have not only addressed several aspects of materials physics, but also their applications. The development of theoretical and computational tools accompanied by novel characterization techniques along with the ability to synthesize these reduced dimensionalities of the carbon family like fullerene, carbon nanotubes, graphene, carbon quantum dots, etc. has significantly improved the understanding of these nanostructures. The ability of computational and theoretical techniques to predict and provide insights into the structure and properties of systems plays a crucial part in substantiating experimental findings. Theoretical and computational modeling of various carbon nanostructures such as fullerene, carbon nanotubes, graphene, and carbon quantum dots will be critically reviewed. The chapter begins with the description of the historical timeline of carbon nanostructures. How the models developed over time have led to the development of carbon nanoforms is reviewed. The impact of theoretical and computational approaches in understanding the physics of these carbon nanostructures is also highlighted.

Keywords Carbon nanostructures · Theoretical and computational modeling · Fullerene · Carbon nanotubes · Graphene

B. Roondhe · S. Saxena (✉)
Department of Metallurgical Engineering and Materials Science, Indian Institute of Technology Bombay, Mumbai 400076, Maharashtra, India
e-mail: sumit.saxena@iitb.ac.in

V. Sharma
Department of Physics, Faculty of Science, The Maharaja Sayajirao University of Baroda, Vadodara 390002, India

© The Author(s), under exclusive license to Springer Nature Singapore Pte Ltd. 2021
A. Hazra and R. Goswami (eds.), *Carbon Nanomaterial Electronics: Devices and Applications*, Advances in Sustainability Science and Technology,
https://doi.org/10.1007/978-981-16-1052-3_7

1 Introduction

Carbon is one of the essential elements in the world; in terms of abundance, it holds the sixth position of typical elements in the universe, fourth in our solar system, and about seventeenth in the Earth's crust [1]. The approximated relative abundance for carbon ranges 180–270 parts per million [2]. It is also noteworthy that the presence of carbon in human beings as an element is only subsequent to oxygen [3] and therefore acquires around 18% of human body weight. One of the remarkable characteristics of carbon is that it can occur in a broad area of metastable phases modeled near ambient environments along with their extensive kinetic stability. Despite the fact that carbon in its elemental form is relatively scarce on the earth's crust [1, 2, 4], it plays a significant role in the ecosystem of the earth. With the ongoing research toward the development of various unique forms of carbon, the current century can be rightly called "The era of carbon allotropes" [5]. Carbon nanoforms or nanostructures comprise various low-dimensional allotropes such as buckminsterfullerene or C_{60}, carbon nanotubes, graphene, poly-aromatic molecules, and carbon quantum dots. The uses of these nanostructures have been explored in different areas like nanoscience, materials science, engineering, and technology [6–12]. Recently, nanotechnology has gathered much attention because of its direct application in developing novel materials comprising significant properties like better directionality, high surface area with flexibility, etc. [13–18]. These properties uncover various applications of carbon nanomaterials design in almost all research domains [9, 19–23]. Ergo, in recent past decades, carbon science has become a trending topic along with its nanoscience discipline.

Carbon is traditionally understood to occur in only two naturally occurring allotropic configurations known as graphite and diamond. Nevertheless, the crystal structure and properties of graphite and diamond are significantly different [24–28]. Chemically, the tendency of carbon atoms to create covalent bonds with other carbon atoms leads to the formation of novel allotropes in the carbon family [29] such as buckminsterfullerene [30, 31], carbon nanotubes [32, 33], and graphene [11]. Although the existence of carbon and its applications has been known to us for centuries, the modern timeline for the development of carbon science is represented in Fig. 1.

A new chapter in the exploration of the carbon family began with the discovery of buckminsterfullerene's ("buckyballs") [30] in the mid-1980s accompanied by the discovery of fullerene nanotubules ("buckytubes") [33]. The breakthrough discovery of these nanostructures triggered increased research efforts in the exploration of carbon materials. Table 1 presents some predictions and discoveries of carbon nanostructures.

The theoretical and computational approach has made significant contributions in the field of carbon nanostructures (graphene, fullerenes, and carbon nanotubes) by offering a framework with predictive structures along with their chemical and physical properties. Computational framework in nanoscience has consistently complemented the experiments for the development of carbon nanostructures with the

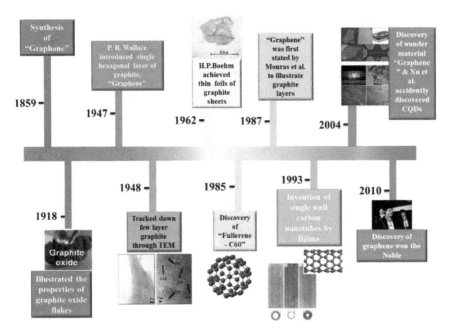

Fig. 1 Timeline of carbon nanostructures

Table 1 Timeline of predictions, discoveries, and observations of carbon nanostructures

Years	Occurrence	Observations
1966	Graphite molecules with hollow-shell were described	Assumption of molecule shape by graphite is made by Jones [34]
1970	A soccer ball-shaped C_{60} molecule is suggested	Osawa [35]
1973	Prediction of stable C_{60} was described	Huckel calculations; closed-shell electronic structure is expected by Bochvar and Galperin [36]
1980	Nanotubes were first observed	Using arc discharge method by Iijima [37]
1985	After several hypotheses, Buckminsterfullerene, C_{60} was discovered	Kroto, Smalley et al. detected C_{60} and C_{70} in the mass spectrum of laser-evaporated graphite [30]
1991	Prediction of hyperfullerenes was made	Curl and Smalley [38]
1993	Single-wall nanotubes were discovered	Using arc process in the presence of iron and cobalt catalytic particles by Iijima et al. [39] and Bethune [40]
2004	Discovery of Graphene	A monolayer graphene was developed using a Scotch tape technique [41]
2004	Discovery of carbon quantum dots	Fluorescent carbon quantum dots were derived accidentally from single-wall carbon nanotubes (SWCNTs) [42]

prediction of their properties. The theoretical approach also provides an under-standing of the reaction and separation mechanisms of carbon nanostructures. Exper-imental methodologies like X-ray diffraction and nuclear magnetic resonance are used for probing and solving the crystal structure of any material. A computa-tional approach can be used alternatively. Several methodologies were developed to deal with the problem of structure prediction. One prominent and effective model comprises investigating material's crystal structure, energy, and thereby choosing the material with the lowest energy as the "best guess" solution. In this context, various methods have been established. Random crystal structure prediction is an easy way that produces random atomic compositions with optimization to stabi-lize those compositions (inside the limits of bond lengths) [43]. While random crystal structure prediction is simplistic, unbiased, and easy to parallelize, it necessi-tates sampling various configurations to achieve better results. Another widespread approach to improve efficiency is evolutionary algorithms [44], which initially starts with a random structure and then enriches guesses with the lowest-energy results with each iteration [45]. In order to improve the results of structural prediction, different algorithms, force statistics, and data mining [46–50] are used to study criteria for crystallization such as in the Inorganic Crystal Structure Database [51]. However, the drawback of data mining methodology is that it is identified by the compounds analogous to previously observed ones, hence, lacking in novel and distinct structural phases. The recent approach for efficient crystal structure prediction involves partial experimental information to apply limitations on symmetry [52]. Every method has its own significance for different applications.

To accomplish electronic structure calculations of carbon nanostructures like fullerenes and model CNTs, many-body empirical potentials, empirical tight-binding molecular dynamics, and local density functional (LDF) means were utilized at begin-ning of the past decade [53, 54]. The Huckel approximation was used to investigate electronic structure for large I_h point group fullerenes [55]. The geometry optimiza-tions of these large fullerenes were also carried by methodologies like molecular mechanics (MM3), semi-empirical methods [56], AM1 [57], PM3 [58], and Semi-Ab Initio Model 1 (SAM1) [59]. The computational strength has also been extensively evolving due to the availability of more powerful computing resources. Consequently, theoreticians are delighted in examining and developing carbon nanostructures past molecular mechanics and semi-empirical methods. An analysis of theoretical and computational approaches utilized to explore different nanostructures of the carbon family is provided in this chapter.

2 Zero Dimensional (0D) Carbon Nanostructures

2.1 Fullerenes

Fullerenes form a hollow cage-like arrangement of carbon atoms comprising solely of hexagon and pentagon rings. Buckminsterfullerene (C_{60}) was the first in the series of developments of such carbon nanostructures [30]. Kroto, Curl, and Smalley were awarded the Nobel Prize in Chemistry in 1996 for this discovery. However, before the experimental realization of these fullerenes, they were first hypothesized by many researchers. In 1966, graphite molecules with hollow-shell were described in the scientific column "Daedalus" [34]. Subsequently, various theoretical hypotheses were made on the capability of 60 carbon atoms with truncated icosahedron [35, 36, 60, 61]. The occurrence of C_{60} was primarily predicted by Osawa in 1970 [35]. These results were later confirmed by mass synthesis of C_{60} by Krätschmer in 1990 using the carbon arc method accompanied by infrared (IR) spectroscopy for structure verification [62]. The aforementioned findings since then sparked widespread novel research for C_{60} along with other fullerene derivatives.

The study for fullerene with the early graphite laser vaporization was initiated and observed by Rohlfing et al. [63]. The carbon clusters formed in the experiments were noticeably bimaximal comprising of even and odd forms of C_n (where $n < = 25$), while only even forms in C_n (where $n > = 40$) relying upon their experimental situations. According to ab initio and various spectroscopic investigations, carbon clusters varying from $n = 2$ to 9 tend to present linear chain structures with single and triplet electronic ground states in odd and even clusters, respectively [63]. Contrary to that, some ab initio studies suggest that even number clusters in the range $n = 2$–8 show cyclic equilibrium structures with lower electronic states [64, 65]. Furthermore, C_n clusters ranging from $n = 10$ to 25 present monocyclic ground state configurations. The above-said conversion from linear chains to monocyclic rings is attributed to the fact that additional bonding associated with ring closure ultimately surpasses the strain energy acquired with the twisting of the polyyne chain to create a ring. Through semi-empirical molecular orbital theory calculations, the transformation point with 10 carbon atoms is predicted [66, 67]. However, according to the intensities in the high-mass region, these carbon nanostructures were indecisive and needed plausible explanations [68]. The photoionization time-of-flight mass spectrum (PI-TOF-MS) of these carbon nanostructures ranging from 1 to 100 atoms is presented in Fig. 2.

Evidently, these elucidations must justify the detail that ion signals of even C_n were observed in the high-mass region. Subsequently, this instantly eliminates a variety of probable configurations for C_n clusters, for instance, fractions of diamond lattice/graphite sheet. These structures tend to present both even and odd peaks of mass by means of linear chains and monocyclic rings. Moreover, it does not exclude the possibilities of other configurations like "carbyne" [63]. The second probable reason consistent with this analysis would be that the second sets of high-mass region carbon clusters are all fullerenes (Fig. 2). This is the well-known fullerene hypothesis and has gathered much attention for the reason that closed cages bypass

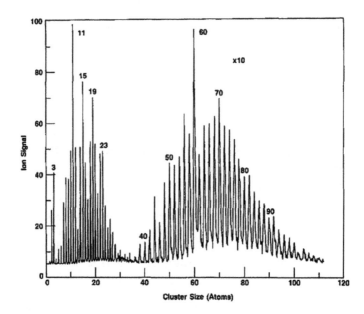

Fig. 2 PI-TOF-MS spectrum (involving the amalgamation of two different spectra) for carbon clusters attained through doubled Nd:YAG vaporizing laser energy (40 mJ) and unfocused ArF ionizing laser energy (1.6 mJ and 193 run). The vertical deflection plate voltage of 300 V is utilized for C_n^+, $1 < n < 30$, leading to the optimization of C_{20}^+ collection while 600 V was utilized for C_{2n}^+, $20 < n < 50$, for the optimization of C_{100}^+. Reproduced with permission from Rohlfing et al. J. Chem. Phys 81, 3322 (1984). Copyright 1984 AIP Publishing

the dangling bonds of edges that are anticipated to destabilize fractions of diamond and graphite lattices [30] and additionally, due to the fact that trivalent cages fulfill the valence necessities of carbon atoms compared to linear chains and monocyclic rings. These qualitative theoretical aspects of the fullerene hypothesis along with electronic structure calculations provided support to the experimentation of C_{60} in 1985.

One of the important investigations performed was the comparison of carbon cages with chains, rings, and toroids along with fractions of infinite diamond and graphite lattices using semi-empirical models [69]. The analysis suggested that cage structures with atoms greater than 25 would be the most stable carbon clusters. Furthermore, the existence of solely pentagonal and hexagonal rings along with the unavailability of adjacent pentagonal rings were conditions for the stability of cage structures [69]. The affinity of fullerenes comprising limited adjacent pentagonal rings was also addressed by Kroto in 1987 using empirical arguments derived from chemical and geodesic rules [70]. The structures studied by Kroto in 1987 are presented in Fig. 3. It is noteworthy that both the aforementioned studies suggested that C_{60} was the smallest fullerene without adjoining pentagonal rings consists of D_{5h} isomer of C_{70}.

Kroto's inference of C_{60} with I_h symmetry as foundational fullerene was supported by Krätschmer et al. [62] in 1990 through four-band IR absorption spectrum and

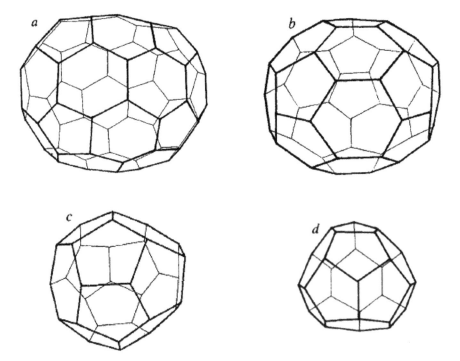

Fig. 3 Structures of fullerenes by Kroto et al. **a** C_{70}, most stable fullerene created by splitting two halves of C_{60} through 10 extra carbon rings, **b** C_{50}, comprising isolated singlet and doublet pentagonal structures, **c** structure of C_{32} with threefold axis, and **d** C_{28} which is a tetrahedral fullerene. Reproduced with permission from Kroto, Nature 329, 529 (1987). Copyright 1987 Springer Nature

latterly in the same year by Taylor et al. [71] through ^{13}C nuclear magnetic resonance (NMR) spectroscopy. Thereafter, several other configurations of fullerenes were synthesized including C_{76} [72], C_{78} [73, 74], and C_{84} [74, 75].

Various configurations of fullerenes are shown in Fig. 4. Each fullerene molecule shows the features of a carbon cage, as each atom is bonded to the other three carbon atoms in the same manner as in graphite [73]. The extensive series of techniques to synthesize fullerenes observed that C_{60} is the most plenteous among fullerenes accompanied by C_{70} [76]. C_{60} with I_h symmetry comprises two C–C bonds with (i) one at the link of two hexagonal rings denoted and (ii) one at the link of pentagonal and hexagonal rings. Contrarily, C_{70} with D_{5h} symmetry consists of eight C–C bonds. It is noteworthy that two pentagonal rings sharing similar C–C bonds are energetically unfavorable. Mathematically, 1812 methods are known to build isomers of 60 carbon atoms, while C_{60} holds its uniqueness and special place with stability due to the fact that all of its pentagonal rings are secluded by its hexagonal rings. The state is known as the "isolated pentagon rule" (IPR) [77]. C_{60} being the smallest member of fullerene family obeying the IPR, C_{62}, C_{64}, C_{66}, and C_{68} fullerenes does not follow the IPR.

Figure 5 presents that the number of IPR isomers is directly proportional to the

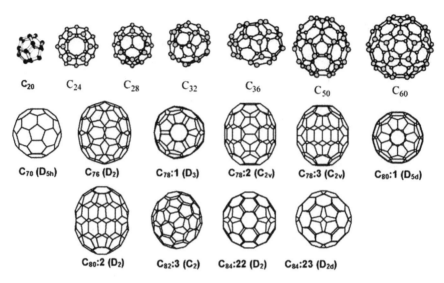

Fig. 4 Structures of fullerenes along with their symmetries. Reprinted with permission from Ref. [73]. Reproduced with permission from Yan et al. Nanoscale 8, 4799 (2016). Copyright 2016 Author(s), licensed under the Creative Commons Attribution 3.0 Unported License

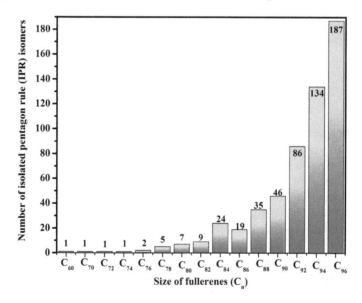

Fig. 5 Size of fullerenes with respect to the number of isolated pentagon rule (IPR) isomers (the details of isomers were taken from Ref. [77])

size of fullerenes. The studies of IPR with possible isomers of fullerenes assisted the experimentalists to identify and characterize them [78–80]. For instance, in C_{78} (consisting five isomers), isomers with C_{2v} and D_3 symmetry were identified using ^{13}CNMR spectra [74]. Theoretical investigations of C_{82} lead to experimental characterization of its three isomers having C_2 symmetry also using ^{13}C NMR spectra [74, 81]. Additionally, several computational investigations were performed since the discovery of fullerenes to thoroughly study their isomers and subsequently to predict the lowest-energy configurations of giant fullerenes [82–89]. Becke, 3-parameter, Lee–Yang–Parr(B3LYP)hybrid functional along with various basis sets were used to examine C_{86} along with its 19 isomers following IPR [87]. Their studies suggested that isomer 17 (C_2 symmetry of C_{86}) is the most stable among them followed by isomer 16 (C_s symmetry of C_{86}). Similarly, several theoretical calculations played a crucial role in predicting accurate lowest-energy structures of the fullerene family [83, 87].

Several theoretical and computational studies in the last decades have been dedicated to exploring C_{60} along with its chemical and physical properties. Theoretical investigations by Fowler and Steer [90] suggested that C_n (n = 60 + 6 k, k = an integer except one) should comprise closed-shell electronic structures. Schmalz et al. showed that the aromaticity of C_{60} is less than that of benzene [69] through resonance circuit theory and Huckel molecular orbital (HMO) theory. The stability occurring through bond delocalization was explained by Amic and Trinajstic [91]. The electronic and vibrational properties of C_{60} were evaluated through the two-dimensional HMO method [92]. Semi-empirical calculations involving overlapping of non-planar π-orbital were also given by the free-electron model in the Coulson–Golubiewski, self-consistent Huckel approximation for the curvature system [93]. The large-scale restricted Hartree–Fock calculations were carried out presenting electron affinity of 0.8 eV and ionization potential to be 7.92 eV with $\Delta H_f = 415$–490 kcal/mol [94–96]. On the basis of ab initio self-consistent field (SCF) theory, the heat of formation was also evaluated by Schulman and Disch [97]. To measure structural parameters, electronic spectra, and oscillator strength, the Pariser–Parr–Pople method and the CNDO/S method (with CI) were used by many researchers [98–101]. The ground and excited states of C_{60} presenting π-bonding character were determined by the tight-binding model using electron–phonon coupling [102]. The primarily vibrational properties of C_{60} were investigated by Newton and Stanton using MNDO theory [103]. It was observed that C_{60} contains four IR active modes because of its high symmetry ("t_{1u}" symmetry) and 10 Raman active modes involving eight "h_g" and two "a_g" symmetries. The 174 vibrational modes of C_{60} contribute to 42 elementary modes with different symmetries. Proceeding to understand magnetic properties of C_{60}, by means of HMO and London theories, the ring current magnetic susceptibility was evaluated with less than 1 ppm shielding because of the termination of the contribution of both diamagnetic and paramagnetic spins [104, 105]. The theory also presented the absence of usual aromatic behavior [104, 105]. Some investigations proposed that the diamagnetic part has been underestimated [106]. Fowler et al. (using coupled Hartree–Fock calculations) in their study proposed that the aforementioned shielding has to be approximately similar as for analogous aromatic structures

[106]. Later on, Haddon and Elser addressed the shielding of fullerenes [104, 105, 107] and reinterpreted the study done by Fowler et al. [106], concluding that their study is inconsistent with the results of small delocalized susceptibility. The chemical shift observed in NMR analysis of C_{60} done by Taylor et al. indicated the presence of aromatic systems; these were confirmed by Fowler and group subsequently [71].

Several theories and computational studies have also been dedicated to exploring doping, defects, functionalization, etc. in fullerenes for their possible applications in antiviral activity, DNA cleavage, photodynamic electron transfer, lightweight batteries, lubricants, nanoscale electrical switches, cancer therapies, and astrophysics [109, 110].

2.2 Carbon Quantum Dots

Carbon quantum dots or carbon dots are relatively newer members among the carbon nanostructure family. These are quasi-spherical nanoparticles involving sp^2/sp^3 amorphous or nanocrystalline forms having size generally <10 nm carrying oxygen/nitrogen groups [111, 112]. Surprisingly, carbon dots were discovered unintentionally in 2004 in an experimental study of carbon nanotubes through electrophoretic fractionation of arc-discharge soot [42]. Carbon dots have gained much attention due to the fact that they possess strong fluorescence with better solubility, biocompatibility, and non-toxicity [113]. However, these fluorescent carbon nanostructures gained significant attention due to improved fluorescence emissions through the surface passivation synthesis approach [114]. The carbon quantum dots along with their STEM and absorption spectra are shown in Fig. 6.

Experimental and theoretical investigations have been used to understand the chemical and physical properties of carbon quantum dots for their applications in various fields like sensing, bio-imaging, nano-medicine, catalysis, optoelectronics, and energy conversion/storage. However, there are considerably rare theoretical studies on carbon quantum dots, and many of them are based on the graphene nanoflakes model [115–120].

Analogous to other quantum dots, the emission of carbon quantum dots is associated to their respective sizes. Carbon quantum dot size <1.2 nm showed UV light emission [121], visible light emissions were reported for quantum dots with size from 1.5 to 3 nm while near-infrared emissions were observed for quantum dots with sizes ~3.8 nm [122]. These observations have also been supported using theoretical calculations. The observation of indirect dependence of the HOMO–LUMO gaps on the size of the carbon quantum dots lead to the conclusion that strong emission of carbon quantum dots is a result of its quantum size rather than carbon–oxygen surface [123]. The photoluminescence mechanism, electronic structures, and frontier molecular orbitals of carbon quantum dots have also been studied using time-dependent density functional theory (TD-DFT) as implemented in Gaussian 09 with B3LYP hybrid functional and the 6-31G(d) basis set [124]. The carbon quantum dots were categorized in two forms: class I representing graphitized carbon core and class II

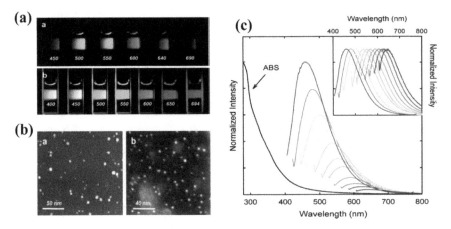

Fig. 6 a Carbon dots attached with PEG1500N in aqueous solution. **b** STEM images of carbon dots. **c** The absorption (ABS) and luminescence emission spectra of carbon dots in an aqueous medium; the graph is plotted with 20 nm increment from longer excitation wavelengths 400 nm on the left and the intensities of emission spectral are normalized to quantum yields (inset is the normalized spectral peaks). Reproduced with permission from Sun et al., J. Am. Chem. Soc. 128, 7756 (2006). Copyright 2006 American Chemical Society

representing disordered carbon core. These classes are depicted in Fig. 7 along with their photoluminescence mechanism.

The study showed that the HOMO–LUMO gap decreases with an increase in the size of class I carbon quantum dots while an opposite trend on the size-dependency of the HOMO–LUMO gap is observed for class II carbon quantum dots. Several studies related to the electronic structure of carbon quantum dots have been explained using molecular orbital (MO) theory [121, 123, 125–127]. In the majority of these reports, carbon quantum dots show n → π* and π → π* transitions because of their well-available transition energies. The π-states of carbon quantum dots are attributed to the sp^2 hybridized carbon in their core, while the n-states are attributed to the functional groups attached. It is found that the energy gap (E$_g$) among π-states reduces consistently with the increase in the number of aromatic rings of carbon quantum dots similar to organic molecules [121, 123]. The electronic properties of amorphous carbon nanodots were explored using semi-empirical molecular–orbital theory using the EMPIRE13 code [128]. Unexpectedly, electronic structures were found to rely weakly on parameters like elemental composition and atomic hybridization. Contrarily, the geometry of sp^2 arrangement describes the band gap of carbon quantum dots. The existence of localized electronic surface states resulting in amphoteric reactivity and near-UV/visible range optical band gaps was predicted [128]. The molecular orbitals, molecular electrostatic potential (MEP), local electron affinity (EA$_L$), and ionization energy (IE$_L$) maps along with excitation energies are depicted in Fig. 8. There have been fewer theoretical studies to understand their optical and electronic mechanisms and in-depth theoretical studies are further expected.

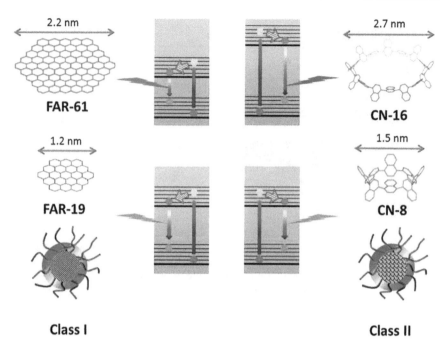

Fig. 7 Photoluminescence (PL) mechanism of class I and class II carbon quantum dots. The number of hexagonal rings is indicated after fused aromatic rings (FARs) and the number of repeating units of cyclo-1,4-naphthylene (CN) is indicated by a number. Reproduced with permission from Zhu et al., J. Mater. Chem. C 1, 580 (2013). Copyright 2013 Royal Society of Chemistry

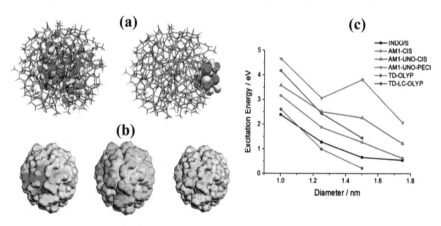

Fig. 8 A 2 nm carbon dot with **a** molecular orbitals; left side presents band-like and right side presents surface states with iso-density surfaces of 0.01 e$^-$ Å^{-3}. **b** Electron iso-density surface maps, MEP (left part) from −50 (blue) to 50 kcal mol^{-1} (red), EA$_L$ (middle part) from −150 (blue) to 5 kcal mol^{-1} (red) and IE$_L$ (right part) from 270 (blue) to 500 kcal mol^{-1} (red). **c** Excitation energies calculated with different methods for different sized carbon dots. Reproduced with permission from Margraf et al., J. Phys. Chem. B 119, 24, 7258 (2015). Copyright 2015 American Chemical Society

3 One Dimensional (1D) Carbon Nanostructures

The first-ever proof for the existence of one-dimensional carbon allotrope was reported in 1993 [40]. Single-walled carbon nanotubes (SWCNT) discovered by Iijima and Bethune lead all scientists for a hunt to utilize this a new form of carbon in many applications for technological advancement like field emission displays, energy storage and energy conversion devices, sensors, hydrogen storage, and semiconductor devices [129–134].

CNT is one of the exceptional inventions which has enriched the field of nanotechnology. It has been consistently studied since the past 20 years due to its potential application in varied areas. The fullerenes discovered by Kroto et al. [30] were the building blocks of the CNTs. CNTs have a variety of physical properties such as stiffness, elasticity, deformation, and tensile strength along with electronic properties showing superconducting, metallic, semiconducting, or insulating behavior.

The discovery of CNTs was reported as a "worm-like" structure long before this tubular form of carbon could be imagined, in 1952 by Radushkevich and Lukyanovich [135]. Dimensionally, SWCNTs are around 1 nm in diameter while their length is in order of a few micrometres. Nevertheless, the size and the shape of nanotubes can vary. The ratio of the diameter and length of the nanotubes, also known as aspect ratio, is typically around 1000 due to which it is generally considered nearly as a one-dimensional structure [136].

The different types of CNTs depend on the number of carbon layers present in them. Monolayered tubes are called single-walled carbon nanotube (SWCNT), while tubes having more than one layer are known as multi-walled carbon nanotubes (MWCNTs). The SWCNTs are generally understood to form by rolling a graphene sheet. Density functional theory calculations have shown the possibility of forming CNTs from bilayer graphene nanoribbons under different pressure conditions depending on the edges of nanoribbons involved [137]. The CNTs are classified into three different types: armchair, zigzag (see Fig. 9), and chiral carbon nanotubes (see Fig. 10). These are formed by rolling graphene sheets along a different axis. The axis of rolling is the chiral vector which is represented by n and m pair (n, m) of indices corresponding to the unit vectors along different directions in the graphene honeycomb crystal lattice sheet. When m = 1, 2,... and n = 0, the nanotube is "zigzag" and if m = n, the nanotube is then termed as "armchair" while the remaining configuration iscalled chiral [136, 138, 139]. Due to the rolling of the sheet into a tube, the symmetry of the plane breaks and forms a new symmetry in a distinct direction of the hexagonal lattice and the axial direction. This develops a peculiar electronic behavior of the nanotube, which is metallic or semiconducting. In the case of the semiconducting tube, its bandgap is sensitive toward its diameter; the small diameter tube has a large band gap while the wide diameter consists of a lower band gap [140]. The diameter of the nanotube thus makes it a conductor with conductivity higher than copper as well as a semiconductor comparable to the potential of silicon. In the structure of a nanotube, every carbon atom is bonded covalently with three nearby carbon atoms with its sp^2 molecular orbital, creating one (the fourth) valence electron free in

(a)

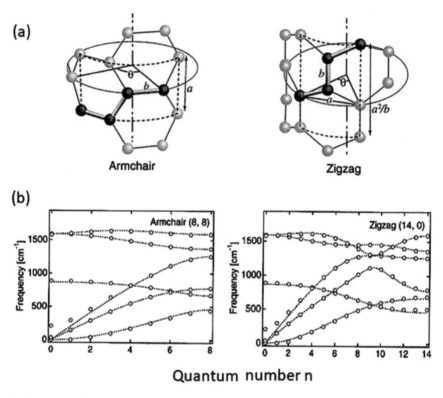

(b)

Fig. 9 **a** Unit cell for two different carbon nanotubes (armchair and zigzag) depicting the primitive azimuthal angle θ (=2π/N). **b** Phonon dispersion curve for armchair and zigzag SWCNT. Reproduced with permission from Maeda et al., Physica B 263–264, 479 (1999). Copyright 1999 Elsevier

every hexagonal unit, which is delocalized over all atoms providing the nanotube its electrical nature. Some CNTs which show metallic nature have the resistivity in the range of 0.34×10^{-4} to 1.0×10^{-4} Ω/cm [141]. The semiconducting CNTs generally show p-type semiconducting behavior [142]. The SWCNTs can also be described as quantum wires due to their ballistic electron transport, while the electronic transport in MWCNTs is quasi-ballistic [143]. Apart from the well-known electronic properties of CNTs, they show equally good mechanical properties as well. The sp^2 carbon–carbon bonds present in the CNTs result in exceptional mechanical properties which were not observed in previously explored material systems. From some previous studies, we get an idea about the stiffness of CNTs, basically in their axial direction [144]. Among all carbon materials, CNTs show extremely high value for Young's modulus (~1TPa) which is even five times higher than steel, and provides a measure of the stiffness of the material [145, 146]. All the studies regarding the mechanical properties of CNTs were first predicted theoretically [53, 147–149]. The transformation from the hexagonal ring of carbon to pentagon–heptagon in CNTs

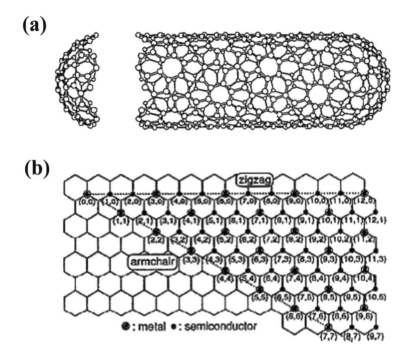

Fig. 10 a Icosahedral C_{140} fullerene-based hemispherical cap covered end chiral fiber with chiral vector $C_h = (10, 5)$. **b** Different probable vectors for the construction of chiral fibers. The two different combinations of circled dots and dots denote the metallic and semiconducting behavior for corresponding chiral fiber constructed. Reproduced with permission from Saito et al. Appl. Phys. Lett. 60, 2204 (1992). Copyright 1992 AIP Publishing

was proposed by Yakobson [150] and Ru [151] when uniaxial tension is applied. DFT calculations suggest that SWCNTs form novel quasi-two-dimensional sheets when subjected to high pressure [152]. In a theoretical study done by Guanghua et al. [153] on the CNTs' mechanical properties, their nature of dependence on diameter is revealed. They found Young's modulus in the range of 0.6–0.7 TPa for nanotubes with diameter >1 nm. The closest agreement with the experimental value of Young's modulus of MWCNTs (1–1.2 TPa) was theoretically calculated by Hernandez et al. [154]. In this study, they also predicted that mechanical properties depend on the diameter of the tube; when the diameter increases, the properties are also enhanced to a certain value and ultimately reach the values corresponding to that of graphene. Calculated values of Young's modulus for individual SWNTs were found in the range from 320 to 1470 GPa [144, 155] while the breaking strength ranged from 13 to 52 GPa [156]. The vibrational properties of CNTs are studied by the normal mode analysis as this technique is standard to understand the dynamics of nanotubes. This technique investigates the harmonic potential analytically for normal mode analysis. The linear combination of Cartesian co-ordinates provides the co-ordinates for normal mode. This method provides a natural description of molecular vibration as it includes the motion of all atoms simultaneously during the vibration.

Apart from the small size, CNTs show quantum effects leading to the low-temperature specific heat and thermal conductivity; CNTs are also of great importance for their thermal properties [149, 157, 158]. The thermal conductivity can be modulated and increased by incorporating different materials with pristine CNTs. The thermal conductivity measured at room temperature for MWCNTs was found to be 3,000 W/K [159], while in a similar study the MWCNTs were found to have thermal conductivities ~200 W/mK higher as compared to the SWCNTs [160]. The main factor which influences the thermal properties is the number of active phonon modes along with a free path of phonon and boundary surface scattering [160–162]. Properties of CNTs are observed to depend on the atomic arrangement, length and diameter of tubes, structural defects, and impurities [163–165].

4 Two-Dimensional (2D) Carbon Nanostructures: Graphene

Graphene is a single atom layer of carbon atoms arranged in a hexagonal honeycomb pattern. It is one of the most studied two-dimensional (2D) materials to date. Figure 11 illustrates a graphene sheet as a 2D building block for different carbon materials in all dimensions such as 0D buckyballs by wrapping up the graphene sheet, 1D nanotube by rolling it, and in 3D graphite by stacking it. Thus, it is known as the mother of

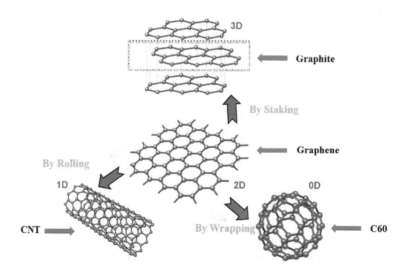

Fig. 11 Graphene sheet is a 2D building block for different carbon materials in all dimensions like 0D buckyballs which is formed by wrapping of graphene sheet, 1D nanotube can be made by rolling it and 3D graphite is formed by stacking it, therefore it is known as the mother of all graphitic forms of carbon material

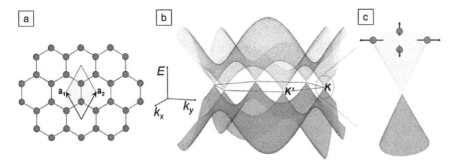

Fig. 12 **a** Unit cell of graphene with the triangular Bravais lattice having lattice vectors a_1 and a_2; unit cell comprises two atoms in the honeycomb lattice. **b** Band structure of graphene calculated by the tight-binding method displaying the pi bands, with only nearest neighbor hopping. The inset E, k_x, and k_y are the energy and the wave vector components in x- and y- directions, respectively. **c** The unique linear dispersion of the band structure near K point with the pseudo-spin vector direction indicated by the arrows. Reproduced with permission from Fuhrer et al. MRS Bulletin 35, 289 (2010). Copyright 2010 Cambridge University Press

all graphitic forms of carbon material. The research has exponentially developed after 2004 when Geim and Novoselov isolated graphene for the first time using the "Scotch Tape" method and characterized it. In the current scenario of the material world, ongoing research is overwhelmed after focusing on characterization, mass production of ultra-thin carbon films including graphene for various applications [166–172].

A unit cell of graphene with the Bravais lattice along with the band structure is shown in Fig. 12. The unique linear dispersion of the band structure near the K point is illustrated by a pseudo-spin direction which is indicated by the arrows. From the past one and half decades, promising applications in the field of corrosion prevention [173], super capacitors [174, 175], long-lasting batteries [176], display panels [177], efficient solar cells [178], desalination [179], and water purification [180–182] have emerged.

The electronic properties of single-layer graphite were investigated by Wallace even before its isolation [183] and introduced the term "graphene" back in 1947. The electronic band structure was investigated theoretically by the tight-binding (TB) approach. The TB approach is more suitable for handling larger systems than the plane waves method, due to its low computational costs. The method was at first described as an interpolation scheme by Slater and Koster [184]. It has been developed comprehensively and now it is a well-established technique to explain the electronic structure of solids [185].

The tight-binding (TB) calculations were performed using the Hamiltonian

$$H = \sum_{il_1\sigma} \epsilon_{l_1} a^\dagger_{il_1\sigma} a_{il_1\sigma} + \sum_{ij} \sum_{l_1,l_2,\sigma} \left(t^{l_1l_2}_{ij} a^\dagger_{il_1\sigma} a_{jl_2\sigma} + H.c. \right) \tag{1}$$

here, the spin σ of the electron is capable of jumping from the orbital l_1 with its onsite energies (ϵ_{l_1}) existing in the ith unit cell to orbital l_2 in the jth unit cell. The hopping interaction strength labeled as $t_{ij}^{l_1 l_2}$ relies on the nature of the orbitals participating as well as on the lattice geometry [184]. Following that, a least-squared error fitting is executed through the alteration of the ϵ's and t's, leading to the calculation of band dispersions at various high-symmetry points. Graphite layer shows semiconducting behavior with zero activation energy at zero temperature, but at higher temperatures due to excitation, the highest bands are filled and show metallic nature. Large anisotropic diamagnetic susceptibility which is greatest across the layers is observed. The study done by Boehm in 1962 provided the concept of single-layer graphite sheet through the reduction of graphite oxide (GO) in dilute sodium hydroxide and also by deflagration of heated GO [186]. To describe the atom intercalation in graphite, effective-mass-approximation differential equations were used at that time for self-consistent screening. At room temperature, graphene displays a strong ambipolar electric field effect between the valence and the conduction bands. This results in ballistic electron transfer at a speed which is slower than light speed and 10–100 times greater than that in silicon chips. Graphene is the thinnest material and is 200 times stronger than steel and harder than diamond but at the same time, it is flexible and transparent [10, 187, 188].

Investigation of the physical properties of graphene reveals that it has a tremendously high optical transparency of up to 97.7%, which makes it a potential material for transparent electrodes for its use in solar cell applications [189]. It also consists of high thermal conductivity of 5000 $Wm^{-1} K^{-1}$ [190], and exceptional mechanical properties like high Young's modulus of 1 TPa [191], and most importantly large specific surface area of 2630 $m^2 g^{-1}$ [192]. Still, there is a need to find a method for the utilization of graphene in many applications and also to guarantee cost-effective production by avoiding some major obstacles. It is a great need to develop a method with the help of which ideally flat graphene membrane without any defects can be achieved. The need to fill the large gap between the theoretical prediction and actual fabrication of graphene is essential. Irreversible agglomerates and the restacking are a key challenge in the synthesis of graphene which need to be addressed.

5 Summary and Outlook

Nanomaterials provide exotic properties, exclusive of the framework of their periodic solid counterparts. Additionally, novel phenomena emerge at the nanoscale level that is not observed in microcrystalline materials. Among all, carbon nanostructures like three dimension (3D—Graphite, diamond), two dimension (2D—graphene), one dimension (1D—carbon nanotubes), and zero dimension (0D—fullerenes and carbon quantum dots) have gained significant attention due to their unique properties. The discovery of C_{60} and carbon quantum dots (0D), CNTs (1D), and graphene (2D) has led to the increased research activity in novel multidisciplinary areas, from synthesis to their theoretical and computational investigations for potential applications. In the present chapter, the theoretical and computational development of carbon nanostructures, specifically on fullerenes, carbon quantum dots, carbon nanotubes, and graphene have been introduced and discussed. The underlying mechanism of size dependency of these carbon cage structures (fullerenes and carbon nanotubes) is essential for modifying their properties according to the potential nanotechnology applications. Computational and theoretical studies have found significant role in predicting and designing their properties accordingly. By the means of powerful supercomputers, performing static and dynamic calculations at high-level ab initio and DFT methodologies is achievable for these carbon nanostructures. Still, the application of futuristic quantum chemical approaches to investigate the structures and properties of large carbon nanostructures (fullerenes, carbon quantum dots, graphene, and CNTs) is a daunting task. The theory of isolated pentagon rule (IRP) in fullerenes chemistry has been discussed. The knowledge on the computational and theoretical aspects of accidentally discovered carbon quantum dots were explored which is still in its developing stage. The theoretical prediction of carbon nanotubes (armchair, zigzag, and chiral) and graphene before their experimental realization is provided. Obtaining insight of the electronic structures along with their chemical and physical properties is still needed for constructing new materials based on carbon-based nanostructures for certain applications. The synergy among theoreticians and experimentalists will expand the applications of carbon nanostructures promptly.

References

1. Zhang Y, Yin Q-Z (2012) Carbon and other light element contents in the Earth's core based on first-principles molecular dynamics. Proc Natl Acad Sci USA 109:19579–19583
2. Allègre CJ, Poirier J-P, Humler E et al (1995) The chemical composition of the Earth. Earth Planet Sci Lett 134:515–526
3. Pace NR (2001) The universal nature of biochemistry. Proc Natl Acad Sci USA 98:805–808
4. Marty B, Alexander CMO, Raymond SN (2013) Primordial origins of Earth's carbon. Rev Mineral Geochem 75:149–181
5. Hirsch A (2010) The era of carbon allotropes. Nat Mater 9:868–871
6. Titirici M-M, White RJ, Brun N et al (2015) Sustainable carbon materials. Chem Soc Rev 44:250–290
7. Loos M (2015) Allotropes of carbon and carbon nanotubes. Elsevier, Amsterdam, The Netherlands
8. Deng J, You Y, Sahajwalla V et al (2016) Transforming waste into carbon-based nanomaterials. Carbon 96:105–115
9. Rodríguez-Reinoso F (1998) The role of carbon materials in heterogeneous catalysis. Carbon 36:159–175
10. Allen MJ, Tung VC, Kaner RB (2010) Honeycomb carbon: a review of graphene. Chem Rev 110:132–145
11. Geim AK, Novoselov KS (2007) The rise of graphene. Nat Mater 6:183–191
12. Novoselov KS, Fal VI, Colombo L et al (2012) A roadmap for graphene. Nature 490:192–200
13. Stankovich S, Dikin DA, Piner RD et al (2007) Synthesis of graphene-based nanosheets via chemical reduction of exfoliated graphite oxide. Carbon 45:1558–1565

14. Zhu Y, Murali S, Stoller MD et al (2011) Carbon-based supercapacitors produced by activation of graphene. Science 332:1537–1541
15. Gadipelli S, Guo ZX (2015) Graphene-based materials: synthesis and gas sorption, storage and separation. Prog Mater Sci 69:1–60
16. Bonaccorso F, Colombo L, Yu G et al (2015) Graphene, related two-dimensional crystals, and hybrid systems for energy conversion and storage. Science 347:1246501–1246509
17. Sun M-J, Cao X, Cao Z (2016) Si(C≡C)4-based single-crystalline semiconductor: diamond-like superlight and super flexible wide-bandgap material for the UV photoconductive device. ACS Appl Mater Interfaces 8:16551–16554
18. Chen Y, Fu K, Zhu S et al (2016) Reduced graphene oxide films with ultrahigh conductivity as Li-Ion battery current collectors. Nano Lett 16:3616–3623
19. Georgakilas V, Tiwari JN, Kemp KC et al (2016) Noncovalent functionalization of graphene and graphene oxide for energy materials, biosensing, catalytic, and biomedical applications. Chem Rev 116:5464–5519
20. Liu J, Cui L, Losic D (2013) Graphene and graphene oxide as new nanocarriers for drug delivery applications. Acta Biomater 9:9243–9257
21. Khadiran T, Hussein MZ, Zainal Z et al (2015) Activated carbon derived from peat soil as a framework for the preparation of shape-stabilized phase change material. Energy 82:468–478
22. Wu Y, Lin Y, Bol AA et al (2011) High-frequency, scaled graphene transistors on diamond-like carbon. Nature 472:74–78
23. Deng J, Li M, Wang Y (2016) Biomass-derived carbon: synthesis and applications in energy storage and conversion. Green Chem 18:4824–4854
24. Ferrari A, Robertson J (2000) Interpretation of Raman spectra of disordered and amorphous carbon. Phys Rev B 61:14095–14107
25. Wei L, Kuo PK, Thomas RL et al (1993) Thermal conductivity of isotopically modified single crystal diamond. Phys Rev Lett 70:3764–3767
26. Titirici M (2013) Sustainable carbon materials from hydrothermal processes. Wiley, Chichester, UK
27. Dai L, Chang DW, Baek J-B et al (2012) Carbon nanomaterials for advanced energy conversion and storage. Small 8:1130–1166
28. Kaneko K, Ishii C, Ruike M et al (1992) Origin of superhigh surface area and microcrystalline graphitic structures of activated carbons. Carbon 30:1075–1088
29. Pang J, Bachmatiuk A, Ibrahim I et al (2016) CVD growth of 1D and 2D sp2 carbon nanomaterials. J Mater Sci 51:640–667
30. Kroto HW, Heath JR, O'Brien SC et al (1985) C60: Buckminsterfullerene. Nature 318:162–163
31. Smalley RE (1991) Great balls of carbon: the Story of Buckminsterfullerene. The Sci 31:22–28
32. Iijima S (2002) Carbon nanotubes: past, present, and future. Phys B 323:1–5
33. Iijima S (1991) Helical microtubules of graphitic carbon. Nature 354:56–58
34. Jones DEH (1966) Hollow molecules. New Sci 32:245
35. Osawa E (1970) Superaromaticity. Kagaku (Kyoto) 25:854–863
36. Bochvar DA, Galperin EG (1973) Hypothetical systems-carbododecahedron, s-icosahedrone and carbo-s-icosahedron. Proc Acad Sci USSR 209:610–612
37. Iijima S (1980) High resolution electron microscopy of some carbonaceous materials. J Microscopy 119:99–111
38. Curl RF, Smalley RE (1991) Fullerenes. Sci Am 265:54–63
39. Iijima S, Ichihashi T (1993) Single-shell carbon nanotubes of 1-nm diameter. Nature 363:603–605
40. Bethune DS, Kiang CH, DeVries MS et al (1993) Cobalt-catalysed growth of carbon nanotubes with single-atomic-layer walls. Nature 363:605–607
41. Novoselov KS, Geim AK, Morozov SV et al (2004) Electric field effect in atomically thin carbon films. Science 306:666–669
42. Xu X, Ray R, Gu Y et al (2004) Electrophoretic analysis and purification of fluorescent single-walled carbon nanotube fragments. J Am Chem Soc 126:12736–12737

43. Pickard CJ, Needs RJ (2011) Ab initio random structure searching. J Phys Condens Matter 23:053201–053223
44. Oganov AR, Glass CW (2006) Crystal structure prediction using *ab initio* evolutionary techniques: principles and applications. J Chem Phys 124:244704–244715
45. Oganov AR, Valle M (2009) How to quantify energy landscapes of solids. J Chem Phys 130:104504–104509
46. Hautier G, Fischer C, Ehrlacher V et al (2011) Data mined ionic substitutions for the discovery of new compounds. Inorg Chem 50:656–663
47. Curtarolo S, Morgan D, Persson K et al (2003) Predicting crystal structures with data mining of quantum calculations. Phys Rev Lett 91:135503–135506
48. Fischer CC, Tibbetts KJ, Morgan D et al (2006) Predicting crystal structure by merging data mining with quantum mechanics. Nat Mater 5:641–646
49. Hautier G, Fischer CC, Jain A et al (2010) Finding nature's missing ternary oxide compounds using machine learning and density functional theory. Chem Mater 22:3762–3767
50. Meredig B, Agrawal A, Kirklin S et al (2014) Combinatorial screening for new materials in unconstrained composition space with machine learning. Phys Rev B 89:094104–094110
51. Bergerhoff G, Hundt R, Sievers R et al (1983) The inorganic crystal structure data base. J Chem Inf Comput Sci 23:66–69
52. Meredig B, Wolverton C (2013) A hybrid computational–experimental approach for automated crystal structure solution. Nat Mater 12:123–127
53. Robertson DH, Brenner DW, Mintmire JW (1992) Energetics of nanoscale graphitic tubules. Phys Rev B 45:12592–12595
54. Zhang BL, Wang CZ, Ho KM et al (1993) The geometry of large fullerene cages: C_{72} to C_{102}. J Chem Phys 98:3095–3102
55. Tang AC, Huang FQ (1995) Electronic structures of giant fullerenes with I_h symmetry. Phys Rev B 51:13830–13832
56. Dewar MJS, Thiel W (1977) Ground states of molecules. 38. The MNDO method. Approximations and parameters. J Am Chem Soc 99:4899–4907
57. Dewar MJS, Zoebisch EG, Healy EF et al (1985) J Am Chem Soc 107:3902–3909
58. Stewart JJP (1989) Optimization of parameters for semiempirical methods I. Method J Comput Chem 10:209–220
59. Dewar MJS, Jie C, Yu J (1993) SAM1; The first of a new series of general purpose quantum mechanical molecular models. Tetrahedron 49:5003–5038
60. Davidson RA (1981) Spectral analysis of graphs by cyclic automorphism subgroups. Theor Chim Acta 58:193–231
61. Schultz HP (1965) Topological organic chemistry. Polyhedranes and Prismanes. J Org Chem 30:1361–1364
62. Krätschmer W, Lamb LD, Fostiropoulos K et al (1990) Solid C60: a new form of carbon. Nature 347:354–358
63. Rohlfing EA, Cox DM, Kaldor A (1984) Production and characterization of supersonic carbon cluster beams. J Chem Phys 81:3322–3330
64. Raghavachari K, Binkley JS (1987) Structure, stability, and fragmentation of small carbon clusters. J Chem Phys 87:2191–2197
65. Parasuk V, Almlof J (1989) The electronic and molecular structure of C6: complete active space self-consistent-field and multireference configuration interaction. J Chem Phys 91:1137–1141
66. Pitzer KS, Clementi E (1959) Large molecules in carbon vapor. J Am Chem Soc 81:4477–4485
67. Hoffmann R (1966) Extended hückel theory—v: Cumulenes, polyenes, polyacetylenes and c_n. Tetrahedron 22:521–538
68. Raghavachari K, Strout DL, Odom GK et al (1993) Isomers of C20. Dramatic effect of gradient corrections in density functional theory. Chem Phys Lett 214:357–361
69. Schmalz TG, Seitz WA, Klein DJ et al (1988) Elemental carbon cages. J Am Chem Soc 110:1113–1127

70. Kroto HW (1987) The stability of the fullerenes Cn, with n = 24, 28, 32, 36, 50, 60 and 70. Nature 329:529–531

71. Taylor R, Hare JP, Abdul-sada AK et al (1990) Isolation, separation and characterisation of the fullerenes C60 and C70: the third form of carbon. J Am Chem Soc Comm 20:1423–1425

72. Ettl R, Chao I, Diederich F et al (1991) Isolation of C76, a chiral (D2) allotrope of carbon. Nature 353:149–153

73. Yan Q-L, Gozin M, Zhao F-Q et al (2016) Highly energetic compositions based on functionalized carbon nanomaterials. Nanoscale 8:4799–4851

74. Kikuchi K, Nakahara N, Wakabayashi T et al (1992) NMR characterization of isomers of C78, C82 and C84 fullerenes. Nature 357:142–145

75. Manolopoulos DE, Fowler PW, Taylor R et al (1992) Faraday communications. An end to the search for the ground state of C84? J Chem Soc Faraday Trans 88:3117–3118

76. Kadish KM, Ruoff RS (eds) (2002) Fullerene: chemistry physics and technology. Wiley, New York

77. Manolopoulos DE, Fowler PW (1992) Molecular graphs, point groups, and fullerenes. J Chem Phys 96:7603–7614

78. Shustova NB, Kuvychko IV, Bolskar RD et al (2006) Trifluoromethyl Derivatives of Insoluble Small-HOMO−LUMO-Gap Hollow Higher Fullerenes. NMR and DFT Structure Elucidation of C_2-$(C_{74}$-$D_{3h})(CF_3)_{12}$, C_s-$(C_{76}$-$T_d(2))(CF_3)_{12}$, C_2-$(C_{78}$-$D_{3h}(5))(CF_3)_{12}$, C_s-$(C_{80}$-$C_{2v}(5))(CF_3)_{12}$, and C_2-$(C_{82}$-$C_2(5))(CF_3)_{12}$. J Am Chem Soc 128:15793–15798

79. Shustova NB, Newell BS, Miller SM et al (2007) Discovering and verifying elusive fullerene cage isomers: structures of C_2-p^{11}-$(C_{74}$-$D_{3h})(CF_3)_{12}$ and C_2-p^{11}-$(C_{78}$-$D_{3h}(5))(CF_3)_{12}$. Angew Chem 46:4111–4114

80. Amsharov KY, Jensen M (2008) A C_{78} fullerene precursor: toward the direct synthesis of higher fullerenes. J Org Chem 73:2931–2934

81. Manolopoulos DE, Fowler PW (1991) Structural proposals for endohedral metal—fullerene complexes. Chem Phys Lett 187:1–7

82. Shao N, Gao Y, Yoo S et al (2006) Search for lowest-energy fullerenes: C_{98} to C_{110}. J Phys Chem A 110:7672–7676

83. Shao N, Gao Y, Zeng XC (2007) Search for lowest-energy fullerenes 2: C_{38} to C_{80} and C_{112} to C_{120}. J Phys Chem C 111:17671–17677

84. Slanina Z, Uhlik F, Yoshida M et al (2000) A computational treatment of 35 IPR isomers of C_{88}. Fullerene Sci Technol 8:417–432

85. Slanina Z, Zhao X, Deota P et al (2000) Relative stabilities of C_{92} IPR fullerenes. J Mol Model 6:312–317

86. Sun G (2003) Assigning the major isomers of fullerene C_{88} by theoretical ^{13}C NMR spectra. Chem Phys Lett 367:26–33

87. Sun G, Kertesz M (2002) ^{13}C NMR spectra for IPR isomers of fullerene C_{86}. Chem Phys 276:107–114

88. Zhao X, Slanina Z, Goto H (2004) Theoretical studies on the relative stabilities of C_{96} IPR fullerenes. J Phys Chem A 108:4479–4484

89. Zhao X, Goto H, Slanina Z (2004) C100 IPR fullerenes: temperature-dependent relative stabilities based on the Gibbs function. Chem Phys 306:93–104

90. Fowler PW, Steer JI (1987) The leapfrog principle: a rule for electron counts of carbon clusters. J Chem Soc Chem Commun 9:1403–1405

91. Amic D, Trinajstic N (1990) On the lack of reactivity of Buckminsterfullerene. A theoretical study. J Chem Soc Perkin Trans 2:1595–1598

92. Coulombeau C, Rassat A (1987) Calculs de propriétés électroniques et des fréquences normales de vibration d'agrégats carbonés formant des polyèdres réguliers et semi-réguliers. J Chim Phys 84:875–882

93. Ozaki M, Takahashi A (1986) On electronic states and bond lengths of the truncated icosahedral C60 molecule. Chem Phys Lett 127:242–244

94. Liithi HP, Almlof J (1987) AB initio studies on the thermodynamic stability of the icosahedral C60 molecule "buckminsterfullerene." Chem Phys Lett 135:357–360

95. Almlof J, Luthi HP (1987) Theoretical methods and results for electronic structure calculations on very large systems. ACS Symp. Ser. 353: (Supercomut. Res. Chem. Chem. Eng.), 35–48

96. Almlof J (1990) Carbon in the Galaxy. In: Tarter JC, Chang S, DeFrees DJ (eds) National Aeronautics and Space Administration Conference Publication Washington, DC, 1990, vol 3061. NASA, USA, p 245

97. Schulman JM, Disch RL (1991) The heat of formation of buckminsterfullerene, C60. J Chem Soc Chem Comm 6:411–412

98. Larsson S, Volosov A (1987) Rosen A (1987) Optical spectrum of the icosahedral C60-"follene-60." Chem Phys Lett 137:501–504

99. Braga M, Larsson S, Rosen A et al (1991) Electronic transition in C60—on the origin of the strong interstellar absorption at 217 NM. Astron Astrophys 245:232–238

100. Kataoka M, Nakajima T (1986) Geometrical structures and spectra of corannulene and icosahedral C_{60}. Tetrahedron 42:6437–6442

101. Lazlo I, Udvardi L (1987) On the geometrical structure and UV spectrum of the truncated icosahedral C60, molecule. Chem Phys Lett 136:418–422

102. Hayden GW, Mele EJ (1987) π bonding in the icosahedral C60 cluster. Phys Rev B 36:5010–5015

103. Newton MD, Stanton RE (1986) Stability of buckminsterfullerene and related carbon clusters. J Am Chem Soc 108:2469–2470

104. Elser V, Haddon RC (1987) Icosahedral C60: an aromatic molecule with a vanishingly small ring current magnetic susceptibility. Nature 325:792–794

105. Elser V, Haddon RC (1987) Magnetic behavior of icosahedral Csub60. Phys Rev A 36:4579–4584

106. Fowler PW, Lazzeretti P, Zanasi R (1990) Electric and magnetic properties of the aromatic sixty-carbon cage. Chem Phys Lett 165:79–86

107. Haddon RC, Elser V (1990) Icosahedral C60 revisited: an aromatic molecule with a vanishingly small ring current magnetic susceptibility. Chem Phys Lett 169:362–364

108. Schmalz TG (1990) The magnetic susceptibility of Buckminsterfullerene. Chem Phys Lett 175:3–5

109. Dresselhaus MS, Dresselhaus G, Eklund PC (1996) Science of fullerenes and carbon nanotubes: their properties and applications. Elsevier, San Diego

110. Lebedeva MA, Chamberlain TW, Khlobystov AN (2015) Harnessing the synergistic and complementary properties of fullerene and transition-metal compounds for nanomaterial applications. Chem Rev. 115:11301–11351

111. Lim SY, Shen W, Gao Z (2015) Carbon quantum dots and their applications. Chem Soc Rev 44:362–381

112. Zhu S, Song Y, Zhao X et al (2015) The photoluminescence mechanism in carbon dots (graphene quantum dots, carbon nanodots, and polymer dots): current state and future perspective. Nano Res 8:355–381

113. Baker SN, Baker GA (2010) Luminescent carbon nanodots: emergent nanolights. Angew Chem Int Ed. 49:6726–6744

114. Sun Y-P, Zhou B, Lin Y et al (2006) Quantum-sized carbon dots for bright and colorful photoluminescence. J Am Chem Soc 128:7756–7757

115. Yamijala SSRKC, Bandyopadhyay A, Pati SK (2014) Electronic properties of zigzag, armchair and their hybrid quantum dots of graphene and boron-nitride with and without substitution: A DFT study. Chem Phys Lett 603:28–32

116. Saidi WA (2013) Oxygen reduction electrocatalysis using N-Doped graphene quantum-dots. J Phys Chem Lett 4:4160–4165

117. Kumar GS, Roy R, Sen D et al (2014) Amino-functionalized graphene quantum dots: origin of tunable heterogeneous photoluminescence. Nanoscale 6:3384–3391

118. Zhao M, Yang F, Xue Y et al (2014) A time-dependent DFT study of the absorption and fluorescence properties of graphene quantum dots. Chem Phys Chem 15:950–957

119. Sk MA, Ananthanarayanan A, Huang L et al (2014) Revealing the tunable photoluminescence properties of graphene quantum dots. J Mater Chem C 2:6954–6960

120. Zarenia M, Chaves A, Farias GA et al (2011) Energy levels of triangular and hexagonal graphene quantum dots: a comparative study between the tight-binding and Dirac equation approach. Phys Rev B 84:245403–245414
121. Li H, He X, Kang Z et al (2010) Water-soluble fluorescent carbon quantum dots and photocatalyst design. Angew Chem Int Ed 49:4430–4434
122. Choudhary RP, Shukla S, Vaibhav K et al (2015) Optical properties of few layered graphene quantum dots. Mater Res Express 2:095024–095028
123. Zhang RQ, Bertran E, Lee S-T (1998) Size dependence of energy gaps in small carbon clusters: the origin of broadband luminescence. Diamond Relat Mater 7:1663–1668
124. Zhu B, Sun S, Wang Y et al (2013) Preparation of carbon nanodots from single chain polymeric nanoparticles and theoretical investigation of the photoluminescence mechanism. J Mater Chem C 1:580–586
125. Park Y, Yoo J, Lim B et al (2016) Improving the functionality of carbon nanodots: doping and surface functionalization. J Mater Chem A 4:11582–11603
126. Hu S, Tian R, Wu L et al (2013) Chemical regulation of carbon quantum dots from synthesis to photocatalytic activity. Chem Asian J 8: 1035–1041
127. Kwon W, Do S, Kim J-H et al (2015) Control of Photoluminescence of carbon nanodots via surface functionalization using para-substituted anilines. Sci Rep 5:12604–12613
128. Margraf JT, Strauss V, Guldi DM et al (2015) The electronic structure of amorphous carbon nanodots. J Phys Chem B 119:7258–7265
129. Ajayan PM, Stephan O, Colliex C et al (1994) Aligned carbon nanotube arrays formed by cutting a polymer resin—nanotube composite. Science 265:1212–1214
130. Saito Y, Hamaguchi K, Hata K et al (1997) Conical beams from open nanotubes. Nature 389:554–555
131. de Heer WA, Châtelain A, Ugarte D (1995) A carbon nanotube field-emission electron source. Science 270:1179–1180
132. Collins PG, Zettl A, Bando H et al (1997) Nanotube nanodevice. Science 278:100–102
133. Nardelli MB, Yakobson BI, Bernholc J (1998) Mechanism of strain release in carbon nanotubes. Phys Rev B 57:R4277-4280
134. Huang JY, Chen S, Ren ZF et al (2006) Real-time observation of tubule formation from amorphous carbon nanowires under high-bias joule heating. Nano Lett 6:1699–1705
135. Radushkevich LV, Lukyanovich VM (1952) The structure of carbon forming in thermal decomposition of carbon monoxide on an iron catalyst. Russian J Phys Chem 26:88–95
136. Saxena S, Tyson TA (2010) Ab initio density functional studies of the restructuring of graphene nanoribbons to form tailored single walled carbon nanotubes. Carbon 48:1153–1158
137. Saito R, Dresselhaus G, Dresselhaus MS (1998) Physical properties of carbon nanotubes. Press, London, Imp. Coll
138. Terrones M (2003) Science and technology of the twenty-first century: synthesis, properties, and applications of carbon nanotubes. Ann Rev Mater Res 33:419–501
139. Zhang M, Li J (2009) Carbon nanotube in different shapes. Mater Today 12:12–18
140. Elliott JA, Sandler JKW, Windle AH et al (2004) Collapse of single-wall carbon nanotubes is diameter dependent. Phys Rev Lett 92:095501–095504
141. Ebbesen TW, Lezec HJ, Hiura H et al (1996) Electrical conductivity of individual carbon nanotubes. Nature 382:54–56
142. Saito R, Fujita M, Dresselhaus G et al (1992) Electronic structure of chiral graphene tubules. Appl Phys Lett 60:2204–2206
143. Delaney P, Di Ventra M, Pantelides ST (1999) Quantized conductance of multiwalled carbon nanotubes. Appl Phys Lett 75:3787–3789
144. Yu MF, Lourie O, Dyer MJ et al (2000) Strength and breaking mechanism of multiwalled carbon nanotubes under tensile load. Science 287:637–640
145. Yu MF, Files BS, Arepalli S et al (2000) Tensile loading of ropes of single wall carbon nanotubes and their mechanical properties. Phys Rev Lett 84:5552–5555
146. Xie S, Li W, Pan Z et al (2000) Mechanical and physical properties on carbon nanotube. J Phys Chem Solids 61:1153–1158

147. Overney G, Zhong W, Tomanek D (1993) Structural rigidity and low frequency vibrational modes of long carbon tubules. Z Phys D 27:93–96
148. Tersoff J (1992) Energies of fullerenes. Phys Rev B 46:15546–15549
149. Sinnott SB, Shenderova OA, White CT et al (1998) Mechanical properties of nanotubule fibers and composites determined from theoretical calculations and simulations. Carbon 36:1–9
150. Yakobson BI (1998) Mechanical relaxation and "intramolecular plasticity" in carbon nanotubes. Appl Phys Lett 72:918–920
151. Ru CQ (2000) Effect of van der Waals forces on axial buckling of a double-walled carbon nanotube. J Appl Phys 87:7227–7231
152. Saxena S, Tyson TA (2010) Interacting quasi-two-dimensional sheets of interlinked carbon nanotubes: a high-pressure phase of carbon. ACS Nano 4:3515–3521
153. Gao G, Çagin T, Goddard WA III (1998) Energetics, structure, mechanical and vibrational properties of single-walled carbon nanotubes. Nanotechnology 9:184–191
154. Hernandez E, Goze C, Bernier P et al (1998) Elastic properties of C and BxCyNz composite nanotubes. Phys Rev Lett 80:4502–4505
155. Yu M-F, Kowalewski T, Ruoff RS (2000) Investigation of the radial deformability of individual carbon nanotubes under controlled indentation force. Phys. Rev. Lett. 85:1456–1459
156. Saeed K, Khan I (2013) Carbon nanotubes–properties and applications: a review. Carbon Lett 14:131–144
157. Ruoff RS, Lorents DC (1995) Mechanical and thermal properties of carbon nanotubes. Carbon 33:925–930
158. Ashcroft NW (1976) Mermin N D (1976) Solid State Physics. Harcourt Brace, Orlando, FL
159. Kim P, Shi L, Majumdar A et al (2001) Thermal transport measurements of individual multiwalled nanotubes. Phys Rev Lett 87:215502–215505
160. Yu C, Shi L, Yao Z et al (2005) Thermal conductance and thermopower of an individual single-wall carbon nanotube. Nano Lett 5:1842–1846
161. Maultzsch J, Reich S, Thomsen C et al (2002) Phonon dispersion of carbon nanotubes. Solid State Commun 121:471–474
162. Ishii H, Kobayashi N, Hirose K (2007) Electron–phonon coupling effect on quantum transport in carbon nanotubes using time-dependent wave-packet approach. Phys E 40:249–252
163. Maeda T, Horie C (1999) Phonon modes in single-wall nanotubes with a small diameter. Phys B 263–264:479–481
164. Kasuya A, Saito Y, Sasaki Y et al (1996) Size dependent characteristics of single wall carbon nanotubes. Mater Sci Eng A 217–218:46–47
165. Popov VN (2004) Theoretical evidence for $T^{1/2}$ specific heat behavior in carbon nanotube systems. Carbon 42:991–995
166. Segal M (2012) Material history: learning from silicon. Nature 483:S43–S44
167. Falcao EHL, Wudl F (2007) Carbon allotropes: beyond graphite and diamond. J Chem Technol Biotechnol 82:524–531
168. Aristov VY, Urbanik G, Kummer K et al (2010) Graphene synthesis on cubic SiC/Si wafers. perspectives for mass production of graphene-based electronic devices. Nano Lett 10:992–995
169. Hernandez Y, Nicolosi V, Lotya M et al (2008) High-yield production of graphene by liquid-phase exfoliation of graphite. Nat Nanotechnol 3:563–568
170. Paredes JI, Villar-Rodil S, Fernández-Merino MJ et al (2011) Environmentally friendly approaches toward the mass production of processable graphene from graphite oxide. J Mater Chem 21:298–306
171. Dikin DA, Stankovich S, Zimney EJ et al (2007) Preparation and characterization of graphene oxide paper. Nature 448:457–460
172. Wang G, Yang J, Park J et al (2008) Facile synthesis and characterization of graphene nanosheets. J Phys Chem C 112:8192–8195
173. Prasai D, Tuberquia JC, Harl RR et al (2012) Graphene: corrosion-inhibiting coating. ACS Nano 6:1102–1108
174. Kiran SK, Shukla S, Struck A et al (2019) Surface enhanced 3D rGO hybrids and porous rGO nano-networks as high performance supercapacitor electrodes for integrated energy storage devices. Carbon 158:527–535

175. Kiran SK, Shukla S, Struck A et al (2019) Surface engineering of graphene oxide shells using Lamellar LDH nanostructures. ACS Appl Mater Interfaces 11:20232–20240

176. Zhao X, Hayner CM, Kung MC et al (2011) In-plane vacancy-enabled high-power Si–graphene composite electrode for Lithium-Ion batteries. Adv Energy Mater 1:1079–1084

177. Z. Radivojevic, et al. (2012) Electrotactile touch surface by using transparent graphene. In: VRIC '12: proceedings of the 2012 virtual reality international conference, association for computing machinery, New York, NY, USA

178. Wang H, Sun K, Tao F et al (2013) 3D honeycomb-like structured graphene and its high efficiency as a counter-electrode catalyst for dye-sensitized solar cells. Angew Chem Int Ed Engl 52:9210–9214

179. Pawar PB, Saxena S, Bhade DK et al (2016) 3D oxidized graphene frameworks for efficient nano sieving. Sci Rep 6:21150–21154

180. Shejale KP, Yadav D, Patil H et al (2020) Evaluation of techniques for the remediation of antibiotic-contaminated water using activated carbon. Mol Syst Des Eng 5:743–756

181. Pandey A, Deb M, Tiwari S et al (2018) 3D oxidized graphene frameworks: an efficient adsorbent for methylene blue. J Mater 70:469–472

182. Pawar PB, Maurya SK, Chaudhary RP et al (2016) Water purification using graphene covered micro-porous, reusable carbon membrane. MRS Adv 1:1411–1416

183. Wallace PR (1947) The band theory of graphite. Phys Rev 71:622–634

184. Slater JC, Koster GF (1954) Simplified LCAO method for the periodic potential problem. Phys Rev B 94:1498–1524

185. Harrison (1980) Electronic structure and the properties of solids: the physics of the chemical bond. W. H, Freeman and Company, San Francisco, p 1980

186. Boehm HP, Clauss A, Fisher GO et al (1962) Das Adsorptionsverhalten sehr dünner Kohlenstoff-Folien. Zeitschrift Fur Anorg Und Allg Chemie 316:119–127

187. Fuhrer MS, Lau CN, MacDonald AH (2010) Graphene: materially better carbon. MRS Bull 35:289–295

188. Singh V, Joung D, Zhai L et al (2011) Graphene based materials: past, present and future. Prog Mater Sci 56:1178–1271

189. Kim KS, Zhao Y, Jang H et al (2009) Large-scale pattern growth of graphene films for stretchable transparent electrodes. Nature 457:706–710

190. Balandin AA, Ghosh S, Bao W et al (2008) Superior thermal conductivity of single-layer graphene. Nano Lett 8:902–907

191. Lee C, Wei X, Kysar JW et al (2008) Measurement of the elastic properties and intrinsic strength of monolayer graphene. Science 321:385–388

192. Zhu Y, Murali S, Cai W et al (2010) Graphene and graphene oxide: synthesis, properties, and applications. Adv Mater 22:3906–3924

Edge State Induced Spintronic Properties of Graphene Nanoribbons: A Theoretical Perspective

Soumya Ranjan Das and Sudipta Dutta

Abstract Low-dimensional carbon-based nanomaterials have generated enormous interest in the scientific community due to their quantum confinement induced novel electronic, magnetic, optical, thermal, mechanical, and chemical properties. The synthesis of two-dimensional graphene has provided a fertile experimental platform for studying these exotic properties and harnessing them for promising carbon-based nanoelectronic devices. This chapter reviews the edge-induced spintronic properties of nanographene ribbons (GNR) from a theoretical perspective and discusses their possible applications in nanoscale devices. The presence of edges bears a crucial impact on the low-energy spectrum of the itinerant Dirac electrons in graphene. Nanoribbons with zigzag edges (ZGNR) possess robust localized edge states near the Fermi energy that induces ferrimagnetic spin polarization along the zigzag edge. In contrast, such localized edge states are absent in nanoribbons with armchair edges (AGNR). We discuss how applying a transverse electric field to ZGNRs, or chemical modification of its edges, can break the spin degeneracy and lead to a half-metallic state which shows spin polarization of the current or the spin-filtering behavior that is very crucial for spintronic device applications. The presence of edges also endows GNRs with peculiar transport properties characterized by the absence of Anderson localization, making them ideal for ultra-low power electronics. Our review highlights the edge states' role as the origin of GNR's diverse physical and chemical properties. It hence has a significant bearing on the future realization of carbon nanomaterial electronics.

S. R. Das · S. Dutta (✉)
Department of Physics, Indian Institute of Science Education and Research (IISER) Tirupati, Tirupati 517507, Andhra Pradesh, India
e-mail: sdutta@iisertirupati.ac.in

S. R. Das
e-mail: soumya.das@students.iisertirupati.ac.in

© The Author(s), under exclusive license to Springer Nature Singapore Pte Ltd. 2021
A. Hazra and R. Goswami (eds.), *Carbon Nanomaterial Electronics: Devices and Applications*, Advances in Sustainability Science and Technology,
https://doi.org/10.1007/978-981-16-1052-3_8

1 Introduction

The ever-increasing demand for high-speed electronic devices and renewable energy has motivated researchers to discover and develop new nanomaterials. This emerging need has led to explosive growth in interest in using carbon-based nanomaterials due to their unique structure and physical properties. Owing to its existence in different allotropic forms, a wide array of bonding nature, and ease of synthesis, carbon, the fourth most abundant element on earth, is essential not only for life on our planet but also in material applications. The manifold electronic properties of carbon materials mainly arise due to the catenation nature of carbon, which enables it to form compounds with different dimensionalities [46, 89]. In fact, carbon-based nanomaterials are composed of sp^2 bonded graphitic carbon and are found in all reduced dimensions. This includes allotropes like zero-dimensional (0D) fullerenes [106], one-dimensional (1D) carbon nanotubes (CNTs) [54, 86, 87], and two-dimensional (2D) graphene [64, 130, 131, 205]. The electronic confinement in reduced length scales in these low-dimensional materials is the origin of many exotic properties which have been at the forefront of condensed matter and materials science research in the last few decades [3, 4, 7, 10, 12, 18, 31, 39, 40, 53, 69, 144, 150, 165, 169].

Graphene, the 2D monolayer of carbon atoms, has emerged as one of the leading systems for fundamental investigations in condensed matter physics and materials science and one of the foremost candidates in carbon-based nanomaterial device applications [64, 130, 131, 205]. Previous theoretical studies have suggested that a perfectly 2D free-standing system at finite temperature is unstable due to quantum fluctuations and thermal agitation [72, 111, 139, 140]. Hence, the experimental realization of graphene was a conjecture until 2004, when Sir Andre Geim's group at Manchester University isolated it by mechanical exfoliation of highly oriented pyrolytic graphite on SiO_2 substrate [130–132]. However, to gain stability, the graphene flakes form ripples that break its perfectly planar structure [56, 122].

The honeycomb lattice structure of graphene consists of two nonequivalent sublattices. The itinerant π-electrons exhibit a unique band structure near the Fermi energy, whose motion is well described by the massless Dirac equation (Weyl equation). Due to its Dirac electron nature and consequent peculiar linear energy dispersion and the presence of a non-zero topological Berry phase [9], graphene provides a platform for extremely unusual 2D electronic properties [11, 29], such as the room temperature anomalous half-integer quantum Hall effect [1, 2, 5, 22, 28, 68, 71, 78, 80, 91, 98, 99, 129, 131, 133, 134, 158, 205], breakdown of the adiabatic Born-Oppenheimer approximation [142], the absence of backward scattering [8–10], Klein tunneling [100], possibility of high T_c superconductivity [27, 67, 79, 90, 172], metal free magnetism [75, 76, 93, 94, 108, 112, 123, 135, 145, 151, 201, 202], and the π-phase shift of Shubnikov-De Haas oscillations [118]. The high charge carrier mobility [21], low spin-orbit coupling, negligible hyperfine interaction, and gate tunability [29] have made graphene an excellent spintronic material [170]. Its high thermal conductivity [16] has also made graphene favorable for next-generation device applications.

With successive miniaturization of graphene, the role of edges on the low-energy π-electrons spectrum, and consequently, the electronic properties of nanographene become crucial [59, 127, 175]. The finite termination of graphene results in quasi 1D ribbons with two basic edge geometries, *armchair* and *zigzag*, that exhibit different electronic properties arising from the differences in their boundary conditions. The armchair edge shows bonding between two atoms from different sublattices, whereas the atoms along the zigzag edge belong to the same sublattice. The ZGNRs show localized edge states near the Fermi energy [59, 127, 174, 175, 186]. However, the ribbons with armchair edges do not show any edge state [46, 186]. The existence of these edge states has been validated by recent scanning tunneling microscopy (STM) [103, 104, 128, 166] and high-resolution angle-resolved photoemission spectroscopy (ARPES) [163] experiments.

The edge states in ZGNRs result in a diverging density of states (DOS) near the Fermi energy. This results in Fermi instability due to the high concentration of localized edge electrons. To circumvent this instability, the spins tend to align in ferrimagnetic order along the zigzag edge, which persists even in very weak electron-electron interactions [59, 174]. Thus, these zigzag edge states are very crucial in inducing magnetism in nanometer-sized graphene systems. The existence of such magnetic states has been theoretically confirmed within density functional theory (DFT) [110, 159], the density matrix renormalization group (DMRG) implementing the Hubbard model [81], and mean-field theory [57, 59, 76, 109, 136, 174, 180, 199, 203]. The edge states are robust to changes in the size and geometry of the nanoscale graphene systems [55, 82, 107]. Recent studies have also shown that these edge states are topological in nature [25, 77] due to non-zero Berry connection and the consequent appearance of non-zero Zak phase, in spite of zero Berry curvature [38, 116, 204]. The functionalities of these systems can be further tailored by structural and chemical modifications of their edges that changes their energy dispersions near the Fermi energy [44, 97, 110, 184, 193].

Recent sophistication in experimental techniques provides various ways of preparing the GNRs of smooth edges with varying widths. These include the top-down approach using lithographic patterning of graphene [34, 73, 167], solution dispersion and sonication of chemically exfoliated graphite [113], and longitudinal unzipping and flattening of carbon nanotubes by chemical and mechanical means [92, 105]. Recent approaches using bottom-up synthesis also provide GNRs on a metal surface with smooth edges of atomic precision. [24, 117, 147]. Current research has also successfully demonstrated a scalable fabrication of GNRs on silicon carbide substrates using self-organized growth, doing away with lithography [161]. Different research groups have also reported zigzag edge state induced transport gap at low-temperatures near the Dirac point [17, 73, 125, 126, 161, 162, 167, 190]. Combining the above experimental techniques has led to a plethora of unique ways to design GNRs with well-controlled widths, edge orientation, and electronic properties.

This chapter reviews the nanoscale edge effects on nanographene systems' electronic, magnetic, and transport properties from a theoretical perspective. We discuss the fundamental aspects of carrier doping and the mechanism of electric-field and chemical modification-induced half-metallicity in GNRs and how they are important for spintronic device applications. We also discuss the peculiar transport properties shown by graphene nanoribbons, characterized by the absence of Anderson localization. Our motivation is to elucidate the critical role of edge states, which enable GNRs to acquire such diverse and interesting physical and chemical properties, and thus can be utilized in future carbon-based nanomaterial device applications.

2 Electronic States of Graphene

We begin by briefly reviewing the π-band dispersion of graphene [188]. In 2D graphene, the sp^2 hybridized carbon atoms form σ bonds with three nearest neighbors with a bond length of 1.42 Å that lead to a planar honeycomb network, as shown in Fig. 1a. The unhybridized p_z orbital on each carbon atom contains one π-electron which governs the electronic states near the Fermi energy. Figure 1a shows the shaded rhombus unit cell, which consists of two atoms from two different sublattices, namely A and B, making graphene a bipartite lattice. Two sublattice points in a bipartite lattice prefer to align the electronic spins in an antiferromagnetic manner, as described by Lieb's theorem [115]. Figure 1b depicts the first Brillouin zone (BZ) of graphene. At the corners of the first BZ, i.e., at **K** and **K'** points, the energy dispersion follow the massless Dirac equation (Weyl equation) [10, 11]. Therefore, these high symmetry points are commonly termed as the Dirac points.

To describe the electronic states of graphene, we use a single-orbital nearest neighbor tight-binding model for the π-electron network [186]. The tight-binding Hamiltonian has the form (assuming $\hbar = 1$) [185]

$$H = -t \sum_{\langle i,j \rangle} \sum_{\sigma} (a_{i,\sigma}^{\dagger} b_{j,\sigma} + h.c.) \tag{1}$$

where $h.c.$ stands for the Hermitian conjugate; the operators $a_{i,\sigma}^{\dagger}(a_{i,\sigma})$ and $b_{i,\sigma}^{\dagger}(b_{i,\sigma})$ create (annihilate) an electron with spin σ ($\sigma = \uparrow, \downarrow$) at the A- and B-sublattice sites, respectively. t is the hopping integral between nearest neighbor carbon atoms, which is estimated to be about 2.75 eV in a graphene system [29]. Note that we have expressed all the energies in this chapter in the unit of t, except the results obtained from DFT calculations.

The Hamiltonian H in Eq. (1) is diagonalized to obtain the energy dispersion relation of graphene, namely [185, 188]

$$E_{\pm}(\mathbf{k}) = \pm t \sqrt{3 + 2\cos\left(\frac{\sqrt{3}k_x a}{2} + \frac{k_y a}{2}\right) + 2\cos\left(\frac{\sqrt{3}k_x a}{2} - \frac{k_y a}{2}\right) + 2\cos(k_y a)} \tag{2}$$

Since each carbon site contributes one π-electron from a singly occupied p_z orbital, only the valence band ($E_-(\mathbf{k})$) gets completely occupied and the conduction band ($E_+(\mathbf{k})$) remains empty. Figure 1c shows the energy dispersion of graphene (in

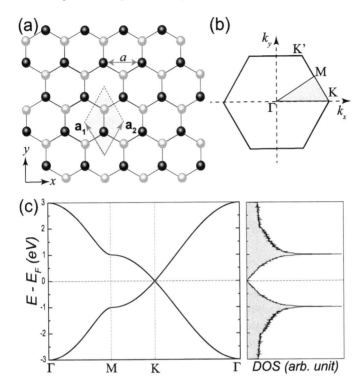

Fig. 1 **a** The honeycomb lattice structure of graphene in real space, composed of two interpenetrating triangular lattices. The A(B)-sublattice sites of the honeycomb lattice are denoted by dark(light) colored circles; $a \approx 2.46$ Å is the lattice constant and $\mathbf{a_1}$, $\mathbf{a_2}$ are the lattice unit vectors, respectively. The shaded rhombus denotes the unit cell. **b** The hexagonal first BZ of graphene, with the high symmetric points. The shaded region denotes the irreducible BZ. The Dirac cones are located at the $\mathbf{K}$ and $\mathbf{K'}$ points. **c** Energy dispersion of graphene in the irreducible first BZ with the corresponding DOS. The energies are presented in the unit of hopping integral (t) and scaled with respect to the Fermi energy (E_F)

the unit of t) with the corresponding DOS. The valence and the conduction bands touch conically at the Dirac points, i.e., at the $\mathbf{K}$ and $\mathbf{K'}$ points (see Fig. 1c), and hence they form a time-reversed pair of opposite chirality [186].

There is a relative phase difference between A and B sublattice sites in the Bloch wave function of graphene, which gives rise to the π-Berry phase around the Dirac points [8, 10] and chiral-dependent Klein tunneling [100] in graphene. The intrinsic phase is preserved in armchair edges but not preserved in case of zigzag edge nanographene. This gives rise to important consequences in their respective electronic properties, which we will discuss in the next section.

The energy spectrum of graphene, as shown in Fig. 1c, is the source of its distinctive electronic transport properties [37]. The unique valley structure, large carrier mobility, and electrostatic-gate tunable Dirac neutrality point are the origin of the anomalous quantum Hall effect, weak localization [64] and supercurrent transport

[79] in graphene. These exotic relativistic properties, coupled with the low intrinsic spin-orbit and hyperfine interactions [85, 171], has enabled graphene to carve its niche in the emerging field of spintronics. In particular, graphene shows a long spin relaxation length at room temperature [41, 88, 170], which enables it to inject, transport, and detect spin signals efficiently within complex multi-terminal device architectures [62, 95]. Recent studies have also shown an enhancement in graphene's spintronic properties when it is coupled with other 2D materials to form van der Waals heterostructures [14, 61, 70, 146]. Hence, the investigation of nanographene's electronic properties and the role of its edge states have become very important for their utilization in spintronic devices. We discuss them in detail in the subsequent sections.

3 Edge States in Graphene Nanoribbons

Terminating the graphene sheet in one direction results in quasi-periodic 1D infinite graphene nanoribbons, where the resulting edge strongly influences its electronic properties. There can be two types of graphene edges: *armchair* and *zigzag*, differing by 30° in their crystallographic orientation. These two edge geometries give rise to distinct π-electronic structures due to different boundary conditions [59]. There appear localized states on zigzag edges, unlike the armchair edge.

The emergence of graphene edge states and their role in the electronic properties of GNRs are theoretically studied in this section by employing a single-orbital tight-binding model for the π-electrons, as we did in the case of graphene [175, 185, 186]. The longitudinal periodicity of GNRs enables us to define the crystal momentum (wavenumber k) and adopt the standard solid-state physics approach to investigate these nano-systems. Figure 2a, b shows the lattice structure of GNRs with the zigzag and armchair edge geometries, respectively. The shaded rectangle in Fig. 2a, b denotes the unit cell of the GNRs. The widths of these GNRs are defined as N_z (N_a) for ZGNRs (AGNRs), where N_z is the number of zigzag lines for ZGNRs and N_a is the number of dimer (two carbon sites) lines for AGNRs. We assume that all the dangling bonds at graphene edges are passivated by hydrogen atoms that do not show any contribution near the Fermi energy. To compare the physical quantities of GNRs of different edge geometries with the same width, we use the definition $W_z = \frac{\sqrt{3}}{2}N_z a + \frac{a}{\sqrt{3}}$ for ZGNRs and $W_a = \frac{1}{2}(N_a + 1)a$ for AGNRs, where a is the lattice constant of graphene (0.246 nm).

Figure 2c relates the BZ of graphene with that of GNRs. The hexagonal first BZ (solid line) for the graphene's rhombus unit cell has been mapped to the rectangular BZ (dashed line) of GNRs. Due to BZ folding in the modified rectangular unit cell, the Dirac points ($\mathbf{K}$ and $\mathbf{K'}$) are now positioned on the line connecting the two high symmetric points Γ and X (denoted by the gray circles).

To find the energy spectrum of the nanoribbons, we diagonalize the tight-binding Hamiltonian H, given by [186]

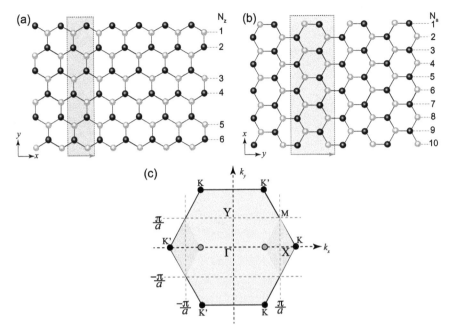

Fig. 2 The lattice structure of **a** zigzag graphene nanoribbons and **b** armchair graphene nanoribbons. The unit cells are denoted by the shaded rectangles, where N_z and N_a are the ribbon widths for zigzag and armchair nanoribbons, depicting the number of zigzag and dimer lines, respectively, along the cross-ribbon directions. Here, we show the ribbons with $N_z = 6$ and $N_a = 10$. **c** The BZ of graphene for rhombus unit cell (solid hexagon) and that for GNRs with rectangular unit cells (dashed rectangles) with the high symmetry points. Due to BZ folding in GNRs, the Dirac points, **K** and **K′**, are mapped onto the gray circles on k_x-axis in the reduced BZ. Note that the mapping of the Dirac point on the k_x-axis ensures the appearance of the same in the 1st BZ of ZGNRs (Γ-X) irrespective of their width, whereas in the case of AGNRs, the Dirac point appears in the 1st BZ (Γ-X) depending on their width

$$H = \sum_{i,j} t_{i,j} c_i^\dagger c_j \tag{3}$$

where $c_i^\dagger$ and c_i are the Fermionic creation and annihilation operators at the ith site of the nanoribbons, respectively, and $t_{i,j}$ is the hopping integral, where

$$t_{i,j} = \begin{cases} -t, & \text{if the sites, } i \text{ and } j \text{ are nearest neighbors} \\ 0, & \text{otherwise} \end{cases}$$

For an armchair nanoribbon with width N_a (Fig. 2b), the energy spectrum is shown in Fig. 3 [185, 186]. Depending on the value of N_a, the armchair edge boundary condition [186] classifies AGNRs into three categories based on their electronic properties, namely $N_a = 3r$ and $N = 3r + 1$, which shows semiconducting behavior and

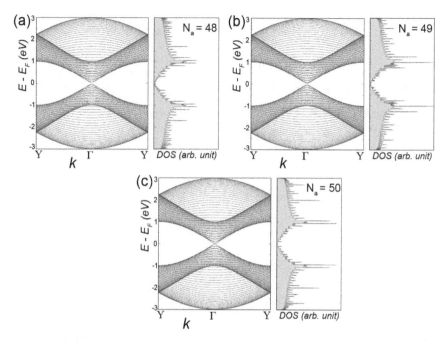

Fig. 3 The energy dispersions and the corresponding DOS of AGNRs with nanoribbon width **a** $N_a = 48$, **b** 49, and **c** 50, respectively. The energies have been scaled with respect to Fermi energy (E_F)

$N = 3r + 2$ which is metallic (r is a positive integer) [22, 59, 127, 159, 175]. This is evident from the direct energy gap at Γ-point for AGNRs in Fig. 3. The energy dispersion of AGNRs is obtained by slicing graphene's band structure, like carbon nanotubes [148, 149]. With an increase in AGNR width, the direct semiconducting band gap decreases and approaches zero for very large width [185]. However, for narrow undoped metallic armchair nanoribbons, an energy gap can be formed by Peierls distortion at low temperatures [60, 82, 124, 159]. Note that the above observations on the width-dependent electronic properties of AGNRs are true only within the tight-binding formalism. First-principles calculations have shown direct semiconducting energy gap at Γ-point for all the three classes of AGNRs [159].

For a zigzag nanoribbon (Fig. 2a), a remarkable characteristic arises in the energy dispersion (see Fig. 4). The top of the valence bands and the bottom of the conduction bands remain degenerate at the high symmetric point X ($ka = \pi$), that is, the BZ boundary (see Fig. 4). These two center bands do not arise from the intrinsic graphene band structure and exhibit flattening with increasing ribbon width within the Dirac point ($ka = 2\pi/3$) and X point; see Fig. 4 [186]. The presence of these partial flat bands in ZGNRs results in a diverging peak in the DOS at the Fermi energy (see Fig. 4). This originates from the localized edge states in ZGNRs and decays exponentially toward the ribbon center [59, 103, 104, 127, 128, 175].

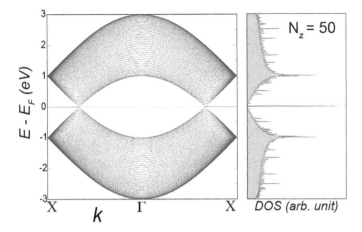

Fig. 4 The energy band structure and the corresponding DOS of ZGNRs with $N_z = 50$. The energies have been scaled with respect to Fermi energy (E_F)

The wave function for AGNRs, obtained by the tight-binding calculations, has a phase difference between A(B) sublattices due to graphene's chiral nature. But the wave functions for ZGNRs for both the extended and localized state are always real and hence have no phase. This difference in wave functions between AGNRs and ZGNRs is related to graphene's pseudospin nature [154] and physical properties like Kohn anomaly [152] and their role in Raman scattering [153].

The edge states that we have studied in this section are robust, and their effect persists even in the absence of a smooth zigzag edge. Any general edge geometry can show a zero-energy edge state, if it is not parallel to the armchair edge. This is analogous to the condition of an unconventional superconductor's zero-energy Andreev bound state [182]. Previous studies have also suggested that due to a considerable imbalance of on-site electronic correlation between two sublattice points in AGNRs, a localized state and a flat band can appear [184]. Thus, the role of electron-electron interaction on the nanoribbons' electronic states is crucial and is discussed in detail in the next section.

4 Spintronic Properties of Zigzag Graphene Nanoribbons

In the previous section, we have discussed how the localized edge states in ZGNRs lead to a sharp peak in the DOS near the Fermi energy and consequent Fermi instability. Lattice distortion arising from the electron-phonon interaction and/or spin polarization due to the electron-electron interaction can eliminate such instability. The non-bonding edge states are unlikely to induce lattice distortion near the zigzag edges [60], which was also confirmed based on first-principles calculations [124, 159].

Thus, we explore the effect of the electronic correlations in GNRs, modeled within the Hubbard Hamiltonian [168], as follows:

$$H_{Hubbard} = -t \sum_{\langle i,j \rangle, \sigma} c^{\dagger}_{i,\sigma} c_{j,\sigma} + U \sum_{i} n_{i,\uparrow} n_{i,\downarrow} \qquad (4)$$

where the operator $c^{\dagger}_{i,\sigma}$ ($c_{i,\sigma}$) creates (annihilates) an electron with spin σ at site i, and the number operator is defined as $n_{i,\sigma} = c^{\dagger}_{i,\sigma} c_{i,\sigma}$. The site indices of GNRs are the same as those defined in Fig. 2a. The first term in the above Hamiltonian is the nearest neighbor tight-binding model that we studied in detail in the previous section (see Eq. (3)), with the addition of the electronic spin. The second term describes the on-site Coulomb interaction U. The Pauli exclusion principle prohibits two electrons with the same spin to occupy a single site. However, two electrons with opposite spins can occupy a single site by expending Coulomb energy ($U > 0$). In the Hubbard model, when $U/t \ll 1$, the electron hopping term dominates over the Coulomb U term, making the system metallic [168]. But if $U/t \gg 1$, the electrons are localized in their respective sites owing to strong on-site Coulomb repulsion, making the system a Mott insulator, where two adjacent sites are antiferromagnetically correlated [115, 168].

The unrestricted Hartree-Fock (HF) approximation (mean-field formalism) is employed to simplify the Hubbard Hamiltonian, [59, 186], which decouples the up- and down-spin sectors. Here, the up (down) spin electron is approximated to experience the potential due to the mean-field of the down (up) spin electron density. Then the expectation values of spin-dependent electron density are determined self-consistently by using the eigenfunctions of Hartree-Fock Hamiltonian [59, 186]. We define the spin density as n_{μ} ($\mu = A$ or B) for each site and the net magnetization has been determined by the following expression [59, 186]

$$m = \mu_B (\langle n_{A(\uparrow)} \rangle - \langle n_{B(\downarrow)} \rangle) \qquad (5)$$

where μ_B is the Bohr magneton.

Figure 5a shows how the magnetization, m, depends on the on-site Coulomb interaction, U, for several sites in ZGNRs with $N_z = 10$. For an arbitrary site, in the limit $U/t \rightarrow 0$, the up-spin and down-spin charge densities are balanced, giving rise to net zero magnetization. An imbalance may occur if we increase U, leading to a finite magnetization in the system. In Fig. 5a, the dashed line is the mean-field solution for 2D graphene. Graphene is a zero-gap semiconductor with zero DOS at the Fermi energy. The dashed line's slope rapidly increases at $U = 2.2t$, consistent with graphene's non-magnetic nature. On the other hand, the large value of magnetization in ZGNRs emerges on the edge carbon atoms, even for a minimal U value. This is because zigzag ribbons have a large DOS at the Fermi energy owing to the edge states. Hence, non-zero magnetization can emerge even for infinitesimally small values of U due to these flat bands, as indicated by the mean-field results and Fig. 5a [186].

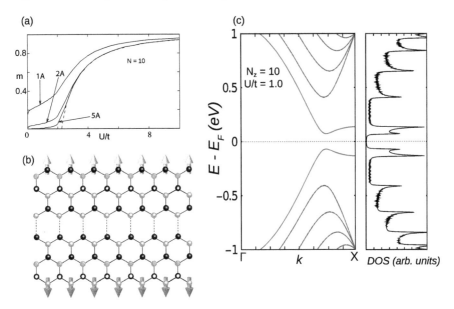

Fig. 5 **a** Dependence of the magnetization m (units of μ_B) on the on-site Coulomb interaction U for ZGNRs with $N_z = 10$ (see Eq. (5)). The mean-field results for the graphene sheet are represented by the extrapolated dashed line. (Adapted with permission from reference [59]. Copyright (1996) by Physical Society of Japan.) **b** The schematic magnetic structure of the same ZGNR system at $U/t = 1.0$. **c** Energy band structure and the corresponding DOS of ZGNR for the same parameter set, calculated using the mean-field Hubbard Hamiltonian. (Adapted with permission from reference [186]. Copyright (2012) by Elsevier)

Figure 5b shows a schematic picture of the real space spin structure for ZGNRs with $N_z = 10$ at $U/t = 1$, calculated using the mean-field Hubbard model [59]. This is also consistent with first-principles calculations [160]. Figure 5b demonstrates that the ground state of ZGNRs has parallel ferrimagnetic spin alignment along the same zigzag edge. In contrast, the spin alignments in the opposite edges are anti-parallel to each other. Note that the honeycomb graphene lattice's bipartite nature gets destroyed at the edges as opposite zigzag edges now belong to different sublattice sites. The spins in the nearest neighbor sites are antiferromagnetically coupled, with a decay of spin density toward the center of the ZGNRs. The vanishing total spin ordering for the ZGNR ground state makes it antiferromagnetic. This anti-parallel spin ordering along the opposite edges is consistent with Lieb's theorem of the half-filled Hubbard model [115].

Figure 5c shows the energy dispersion and the corresponding DOS for the ZGNR systems with the same parameter set. Comparing the energy dispersion with that from Fig. 4, we see that in nanographene with zigzag edges, the inclusion of a finite on-site Coulomb interaction U makes the system spin-polarized and leads to the opening of a band gap near the Fermi energy. Consequently, the DOS shows a splitting into two peaks at the Fermi energy (see Fig. 5c). This peak splitting of edge states has been reported in recent experiments for chiral GNRs at low temperatures using scanning tunneling spectroscopy [119, 166]. This is due to the occurrence of edge spin polarization. However, the AGNRs do not show such magnetic properties as shown in Fig. 5. Similar magnetic behavior in ZGNRs is also shown around a lattice vacancy, defects, or rough edges [52, 109, 202]. The edge reconstruction of semiconducting ZGNRs on one side also results in magnetic and metallic nature [45]. Note that more accurate quantum many-body configuration interaction (CI) methods [164] for numerical analysis of the Hubbard Hamiltonian suggest that, although the electron density prefers to localize on the edges, the GNRs prefer to mix both up and down spin densities throughout the lattice [42]. As such, the long-range spin order derived from first-principles and mean-field theory is spurious [143]. Semi-empirical CI calculations have also shown that, with enhanced correlation strength, the charge gap increases with a decrease in the spin gap in ZGNRs, resulting in insulating behavior, making magnetic excitations easier [51]. This has enormous potential in spin-filtering and spin transport applications, where the spin degrees of freedom can be exploited for high-speed and low-power computing devices [14, 195, 207]. Hence, in the subsequent sections, we build on our theoretical observations of the edge states in nanographene and give an overview of their important role in spintronic and transport applications [14, 207].

5   Effect of Carrier Doping in ZGNRs

In the preceding section, we have discussed how nanographene's magnetic properties arise due to their zigzag edges. In fact, the edge magnetic states in ZGNRs are very sensitive to carrier doping [48], which plays a significant role in tailoring their electronic properties [20, 30, 33, 121, 141, 191, 200].

We have taken non-periodic finite quantum dot systems with long, smooth zigzag edges as they inherently include scattering events connecting two different k-states, arising from electronic correlations. The observations match with the periodic ZGNR systems for a similar level of theoretical calculation [48]. The finite size ZGNRs have been modeled within Hubbard Hamiltonian (see Eq. (4)) with on-site Coulomb correlation U. Here, we have considered $U/t = 1.0$ to ensure minimal bulk spin polarization. The semi-empirical many-body CI method [164] has been used for the numerical analysis of the Hubbard Hamiltonian. Here, n_h is the number of holes doped in the nanoribbon quantum dot. To investigate the conduction and magnetic properties, we calculate the charge gap (Δ_c) and spin gap (Δ_s), respectively, using [48, 50, 81, 114]

$$\Delta_c = E_0(N_e + 1, S_0) + E_0(N_e - 1, S_0) - 2E_0(N_e, S_0) \qquad (6)$$

and,

$$\Delta_s = E_0(N_e, S_0 + 1) - E_0(N_e, S_0) \qquad (7)$$

where $E_0(N_e, S_0)$ is the lowest energy in the subspace of N_e (= $N_{site} - n_h$) electrons with lowest S_z^{tot} (z-component of the total spin). Note that N_{site} denotes the number of sites of a finite size ZGNR system. Δ_c indicates the energy difference between charging and discharging, whereas Δ_s refers to the energy required to flip a spin, as evident from Eqs. 6 and 7, respectively [48]. Figure 6a, b shows the many-body charge-gap and the spin-gap nature, respectively, for finite-size ZGNR quantum dots with varying inverse length (L), that is, the number of unit cells for a fixed width ($N_z = 6$). We see that both Δ_c and Δ_s decrease upon hole doping. This indicates that hole-doped finite-size ZGNRs become magnetic and conducting. The spin gap is very low even in a half-filled system with no hole doping ($n_h = 0$). This suggests that the magnetic state is easily accessible [48].

Figure 6c shows the spin-spin correlation $\langle S_0^z S_j^z \rangle$ for a finite-size ZGNR quantum dot with long and smooth zigzag edges, with varying hole doping [48]. The spin-spin correlation is calculated between an atom located at "0" edge site (depicted in the middle panel) with all other atoms along the same edge (top panel) and opposite edge (bottom panel) denoted by j. Since the spins show ferrimagnetic alignment along the same zigzag edge (top panel), the spin-spin correlation is positive. However, this positive spin-spin correlation shows a slow power-law decay with an increase in hole doping, since the edge electrons gradually vanish. This makes the zigzag edge magnetism in ZGNRs quite robust even after hole doping [187]. On the other hand, the spin-spin correlation function, as shown in the bottom panel of Fig. 6c, is negative, indicating that the spins at opposite edges are antiferromagnetically coupled. This antiferromagnetic coupling drastically decreases with hole doping, making the spin excitation energy almost zero that results in a magnetic ground state [48].

Hole-doped ZGNRs have a magnetic ground state with a reduced charge gap. This suggests that these systems behave as magnetic conductors with an excess of one kind of spin in the charge carriers. This unique property can be utilized in spin-filtering devices by injecting carriers through gate electrodes, as shown by the schematic device arrangement in Fig. 7. The antiferromagnetic and semiconducting ZGNR (see Fig. 7a) can be tailored to be used as conducting spin filters by inducing holes through the application of a gate voltage (see Fig. 7b). Thus, this proposed experimental setup can achieve spin-dependent magneto-transport [48].

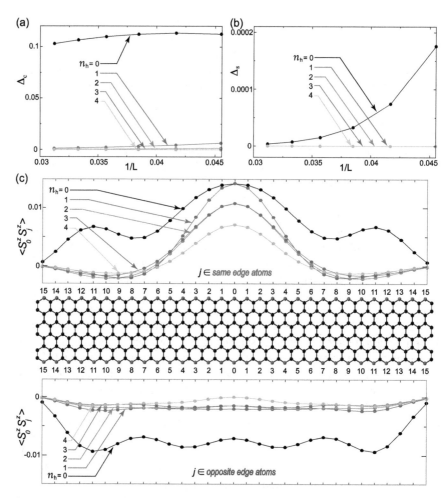

Fig. 6 **a** The charge gap (Δ_c) and **b** spin gap (Δ_s) for finite ZGNR constriction ($N_z = 6$) as a function of inverse length (L), that is, the number of unit cells, for different number of hole (n_h) doping. **c** The spin-spin correlation function ($\langle S_0^z S_j^z \rangle$), calculated from the "0th" atom, located at the middle of the edge of the ZGNR structure (middle panel) with all the edge atoms (j), along the same edge (top panel) and along the opposite edge (bottom panel) for varying number of hole (n_h) doping. Here, $U/t = 1.0$. (Adapted with permission from reference [48]. Copyright (2012) by Springer Nature)

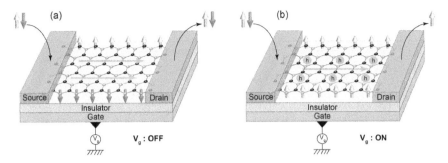

Fig. 7 Schematic of possible spintronic device applications of ZGNRs. **a** In the absence of gate bias (V_g), the ZGNRs show semiconducting and antiferromagnetic ground state with anti-parallel spin orientation between opposite edges. **b** Hole doping induced by the gate electrode turns the ZGNR metallic and magnetic, enabling it to be used in spin filter-based devices

6 Electric Field-Induced Half-Metallicity in ZGNR

ZGNRs are theoretically predicted to show half-metallic properties on the application of an external electric field along the cross-ribbon direction [43, 46, 160]. This property is very crucial for spintronics. In half-metallic materials, the spin degeneracy near the Fermi energy is lifted. Unlike conventional metals and semiconductors, half-metals show metallicity for electrons with one spin orientation and semiconducting or insulating nature for the other spin channel. This is shown schematically by the DOS in Fig. 8a. Half-metallic materials were previously observed in Heusler compounds and manganese perovskites [101, 138]. Being a source of spin-polarized current, their unique spin-filtering capability makes them vital for spintronic device applications. Half-metallicity can also be exploited to enhance the efficiency of magnetic data reading and consequent memory storage devices. In the case of ZGNRs, the external electric field lifts the spin degeneracy, and a half-metallic state is induced in the system, as shown in Fig. 8b, c [160].

We show a theoretical setup for the electric-field-induced half-metallic device based on ZGNRs in Fig. 9. The electric voltage V is applied in the transverse direction of ZGNRs (see Fig. 2a); thus, the cross-ribbon electric field E is given by $E = \frac{V}{W_z}$. The potential term V due to the transverse electric field is added in the mean-field Hubbard Hamiltonian, as described in the previous section. Here, the Hubbard Hamiltonian also includes the second nearest neighbor hopping terms (t'). The inclusion of t' helps to reproduce the energy dispersion as obtained from the DFT calculations with the local spin density approximation (LSDA) [160]. It does not incur any change in the qualitative observations except breaking the particle-hole symmetry of the energy spectrum. However, the mirror symmetry of the magnetization remains preserved. The energy band structure based on this mean-field Hubbard model with $t'/t = 0.1$ and $U/t = 1.0$ is shown in Fig. 10a. It qualitatively agrees with the energy dispersion obtained by Son et al. using the DFT-LSDA approach [160]. Here, the

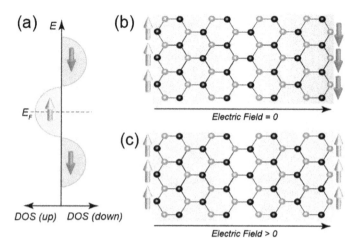

Fig. 8 **a** Half-metallic DOS. The left (right) side shows the DOS for up- (down-) spins, where the degeneracy between the two spin states is lifted. Only the up-spin states can contribute to electronic conduction as they possess a finite DOS near the Fermi energy (E_F). The down-spin states, however, are either semiconducting or insulating. **b** and **c** show the schematic representation of the spatial spin distribution of the valence band states without and with the application of an external electric field along the cross- ribbon direction, respectively

Fig. 9 The schematic representation of ZGNRs with the application of an external transverse electric field along the cross-ribbon width direction, using the source-drain bias voltage $\pm V$

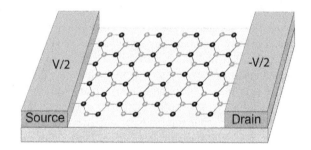

up-spin and down-spin bands are degenerate. Figure 10b, c shows the energy dispersion and the corresponding DOS with a gradual increase in the transverse electric field. We see that the spin degeneracy is lifted on the application of the applied electric field. The up-spins' energy gap decreases with an increase in the electric field, while that for the down-spins remains open. Moreover, the decreasing up-spin band gap transitions from direct to indirect as the electric field increases. Consequently, only the up-spins become conducting near the Fermi energy, making the system half-metallic under the external electric field's application.

These observations are consistent with varying ribbon width (see Fig. 11). However, the semiconducting degenerate band gap and the critical electric field strength that is required to get half-metallicity decrease with an increase in ribbon width, as observed within first-principles-based calculations (see Fig. 11) [43, 160]. The semiconductor to half-metal transition becomes faster as the electrostatic poten-

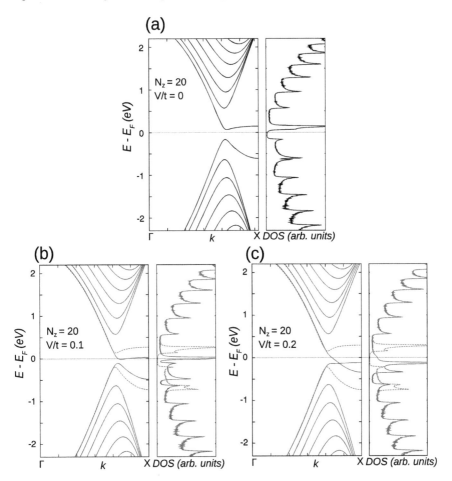

Fig. 10 **a** Energy dispersion and corresponding DOS of ZGNRs with $N_z = 20$ without the application of the external electric field. The up-spin and down-spin states are degenerate. The band structures and DOS in case of an applied electric field are shown in **b** $V/t = 0.1$ and **c** $V/t = 0.2$. In **b** and **c**, the solid blue and the red dashed lines denote the up-spin and down-spin states, respectively. Here, the ZGNRs are modeled within mean-field Hubbard Hamiltonian with on-site Coulomb interaction ($U/t = 1$) and the hopping between the next nearest neighbor sites ($t'/t = 0.1$). (Adapted with permission from reference [186]. Copyright (2012) by Elsevier)

tial difference between opposite edges becomes larger with an increase in width [46, 160].

Recent studies have shown that the phenomenon of half-metallicity of ZGNRs under the influence of a transverse electric field is universal and is well preserved even for finite and extremely short ribbon widths [83]. In the next two sections, we discuss the effect of selective edge doping and chemical modifications in ZGNRs to achieve half-metallicity even without applying an external electric field and the consequent spintronic properties of GNRs that can be used for device applications.

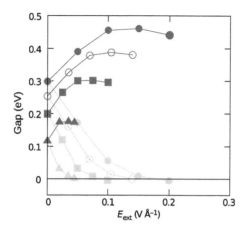

Fig. 11 Spin-polarized band gap as a function of the applied transverse electric field in ZGNRs with $N_z = 8, 11, 16$, and 32 (filled circles, open circles, squares, and triangles, respectively). The results have been obtained using DFT. The light yellow (dark green) color denotes the up- (down-) spin states, respectively. At zero external bias, the up- and down-spin gaps are degenerate. The transverse electric field removes the spin degeneracy, resulting in an increase in the gap for the down-spins, while reducing the up-spin gap. (Adapted and suitably modified with permission from reference [160]. Copyright (2006) by Springer Nature)

7   Effect of Chemical Modifications in ZGNRs

The magnetic and conducting states of ZGNRs can be modified by selectively replacing all the edge carbon atoms with boron (B) or nitrogen (N) atoms [43, 186]. The ground state of ZGNRs with hydrogen-passivated edges is antiferromagnetic and is stabler than the ferromagnetic state by a small margin of meV order [43, 160]. Instead of hydrogen passivation, if we replace the edge carbon atoms with B or N atoms, the electronic properties of ZGNRs change drastically. The introduction of B or N atoms dopes the system as it introduces holes or electrons in GNRs, respectively. In addition to that, they can also scatter the charge carriers or can distort the lattice [141]. The replacement of edge carbon atoms by B results in linear band dispersion near Fermi energy, and hence makes the ZGNRs metallic [186]. The B atoms, having one less electron as compared to carbon atoms, make the system hole-doped and lead to a ferromagnetic ground state where the spins along the same and opposite edges tend to align parallelly [43, 186]. Similar Nagaoka ferromagnetism has been proposed in previous theoretical works for strongly correlated half-filled low-dimensional systems in the presence of holes [157]. Moreover, this result agrees with the hole doping-induced magnetic and metallic ground states of ZGNRs discussed in Sect. 5 [48].

From the DOS analysis, as shown in Fig. 12, it is clear that, even without the transverse electric field, B-doped ZGNRs show half-metallicity [43]. This half-metallic ground state remains robust under forward or reverse bias. Note that the gap for

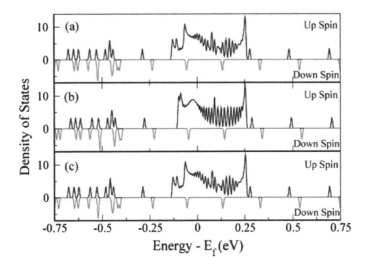

Fig. 12 Up- and down-spin DOS of ZGNR ($N_z = 8$) with both the edge atoms replaced by electron-deficient boron atoms in the presence of transverse electric fields, **a** -0.2, **b** 0.0, and **c** 0.2 V Å^{-1}. The energy is scaled with respect to the Fermi energy (E_F). (Adapted with permission from reference [43]. Copyright (2008) by American Chemical Society)

minority spin channel ($\approx$0.2 eV) is much higher than the room temperature (0.025 eV) and remains unaffected by the external field (see Fig. 12). Since the system is ferromagnetic with a similar spin polarization of opposite edges, the polarity of the external transverse electric field does not affect the spin polarization of conducting electrons. The projected DOS (pDOS) analysis for B-doped ZGNRs also shows that the edge B atoms significantly contribute to the majority spin DOS at Fermi energy [43]. Thus on B-doping, ZGNRs become half-metallic irrespective of width and the external field strength, and this ground state remains robust even at room temperature. This immensely increases the potential of these nanoribbon systems to be used as spintronic devices [43].

Replacing the edge carbon atoms with nitrogen atoms makes the ZGNRs electron-doped and the ground state antiferromagnetic with anti-parallel spin alignment between opposite edges. However, relatively small electron accumulation along the edges makes the ZGNRs weakly metallic, sustaining the transverse electric field [43]. Similar pDOS analysis shows that edge N atoms are dominant contributors to the DOS at the Fermi energy [43].

Instead of selectively doping the edge carbon atoms, other effective chemical modification schemes enhance the device applicability of GNRs by retaining the system's intrinsic half-metallicity without introducing any additional charge carriers [44]. This can be done by replacing two consecutive carbon (electronic configuration: $1s^2 2s^2 2p^2$) atoms in ZGNRs with one nitrogen (electronic configuration: $1s^2 2s^2 2p^3$) atom and one boron (electronic configuration: $1s^2 2s^2 2p^1$) atom, by substituting the

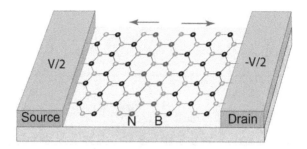

Fig. 13 The schematic representation of a modified ZGNR system ($N_z = 8$) in the presence of external transverse source-drain bias. Here, the two zigzag carbon chains in the middle are replaced by isoelectronic boron-nitrogen chains. The doping concentration increases gradually from the middle toward the edge, as indicated by the arrows

zigzag carbon chains with zigzag boron-nitrogen chains, starting from the middle of the ZGNRs with progressive replacement toward edges (see Fig. 13). Clearly, adding B-N chemical dopants does not result in any addition of extra electrons and holes, and thus the doped system remains isoelectronic with pristine hydrogen passivated ZGNR (Fig. 13). The system depicted in Fig. 13 shows that the chemical modification of ZGNRs results in a polyborazine structure. Replacement of all carbon zigzag chains by the isoelectronic B-N chains finally results in zigzag boron nitride nanoribbons (ZBNNR). The ZBNNR show an antiferromagnetic ground state with a wide band gap insulating behavior, in spite of being isoelectronic with ZGNRs behavior [44, 137, 206].

Figure 14 presents the DOS of the ZGNR system ($N_z = 8$) with the replacement of ($a1$) 2, ($b1$) 4, and ($c1$) 6 middle zigzag carbon chains with B-N chains. The ZGNR system with two zigzag B-N chains at the middle shows an asymmetric energy gap for up and down spin channels. Replacement of four middle zigzag carbon chains by B-N chains results in half-metallic behavior with one conducting spin channel, leaving the other spin channel semiconducting. However, the system with two zigzag carbon chains on either sides and all the six zigzag B-N chains at the middle exhibits isolated DOS peaks, a signature of nondispersive bands and localized electrons [44].

Unlike ZGNRs, the electrons in the ZBNNRs do not show any spin ordering. The spin density plot in the half-metallic case (in Fig. 14(b2)) suggests the breaking of bipartite symmetry when carbon gets attached to B or N. The Lewis acid character of boron results in an electron transfer from carbon to boron. Similarly, nitrogen atoms donate the electrons to adjacent carbon atoms, due to their Lewis base character. Consequently, this creates a potential gradient across the ribbon width that has a striking resemblance with the case of undoped hydrogen passivated ZGNRs under a transverse electric field [160], thus effectively inducing an intrinsic half-metallicity in the modified ZGNR systems [44].

Note that, except the system with two zigzag carbon chains in either edges with all B-N chains in the middle, all other systems are made by stitching the polyacene and polyborazine units that are constituents of ZGNRs and ZBNNRs, respectively.

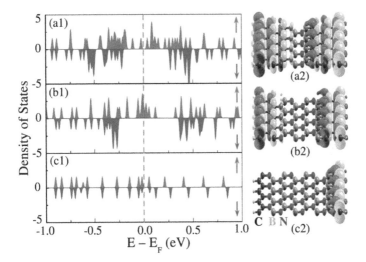

Fig. 14 The DOS plot for the ZGNRs ($N_z = 8$), with (a1) 2, (b1) 4, and (c1) 6 zigzag boron-nitrogen chains at the middle. The energy is scaled with respect to the Fermi energy (E_F). (a2), (b2), and (c2) show their respective spin density profiles. (Adapted with permission from reference [44]. Copyright (2009) by American Physical Society)

That is why, this system shows very distinct spin localization along only one edge unlike the other systems (see Fig. 14(c2)). The half-metallic behavior, as observed in the system with middle B-N chains that are sandwiched between two polyacene units on either sides, remains unchanged for any ribbon width. This unique behavior arises mainly from the carbon atoms that are attached to the boron atoms and subsequent polyacene units on the same side, as evident from the pDOS analysis. This intrinsic half-metallicity remains robust against a large external electric field with the semiconducting gap for one kind of spin exceeding the room temperature [44]. These results show that chemical modification of graphene nanoribbons can open new avenues to study intrinsic half-metallicity at room temperature for advanced spintronic device applications.

8 Transport Properties of Graphene Nanoribbons

We have discussed how the edge states give rise to nanographene's unusual electronic properties. We have also discussed how half-metallic conduction and other spintronic properties that emerge in ZGNRs can be utilized for carbon-based nano-material devices. The electronic transport properties in doped and gated graphene nanostructures have also generated significant interest in the scientific community [37]. Nanojunctions made of graphene can show zero-conductance Fano resonances, having potential applications in current control and switching devices [176, 177,

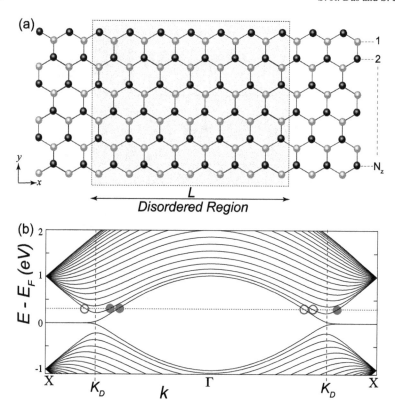

Fig. 15 a Structure of disordered ZGNR. The shaded region with length L denotes the disordered region with randomly distributed impurities. **b** Energy dispersion of ZGNRs with $N_z = 10$. The vertical dashed lines at K_D show the locations of valleys in the energy dispersion, originating from the Dirac (**K, K'**) points of graphene. The red-filled (blue-unfilled) circles denote the right- (left-) going open channel at energy slightly above E_F. The degeneracy between these two channels is lifted in the left (right) valley because of one excess right (left)-going mode. This breaks the time-reversal symmetry under the intravalley scattering

179, 197]. Previous studies have shown that the localized edge states of disordered nanographene exhibit peculiar charge transport that transitions between variable range hopping at low temperatures and thermally activated transport at higher temperatures [74]. The inclusion of spin-orbit interaction in ZGNRs can also be used as a model to study the quantum spin Hall effect [99] and topological insulators [77]. Recent experiments have also obtained signatures of edge-mode transport in atomic-scale constrictions in single and multilayer nanographene [102].

Hence, in this section, we will see how the edge boundary conditions are intricately linked to the electronic transport properties of graphene ribbons. In particular, we will focus on disorder effects on the electronic transport properties of zigzag graphene nanoribbons. As shown in Fig. 15a, we have a disordered region of length L in ZGNRs with width N_z. In Fig. 15b, we show the energy dispersion for the disordered ZGNRs with $N_z = 10$. We have already discussed through our tight-binding calculations in the previous section that the ZGNRs show metallic behavior at finite doping because of the partial flat band at the Fermi energy induced by the edge states, irrespective of their width. To probe the role of the edge states in the electronic transport properties of ZGNRs, we treat them as a new class of quantum wires. In general, the scattering matrix (S) formalism is used to describe the electron scattering in a quantum wire [19]. The relation between the scattering wave O and the incident wave I can be represented as [182, 183],

$$\begin{pmatrix} O_L \\ O_R \end{pmatrix} = S \begin{pmatrix} I_L \\ I_R \end{pmatrix} = \begin{pmatrix} r & t' \\ t & r' \end{pmatrix} \begin{pmatrix} I_L \\ I_R \end{pmatrix} \tag{8}$$

where L (R) denote left (right) lead, while r (r') and t (t') represent reflection and transmission matrices, respectively. In the quantum wire regime, the phase coherence length and electronic mean free path are larger than the system size, and hence the electronic conductance of the mesoscopic systems can be evaluated by the Landauer-Büttiker formula [23] by using the S-matrix as follows [182, 183]:

$$G(E) = \frac{e^2}{\pi\hbar} Tr(tt^\dagger) = \frac{e^2}{\pi\hbar} g(E) \tag{9}$$

We express the electronic conductance in the unit of conductance quanta ($\frac{e^2}{\pi\hbar}$), i.e., as dimensionless conductance $g(E)$.

For a pure system, the transmission probability is unity in the clean limit. Hence, the number of conducting channels can give dimensionless conductance at zero temperature. As shown in Fig. 15b, there is always one excess left-going channel in the right valley and one excess right-going channel in the left valley within the energy window of $|E - E_F| \leq 1$. However, the number of right-going and left-going channels in the whole system remains the same, maintaining the overall time-reversal symmetry. However, there is always one excess channel in one direction at each valley, which we call the chiral mode. Using this information, we can write the dimensionless conductance as $g(E) = 2n + 1$ for ZGNRs, where $n = 0, 1, 2, ...$ [182].

Figure 15a shows that the impurities are randomly distributed with a density n_{imp} in the disordered region of the nanoribbons. The large momentum difference between the two valleys requires short-range impurities (SRI) with a range smaller than the lattice constant for intervalley scattering, whereas long-range impurities (LRI) are the cause of intravalley scattering [10]. We expect random impurities will cause Anderson localization [6], which will make the electrons localized over a certain region, rather than spread out over the whole system. Hence, the conductance will decay exponentially with an increase in the disorder length L and vanish in the limit of $L \to \infty$. However, if we look at Fig. 16a, the average dimensionless conductance $\langle g \rangle$ for different incident energies E (Fermi energies) shows a typical localization effect that gradually decreases with growing L. However, in the limit $L \to \infty$, $\langle g \rangle$ converges to 1 for LRIs, as shown in Fig. 16a. This indicates a single perfectly con-

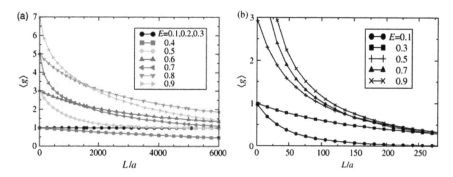

Fig. 16 Dependence of disorder length (L) on the average dimensionless conductance $\langle g \rangle$ for ZGNRs with $N_z = 10$, in the **a** absence and **b** presence of intervalley scattering. The potential strength and the impurity density are considered to be $u = 1.0$ and $n_{imp} = 0.1$, respectively. (Adapted with permission from reference [183]. Copyright (2009) by Institute of Physics)

ducting channel in ZGNRs related to the chiral propagating mode that we have discussed before. Thus, ZGNRs with long-range impurities do not exhibit Anderson localization [181]. However, in the presence of SRIs, the dominance of intervalley scattering suppresses the perfectly conducting channel (see Fig. 16b). Such phenomena can occur in large impurity concentrations in the disordered region or by edge-roughness that alters the ribbon width considerably over a short length. This is largely due to the localization of electrons and the consequent rapid decay of average conductance [181]. In such a situation, we observe perfect backward scattering due to zero-conductance Fano resonances in the low-energy transport regime [178].

Owing to the nontrivial Berry π-phase for the single-channel linear band originating from the low-energy Dirac dispersion, the AGNRs show a perfectly single conducting channel in the presence of LRIs [155]. Note that, unlike ZGNRs, the two valleys are not well separated in AGNRs.

Thus, we see how the perfectly conducting channel can be the origin of a robust electron transport mechanism in ZGNRs [181, 183]. Recent experiments have utilized this exceptional room temperature ballistic transport mechanism of ZGNRs to design ultra-low power consumption nanoelectronic devices [17].

9 Conclusion and Future Outlook

This chapter has reviewed the recent studies in the exciting field of quasi 1D graphene nanoribbons, given its important role in carbon-based nanomaterial device applications. In particular, we have reviewed the theoretical investigations on the nanoscale edge effects on graphene's electronic properties. The presence of edges crucially affects the electronic states of nanographene. We have discussed two typical graphene edges in detail, namely *armchair* and *zigzag* and have shown how they exhibit completely different electronic properties. The properties of Dirac fermions remain intact in the case of armchair edge, while the zigzag edge induces localized edge states that give rise to a large peak in the DOS at the Fermi energy. The resulting Fermi instability leads to a ferrimagnetic spin polarization along the zigzag edge. Thus, ZGNRs show magnetic properties that are absent in pristine two-dimensional graphene. We have shown how these magnetic states respond to carrier doping. Such metal-free magnetic states can be exploited to realize carbon-based spintronic devices.

Under the influence of the external transverse electric field, the semiconducting ZGNRs also exhibit half-metallic behavior. A similar property can be achieved by selective doping of edge carbon atoms in ZGNR by borons and chemically modifying ZGNRs by B-N pair replacements. These observations pave novel ways to control the spin-polarized current in quasi-one-dimensional carbon systems.

The edge states in ZGNRs can induce a perfectly conducting channel with the absence of Anderson localization. We have discussed this peculiar spin transport behavior of nanoribbons that will help in designing ultra-low power consumption electronics based on carbon nanomaterials.

Nanomaterials based on low-dimensional systems are an emerging field, and rapid progress is made almost every day. Hence, it is challenging to give a comprehensive review of this fast-evolving topic. Thus, we have merely scratched the surface of this rapidly developing field in materials science and have not included many significant aspects where cutting-edge research is currently going on for spintronic device applications.

We have skipped discussing lattice defects or vacancies, but studies have shown that they can also be a source of magnetism in nanographene [201]. The spin-orbit interaction also plays an important role, which can be tuned by hydrogenation to induce the spin Hall effect [15]. The unconventional electronic properties in graphene have also motivated researchers to go beyond the conventional carbon allotropes and study spintronics and unusual properties in other low-dimensional materials [14], like monolayers of hexagonal boron nitride (hBN) [194], phosphorene [196], silicene [58, 173], BC_3 [47, 49, 96], transition metal dichalcogenides (TMDs) [120, 189], among others [63]. The proximity-induced SOC enhancement [192] and optical spin injection [13, 65] have made heterostructures made of graphene and TMDs very appealing for spintronics and optospintronics. Experimental and theoretical works on atomic-scale self-assembled TMD heterojunctions formed by controlled oxidation have also shown how these materials can be exploited for advanced electronic, optoelectronic, and valleytronic device applications [35, 198]. Such laterally stitched or vertically stacked Van der Waals heterostructures made from different 2D materials have also become very popular and are actively exploited in next-generation spin filters and spin-rectifying devices [36].

Very recently, experiments in exchange anisotropy-induced 2D magnetism in layered materials like CrI_3 [84] and Cr_2Ge2Te_6 [66], etc., have opened a new area of research in condensed matter and materials science to utilize their magneto-electric and magneto-optical effects and interface them with other low-dimensional materials to form heterostructures and engineer new physics. The recent surge in research to unravel the unusual flat band topology, high T_c superconducting phases, and emergent magnetism observed in twisted bilayer low-dimensional systems has also been very promising for electronic applications [26, 27, 32, 156]. We hope that this continued and sustained way to explore the "plenty of room" in these nanoscale low-dimensional systems will aid in their realization of next-generation nanomaterial devices.

Acknowledgements SRD and SD thank IISER Tirupati for Intramural Funding and Science and Engineering Research Board (SERB), Dept. of Science and Technology (DST), Govt. of India, for Early Career Research (ECR) award grant (ECR/2016/000283).

References

1. Abanin DA, Lee PA, Levitov LS (2006) Spin-filtered edge states and quantum Hall effect in graphene. Phys Rev Lett 96:176803
2. Abanin DA, Novoselov KS, Zeitler U, Lee PA, Geim AK, Levitov LS (2007) Dissipative quantum Hall effect in graphene near the Dirac point. Phys Rev Lett 98:196806
3. Ajayan PM (1999) Nanotubes from carbon. Chem Rev 99:1787–1800
4. Ajiki H, Ando T (1993) Electronic states of carbon nanotubes. J Phys Soc Jpn 62:1255–1266
5. Alicea J, Fisher MPA (2006) Graphene integer quantum Hall effect in the ferromagnetic and paramagnetic regimes. Phys Rev B 74:075422
6. Anderson PW (1958) Absence of diffusion in certain random lattices. Phys Rev 109:1492–1505
7. Ando T (1997) Excitons in carbon nanotubes. J Phys Soc Jpn 66:1066–1073
8. Ando T, Nakanishi T (1998) Impurity scattering in carbon nanotubes—absence of back scattering. J Phys Soc Jpn 67:1704–1713
9. Ando T, Nakanishi T, Saito R (1998) Berry's phase and absence of back scattering in carbon nanotubes. J Phys Soc Jpn 67:2857–2862
10. Ando T (2005) Theory of electronic states and transport in carbon nanotubes. J Phys Soc Jpn 74:777–817
11. Ando T (2008) Physics of graphene: zero-mode anomalies and roles of symmetry. Prog Theor Phys Suppl 176:203–226
12. Avouris P (2002) Molecular electronics with carbon nanotubes. Acc Chem Res 35:1026–1034
13. Avsar A, Unuchek D, Liu J, Sanchez OL, Watanabe K, Taniguchi T, Özyilmaz B, Kis A (2017) Optospintronics in graphene via proximity coupling. ACS Nano 11:11678–11686
14. Avsar A, Ochoa H, Guinea F, Özyilmaz B, van Wees BJ, Vera-Marun IJ (2020) Colloquium : spintronics in graphene and other two-dimensional materials. Rev Mod Phys 92:021003
15. Balakrishnan J, Kok Wai Koon G, Jaiswal M, Castro Neto AH, Özyilmaz B (2013) Colossal enhancement of spin- orbit coupling in weakly hydrogenated graphene. Nat Phys 9:284–287
16. Balandin AA, Ghosh S, Bao W, Calizo I, Teweldebrhan D, Miao F, Lau CN (2008) Superior thermal conductivity of single-layer graphene. Nano Lett 8:902–907
17. Baringhaus J, Ruan M, Edler F, Tejeda A, Sicot M, Taleb-Ibrahimi A, Li AP, Jiang Z, Conrad EH, Berger C, Tegenkamp C, de Heer WA (2014) Exceptional ballistic transport in epitaxial graphene nanoribbons. Nature 506:349–354
18. Baughman RH (2002) Carbon nanotubes-the route toward applications. Science 297:787–792
19. Beenakker CWJ (1997) Random-matrix theory of quantum transport. Rev Mod Phys 69:731–808
20. Biel B, Blase X, Triozon F, Roche S (2009) Anomalous doping effects on charge transport in graphene nanoribbons. Phys Rev Lett 102:096803
21. Bolotin KI, Sikes KJ, Jiang Z, Klima M, Fudenberg G, Hone J, Kim P, Stormer HL (2008) Ultrahigh electron mobility in suspended graphene. Solid State Commun 146:351–355

22. Brey L, Fertig HA (2006) Edge states and the quantized Hall effect in graphene. Phys Rev B 73:195408
23. Büttiker M, Imry Y, Landauer R, Pinhas S (1985) Generalized many-channel conductance formula with application to small rings. Phys Rev B 31:6207–6215
24. Cai J, Ruffieux P, Jaafar R, Bieri M, Braun T, Blankenburg S, Muoth M, Seitsonen AP, Saleh M, Feng X, Müllen K, Fasel R (2010) Atomically precise bottom-up fabrication of graphene nanoribbons. Nature 466:470–473
25. Cao T, Zhao F, Louie SG (2017) Topological phases in graphene nanoribbons: junction states, spin centers, and quantum spin chains. Phys Rev Lett 119:076401
26. Cao Y, Fatemi V, Demir A, Fang S, Tomarken SL, Luo JY, Sanchez-Yamagishi JD, Watanabe K, Taniguchi T, Kaxiras E, Ashoori RC, Jarillo-Herrero P (2018a) Correlated insulator behaviour at half-filling in magic-angle graphene superlattices. Nature 556:80–84
27. Cao Y, Fatemi V, Fang S, Watanabe K, Taniguchi T, Kaxiras E, Jarillo-Herrero P (2018b) Unconventional superconductivity in magic-angle graphene superlattices. Nature 556:43–50
28. Castro Neto AH, Guinea F, Peres NMR (2006) Edge and surface states in the quantum Hall effect in graphene. Phys Rev B 73:205408
29. Castro Neto AH, Guinea F, Peres NMR, Novoselov KS, Geim AK (2009) The electronic properties of graphene. Rev Mod Phys 81:109–162
30. Cervantes-Sodi F, Csányi G, Piscanec S, Ferrari AC (2008) Edge-functionalized and substitutionally doped graphene nanoribbons: Electronic and spin properties. Phys Rev B 79:677–732
31. Charlier JC, Blase X, Roche S (2007) Electronic and transport properties of nanotubes. Rev Mod Phys 79:677–732
32. Chen G, Sharpe AL, Fox EJ, Zhang YH, Wang S, Jiang L, Lyu B, Li H, Watanabe K, Taniguchi T, Shi Z, Senthil T, Goldhaber-Gordon D, Zhang Y, Wang F (2020) Tunable correlated Chern insulator and ferromagnetism in a moiré superlattice. Nature 579:56–61
33. Chen W, Chen S, Qi DC, Gao XY, Wee ATS (2007a) Surface transfer p-type doping of epitaxial graphene. J Am Chem Soc 129:10418–10422
34. Chen Z, Lin YM, Rooks MJ, Avouris P (2007b) Graphene nano-ribbon electronics. Physica E Low Dimens Syst Nanostruct 40:228–232
35. Das SR, Wakabayashi K, Yamamoto M, Tsukagoshi K, Dutta S (2018) Layer-by-layer oxidation induced electronic properties in transition-metal dichalcogenides. J Phys Chem C 122:17001–17007
36. Das SR, Dutta S (2019) Spin filtering and rectification in lateral heterostructures of zigzag-edge BC3 and graphene nanoribbons: implications for switching and memory devices. ACS Appl Nano Mater 2:5365–5372
37. Das Sarma S, Adam S, Hwang EH, Rossi E (2011) Electronic transport in two-dimensional graphene. Rev Mod Phys 83:407–470
38. Delplace P, Ullmo D, Montambaux G (2011) Zak phase and the existence of edge states in graphene. Phys Rev B 84:195452
39. Dresselhaus M (1995) Carbon nanotubes. Carbon 33:871–872
40. Dresselhaus MS, Dresselhaus G, Eklund PC (1996) Science of fullerenes and carbon nanotubes. Academic Press, San Diego
41. Drögeler M, Franzen C, Volmer F, Pohlmann T, Banszerus L, Wolter M, Watanabe K, Taniguchi T, Stampfer C, Beschoten B (2016) Spin lifetimes exceeding 12 ns in graphene nonlocal spin valve devices. Nano Lett 16:3533–3539
42. Dutta S, Lakshmi S, Pati SK (2008) Electron-electron interactions on the edge states of graphene: a many-body configuration interaction study. Phys Rev B 77:073412
43. Dutta S, Pati SK (2008) Half-metallicity in undoped and boron doped graphene nanoribbons in the presence of semilocal exchange-correlation interactions. J Phys Chem B 112:1333–1335
44. Dutta S, Manna AK, Pati SK (2009) Intrinsic half-metallicity in modified graphene nanoribbons. Phys Rev Lett 102:096601
45. Dutta S, Pati SK (2010a) Edge reconstructions induce magnetic and metallic behavior in zigzag graphene nanoribbons. Carbon 48:4409–4413

46. Dutta S, Pati SK (2010b) Novel properties of graphene nanoribbons: a review. J Mater Chem 20:8207–8223
47. Dutta S, Wakabayashi K (2012a) Anomalous energy-gap behaviour of armchair BC3 ribbons due to enhanced π-conjugation. J Mater Chem 22:20881
48. Dutta S, Wakabayashi K (2012b) Tuning charge and spin excitations in zigzag edge nanographene ribbons. Sci Rep 2:519
49. Dutta S, Wakabayashi K (2013a) Edge state induced metallicity in zigzag BC3 ribbons. J Mater Chem C 1:4854
50. Dutta S, Wakabayashi K (2013b) Interacting spins and holes in zigzag edge nanographene. AIP Conf Proc 1566:153–154
51. Dutta S, Wakabayashi K (2014) Spin and charge excitations in zigzag honeycomb nanoribbons: effect of many body correlation. Jpn J Appl Phys 53:06JD01
52. Dutta S, Wakabayashi K (2015) Magnetization due to localized states on graphene grain boundary. Sci Rep 5:11744
53. Ebbesen TW (1994) Carbon nanotubes. Annu Rev Mater Sci 24:235–264
54. Ebbesen TW, Ajayan PM (1992) Large-scale synthesis of carbon nanotubes. Nature 358:220–222
55. Ezawa M (2007) Metallic graphene nanodisks: Electronic and magnetic properties. Phys Rev B 76:245415
56. Fasolino A, Los JH, Katsnelson MI (2007) Intrinsic ripples in graphene. Nat Mater 6:858–861
57. Fernández-Rossier J, Palacios JJ (2007) Magnetism in graphene nanoislands. Phys Rev Lett 99:177204
58. Fleurence A, Friedlein R, Ozaki T, Kawai H, Wang Y, Yamada-Takamura Y (2012) Experimental evidence for epitaxial silicene on diboride thin films. Phys Rev Lett 108:245501
59. Fujita M, Wakabayashi K, Nakada K, Kusakabe K (1996) Peculiar localized state at zigzag graphite edge. J Phys Soc Jpn 65:1920–1923
60. Fujita M, Igami M, Nakada K (1997) Lattice distortion in nanographite ribbons. J Phys Soc Jpn 66:1864–1867
61. Garcia JH, Vila M, Cummings AW, Roche S (2018) Spin transport in graphene/transition metal dichalcogenide heterostructures. Chem Soc Rev 47:3359–3379
62. Gebeyehu ZM, Parui S, Sierra JF, Timmermans M, Esplandiu MJ, Brems S, Huyghebaert C, Garello K, Costache MV, Valenzuela SO (2019) Spin communication over 30 μm long channels of chemical vapor deposited graphene on SiO 2. 2D Mater 6:034003
63. Geim AK, Grigorieva IV (2013) van der Waals heterostructures. Nature 499:419–425
64. Geim AK, Novoselov KS (2007) The rise of graphene. Nat Mater 6:183–191
65. Gmitra M, Fabian J (2015) Graphene on transition-metal dichalcogenides: A platform for proximity spin-orbit physics and optospintronics. Phys Rev B 92:155403
66. Gong C, Li L, Li Z, Ji H, Stern A, Xia Y, Cao T, Bao W, Wang C, Wang Y, Qiu ZQ, Cava RJ, Louie SG, Xia J, Zhang X (2017) Discovery of intrinsic ferromagnetism in two-dimensional van der Waals crystals. Nature 546:265–269
67. González J, Perfetto E (2007) Cooper-pair propagation and superconducting correlations in graphene. Phys Rev B 76:155404
68. Guinea F, Katsnelson MI, Geim AK (2010) Energy gaps and a zero-field quantum Hall effect in graphene by strain engineering. Nat Phys 6:30–33
69. Guldi DM, Prato M (2000) Excited-state properties of C^{60} fullerene derivatives. Acc Chem Res 33:695–703
70. Gurram M, Omar S, van Wees BJ (2018) Electrical spin injection, transport, and detection in graphene-hexagonal boron nitride van der Waals heterostructures: progress and perspectives. 2D Mater 5:032004
71. Gusynin VP, Sharapov SG (2005) Unconventional integer quantum Hall effect in graphene. Phys Rev Lett 95:146801
72. ter Haar D (2014) Collected papers of L.D. Landau. Elsevier Science, Saint Louis
73. Han MY, Özyilmaz B, Zhang Y, Kim P (2007) Energy band-gap engineering of graphene nanoribbons. Phys Rev Lett 98:206805

74. Han MY, Brant JC, Kim P (2010) Electron transport in disordered graphene nanoribbons. Phys Rev Lett 104:056801
75. Harigaya K (2001) The mechanism of magnetism in stacked nanographite: Theoretical study. J Phys: Condens Matter 13:1295–1302
76. Harigaya K, Enoki T (2002) Mechanism of magnetism in stacked nanographite with open shell electrons. Chem Phys Lett 351:128–134
77. Hasan MZ, Kane CL (2010) Colloquium: topological insulators. Rev Mod Phys 82:3045–3067
78. Hasegawa Y, Kohmoto M (2006) Quantum Hall effect and the topological number in graphene. Phys Rev B 74:155415
79. Heersche HB, Jarillo-Herrero P, Oostinga JB, Vandersypen LMK, Morpurgo AF (2007) Bipolar supercurrent in graphene. Nature 446:56–59
80. Herbut IF (2007) Theory of integer quantum Hall effect in graphene. Phys Rev B 75:165411
81. Hikihara T, Hu X, Lin HH, Mou CY (2003) Ground-state properties of nanographite systems with zigzag edges. Phys Rev B 68:035432
82. Hod O, Peralta JE, Scuseria GE (2007) Edge effects in finite elongated graphene nanoribbons. Phys Rev B 76:233401
83. Hod O, Barone V, Scuseria GE (2008) Half-metallic graphene nanodots: A comprehensive first-principles theoretical study. Phys Rev B 77:035411
84. Huang B, Clark G, Navarro-Moratalla E, Klein DR, Cheng R, Seyler KL, Zhong D, Schmidgall E, McGuire MA, Cobden DH, Yao W, Xiao D, Jarillo-Herrero P, Xu X (2017) Layer-dependent ferromagnetism in a van der Waals crystal down to the monolayer limit. Nature 546:270–273
85. Huertas-Hernando D, Guinea F, Brataas A (2006) Spin-orbit coupling in curved graphene, fullerenes, nanotubes, and nanotube caps. Phys Rev B 74:155426
86. Iijima S (1991) Helical microtubules of graphitic carbon. Nature 354:56–58
87. Iijima S, Ichihashi T (1993) Single-shell carbon nanotubes of 1-nm diameter. Nature 363:603–605
88. Ingla-Aynés J, Guimarães MHD, Meijerink RJ, Zomer PJ, van Wees BJ (2015) $24 - \mu$m spin relaxation length in boron nitride encapsulated bilayer graphene. Phys Rev B 92:201410
89. Jariwala D, Sangwan VK, Lauhon LJ, Marks TJ, Hersam MC (2013) Carbon nanomaterials for electronics, optoelectronics, photovoltaics, and sensing. Chem Soc Rev 42:2824–2860
90. Jiang Y, Yao DX, Carlson EW, Chen HD, Hu J (2008) Andreev conductance in the $d + id'$-wave superconducting states of graphene. Phys Rev B 77:235420
91. Jiang Z, Zhang Y, Stormer HL, Kim P (2007) Quantum Hall states near the charge-neutral Dirac point in graphene. Phys Rev Lett 99:106802
92. Jiao L, Zhang L, Wang X, Diankov G, Dai H (2009) Narrow graphene nanoribbons from carbon nanotubes. Nature 458:877–880
93. Jung J, MacDonald AH (2009) Carrier density and magnetism in graphene zigzag nanoribbons. Phys Rev B 79:235433
94. Jung J, Pereg-Barnea T, MacDonald AH (2009) Theory of interedge superexchange in zigzag edge magnetism. Phys Rev Lett 102:227205
95. Kamalakar MV, Groenveld C, Dankert A, Dash SP (2015) Long distance spin communication in chemical vapour deposited graphene. Nat Commun 6:6766
96. Kameda T, Liu F, Dutta S, Wakabayashi K (2019) Topological edge states induced by the Zak phase in A 3 B monolayers. Phys Rev B 99:075426
97. Kan EJ, Li Z, Yang J, Hou JG (2008) Half-metallicity in edge-modified zigzag graphene nanoribbons. J Am Chem Soc 130:4224–4225
98. Kane CL, Mele EJ (2005a) Z_2 Topological order and the quantum spin Hall effect. PhysRev Lett 95:146802
99. Kane CL, Mele EJ (2005b) Quantum spin Hall effect in graphene. Phys Rev Lett 95:226801
100. Katsnelson MI, Novoselov KS, Geim AK (2006) Chiral tunnelling and the Klein paradox in graphene. Nat Phys 2:620–625
101. Katsnelson MI, Irkhin VY, Chioncel L, Lichtenstein AI, de Groot RA (2008) Half-metallic ferromagnets: from band structure to many-body effects. Rev Mod Phys 80:315–378

102. Kinikar A, Phanindra Sai T, Bhattacharyya S, Agarwala A, Biswas T, Sarker SK, Krishna-murthy HR, Jain M, Shenoy VB, Ghosh A (2017) Quantized edge modes in atomic-scale point contacts in graphene. Nat Nanotechnol 12:564–568

103. Kobayashi Y, Ki Fukui, Enoki T, Kusakabe K, Kaburagi Y (2005) Observation of zigzag and armchair edges of graphite using scanning tunneling microscopy and spectroscopy. Phys Rev B 71:193406

104. Kobayashi Y, Ki Fukui, Enoki T, Kusakabe K (2006) Edge state on hydrogen-terminated graphite edges investigated by scanning tunneling microscopy. Phys Rev B 73:125415

105. Kosynkin DV, Higginbotham AL, Sinitskii A, Lomeda JR, Dimiev A, Price BK, Tour JM (2009) Longitudinal unzipping of carbon nanotubes to form graphene nanoribbons. Nature 458:872–876

106. Kroto HW, Heath JR, O'Brien SC, Curl RF, Smalley RE (1985) C60: Buckminsterfullerene. Nature 318:162–163

107. Kudin KN (2008) Zigzag graphene nanoribbons with saturated edges. ACS Nano 2:516–522

108. Kumazaki H, Hirashima S, D, (2007) Nonmagnetic-defect-induced magnetism in graphene. J Phys Soc Jpn 76:064713

109. Kumazaki H, Hirashima S, D, (2008) Local magnetic moment formation on edges of graphene. J Phys Soc Jpn 77:044705

110. Kusakabe K, Maruyama M (2003) Magnetic nanographite. Phys Rev B 67:092406

111. Landau LD, Lifšic EM, Pitaevskij LP, Landau LD (2008) Statistical Physics, Part 1, 3rd edn. Course of Theoretical Physics, Elsevier [u.a.], Amsterdam [u.a]

112. Li W, Zhao M, Xia Y, Zhang R, Mu Y (2009) Covalent-adsorption induced magnetism in graphene. J Mater Chem 19:9274–9282

113. Li X, Wang X, Zhang L, Lee S, Dai H (2008) Chemically derived, ultrasmooth graphene nanoribbon semiconductors. Science 319:1229–1232

114. Lieb EH, Wu FY (1968) Absence of Mott transition in an exact solution of the short-range, one-band model in one dimension. Phys Rev Lett 20:1445–1448

115. Lieb EH (1989) Two theorems on the Hubbard model. Phys Rev Lett 62:1201–1204

116. Liu F, Yamamoto M, Wakabayashi K (2017) Topological edge states of honeycomb lattices with Zzero Berry curvature. J Phys Soc Jpn 86:123707

117. Liu J, Li BW, Tan YZ, Giannakopoulos A, Sanchez-Sanchez C, Beljonne D, Ruffieux P, Fasel R, Feng X, Müllen K (2015) Toward cove-edged low band gap graphene nanoribbons. J Am Chem Soc 137:6097–6103

118. Luk'yanchuk IA, Kopelevich Y (2004) Phase analysis of quantum oscillations in graphite. Phys Rev Lett 93:166402

119. Magda GZ, Jin X, Hagymási I, Vancsó P, Osváth Z, Nemes-Incze P, Hwang C, Biró LP, Tapasztó L (2014) Room-temperature magnetic order on zigzag edges of narrow graphene nanoribbons. Nature 514:608–611

120. Mak KF, Lee C, Hone J, Shan J, Heinz TF (2010) Atomically thin MoS 2: A new direct-gap semiconductor. Phys Rev Lett 105:136805

121. Martins TB, Miwa RH, da Silva AJR, Fazzio A (2007) Electronic and transport properties of boron-doped graphene nanoribbons. Phys Rev Lett 98:196803

122. Meyer JC, Geim AK, Katsnelson MI, Novoselov KS, Booth TJ, Roth S (2007) The structure of suspended graphene sheets. Nature 446:60–63

123. Min H, Borghi G, Polini M, MacDonald AH (2008) Pseudospin magnetism in graphene. Phys Rev B 77:041407

124. Miyamoto Y, Nakada K, Fujita M (1999) First-principles study of edge states of H-terminated graphitic ribbons. Phys Rev B 59:9858–9861

125. Miyazaki H, Odaka S, Sato T, Tanaka S, Goto H, Kanda A, Tsukagoshi K, Ootuka Y, Aoyagi Y (2008) Coulomb blockade oscillations in narrow corrugated graphite ribbons. Appl Phys Express 1:024001

126. Molitor F, Güttinger J, Stampfer C, Graf D, Ihn T, Ensslin K (2007) Local gating of a graphene Hall bar by graphene side gates. Phys Rev B 76:245426

127. Nakada K, Fujita M, Dresselhaus G, Dresselhaus MS (1996) Edge state in graphene ribbons: nanometer size effect and edge shape dependence. Phys Rev B 54:17954–17961
128. Niimi Y, Matsui T, Kambara H, Tagami K, Tsukada M, Fukuyama H (2005) Scanning tunneling microscopy and spectroscopy studies of graphite edges. Appl Surf Sci 241:43–48
129. Nomura K, MacDonald AH (2006) Quantum Hall ferromagnetism in graphene. Phys Rev Lett 96:256602
130. Novoselov KS, Geim AK, Morozov SV, Jiang D, Zhang Y, Dubonos SV, Grigorieva IV, Firsov AA (2004) Electric field effect in atomically thin carbon films. Science 306:666–669
131. Novoselov KS, Geim AK, Morozov SV, Jiang D, Katsnelson MI, Grigorieva IV, Dubonos SV, Firsov AA (2005a) Two-dimensional gas of massless Dirac fermions in graphene. Nature 438:197–200
132. Novoselov KS, Jiang D, Schedin F, Booth TJ, Khotkevich VV, Morozov SV, Geim AK (2005b) Two-dimensional atomic crystals. Proc Nat Acad Sci 102:10451–10453
133. Novoselov KS, McCann E, Morozov SV, Fal'ko VI, Katsnelson MI, Zeitler U, Jiang D, Schedin F, Geim AK (2006) Unconventional quantum Hall effect and Berry's phase of 2π in bilayer graphene. Nat Phys 2:177–180
134. Novoselov KS, Jiang Z, Zhang Y, Morozov SV, Stormer HL, Zeitler U, Maan JC, Boebinger GS, Kim P, Geim AK (2007) Room-temperature quantum Hall effect in graphene. Science 315:1379
135. Palacios JJ, Fernández-Rossier J, Brey L (2008) Vacancy-induced magnetism in graphene and graphene ribbons. Phys Rev B 77:195428
136. Palacios JJ, Fernández-Rossier J, Brey L, Fertig HA (2010) Electronic and magnetic structure of graphene nanoribbons. Semicond Sci Technol 25:033003
137. Park CH, Louie SG (2008) Energy gaps and Stark effect in boron nitride nanoribbons. Nano Lett 8:2200–2203
138. Park JH, Vescovo E, Kim HJ, Kwon C, Ramesh R, Venkatesan T (1998) Direct evidence for a half-metallic ferromagnet. Nature 392:794–796
139. Peierls RE (1934) Bemerkungen über Umwandlungstemperaturen. Helv Phys Acta 7:81
140. Peierls RE (1935) Quelques propriétés typiques des corps solides. Ann Inst Henri Poincaré 5:177–222
141. Peres NMR, Klironomos FD, Tsai SW, Santos JR, dos Santos JMBL, Neto AHC (2007) Electron waves in chemically substituted graphene. EPL 80:67007
142. Pisana S, Lazzeri M, Casiraghi C, Novoselov KS, Geim AK, Ferrari AC, Mauri F (2007) Breakdown of the adiabatic Born-Oppenheimer approximation in graphene. Nat Mater 6:198–201
143. Pitaevskii L, Stringari S (1991) Uncertainty principle, quantum fluctuations, and broken symmetries. J Low Temp Phys 85:377–388
144. Prato M (1997) [60]Fullerene chemistry for materials science applications. J Mater Chem 7:1097–1109
145. Radovic LR, Bockrath B (2005) On the chemical nature of graphene edges: Origin of stability and potential for magnetism in carbon materials. J Am Chem Soc 127:5917–5927
146. Roche S, Åkerman J, Beschoten B, Charlier JC, Chshiev M, Dash SP, Dlubak B, Fabian J, Fert A, Guimarães M, Guinea F, Grigorieva I, Schönenberger C, Seneor P, Stampfer C, Valenzuela SO, Waintal X, van Wees B (2015) Graphene spintronics: The European Flagship perspective. 2D Mater 2:030202
147. Ruffieux P, Wang S, Yang B, Sanchez-Sanchez C, Liu J, Dienel T, Talirz L, Shinde P, Pignedoli CA, Passerone D, Dumslaff T, Feng X, Müllen K, Fasel R (2016) On-surface synthesis of graphene nanoribbons with zigzag edge topology. Nature 531:489–492
148. Saito R, Fujita M, Dresselhaus G, Dresselhaus MS (1992a) Electronic structure of chiral graphene tubules. Appl Phys Lett 60:2204–2206
149. Saito R, Fujita M, Dresselhaus G,Dresselhaus MS (1992b) Electronic structure of graphene tubulesbased on C_{60}. Phys Rev B 46:1804–1811
150. Saito R, Dresselhaus G, Dresselhaus MS (1998) Physical properties of carbon nanotubes. Published by Imperial College Press and distributed by World Scientific Publishing Co

151. Sasaki K, Saito R (2008) Magnetism as a mass term of the edge states in Graphene. J Phys Soc Jpn 77:054703

152. Sasaki KI, Yamamoto M, Murakami S, Saito R, Dresselhaus MS, Takai K, Mori T, Enoki T, Wakabayashi K (2009) Kohn anomalies in graphene nanoribbons. Phys Rev B 80:155450

153. Sasaki KI, Saito R, Wakabayashi K, Enoki T (2010a) Identifying the orientation of edge of graphene using G band Raman spectra. J Phys Soc Jpn 79:044603

154. Sasaki KI, Wakabayashi K (2010b) Chiral gauge theory for the graphene edge. Phys Rev B 82:035421

155. Sasaki KI, Wakabayashi K, Enoki T (2010c) Berry's phase for standing waves near graphene edge. New J Phys 12:083023

156. Sharpe AL, Fox EJ, Barnard AW, Finney J, Watanabe K, Taniguchi T, Kastner MA, Goldhaber-Gordon D (2019) Emergent ferromagnetism near three-quarters filling in twisted bilayer graphene. Science 365:605–608

157. Shastry BS, Krishnamurthy HR, Anderson PW (1990) Instability of the Nagaoka ferromagnetic state of the U =∞ Hubbard model. Phys Rev B 41:2375–2379

158. Sheng L, Sheng DN, Haldane FDM, Balents L (2007) Odd-integer quantum Hall effect in graphene: Interaction and disorder effects. Phys Rev Lett 99:196802

159. Son YW, Cohen ML, Louie SG (2006a) Energy gaps in graphene nanoribbons. Phys Rev Lett 97:216803

160. Son YW, Cohen ML, Louie SG (2006b) Half-metallic graphene nanoribbons. Nature 444:347–349

161. Sprinkle M, Ruan M, Hu Y, Hankinson J, Rubio-Roy M, Zhang B, Wu X, Berger C, de Heer WA (2010) Scalable templated growth of graphene nanoribbons on SiC. Nat Nanotechnol 5:727–731

162. Stampfer C, Güttinger J, Hellmüller S, Molitor F, Ensslin K, Ihn T (2009) Energy gaps in etched graphene nanoribbons. Phys Rev Lett 102:056403

163. Sugawara K, Sato T, Souma S, Takahashi T, Suematsu H (2006) Fermi surface and edge-localized states in graphite studied by high-resolution angle-resolved photoemission spectroscopy. Phys Rev B 73:045124

164. Szabo A, Ostlund NS (1996) Modern quantum chemistry: introduction to advanced electronic structure theory. Dover Publications, Mineola, N.Y

165. Tans SJ, Verschueren ARM, Dekker C (1998) Room-temperature transistor based on a single carbon nanotube. Nature 393:49–52

166. Tao C, Jiao L, Yazyev OV, Chen YC, Feng J, Zhang X, Capaz RB, Tour JM, Zettl A, Louie SG, Dai H, Crommie MF (2011) Spatially resolving edge states of chiral graphene nanoribbons. Nat Phys 7:616–620

167. Tapasztó L, Dobrik G, Lambin P, Biró LP (2008) Tailoring the atomic structure of graphene nanoribbons by scanning tunnelling microscope lithography. Nat Nanotechnol 3:397–401

168. Tasaki H (1998) The Hubbard model—an introduction and selected rigorous results. J Phys: Condens Matter 10:4353–4378

169. Thostenson ET, Ren Z, Chou TW (2001) Advances in the science and technology of carbon nanotubes and their composites: A review. Compos Sci Technol 61:1899–1912

170. Tombros N, Jozsa C, Popinciuc M, Jonkman HT, van Wees BJ (2007) Electronic spin transport and spin precession in single graphene layers at room temperature. Nature 448:571–574

171. Trauzettel B, Bulaev DV, Loss D, Burkard G (2007) Spin qubits in graphene quantum dots. Nat Phys 3:192–196

172. Uchoa B, Castro Neto AH (2007) Superconducting states of pure and doped graphene. Phys Rev Lett 98:146801

173. Vogt P, De Padova P, Quaresima C, Avila J, Frantzeskakis E, Asensio MC, Resta A, Ealet B, Le Lay G (2012) Silicene: compelling experimental evidence for graphenelike two-dimensional silicon. Phys Rev Lett 108:155501

174. Wakabayashi K, Sigrist M, Fujita M (1998) Spin wave mode of edge-localized magnetic states in nanographite zigzag ribbons. J Phys Soc Jpn 67:2089–2093

175. Wakabayashi K, Fujita M, Ajiki H, Sigrist M (1999) Electronic and magnetic properties of nanographite ribbons. Phys Rev B 59:8271–8282
176. Wakabayashi K, Sigrist M (2000) Zero-conductance resonances due to flux states in nanographite ribbon junctions. Phys Rev Lett 84:3390–3393
177. Wakabayashi K (2001) Electronic transport properties of nanographite ribbon junctions. Phys Rev B 64:125428
178. Wakabayashi K (2002) Numerical study of the lattice vacancy effects on the single-channel electron transport of graphite ribbons. J Phys Soc Jpn 71:2500–2504
179. Wakabayashi K, Aoki T (2002) Electrical conductance of zigzag nanographite ribbons with locally applied gate voltage. Int J Mod Phys B 16:4897–4909
180. Wakabayashi K, Harigaya K (2003) Magnetic structure of nano-graphite Möbius ribbon. J Phys Soc Jpn 72:998–1001
181. Wakabayashi K, Takane Y, Sigrist M (2007) Perfectly conducting channel and universality crossover in disordered graphene nanoribbons. Phys Rev Lett 99:036601
182. Wakabayashi K, Takane Y, Yamamoto M, Sigrist M (2009a) Edge effect on electronic transport properties of graphene nanoribbons and presence of perfectly conducting channel. Carbon 47:124–137
183. Wakabayashi K, Takane Y, Yamamoto M, Sigrist M (2009b) Electronic transport properties of graphene nanoribbons. New J Phys 11:095016
184. Wakabayashi K, Okada S, Tomita R, Fujimoto S, Natsume Y (2010a) Edge states and flat bands of graphene nanoribbons with edge modification. J Phys Soc Jpn 79:034706
185. Wakabayashi K, Ki Sasaki, Nakanishi T, Enoki T (2010b) Electronic states of graphene nanoribbons and analytical solutions. Sci Technol Adv Mater 11:054504
186. Wakabayashi K, Dutta S (2012) Nanoscale and edge effect on electronic properties of graphene. Solid State Commun 152:1420–1430
187. Wakabayashi K (2018) Graphene nanotechnology. Materials Nanoarchitectonics, John Wiley & Sons Ltd, chap 5:109–123
188. Wallace PR (1947) The band theory of graphite. Phys Rev 71:622–634
189. Wang QH, Kalantar-Zadeh K, Kis A, Coleman JN, Strano MS (2012) Electronics and opto-electronics of two-dimensional transition metal dichalcogenides. Nat Nanotechnol 7:699–712
190. Wang X, Ouyang Y, Li X, Wang H, Guo J, Dai H (2008) Room-temperature all-semiconducting sub-10-nm graphene nanoribbon field-effect transistors. Phys Rev Lett 100:206803
191. Wang X, Li X, Zhang L, Yoon Y, Weber PK, Wang H, Guo J, Dai H (2009) N-doping of graphene through electrothermal reactions with Ammonia. Science 324:768–771
192. Wang Z, Ki D, Chen H, Berger H, MacDonald AH, Morpurgo AF (2015) Strong interface-induced spin- orbit interaction in graphene on WS2. Nat Commun 6:8339
193. Wassmann T, Seitsonen AP, Saitta AM, Lazzeri M, Mauri F (2008) Structure, stability, edge states, and aromaticity of graphene ribbons. Phys Rev Lett 101:096402
194. Watanabe K, Taniguchi T, Kanda H (2004) Direct-bandgap properties and evidence for ultra-violet lasing of hexagonal boron nitride single crystal. Nat Mater 3:404–409
195. Wolf SA, Chtchelkanova AY, Treger DM (2006) Spintronics-A retrospective and perspective. IBM J Res & Dev 50:101–110
196. Xia F, Wang H, Jia Y (2014) Rediscovering black phosphorus as an anisotropic layered material for optoelectronics and electronics. Nat Commun 5:4458
197. Yamamoto M, Wakabayashi K (2009) Control of electric current by graphene edge structure engineering. Appl Phys Lett 95:082109
198. Yamamoto M, Dutta S, Aikawa S, Nakaharai S, Wakabayashi K, Fuhrer MS, Ueno K, Tsuk-agoshi K (2015) Self-limiting layer-by-layer oxidation of atomically thin WSe^2. Nano Lett 15:2067–2073
199. Yamashiro A, Shimoi Y, Harigaya K, Wakabayashi K (2003) Spin- and charge-polarized states in nanographene ribbons with zigzag edges. Phys Rev B 68:193410
200. Yan Q, Huang B, Yu J, Zheng F, Zang J, Wu J, Gu BL, Liu F, Duan W (2007) Intrinsic current-voltage characteristics of graphene nanoribbon transistors and effect of edge doping. Nano Lett 7:1469–1473

201. Yazyev OV, Helm L (2007) Defect-induced magnetism in graphene. Phys Rev B 75:125408
202. Yazyev OV (2008) Magnetism in disordered graphene and irradiated graphite. Phys Rev Lett 101:037203
203. Yoshioka H (2003) Spin excitation in nano-graphite ribbons with zigzag edges. J Phys Soc Jpn 72:2145–2148
204. Zak J (1989) Berry's phase for energy bands in solids. Phys Rev Lett 62:2747–2750
205. Zhang Y, Tan YW, Stormer HL, Kim P (2005) Experimental observation of the quantum Hall effect and Berry's phase in graphene. Nature 438:201–204
206. Zhang Z, Guo W (2008) Energy-gap modulation of BN ribbons by transverse electric fields: First-principles calculations. Phys Rev B 77:075403
207. Žutić I, Fabian J, Das Sarma S (2004) Spintronics: Fundamentals and applications. Rev Mod Phys 76:323–410

Carbon Nanotube Field-Effect Transistors (CNFETs): Structure, Fabrication, Modeling, and Performance

Navneet Gupta and Ankita Dixit

Abstract The problems associated with attempting to scale down traditional metal oxide field-effect transistors (MOSFET) have led researchers to look into CNT-based field-effect transistors (CNFETs), as an alternative. Though the scaling of MOSFET has been the driving force toward the technological advancement, but due to continuous scaling, various secondary effects which include short channel effects, high leakage current, excessive process variation, and reliability issues degrade the device performance. On the other hand, CNFETs are not subjected to the scaling problems. The operation principle of the CNFET is similar to traditional MOSFET but the conduction phenomena are different. The traditional MOSFETs are based on the drift and diffusion phenomena in which channel length is very large as compared to mean free path of charge carriers whereas the CNFETs are based on ballistic transport conduction mechanism, in which channel length is very small as compared to mean free path of charge carriers. In CNFET, electrons are injected from source to drain and transported through the nanotubes without scattering. Due to ballistic transport the nanotubes act as a perfect conductor for electrons such that the full quantum information of these electrons (momentum, energy, spin) can be transferred without losses. The channel current in CNFETs depends on gate voltage, number of nanotubes in channel, dielectric material and its thickness, and diameter and chirality of carbon nanotubes. So in this chapter we shall discuss different device structures of CNFET, steps involved in the fabrication of CNFETs, advantages and limitations of various methods involved in the synthesis of CNTs, conduction models, and performance parameters.

N. Gupta (✉) · A. Dixit
Department of Electrical and Electronics Engineering, Birla Institute of Technology and Science, Pilani, Rajasthan 333031, India
e-mail: ngupta@pilani.bits-pilani.ac.in

© The Author(s), under exclusive license to Springer Nature Singapore Pte Ltd. 2021   199
A. Hazra and R. Goswami (eds.), *Carbon Nanomaterial Electronics: Devices and Applications*, Advances in Sustainability Science and Technology,
https://doi.org/10.1007/978-981-16-1052-3_9

1 Introduction

Over the years, silicon metal-oxide-semiconductor field-effect transistors (Si-MOSFETs) have come down to a size of 10 nm, but further scaling has proved to be challenging due to short channel effects (SCE) such as drain-induced barrier lowering (DIBL), velocity saturation, and hot carrier generation, which degrade the performance of the device in terms of high power consumption during turning off the device [1, 2].

Carbon nanotube field-effect transistors (CNFETs), which uses carbon nanotubes (CNTs) as the channel material, have come up as a preferred alternative over the Si-MOSFETs [3]. CNTs have high carrier velocity and possess near-ballistic transport property which make it highly suitable for FET devices [4, 5]. Moreover, owing to their unique physical properties in addition to their remarkable electronic, mechanical, optical, vibrational, and thermal properties [6–10], CNTs can overcome SCE and also provide a better gate control than Si-MOSFET in nanoscale region as a result of their significantly thinner structure. These attributes have made CNTs a preferred option for various other potential applications as well [11–15].

CNTs are the backbone of carbon nanotube field-effect transistors (CNFETs). It was discovered by S. Iijima in 1991 [16]. CNTs can be divided into two categories on the basis of number of walls and chirality. Based on number of walls, it is further classified into two types, i.e., single-wall CNTs and multiwall CNTs. On the basis of chirality (n, m) and chiral angle (θ), CNTs can be categorized into three different types: zigzag $(n, 0)$, $\theta = 0°$ [17]; armchair (n, n), $\theta = 30°$; and helical CNT (n, m), $0° < \theta < 30°$ [18]. Figure 1 shows the schematic illustration of these different types of CNT configurations.

The diameter (d) of CNT depends on the chiral vector of CNT. The diameter and energy bandgap (E_g) of CNT can be given as [19] follows:

$$d = \frac{\sqrt{3}a_{c-c}}{\pi}\sqrt{n^2 + nm + m^2} \tag{1}$$

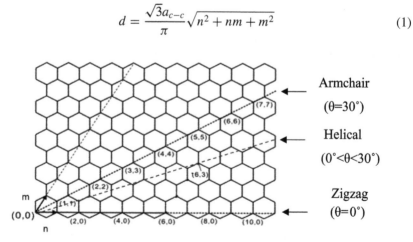

Fig. 1 Schematic illustration of carbon nanotube configurations

Table 1 Diameter and energy bandgap for various CNTs

Types of CNT	Diameter (nm)	Energy band gap (eV)	Nature of CNT	
Achiral CNT	CNT (6, 0)	0.4698	0	Metallic
	CNT (7, 0)	0.548128	1.55438	Semiconductor
	CNT (7, 7)	0.94938	0	Metallic
Chiral CNT	CNT (8, 5)	0.8891	0	Metallic
	CNT (5,7)	0.8173	1.0424	Semiconductor

$$E_g = \begin{cases} \frac{2a_{c-c}t}{d}, n - m \neq 3i \\ 0, n - m = 3i \end{cases} \tag{2}$$

where a_{c-c} is the carbon bond length (1.42 Å), t is the tight binding energy (~ 3 eV), and "i" is an integer. CNTs with $n-m \neq 3i$ are semiconductor, otherwise metallic in nature. Specifically, armchair CNTs are metallic, while zigzag carbon nanotubes are either semiconductor or metallic. Table 1 shows the calculated values of diameter and energy bandgap using Eq. (1) and (2) for achiral and chiral CNT. It was also observed that energy bandgap is inversely proportional to diameter of CNT.

CNFET is a three-terminal device with source, drain, and gate like MOSFET. The performance of CNFET depends on drain current (I_d) which is controlled by gate-to-source voltage (V_{gs}) and drain-to-source voltage (V_{ds}). Apart from diameter and chirality of CNT, other physical parameters that affect the drain current are contact between CNT and source/drain (S/D), gate dielectric material, channel length, and thickness of gate dielectric [20]. It was observed that a small diameter of CNT leads to high value of leakage current which in turn reduces the mobility. It was also reported that for the diameter (d) of 1 nm, current through zigzag CNT is less as compared to armchair CNT [21]. In CNFET, threshold voltage (V_t) depends on temperature and channel length [22]. As the channel length increases from 10 nm to 20 nm threshold voltage decreases rapidly in CNFET, whereas for MOSFET threshold voltage increases sharply when channel length increases from 10 nm to 20 nm [23].

High-k dielectric material plays very important role in the performance of CNFET which allows the charge injection into transistor channel and reduces the leakage current [24]. As CNTs are free from dangling bond, electrolytic dielectrics can also be used as gate in CNFET device [25]. It was observed that for CNFET with high-k dielectric the on-current (I_{on}) per unit width is approximately three times higher than the on-current for MOSFET at gate voltage (V_g) of 0.6 V [26].

This chapter provides an overview on CNFET. The device structure and fabrication steps for two different structures of CNFET are described in Sects. 2 and 3, respectively. Section 4 explains various conduction models for CNFET proposed by various research groups. Different parameters that affect the performance of CNFET are described in Sect. 5. Finally, the possible applications of CNFET are explained in the last section.

2 Device Structure

CNFET device structure can be classified into two categories: *planar structure* CNFET and *Gate-All-Around (GAA) structure* CNFET [27]. Figure 2 shows the schematic diagram of both these CNFETs.

There are basically four different configurations of planar structure of CNFET and each structure has its own advantages and limitations. Figure 3 (a) shows the single gate planar structure of CNFET with heavily doped substrate as a back gate. This device structure is easy to fabricate but suffer from low drain current (I_d) which can be increased by using a double gate structure of CNFET, as shown in Fig. 3(b). In this double gate device structure, top gate is used to control the charge density in the channel and back gate is used for electrostatic doping in the extension region of the CNT in channel. Another structure of CNFET as shown in Fig. 3(c) is almost similar to the device structure in Fig. 3(b) except that the CNT is doped. This device structure

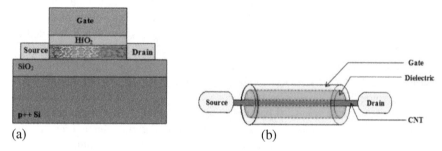

Fig. 2 Schematic view of planar CNFET and GAA-CNFET

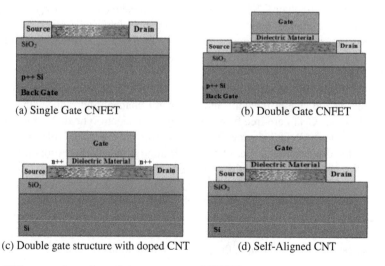

(a) Single Gate CNFET   (b) Double Gate CNFET

(c) Double gate structure with doped CNT   (d) Self-Aligned CNT

Fig. 3 Different configurations of planar structure of CNFET

provides the low resistance path between CNT and S/D contacts [28, 29]. A self-aligned device structure of CNFET is shown in Fig. 3(d) in which there is no extension region. This device structure is used to reduce the parasitic capacitance between gate and source and between gate and drain, which improves the performance of the device and also reduces the size of the device.

In order to provide ohmic contact, CNFET has been fabricated with different source/drain materials like palladium [30], Ti/Au [31], Cr/Pd/Au [32], and Cr/Au [33, 34]. The next generation of CNFET was developed with GAA-CNFET to increase the performance of transistors. Z. Chen et al. [35] fabricated GAA-CNFET with dielectric layer of Al_2O_3 and Pd is used as a source/drain material and the segments of length of 100 nm were designed between gate and extended source/drain. A.D. Franklin et al. [36] fabricated the GAA-CNFET with HfO_2 as high-k gate dielectric material and observed that n-type and p-type CNFETs can be realized by using HfO_2 and Al_2O_3 dielectric materials, respectively [37].

3   Fabrication of CNFET

The fabrication steps of CNFET are almost similar to that of MOSFET except an additional step of CNT deposition. In order to fabricate CNFET, we need first to synthesis CNT and then follow steps to fabricate the device.

3.1   Synthesis of CNT

Various synthesis methods such as arc discharge method, LASER ablation, electrolysis, hydrothermal or sono-chemical, and chemical vapor deposition (CVD) have been developed to synthesize the CNT. Each method has its own advantages and limitations. The comparison among these synthesis methods is given in Table 2.

3.2   Device Fabrication

The CNFETs have been fabricated with both planar and GAA structures. The fabrication steps of CNFET include doping, masking, etching, lithography, and metal deposition process. Let us discuss these steps for planar and GAA structures in next section.

3.2.1   Fabrication Steps for Planar CNFET

The fabrication process for planar CNFET is illustrated in Fig. 4. First, silicon wafer

Table 2 Comparison among synthesis methods of CNTs

Synthesis methods	Description		
	Advantages	Limitations	References
Arc discharge	Both single-wall and multiwall CNTs with length up to 50 μm can be synthesized	Cooling is required for electrodes and chamber and required vacuum chamber and high current power supply	[38]
LASER ablation	Primarily used for the synthesis of SWCNTs with controllable diameter	High temperature is required and this method is expensive	[39]
Electrolysis	Inexpensive and use green house gas (CO_2)	Synthesizes only MWNTs	[40]
Hydrothermal	Required low temperature, does not require any hazardous gas such as H_2	Synthesizes only MWNTs	[41, 42]
Chemical vapor deposition	Low temperature required (<800 °C). Patterned nanotube can be grown with controllable diameter	Catalyst is used to grow the nanotube and also used to control the diameter of CNT	[43]

is oxidized using wet thermal oxidation. Then Mask-1 is used to deposit the source and drain with specific channel length. Cr/Au can be used for metalization. Mask-2 is used for the dielectrophoresis (DEP) method. The AC voltage of 10 V with 1 MHz frequency was applied for the DEP method. After the alignment of CNT between source and drain, gate dielectric can be deposited using e-beam evaporation. In the final step, an aluminum thin film was deposited for the contact of source, drain, and gate.

3.2.2 Fabrication Steps for GAA-CNFET

Figure 5 shows the fabrication steps for GAA-CNFET [44]. First, CNT is deposited at selected area on Si substrate. After the metallization of source and drain, silicon layer is etched to form a recess. Then the spacer material (i.e., dielectric material) is deposited using either atomic layer deposition (ALD) or chemical vapor deposition (CVD) process to provide an insulating layer between gate and S/D contact. After deposition of spacer material, a high-k dielectric material is deposited on exposed surface including channel (CNT).The gate terminal is formed by metallization using palladium (Pd) or tungsten (W) material on gate dielectric material to control the current between source and drain.

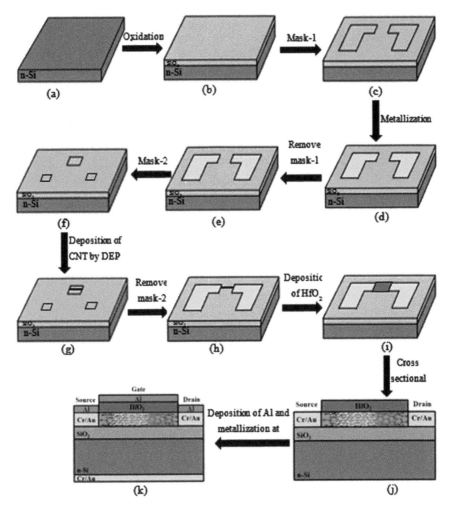

Fig. 4 Fabrication steps for planar CNFET

4 Conduction Models of CNFET

Conduction models are used to analyze the electrical behavior and performance of CNFET. The device model for CNFET is classified into three broad categories: transport model, charge model, and finite element method (FEM) model [45]. Figure 6 shows further classification of transport models for CNFET.

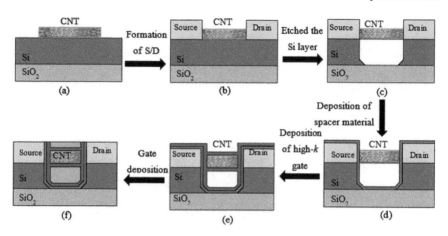

Fig. 5 Fabrication steps for gate-all-around CNFET

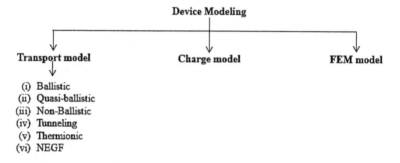

Fig. 6 Classifications of conduction models for CNFET

4.1 Transport Mode

Various transport models for CNFET have been proposed based on different mechanisms behind the conduction phenomena. Following sub-section explains these models in detail.

4.1.1 Ballistic Model

Ballistic transportation means that there is no resistance in the path of charge carriers of the device. It also represents that there is no loss of energy in charge carriers due to scattering. In ballistic transportation, mean free path is greater than the channel length. I. Bejenari, M. Schröter, and M. Claus developed a new analytical model to the drain current in 1-D ballistic Schottky barrier transistor [46]. This analytical model is based on Wentzel–Kramers–Brillouin approximation. An analytical solution for

Landauer integral was derived by using the Fermi–Dirac distribution function and transmission probability, to overcome the limitations of existing models. This model decreases the evaluation time as it is free from numerical integration. The drain current equation for ballistic model is given as follows:

$$I_d = \frac{4q}{h} k_B T \ln \left\{ \frac{1 + \exp(-\phi_b/k_B T)}{1 + \exp[-(\phi_b + q V_{ds})/k_B T]} \right\} \tag{3}$$

where k_B is Boltzmann constant, T is temperature, q is the electronic charge, ϕ_b is the barrier height, h is Planck's constant, and V_{ds} is the voltage between source and drain.

4.1.2   Quasi-Ballistic Model

Quasi-ballistic model refers to the model where the transportation of charge carriers has small amount of scattering in their path. C-S Lee, A. D. Franklin, W. Haensch, and H-S P. Wong [47] developed a model for the quasi-ballistic CNFET based on virtual source model. In this model, it has been considered that the carrier effective mobility and velocity of charge carriers depend on the diameter of CNT and gate capacitance also depends on quantum capacitance which occurs due to the CNT. The drain current equation for quasi-ballistic model is given as follows:

$$I_d = \frac{4q}{h} k_B T \ln \left[\frac{1 + \exp\left(\frac{\psi_s - E_g/2q}{k_B T}\right)}{1 + \exp\left(\frac{\psi_s - E_g/2q - V_{ds}}{k_B T/q}\right)} \right] \tag{4}$$

where k_B is Boltzmann constant, T is temperature, q is the electronic charge, ψ_s is the surface potential, h is Planck's constant, and V_{ds} is the voltage between source and drain.

4.1.3   Non-Ballistic Model

In non-ballistic model, there is a resistance for the charge carriers in the transportation path between source and drain. For non-ballistic transportation, electron scattering by acoustic and optical phonons in the channel is also considered. I. Bejenari et al. had developed a novel analytical model for the study of drain current in 1-D non-ballistic CNFET transistor [48]. In this model, a piece-wise approximation of Fermi–Dirac distribution and transmission probability has been used for the derivation of drain current equation in the Landauer–Büttiker formalism.

4.1.4   Tunneling-Based Model

Tunneling is a process that allows the charge carriers to move from one terminal to other through the barrier. It is the main reason for the leakage current. S. Frégonèse, C. Maneux, and T. Zimmer [49] have proposed a compact model in CNFET for the implementation of tunneling phenomena. The CNFET has been analyzed for the circuit configuration with and without and band-to-band tunneling (BTBT) which is having a major impact on the figure of merit of the circuit. The tunneling current probability and the current density is a function of CNT diameter, oxide thickness, and gate voltage (V_g).

4.1.5   Thermionic Model

The process in which electrons are emitted from heated source is known as thermionic emission. In CNFET, thermionic emission is due to the electric field between source and drain. The thermionic emission required higher energy than the energy required for tunneling [50].

4.1.6   NEGF-Based Model

Non-equilibrium Green's function (NEGF) formalism is a matrix-based computational process which includes Hamiltonian matrix and self-energy matrix but it is too computationally expensive for device optimization and circuit simulation [51–53].

4.2   Charge Model

In charge model for GAA structure of CNFET, potential and carrier density can be solved by Poisson–Schrodinger equations. The analytical expressions for density of state and drain current can be derived for CNFET, which has a SPICE–compatible model [54].

4.3   Finite Element Method (FEM) Model

The finite element method (FEM) has an attractive feature to handle complicate geometries (boundaries) with relative ease. In this model, a large problem is subdivided into small problems called finite elements. The equations of these models are assembled to obtain the solution to the entire problem. A 3-D model of the CNFET was developed for evaluating the circuit performance in terms of ON-current, speed, and power [55]. This model was developed for the multichannel gate-all-around structure with k_1 as a dielectric constant for gate oxide and k_2 as a dielectric constant for the substrate.

5 Performance Parameters of CNFET

5.1 Impact of Gate Dielectric Materials

The performance of CNFET is affected by gate dielectric materials. CNFETs have been studied by various gate dielectric materials such as SiO_2, Al_2O_3, La_2O_3, HfO_2, and ZrO_2 [24]. The gate dielectric materials with relative dielectric constant of more than 7 (i.e., $\varepsilon_r > 7$) are referred as high-k dielectric materials and those materials with relative dielectric constant less than 3.9 (i.e., $\varepsilon_r < 3.9$) are referred as low-k dielectric materials [56]. As the gate dielectric thickness reduces, the leakage current increases; hence, it is necessary to have high-k gate dielectric material. The analysis shows that for improved performance of CNFET, La_2O_3 material is the best possible gate dielectric material followed by HfO_2 [24]

5.2 Impact of Gate Dielectric Thickness

Due to ballistic transport effects, direct tunneling from the source to the drain will be most detrimental to low-voltage transistor operation because it not only degrades the subthreshold swing (SS) but also increases the leakage current in the subthreshold region. A.Shaukat et al. [20] studied the effect of the thickness of SiO_2 on CNFET. As the thickness of gate dielectric material reduces, transconductance and voltage gain increase and subthreshold swing reduces; **however,** DIBL slightly increases. For different structures, it has been observed that the GAA provides better performance as compared to planar CNFET.

5.3 Effect of Channel Length

In CNFET, channel length also plays an important role. When the channel length is smaller than the mean free path, ballistic transportation occurs in CNTs. If the device is scaled down, electric field is increased between source **and drain,** which reduced the barrier for charge carriers.

It has been observed that when channel length reduces, subthreshold swing also reduces and GAA-CNFET provides high switching speed due to less subthreshold swing. (The subthreshold swing is reciprocal of the subthreshold slope.)

5.4   Effect of Source and Drain Materials

The performance of CNFET also depends on source/drain materials. The metal/CNT contact reacts with atmospheric gases as they change the metal work function, which causes the changes in the barrier height [57]. Earlier, Nobel metals like gold (Au) and platinum (Pt) were used because of high conductivity but they provide high parasitic resistance in megaohms. Therefore, at present, many researchers are using Pd, Ti, Al, and Cr metal as a S/D material [34–36, 58]. Al contacts provide SB barrier height for holes due to low work function whereas CNFET shows ambipolar behavior with Ti because it provides equal barrier height for both holes and electrons. Pd is mostly used because it provides less contact resistance which reduces the leakage current and enhances the performance of the CNFET.

6   Potential Applications

CNFETs have many applications due to the unique properties of CNTs. CNTs have distinguished electrical, chemical, mechanical, and thermal properties besides having high tensile strength. There is no boundary scattering in CNTs due to hollow cylindrical structure. It can transport a large amount of charge carriers as compared to silicon. The CNFETs have been used for nanosensors, electronic devices, and circuit applications and RF applications [59–63]. Due to remarkable properties of CNTs, it is suitable for low power, highly sensitive, and low-cost sensors. CNFET is also used for many digital circuit applications such as a quaternary logic gates and arithmetic circuits (half adder and full adder) using voltage divider. An electronic circuit based on memristor with ternary inverters (T-Inverters) using CNFET has also been proposed [62]. CNFET also has the potential to be used in RF devices which enhances the performance parameters (frequency band, power consumption, and the size of units) of device.

7   Conclusion and Outlook

This chapter provides an overview of the CNFETs which summarizes research advancement of CNFET from three aspects: (1) geometrical structure, (2) fabrication, and (3) device modeling of the CNFET. This chapter covers the information of the performance parameters of CNFET and describes the device applications. In nanoscale regime, CNFET is better than MOSFET due to CNT's remarkable and unique properties. The structure of CNFET device also plays an important role in performance of the device. GAA-CNFET has better gate control over the planar CNFET. It is also evident that the transconductance in GAA-CNFET is remarkably high when compared to the planar CNFET. GAA-CNFET has less leakage current and high drain current than planar structure. The performance of CNFET also depends on the gate dielectric material and its thickness, channel length, and source and drain materials. Working of CNFET can be explained by using mainly three models, i.e., transport, charge, and FEM models. As the size is compact and performance is better than MOSFET, the CNFET is used for nanosensor applications, bio-sensors, gas sensors, electronic device applications, and RF applications.

Acknowledgements The authors would like to thank the Defence Research and Development Organisation (DRDO) for funding the work reported in this paper (ERIP/ER/DG-MED&OS/990416502/M/01/1657).

References

1. Xu L, Qiu C, Zhao C, Zhang Z, Peng L (2019) Insight into ballisticity of room-temperature carrier transport in carbon nanotube field-effect transistors. IEEE Trans Electron Devices 66(8):3535–3540
2. Obite F, Ijeomah G, Bassi JS (2013) Carbon nanotube field effect transistors: toward future nanoscale electronics. Int J Comput Appl 41:149–164
3. Najari M, Frégonèse S, Maneux C, Mnif H, Masmoudi N, Zimme T (2011) Schottky barrier carbon nanotube transistor: compact modeling, scaling study, and circuit design applications. IEEE Trans Electron Devices 58(1)
4. Javey A, Guo J, Wang Q, Lundstrom M, Dai H (2003) Ballistic carbon nanotube field-effect transistors. Nature 424(6949):654–657
5. Natori K (2005) Kimura Y and Shimizu, Characteristics of a Carbon nanotube field-effect transistor analysed as a ballistic nanowire field-effect transistor. J Appl Phys 97(3):
6. de Heer WA, Chatelain A, Ugaarte D (1995) A carbon nanotubefield-emission electron source. Science 270:1179–1180
7. Yu MF, Files BS, Arepalli S, Ruoff RS (2000) Tensile loading ofropes of single wall carbon nanotubes and their mechanicalproperties. Phys Rev Lett 84:5552–5555
8. Hone J, Batlogg B, Benes Z, Johnson AT, Fischer JE (2000) Quantized phonon spectrum of single-wall carbon nanotubes. Science 289:1730–1733
9. Popov VN (2004) Carbon nanotubes: properties and application. Mater Sci Eng R Rep 43(3)
10. Zhao H, Zhang Y, Bradford PD, Zhou Q, Jia Q, Yuan FG, Zhu Y (2010) Carbon nanotube yarn strain sensors. Nanotechnology 21(30)
11. Robinson JA, Snow ES, Bădescu SC, Reinecke TL, Perkins FK (2006) Role of defects in single-walled carbon nanotube chemical sensors. Nano Lett 6(8)
12. Wook CY, Oh JS, Yoo SH, Choi HH, Yoo K-H (2007) Electrically refreshable carbon-nanotube-based gas sensors. Nanotechnology 18(43)
13. Lee NS, Chung DS, Han IT, Kang JH, Choi YS, Kim HY, Park SH et al (2001) Application of carbon nanotubes to field emission displays. Diam Relat Mater 10(2)
14. Bianco A, Kostarelos K, Prato M (2005) Applications of carbon nanotubes in drug delivery. Curr Opin Chem Biol 9(6)
15. Ayala P, Arenal R, Rümmeli M, Rubio A, Pichler T (2010) The doping of carbon nanotubes with nitrogen and their potential applications. Carbon 48(3)
16. Iijima S (1991) Helical microtubules of graphitic carbon. Nature 354:56–58
17. Gu¨lseren O, Yildirim T, Ciraci S (2002) Systematic Ab- initio study of curvature effects in carbon nanotubes, vol 65, pp 153405(1–4). The American Physical Society
18. Chaudhury S, Sinha SK (2019) Carbon nanotube and nanowires for future semiconductor devices applications. Nanoelectronics, 375–398. https://doi.org/10.1016/b978-0-12-813353-8.00014-2

19. Krueger A (2010) Carbon materials and nanotechnology. Wiley-VCH, Great Britain, pp 126–127
20. Shaukat A, Umer R, Isla N (2017) Impact of dielectric material and oxide thickness on the performance of carbon nanotube field effect transistor. In IEEE International Conference on Nanotechnology. Pittsburgh, USA, pp 250–254
21. O'Connor I, Liu J, Gaffiot F, Prégaldiny F, Lallement C, Maneux C, Goguet J, Frégonèse S, Zimmer T, Anghel L, Dang TT, Leveugle R (2007) —CNTFET modeling and reconfigurable logic-circuit design. ‖ IEEE Trans. Circuit Syst 54:2365–2379
22. Tans SJ, Verschueren ARM, Dekker C (1998) Room temperature transistor based on a single carbon nanotube. Nature 393:49–52
23. Sinha SK, Chaudhury S (2014) Advantage of CNTFET characteristics over MOSFET to reduce leakage power. In: International conference on devices, circuits and systems (ICDCS), Combiatore, pp 1–5
24. Dixit A, Gupta N (2019) Investigation on gate dielectric material using different optimization techniques in carbon nanotube field effect transistors (CNFETs) J Micromech Microeng 29:094002. IOP Publishing Ltd
25. Moaiyeri MH et al (2011) Efficient CNTFET-based ternary full adder cells for nanoelectronics. Nano-Micro Lett 3.1:43–50
26. Guo J, Datta S, Lundstrom M, Brink M, McEuen P, Javey A, Dai H, Kim H, McIntyre P (2002) Assessment of silicon MOS and carbon nanotube FET performance limits using a general theory of ballistic transistors. IEEE Elec Devices Meet, 29.3.1–29.3.4
27. Saiphani Kumar G, Singh A, Raj B (2017) Design and analysis of a gate-all-around CNTFET-based SRAM cell. J Comput Electron 17(1):138–145. https://doi.org/10.1007/s10825-017-1056-x
28. Dobson Peter (2012) Carbon Nanotube and Graphene Device Physics by H.-S. Philip Wong and Deji Akinwande. Contemp Phys CONTEMP PHYS 53:1–1. https://doi.org/10.1080/00107514.2012.689336
29. Sinha S, Balijepalli A, Cao Y (2009) Compact model of carbon nanotube transistor and interconnect. IEEE Trans Electron Devices 56:2232–2242
30. Jejurikar S, Casterman D, Pillai PB, Petrenko O, De Souza MM, Tahraoui A, Durkan C, Milne WI (2010) Anomalous n-type electrical behaviour of Pd-contacted CNTFET fabricated on small-diameter nanotube. Nanotechnology 21(21):1–7
31. Raman Pillai SK, Chan-Park MB (2012) High-performance printed carbon nanotube thin-film transistors array fabricated by a nonlithography technique using hafnium oxide passivation layer and mask. ACS Appl Mater Interfaces 4(12):7047–7054
32. Zhang X et al (2013) Fabrication of hundreds of field effect transistors on a single carbon nanotube for basic studies and molecular devices. J Vac Sci Technol B, Nanotechnol Microelectron Mater Process Meas Phenom 31.6: 06FI01
33. Xiao Z et al (2016) The fabrication of carbon nanotube electronic circuits with dielectrophoresis. Microelectron Eng 164: 123–127
34. Agarwal PB, Singh AK, Agarwal A (2018) Stable Metal-CNT contacts using shadow mask technique for CNTFET. AIP Conf Proc 1989: https://doi.org/10.1063/1.5047720
35. Chen Z et al (2008) Externally assembled gate-all-around carbon nanotube field-effect transistor. IEEE Electron Device Lett 29.2: 183–185
36. Franklin AD et al (2012) Scalable and fully self-aligned n-type carbon nanotube transistors with gate-all-around. 2012 Int Electron Devices Meet. IEEE
37. Franklin AD et al (2013) Carbon nanotube complementary wrap-gate transistors. Nano Lett 13.6: 2490–2495
38. Collins PG (2000) Nanotubes for electronics. Sci Am 283(6): 67–69. Bibcode:2000SciAm.283f..62C. https://doi.org/10.1038/scientificamerican1200-62. PMID 11103460
39. Chrzanowska J et al (2015) Synthesis of carbon nanotubes by the laser ablation method: effect of laser wavelength. Phys Status Solidi (b) 252.8: 1860–1867

40. Ren J, Li F-F, Lau J, González-Urbina L, Licht S (2015) One-pot synthesis of carbon nanofibers from CO2. Nano Lett 15(9):6142–6148
41. Katoh R, Tasaka Y, Sekreta E, Yumura M, Ikazaki F, Kakudate Y, Fujiwara S (1999) Sono chemical production of a carbon nanotube. Ultrason Sonochem 6(4):185–187
42. Manafi S, Nadali H, Irani HR (2008) Low temperature synthesis of multi-walled carbon nanotubes via a sono chemical/hydrothermal method. Mater Lett 62(26):4175–4176
43. De Jonge N, Bonard J-M (2004) Carbon nanotube electron sources and applications. Philos Trans Royal Soc A Math Phys Eng Sci 362(1823):2239–2266. https://doi.org/10.1098/rsta.2004.1438
44. Javey A, Guo J, Farmer DB, Wang Q, Yenilmez E, Gordon RG, ... Dai H (2004) Self-aligned ballistic molecular transistors and electrically parallel nanotube arrays. Nano Lett 4(7)
45. Franklin AD, Koswatta SO, Smith JT (2015) Gate-all-around carbon nanotube transistor with selectively doped spacers. U.S. Patent No. 9,000,499. 7 Apr. 2015
46. Bejenari I, Schroter M, Claus M (2017) Analytical drain current model of 1-D ballistic schottky-barrier transistors. IEEE Trans Electron Devices 64(9):3904–3911. https://doi.org/10.1109/ted.2017.2721540
47. Lee C-S, Pop E, Franklin AD, Haensch W, Wong H-SP (2015) A compact virtual-source model for carbon nanotube FETs in the Sub-10-nm regime—Part I: intrinsic elements. IEEE Trans Electron Devices 62(9):3061–3069. https://doi.org/10.1109/ted.2015.2457453
48. Bejenari I, Schroter M, Claus M (2017) Analytical drain current model for non-ballistic Schottky-Barrier CNTFETs. 2017 47th European Solid-State Device Research Conference (ESSDERC)
49. Fregonese Sé, Maneux C, Zimmer T (2009) Implementation of tunneling phenomena in a CNTFET compact model. IEEE Trans Electron Devices 56(10); 2224–2231. https://doi.org/10.1109/ted.2009.2028621
50. Perello DJ, Lim SC, Chae SJ, Lee I, Kim MJ, Lee YH, Yun M (2011) Thermionic field emission transport in carbon nanotube transistors. ACS Nano 5(3); 1756–1760. https://doi.org/10.1021/nn102343k
51. Datta (2004) The NEGF method: capabilities and challenges. In: 2004 Abstracts 10th international workshop on computational electronics, West Lafayette, IN, USA, pp 61–62, https://doi.org/10.1109/iwce.2004.1407323
52. Xinghui L, Junsong Z, Zhong Q, Fanguang Z, Jiwei W, Chunhua G (2011) Study on transport characteristics of CNTFET based on NEGF theory. In: 2011 IEEE international conference of electron devices and solid-state circuits. https://doi.org/10.1109/edssc.2011.6117655
53. Deyasi A, Sarkar A (2018) Analytical computation of electrical parameters in GAAQWT and CNTFET with identical configuration using NEGF method. Int J Electron, 1–16. https://doi.org/10.1080/00207217.2018.1494339
54. Dokania V, Islam A, Dixit V, Tiwari SP (2016) Analytical modeling of wrap-gate carbon nanotube FET with parasitic capacitances and density of states. IEEE Trans Electron Devices 2016:1–6. http://dx.doi.org/10.1109/TED.2016.2581119
55. Rakibul Karim Akanda Md, Khosru Quazi DM (2013) FEM model of wraparound CNTFET with multi-CNT and its capacitance modeling. IEEE Trans Electron Dev, 60(1):97–102
56. Teja KBR, Gupta N (2019) Hybrid bilayer gate dielectric-based organic thin film transistors. Bull Mater Sci 42.1: 2
57. Franklin AD et al (2012) Sub-10 nm carbon nanotube transistor. Nano Lett 12(2):758–762
58. Sinha SK, Chaudhury S (2015) Effect of device parameters on carbon nanotube field effect transistor in nanometer regime. J Nano Res 36:64–75
59. Ghodrati M, Farmani A, Mir A (2019) Nanoscale Sensor-based tunneling carbon nanotube transistor for toxic gases detection: a first-principle study. IEEE Sens J 1–1. https://doi.org/10.1109/jsen.2019.2916850
60. Hosseingholipourasl A, Ariffin SHS, Koloor SSR, Petru M, Hamzah A (2020) Analytical prediction of highly sensitive CNTFET based sensor performance for detection of gas molecules. IEEE Access 1–1. https://doi.org/10.1109/access.2020.2965806

61. Ebrahimi SA et al (2016) Efficient CNTFET-based design of quaternary logic gates and arithmetic circuits. Microelectronics J 53: 156–166

62. Soliman Nancy S, Fouda Mohammed E, Radwan Ahmed G (2018) Memristor-CNTFET based ternary logic gates. Microelectron J 72:74–85

63. Taghavi A, Carta C, Meister T, Ellinger F, Claus M, Schroter M (2017) A CNTFET Oscillator at 461 MHz. IEEE Microw Wirel Compon Lett 27(6):578–580

Carbon Nanomaterials for Emerging Electronic Devices and Sensors

Venkatarao Selamneni, Naveen Bokka, Vivek Adepu, and Parikshit Sahatiya

Abstract Over the last two decades, carbon nanomaterials including two-dimensional graphene, one-dimensional carbon nanotubes (CNTs), and zero-dimensional carbon quantum dots, fullerenes have gained tremendous attention from researchers due to their unique optical, electronic, mechanical, chemical, and thermal properties. Furthermore, to enhance the properties of pristine carbon nanomaterials, their hybrid materials have been synthesized. Even though tremendous advancement in carbon nanomaterials-based electronic devices and sensors has been achieved, a few challenges need to be addressed before the commercialization of carbon nanomaterials-based devices. Apart from the improvements, the device to device variations, and extrinsic factors like dielectric layers, metal contact resistance remain an issue. Strategies such as chemically tuning and enhancing the properties of carbon nanomaterials are important for the further improvement of carbon nanomaterial-based device performance. This chapter focuses on understanding the basic electronic properties of graphene, CNT. and carbon quantum dots/fullerenes and their applications in electronic devices (field-effect transistors, diodes, etc.), optoelectronics, and various chemical and physical sensors.

Keywords Carbon nanomaterials · Fullerenes · Carbon quantum dots · Carbon nanotubes · Graphene · CNTFET · GFET · Chemical sensors · Physical sensors · Photodetectors

1 Introduction

Solid materials with at least one dimension restricted to the nanoscale (1–100 nm) are called nanomaterials. Materials whose size range in micro-meter have their properties almost similar to that of the bulk materials, but on the other hand, nanomaterials have unique and different properties such as higher surface to volume ratio,

V. Selamneni · N. Bokka · V. Adepu · P. Sahatiya (✉)
Department of Electrical and Electronics Engineering, Birla Institute of Technology and Science
Pilani Hyderabad Campus, Hyderabad 500078, India
e-mail: parikshit@hyderabad.bits-pilani.ac.in

© The Author(s), under exclusive license to Springer Nature Singapore Pte Ltd. 2021   215
A. Hazra and R. Goswami (eds.), *Carbon Nanomaterial Electronics: Devices and Applications*, Advances in Sustainability Science and Technology,
https://doi.org/10.1007/978-981-16-1052-3_10

improved electrical, thermal, mechanical, optical properties, and with minimal imperfections which are completely diverse from bulk counterparts. In 1985 first carbon material named buckminsterfullerene (C_{60}) was synthesized and then later, various other carbon materials came into existence. Over the years, carbon nanomaterials such as fullerenes, nanodiamonds, nanotubes, nanofibers, and graphene have got the utmost prominence for their utilization in various applications. We must understand these different carbon nanomaterials before a detailed discussion regarding different applications.

1.1 Fullerene

Carbon atoms organized in hexagonal and pentagonal rings with closed hollow cages are known as fullerenes. Cylindrical/ ellipsoidal, the spherical arrangement of molecules that entirely contain carbon atoms falls under the category of fullerenes. Carbon nanotubes/bucky tubes are cylindrical-shaped fullerenes, and spherical-shaped fullerene is known as buckyballs.

Spherical Fullerene

This class of carbon materials is zero-dimensional in which all the dimensions of the molecule are restricted to the nanoscale. C_n is the chemical formula of the spherical fullerene where the number of carbon atoms existing in a fullerene molecule is denoted by n in C_n. These are classified into C_{60}, C_{70}, $C_{76,}$ $C_{84,}$ and this sequence continues till gigantic [1] (where the number of carbon atoms greater than 100) and fullerenes of onion type [2] (where concentric shells are combined to form a hollow shape structure). Among all the fullerene structures, buckminsterfullerene-C_{60} is the most stable and prominent structure. It is the carbon molecule in which pentagonal faces are out-of-the-way with each other with a diameter of approximately 0.683 nm. Fullerenes are found as less stable dynamically but considered as chemically more stable in comparison to graphite. These are applied in various applications such as photodetectors [3], solar cells [4], field-effect transistors [5], etc (Fig. 1).

1.1.1 Carbon Nanotubes

Carbon atoms of sp^2 hybridization must be bent to form fullerenes of cylindrical structures knows as carbon nanotubes. During the thermal decomposition of various hydrocarbons, it was found that there is a possibility of the formation of carbon filaments before the invention of a transmission the electron microscope (TEM) [6]. Later after the invention of TEM, it is confirmed that tubular structures like that of nanotubes came into existence from various reports issued in the twentieth century. The carbon filaments of nano size and tubular form using TEM images in 1952 were stated by Radushkevich and Lukyanovich [7]. The first CNT was grown without the usage of any catalyst in a report published by Ijima in 1991 [8]. Different methods such as chemical vapor deposition [9, 10], laser ablation [11], arc discharge [12], etc., are employed for the production of CNTs. CNTs have an analogous structure as that of three-Dimensional graphite because it is derived from the rolled monolayer layer of graphite structure. If the tube wall of carbon nanotube is formed from only one-layer, multiple layers of graphite, then they are known as SWCNTs (single-walled carbon nanotube) and MWCNTs (multi-walled carbon nanotubes), respectively. Intertube spacing in MWCNTs is 0.34 nm almost similar to that of the interlayer distance between carbon atoms of graphite structure [13]. In CNTs, only one dimension of the structure is restricted to the nanoscale, therefore it is considered as a one-dimensional nanomaterial with several hundred nanometers of diameter [14] and length up to several centimeters [15]. Chiral vector (C_h) represents the structure of the SWCNTs. Two integers (n,m) and base vectors a1, a2 will define the chiral vector of the CNTs [16, 17]. These integers of chiral vector classify SWCNTs into the armchair tubes (when m $=$ n i.e., (n, n), zig-zag tubes (when m $=$ 0, i.e., (n, 0)) and for other possible (n, m) values of integers, they are chiral tubes. The information about the chiral angle and the diameter of the nanotubes is known by integer indices (n, m) [18]. The electrical properties of SWCNTs are proportionate to the chirality of nanotubes. When m–n/3 is an integer then nanotube (SWCNT) is considered as metallic or else for all other conditions it is a semiconductor. From various reports, it is confirmed that CNTs are unique carbon nanomaterials with improved electrical, thermal, mechanical, and various other properties for utilizing them in different applications such as flat panel displays [19], sensing devices [20, 21], Li-ion batteries [22], fuel cells [23, 24], etc (Fig. 2).

1.2   Carbon Nanofibers

Carbon nanofibers are a different class of carbon nanomaterial in which the filament-like structure of graphite completely different from nanotubes is aligned in the monolayer graphitic planes. Nanofibers are those in which layers of graphite are organized at an angle (herringbone form) or perpendicular to the plane (stacked form) of the fiber axis [27]. Subjecting hydrocarbons in gaseous form to high temperatures in the presence of catalyst will result in carbon nanofibers. Factors influencing the carbon nanofibers are catalyst material, the temperature used for the synthesis, reactant gas composition. These carbon nanofibers have exceptional mechanical properties such as high mechanical strength which is used in nanocomposite preparation for various applications [28] (Fig. 3).

1.3   Graphene

Graphene is a two-dimensional carbon nanomaterial that is formed from single or various monolayers of graphite. Carbon atoms of sp^2 hybridization are tightly crammed with a bond length of 0.142 nm, in a honeycomb crystal lattice. An only sheet of graphene is termed single-layer graphene (SLG) and multiple sheets super-imposed upon one another with an interplanar spacing of 0.335 nm are called few-layered graphene (FLG). Graphene is considered a basic building block of carbon nanomaterials. Zero-dimensional (spherical fullerene) and one-dimensional (nanotubes) carbon nanomaterial can be easily prepared from graphene by wrap-ping layers in a specific direction of the material. During 1990 and 2004, using the mechanical exfoliation technique several attempts are made for producing thin films of graphite which resulted in the making of nothing less than numerous tens of layers [30]. In 2004, Novoselov and Geim succeeded in the production of graphene in the order of single-atom thickness from bulk graphite structure using a technique known as the micro-mechanical cleavage [31]. Most of the reported methods such as mechanical exfoliation [31], chemical vapor deposition [32], oxidation of graphite [33], liquid-phase exfoliation [34], etc., have challenges in controlling the size, shape, and other parameters during the synthesis of graphene. Apart from all these hardships in synthesis, still, graphene has been utilized in different applications owing to its remarkable mechanical, thermal, electrical, and other properties [35–38] (Fig. 4).

Different properties of the above-discussed carbon-based nanomaterials such as electrical, elastic, mechanical, etc., are shown in Table 1.

1.4 Carbon Quantum-Dots

A quantum dot is a zero-dimensional nanomaterial relative to bulk materials. It is a fluorescent semiconducting nanocrystal that consists of elements from the periodic table from II to VI, III to V, or IV to VI groups (e.g., CdTe) [40]. The size of these nanomaterials is in the range of 2–10 nm in diameter and contain 200–10,000 atoms approximately [41]. Quantum dots are highly photostable due to the quantum confinement effect with broad absorption, symmetric, and narrow emission spectra. Size, surface chemistry, and chemical composition of nanomaterials are important for the emitted light and can be adjusted to visible and near-infrared regions from the UV region. Electron energy levels of the bulk form of semi-conductor are continuous and these energy levels are distinct due to the quantum confinement effect at the nanoscale. Coulomb stated that electron–hole pair is created due to jumping off an electron from the valency band to the conduction band as exciton and physical confinement of electrons in 3D leads to quantum confinement [42, 43]. An electron that is vaulted into the CB (conduction band) will fall back to the VB (valence band) resulting in radiation (electromagnetic) which is diverse in comparison to actual inducement. The frequency with which it is emitted is allegedly attributed to fluorescence which depends on the size of the bandgap that can be tuned by altering the size and surface chemistry of the quantum dot. Therefore, it is concluded that the smaller the quantum dot higher will be its bandgap.

Table 1 Properties of carbon-based nanomaterials. Reprinted with permission from [39]. Copyright © 2017 Elsevier Ltd. All rights reserved

Property	Fullerene	Single-wall carbon nanotube (SWCNT)	Double-wall carbon nanotube (DWCNT)	Multi-wall carbon nanotube (MWCNT)	Graphene	Carbon-based nanofiber
Density (g/cm^3)	1.7	1.3–1.5	1.5	1.8–2.0	N.A	2.25
Electrical conductivity (S/m)	10^{-3}	10^6	10^6	10^6	6×10^5	10^3
Distinctive diameter (nm)	0.7	1	5	20	N.A	50–500
Tensile strength (GPa)	N.A	50–500	23–63	10–60	130	3–7
Thermal conductivity in ambient temperature conditions (W/m–K)	0.4	6000	3000	2000	5000	1900
Elongation at break (%)	N.A	5.8	28	N.A	20	05–2.5
Thermal solidity (in the air)	600	>700	>700	>700	450–650	N.A
Specific surface area (m^2/g)	42–85	10–20	10–20	10–20	2675	50
Elastic modulus (TPa)	N.A	1	N.A	0.3–1	1	0.5

NA—-Not Available

Carbon-based quantum dots are a class of this zero-dimensional nanomaterial that consists of graphene quantum dots and carbon-quantum dots with a size of less than 10 nm. Carbon-based quantum dots because of their durable luminescence and admirable solubility have got great attention to be called carbon-based nano lights [43, 44]. They are synthesized using various methods such as laser ablation [45], chemical ablation [46, 47], electrochemical carbonization [48, 49], hydrothermal/solvothermal [50, 51], etc. These types of quantum dots have exceptional electronic properties which lead to chemiluminescence and electro-luminescence made potential applications in sensors, photodetectors, catalysis, etc. Over the years, 0D, 1D, 2D carbon nanomaterials (Fullerene, nanotubes, graphene) have got a wide range of scope due to their exceptional electronic, thermal, optical, chemical, and mechanical properties [52]. Electronic devices such as CNTFET, RF transistors fabricated using graphene, CNT-based digital logic circuits, etc., have improved their performance due to usage of carbon-nanomaterials during their device fabrication. Various physical and chemical sensors such as pressure sensors, strain sensors, electrochemical, biosensors, humidity sensors, etc., have improved their sensing properties by using pristine as well as hybrid carbon nanomaterials. In this chapter, a brief description of pristine, hybrid carbon nanomaterials applications in different electronic, optoelectronic, and sensing devices is discussed.

2 Carbon Nanomaterials-Based Electronic Devices

The exploitation of carbon nanomaterials as electronic materials with high mobility is done due to the extreme delocalization of its sp^2 hybridized electronic structure. Furthermore, in semiconducting CNTs bandgap and diameter are dependent on its atom arrangement called chirality. Therefore, the tunable bandgap of CNTs offers unique opportunities to customize optoelectronic properties. In comparison to conventional semiconducting materials, carbon nanomaterials are found as probable replacements in various electronic applications based on the above reasons [53]. This section provides recent developments in carbon nanomaterials-based electronic devices.

2.1 CNTs Materials in Digital Electronics Applications

In field-effect transistors, semiconducting CNTs are used as capable materials for the channel. If current flows using a channel that is made of CNT material which is in between the drain and source terminals of the FET device, then it is called CNTFET which has three terminals (Fig. 5a). In 1991, the MWCNT was the first CNT to be discovered; but after 2 years, SWCNT was synthesized by researchers [8]. The research on the development of CNT FETs using SWCNTs was increased significantly compared to MWCNTs due to its better tenability of bandgap. CNTFETs are classified into four types: (1) Conventional CNTFETs, (2) Schottky barrier CNTFETs, (3) Tunnel CNTFETs, and (4) Partially gated CNTFETs. In conventional CNTFETs, doped CNT is used as a channel that is similar to the conventional MOSFETs. Undoped CNTs are used in the Schottky barrier CNTFETs which creates the Schottky barrier with metals at source and drain. Tunnel CNTFETs consist of oppositely doped drain and source regions, and partially gated CNTFETs of partial gate region covering channel. Minimal parasitic capacitances, small switching energies are observed due to smaller value of capacitance (i.e. < 0.05 aF/nm) of CNTs and atomically smooth surface of CNTs decreases the carries scattering and scattering at small-angle carries which outcomes in only onward and backward scattering are removed by the 1D structure of CNTs [54, 55]. In comparison to other semiconductors with high mobility, for example, InSb, CNTs have shown high minimal field mobility of greater than 100,000 cm^2/Vs and current densities of 10^8 A/cm^2 at room temperature conditions [56, 57]. Also, CNTs have few limitations in nanoelectronic applications, particularly the contact resistance between one-dimensional CNT and metal electrodes has a lower limit of ~6.45 kΩ [58]. Further, the contact resistance increases due to the Schottky contacts between CNTs and metals. The difference in CNT Fermi level and metalwork function can be accustomed to allow both FETs of p-type and n-type due to the ambipolar nature of CNTFETs intrinsically. Without tuning of electrode work function or internal doping, generally, CNTFETs are p-type in ambient conditions due to oxidation [59, 60].

The first CNT FET was demonstrated in 1998 (Fig. 5b) [61]. These Initial studies sparked substantial interest in transport phenomena and electronic devices based on CNTs for example SET(single-electron transistors) and wired like structures at the quantum level (quantum wires) [62], ballistic transistors [63], Luttinger liquid behavior [64], and ambipolar FETs [65]. In the early research, Snow et al. studied the transport properties of fabricated thin film transistor s(TFT) using as-grown random CNTs [66]. The on/off ratio of 10^5 and field-effect mobility of 10 cm^2/Vs· are seen in CNT-based TFTs. The study revealed the trade-offs between field-effect mobility and on–off ratio. From the simulation studies, it was deep-rooted that percolation effects dominate the transport of charge in CNT FETs. Percolation effect displayed power-law relation channel resistance with channel width and length, CNT alignment, CNTs network density. Due to two reasons, CNT TFTs field-effect mobility (i.e., (<100 cm^2/Vs) is found lower than that of SWCNT (i.e. >10,000 cm^2/V s). First, contact resistance of CNT-CNT in percolating network [67]. The second reason being the overestimation of the random network CNTs capacitance. Numerical methods were used to suppress the consequence of metallic CNTs for overcoming the on/off ratio of digital circuits. Collins et al. reported a selective removal of metallic CNTs using Joule heating [68]. This technique is successful particularly for CNTs which are grown using the CVD method, where all CNTs bridge the drain and source gap. The drawback of this technique is, it causes collateral damage to adjacent nanotubes.

DGU (Density gradient ultracentrifugation) centrifuged semiconducting CNT inks-based devices showed larger values of on/off ratios, current densities, field-effect mobilities [69]. Simple assembly methods such as transfer printing, dip coating, and drop-casting mostly use these semiconducting CNT inks. The density of these semiconducting CNTs could be improved without affecting the on–off ratio. Apart from DGU-sorted CNTs, methods that are based on gels also showed a higher value of on/off ratio which is greater than 10^4 [70]. Roberts et al. fabricated low-voltage TFTs with semiconducting CNTs, with effective device mobility of 13.4 cm^2/Vs, on/off ratio of >1000, and a low subthreshold swing (130 mV/dec) [71] (Fig. 5).

2.2 Carbon Nanotubes for Radiofrequency Transistors

Due to high saturation velocity and carrier mobility, the low intrinsic capacitance of CNTs, they have been significantly used as a material for the channel which is used to fabricate high-performance radio-frequency (RF) transistors [73]. CNT-based RF transistors showed extraordinary linearity that is desirable in RF devices. Steiner et al. with the help of aligned semiconducting CNTs fabricated a radio frequency transistor. For a 100 nm gate length, power gain cut-off frequency of 7 GHz (153 GHz, intrinsically) and current gain cut-off frequencies of 15 GHz (30 GHz, intrinsically) output current was saturated [74]. Cao et al. studied the RF performance of fabricated CNT transistors based on high-density, aligned, polyfluorene-sorted semiconducting CNTs with a self-aligned T-shaped gate (Fig. 6) [75]. The fabricated RF transistor exhibited excellent DC with transconductance of 310 μS/μm, output normalized resistance >100 kΩ.μm, and both current gain cut-off frequency and oscillation frequency were greater than 70 GHz. Further, the transistor displayed good linearity with 22 dBm of third-order intercept point (IIP3) and 14 dBm of P_{1dB} (gain compression point). Zhong et al. demonstrated an RF transistor depends on arbitrarily oriented semiconducting CNT films [76]. The maximum transconductance of 0.38 mS/μm was achieved by reducing the CNT/CNT junctions and at 103 GHz (281 GHz, intrinsically) current gain cut-off frequency, up to 107 GHz (190 GHz, intrinsically) power gain cut-off frequency.

2.3 Graphene for Digital Electronics

Graphene has been used to explore its applications in digital electronics due to its remarkable field-effect mobility [77]. The better scaling of the device channel is permitted due to its two-dimensional structure [78]. Unlike CNTs, the device's high conductance is attributed to their lower values of contact resistance. But, very low on–off ratio (5–10) and due to the presence of zero bandgap in graphene, it is not used for applications such as digital electronics wherein the anticipated on/off ratio is very high (i.e., in the range of 10^4–10^6). Even at zero carrier concentration, graphene shows minimum conductivity, hence it cannot be turned off completely [79]. The two most successful approaches used for bandgap engineering in graphene are (i) Due to quantum confinement effect in graphene nanoribbons [78], the bandgap could be opened in graphene and (ii) By changing the symmetry in K and K′ carbon atoms of honeycomb lattice structure [35]. Therefore, several fundamental material issues need to be addressed before considering graphene for digital electronics.

2.4   Graphene for Radio-frequency Transistors

Over the years, owing to superior electronic properties mainly the high mobility of charge carriers, graphene has gained significant attention as a channel material for RF electronics [78]. Lin et al. described a breakthrough study of RF devices fabricated on SiC using graphene, which has 240 nm of gate length [80]. Despite moderate carrier mobility (1500 cm^2/Vs), a high operating frequency (f_T) of 100 GHz was achieved. Badmaev et al. developed graphene transistors by a scalable fabrication technique using a T-shaped gate structure (Fig. 7a) [81]. They successfully scaled down the channel length to 110 nm. Guo et al. developed a graphene transistor on silicon–carbon substrate using C-face epitaxial graphene (Fig. 7b) [82]. Significant improvement in the transistor was observed over Si-face epitaxial graphene transistor and a value of 70 Giga-Hertz maximum power gain cut-off frequency was achieved. Self-aligned contacts and high k dielectric T-gate contributed to the maximum frequency value.

3   Carbon Nanomaterials Hybrid-Based Electronic Devices

Hybrids of 0D, 1D, and 2D carbon nanomaterials provide the tremendous potential to obtain next-generation scalable and high-performance devices. Graphene has made significant scientific attention due to its number of uses, such as the conversion of energy and storage, optics, sensors, and electronic devices [83–85]. The electronic device applications of graphene were less due to its gapless nature [86]. Alternatively, chemical changes in various reactive oxygen groups of graphene oxide (GO), enable the development of functional materials used in various applications. Feng et al. [87] fabricated a semi-transparent graphene p–n junction diode, obtained by coating two layers of oppositely charged GO layers, and it has carbon nanotubes electrodes. The current rectification of this diode occurs because the tunneling conduction electrons are governed by an internal field due to this device different from other diodes. The current rectification occurrence in graphene p–n junctions significantly extends the class of materials and physical processes that can be used to design electronic components. This graphene diode can be utilized for simple logic operations such as AND and OR logic gates. Figure 8a shows a high output potential for the AND logic gate if both switches A and B are given positive potential. Unlike the AND gate, OR logic gate has a high output potential, due to either switch A and B, a high potential is triggered as observed in Fig. 8b. Deep et al. [88] reported gate-tunable p–n heterojunction diode by the embedding of n-type single-layer molybdenum disulfide MoS$_2$ and p-type SWCNTs semiconductors. The heterojunction was formed because these two semiconductors were stacked vertically. The wide range of charge transport is achieved by tuning the heterojunction by the applied gate bias voltage. Under various gate biases, the device's output graphs are shown in Fig. 8c. The p–n

Fig. 1 C$_{60}$ molecular structure

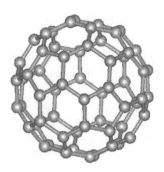

heterojunction diode often exhibits antiambipolar behavior when handled as a three-terminal device. Figure 8d illustrates junction transfer characteristics, and the green line shows an unusual gate voltage dependence, indicates antiambipolar behavior.

In living systems, metal ion analysis is very important to understand metallic element stabilization and associated diseases [89]. Currently, conventional techniques used for identification and their applications are limited due to time consumption, require sophisticated and costly equipment, and inadequate real-time monitoring. To overcome those limitations, Fan et al. [90] demonstrated the identification of Cu^{2+} ions by functional CQDs modified gate electrodes with the solution-gated graphene transistors (SGGTs). The sensing mechanism of the sensor was channel current varies with the electrical double layer (EDL) capacitance change due to interaction between CQDs and Cu^{2+} ions. Compared to conventional detection methods, the combination of CQDs with SGGTs demonstrates Cu^{2+} detection with a minimum concentration range (1×10^{-14} M). It shows quick response time in seconds. The schematic and working of the SGGT-based Cu^{2+} ion device are shown in Fig. 9 [10].

Over the last few years, research is gaining momentum on new elastic semiconductor materials as they can be integrated into modern, flexible, portable, and handheld consumer electronics [91–93]. So far, carbon nanotubes-based flexible electronics such as flexible and transparent transistors have been fabricated on PET and polymer substrates [94, 95]. Different techniques like floating-catalyst CVD, roll-to-roll transfer, and gas-phase filtration technique employed to fabricate TFT and integrated circuits on flexible transparent substrates [96, 97]. In spite of the significant advances in flexible electronics carbon nanotube and graphene have some obstacles remain. The stretchable transistors need less contact resistance with the channels, electrodes with excellent electrical conductivity and an active channel with a large on–off ratio, both of which point to a cooperation strategy between graphene and CNTs. Electronic device performance-enhanced beyond the flexible electronics by hybrid-based graphene/CNTs [98, 99]. Tung et al. [100] reported solution-based, low-cost scalable, and flexible graphene/CNTs hybrid films show excellent conductivity at high optical transmittance of 86%. This film does not include surfactants to maintain the mechanical and electronic properties of both components. The flexibility of film better than the transparent rigid inorganic crystal structure of ITO suggests maximum compatibility with flexible substrates.

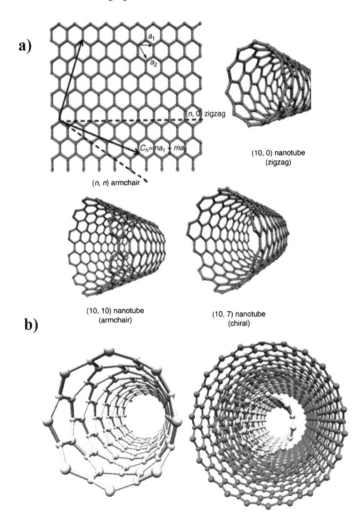

Fig. 2 **a** Schematic representation for different types of CNTs. Reprinted with permission from [25]. Copyright © 2018 Elsevier Inc. All rights reserved **b** Single-walled carbon nanotubes and Multi-walled carbon nanotubes. Reprinted with permission from [26] Copyright @ 202 Springer Nature

4 Carbon Nanomaterials Based Sensors

Based on the types of measurand, sensors are roughly categorized as physical and chemical sensors. Parameters such as strain, temperature, pressure, force, displacement, position, flow rate, etc., mostly are detected using physical sensors As explained in the introduction, CNTs have excellent mechanical properties with Young's modulus value of 1 TPa approximately for single-walled CNTs [101].

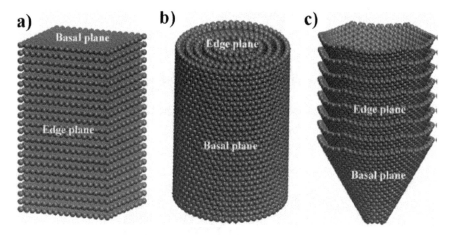

Fig. 3 **a–c** schematic representation of different carbon nanofiber structures. Reprinted with permission from [29] Copyright © 2012 Elsevier Ltd.

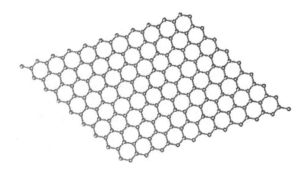

Fig. 4 Schematic illustration of 2D layered graphene structure

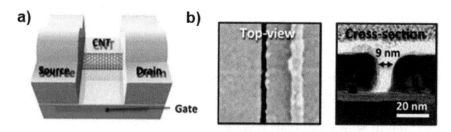

Fig. 5 **a** CNT FET with bottom gate schematic **b** Bottom-gate CNT FET with sub-10 nm channel SEM image from top-view and TEM image of the cross-sectional view. Reprinted with permission from [72]. Copyright © 2012, American Chemical Society

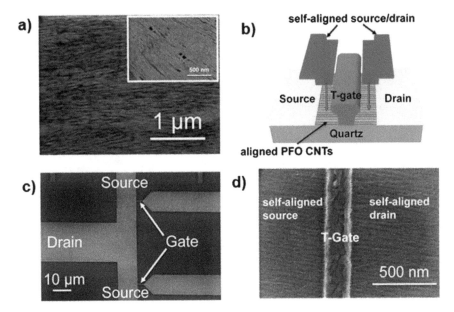

Fig. 6 a SEM graph of Polyfluorene-organized CNT film on a quartz substrate. Approximately 40 nanotubes/μm packing density is seen **b** Schematic of the self-aligned transistor structure with T-shaped gate **c** SEM micrograph of the fabricated self-aligned T-shaped gate transistor **d** High-magnification SEM micrograph of the channel area. Reprinted with permission from [75]. Copyright © 2016, American Chemical Society

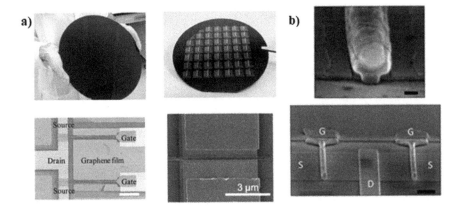

Fig. 7 a 2-inch Silicon wafer with large area graphene (CVD grown). Graphene transistors on Silicon wafer. SEM image (color) of dual-gate graphene transistor (scale bar 10 μm). High-magnification SEM image showing the active area of the transistor. Reprinted with permission from [81]. Copyright © 2012, American Chemical Society **b** SEM graph (scale bar 100 nm) of the device with 100 nm T-gate with a trilayer resist and drain and source contact metal aligned to the gate foot and a dual-gate graphene FET on C-face SiC (with scale bar 1 μm) Reprinted with permission from [82]. Copyright © 2013, American Chemical Society

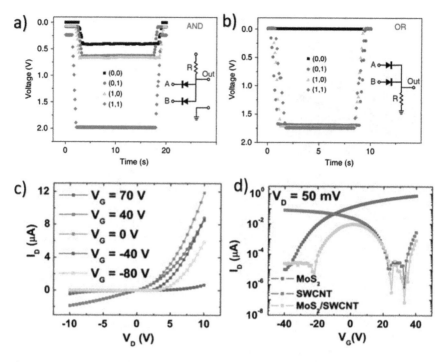

Fig. 8 **a, b** Graphene p-n diode and resistor-based AND and OR logic gates, Reprinted with permission from [87] Copyright © 2018, Springer Nature **c** output characteristics with various gate voltages **d** Transfer characteristics of the p–n junction. Reprinted with permission from [88] Copyright © 2013, National Academy of Sciences

Wong et al. reported an experimental value of 1.28 ± 0.5 TPa for MWCNTs [102]. Also, without any deformation, CNTs showed a high tensile strength (up to 40%).

4.1 Carbon Nanomaterial-Based Pressure Sensor

A pressure sensor is a sensor which converts mechanical displacement into an electrical signal [103–106]. In the last two decades, prominent research has been done in the fabrication of highly sensitive pressures sensor using novel carbon nanomaterials, CNT, and graphene. Zhan et al. reported a wearable and flexible pressure sensor based on SWCNTs/paper through a highly scalable and cost-effective approach with 2.2 kPa^{-1} sensitivity value in a broad range of 35 Pa–2.5 kPa and 1.3 kPa^{-1} sensitivity in 2500–11,700 Pa [107]. The sensor was fabricated by impregnating SWCNTs into paper and sandwiched between a PDMS film and PI sheet that is decorated with interdigitated gold (Au) electrodes (Fig. 10).

Sahatiya et al. demonstrated a flexible, biodegradable pressure sensor by sandwiching MWCNTs between PI substrate and cellulose paper. Multi-walled CNTs

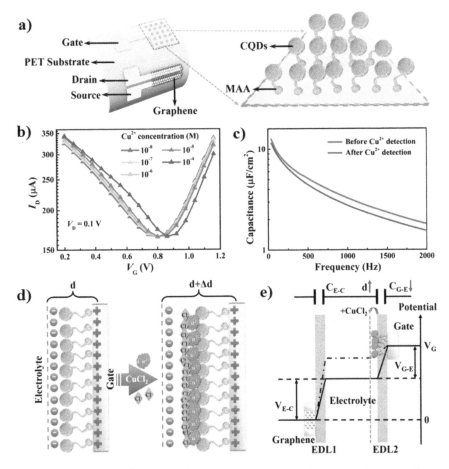

Fig. 9 **a** SGGT-based Cu^{+2} ion sensor **b** Transfer curves of SGGT **c** frequency and gate capacitance relationship before and after Cu^{+2} ions are added **d** Schematic illustration of the EDL capacitance variation mechanism **e** Potential drops on the surface of graphene channel and gate, across the two EDLs. Reprinted with permission from [90]. Copyright © 2020, American Chemical Society

were deposited on the substrate (PI) by roll pin and pre-compaction mechanical pressing technique (Fig. 11a) [108]. Sensitivity value of 0.549 kPa^{-1} was seen by a fabricated pressure sensor and a response time of lesser than 32 ms. Furthermore, the fabricated pressure sensor was used as a touchpad and electronic skin application. Graphene was widely used as active as well as inactive material for pressure sensor applications. Although graphene no piezoelectric properties, it was used as electrodes to support piezoelectric material and also used as an additive to piezoelectric polymer like PVDF to improve piezoelectricity [109]. Furthermore, due to graphene's excellent mechanical and electronic properties, it has been extensively used as an active material for the piezoresistive pressure sensor as elucidated in the introduction. Zhu et al. developed a piezoresistive pressure sensor using graphene

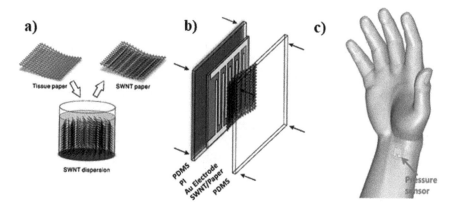

Fig. 10 Pressure sensors fabrication process illustration. **a** The SWNTs were dip-coated on tissue paper **b** Single-wall nanotube/tissue paper is accumulated onto gold electrodes on a PI substrate; encapsulated using PDMS layers which provides mechanical support and **c** To sense heart pulse human wrist is mounted with a pressure sensor Reprinted with permission from [107]. Copyright © 2017, American Chemical Society

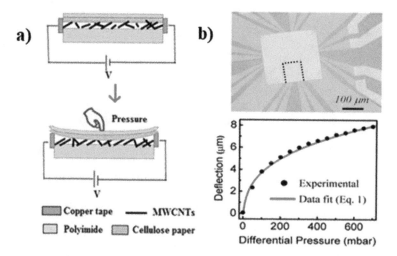

Fig. 11 **a** Ultrasensitive touchpad schematic which is done by sandwiching MWCNTs between polyimide and cellulose paper [108] **b** Silicon nitride square membrane with graphene piezoresistors optical microscope image (the dotted lines indicate the device used) and SiN membrane with deflection v/s differential pressure of 100 nm. Reprinted with permission from [110] Copyright © 2013, American Institute of Physics

on the silicon nitride membrane showed a dynamic range of 0–700 mbar [110]. In this strain sensor, multilayer polycrystalline was fabricated using a chemical vapor deposition method (Fig. 11b). Yao et al. developed a pressure sensor using a flexible graphene/polyurethane sponge using a fracture microstructure [111]. PU sponge which is available commercially was coated with graphene oxide using the dip-coating technique and then it was reduced to improve thermal conductivity using hydrogen iodide, further softened with hydrothermal treatment. The fabricated sensor exhibited a sensitivity of 0.26 kPa^{-1} (0–2 kPa range) and 0.03 kPa^{-1} (2–10 kPa). The fractured microstructure improved the sensitivity when compare with rGO/PU sponge sensor without fractured microstructure. Jian et al. fabricated a graphene/CNT hybrid-based highly sensitive pressure sensor [112]. Because of the synergistic effect in the hybrid, the fabricated pressure sensor exhibited a high sensitivity of 19.8 kPa^{-1} and a very small detection limit around 0.6 Pa. Tian et al. reported a pressure sensor that is sensitive to a 0–50 kPa wide pressure range using laser scriber graphene with a foam-like structure. The sensitivity value of 0.9 kPa^{-1}was seen using this pressure sensor [113].

4.2 Carbon Nanomaterials-Based Strain Sensor

Lee et al. presented a fully microfabricated strain sensor by SWCNT film on PI film. Polyimide was prepared by spin coating glass wafer and photolithography was used for patterning the electrode [114]. Later, by using a spray coating method, 280-nm-thick CNT film was deposited on PI film. The fabricated device showed a linear relationship between the applied strain and resistance over a range from 0 to 400 microstrain and a high gauge factor of 60. Dharap et al. demonstrated a strain sensor using pristine-SWCNTs film using vacuum filtration by 0.2 mm Teflon membrane [115]. The SWCNTs thin film was attached to a brass specimen and the results showed that the relation between the applied strain and change in the voltage across the film is linear. In the last few years, MWCNTs have attracted the researchers' interest due to relatively higher purity, cost-effectiveness, and superior electronic properties in comparison with SWCNTs. Zhang and co-workers have fabricated strain sensors based on silver nanoparticles and MWCNTs on flexible PDMS substrates (Fig. 12) [116]. The sensor exhibited gauge factors that are tunable in the range of 2.1–39.8 and stretchability of 95.6%. Further, the sensitivity and linearity of the sensor could be adjusted by Ag NPs concentration.

Zhao and co-authors demonstrated the very sensitive strain sensor using nanographene films on the mica substrate [117]. Fabricated sensor achieved a gauge factor of more than 300. Li et al. demonstrated a strain sensor using graphene woven fabrics (GWFs) on the PDMS substrate [118]. SWFs were prepared using a woven copper mesh by a CVD process and later copper was etched away. After placing GWFs on PDMS, contacts were taken using silver paste (Fig. 13). The fabricated sensor achieved an approximate value of 10^3 gauge factors for 2–6% strains and ~10^6 for greater than 7% strains. The ultrahigh sensitivity of the device was ascribed to mesh configuration and fracture behavior of graphene micro ribbons.

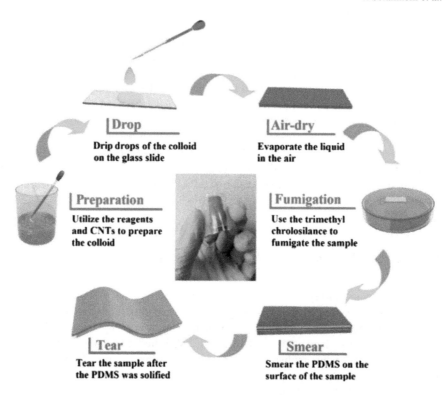

Fig. 12 CNTs/PDMS strain sensor fabrication process which is modified using silver nanoparticles. The as-prepared product is illustrated in the center. Reprinted with permission from [116]. Copyright © 2015, Elsevier

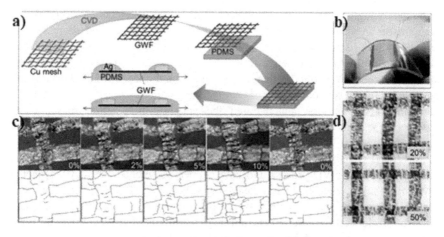

Fig. 13 **a** Schematic illustration of GWF/PDMS fabrication **b** Wired samples macroscopic optical image **c** Establishment of crack and their evolution in GWF with various strain optical images, and schematic illustration **d** large strains GWF Optical images (20 and 50%) Reprinted with permission from [118]. Copyright © 2012, Springer Nature

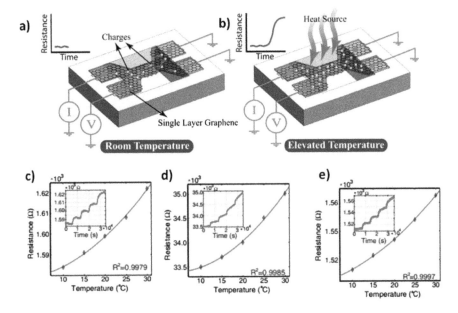

Fig. 14 a Graphene sensors schematic illustration. With the temperature rises from room temperature to a high temperature **b**, Mobility will be decreased due to a proportionate increase in the number of electrons. A resistance varies in terms of temperature and time for the three different graphene sensors: **c** on a SiO_2/Si substrate, **d** on a SiN membrane **e** the deferred graphene sensor. Reprinted with permission from [119]. Copyright © 2017, Springer Nature

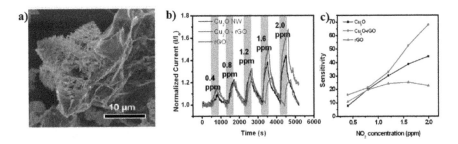

Fig. 15 a Scanning electron microscope image of rGO/Cu_2O nanowire mesocrystal, **b** Dynamic response of different hybrid carbon nanomaterials under the exposure of NO_2 gas with increased ppm levels, **c** sensitivity of three devices (Cu_2O, Cu_2O-rGO, and rGO) are plotted. Reprinted with permission from [145]. Copyright © 2012, American Chemical Society

4.3 Carbon Nanomaterials-Based Temperature Sensor

Temperature sensor is a crucial part in several major applications, such as health monitoring, artificial electronic skin, etc. Resistive-based temperature detectors are the most widely used sensor due to their high accuracy, stability, fast response.

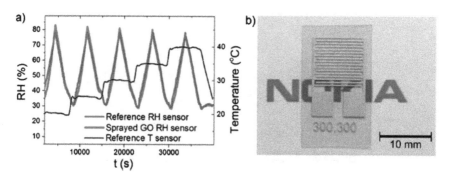

Fig. 16 **a** Sensing performance of GO RH sensor of 15 nm thick compared with commercially fabricated RH sensor with high-performance **b** Image of GO film sprayed on Ag electrodes. Reprinted with permission from [158]. Copyright © 2013, American Chemical Society

Furthermore, thermal sensors, mercury thermometers, and infrared temperature sensors are also widespread. In the past few years, carbon-based sensitive materials including graphene, CNTs, and carbon fiber have gained more attention for temperature sensor applications due to their extraordinary mechanical and electronic properties compared to ceramics, metals, metal oxides. Davaji et al. developed a temperature sensor based on single-layer graphene on three dissimilar substrates, a silicon nitride membrane, a silicon/silicon dioxide-based substrate, and a suspended architecture (Fig. 14) [119]. The fabricated temperature sensors acted as a resistive type, and the resistance change was explained using electron–phonon scattering and electron mobility–temperature relationship ($\sim T^{-4}$). From the analysis, the team revealed that the silicon nitride membrane-based sensor shown the highest sensitivity because of a smaller value thermal mass and it has shown the lower value of sensitivity for the sensor on SiO_2/Si. Furthermore, the sensor on silicon nitride showed improved mechanical stability in comparison to the deferred graphene sensor. Sahatiya et al. demonstrated flexible wearable temperature sensors based on flakes of graphene and solar exfoliated reduced graphene oxide (SrGO) on the PI substrate [120]. Both graphene flakes and SrGO based temperature sensors discovered a negative temperature coefficient that was comparable with commercial temperature sensors. This research has clearly shown that graphene on a flexible PI substrate could be used effectively as wearable temperature sensing applications.

4.4 Carbon-Based Nanomaterials for Chemical Sensors

A device that converts the existence of target compounds into a computable quantity is called a chemical sensor [121]. Gas sensors, vapor sensors, humidity sensors, etc., used for the detection of target compounds are different types of chemical sensors [122–126]. A detailed explanation of carbon-based nanomaterials used to detect target compounds is depicted below.

4.4.1 Carbon-Based Nanomaterials for Gas Sensors

The measurement and existence of concentration of a specific gas in the earth's surrounding atmosphere such as natural fumes, dangerous gases, etc., are detected by using a device known as gas sensor [127]. These sensors are used for monitoring the environment [128], control of hazardous gas emissions, production control in cultivation and health diagnostics, etc. [129–131]. Over the years, researchers have utilized carbon nanomaterials like graphene, GO, rGO, CNTs, etc., and found them as efficient candidates for gas sensing applications. Chemiresistors, field-effect transistors, Schottky diode, etc., are different configurations of device architecture for gas sensing [127]. Noveselov and his co-workers utilized graphene for the first time for fabricating gas sensors in 2007 to detect NO_2 gas [132] graphene-based sensing devices are fabricated using graphite flakes which are exfoliated micro-mechanically and supported on Si substrate. These fabricated devices have shown remarkable sensitivity for detecting individual gas molecules. Later in 2009, Dan and his co-workers proved that the pristine graphene-based device fabricated by Nove Selo has improved its sensitivity by forming a contamination layer on the graphene on Si substrate using an electron-beam lithography technique [133]. This layer helps to detect various other individual molecules such as octanoic acid, H2O, nonanal, and various other gas vapors which are down to ppm level. Balandin and his co-workers also employed an electron beam lithography technique for fabricating the FET device using pristine graphene and with the help of low-frequency noise of the device they detected various organic vapors [134].

Later, CVD technique was used for fabricating a sensing device using graphene sheets in the layered form with exceptional quality on Si substrate to detect gases such as ammonia [135, 136], carbon dioxide, nitrous oxide [137, 138], and also used to detect humidity [139]. Gautam et al. deposited graphene monolayers on the Si support to detect CH_4, H_2, and NH_3 [135]. Improved sensitivity is observed using this device by operating it at a temperature range between 150 and 200 °C but there is no enhancement in response time [135]. Choi et al. fabricated a sensing device where SLG (single-layer graphene) is used as an active component and BLG (bi-layer graphene) is used as a heating element that can heat up to 200 °C [140]. This device achieved a response in the range of 39% to 41 ppm of NO_2 and further, the existence of the element for heating facilitated to upsurge the retrieval time of the device. In the future, a lot of modifications (such as the incorporation of defects and impurities) should be done to pristine graphene to improve its selectivity, sensitivity, and durability of the fabricated sensing device. Graphene oxide is also used for sensing various gases owing to the presence of oxygen-containing functional groups in its structure. To improve the sensing activity of the graphene oxide device it is further reduced in an atmosphere such as hydrogen (Via thermal route) to obtain rGO. In 2008, Robinson and his co-workers first used rGO based devices for detecting various gases [141]. Initially using the spin coating method GO is deposited on the Si substrate and later reduced with reducing agent vapors of hydrazine to fabricate rGO-based sensing device at 100 °C. Manohar et al. fabricated rGO-based flexible printed sensing device which is prepared using an ascorbic acid reducing agent [142]. This

Table 2 Different carbon-based nanomaterials gas sensing performance for detecting various gases

Material	Detected gases	Response/ Retrieval time	Detection limit	Sensitivity (response x ppm^{-1})	Reference
Single Layer graphene / Bilayer graphene (CVD)	Nitrous oxide	95 s/11 s	–	0.98%	[140]
Single Layer graphene (CVD) patterned	Nitrous oxide, Ammonia	89 s/579 s	–	2.6% (Nitrous oxide)	[139]
PANI/reduced graphene oxide	Hydrogen	2 min/3 min	–	0.0016%	[146]
CNTs/graphene oxide	Nitrous oxide	1 h/3 h	0.5 ppm	2%	[147]
In$_2$O$_3$/ reduced graphene oxide	Nitrous oxide	3 min/4 min	–	12.2%	[148]
ZnO Ncryst/ reduced graphene oxide	Methane	1 min/10 s	100 ppm	0.05%	[149]

device was fabricated using the inkjet printing technique by dissolving rGO in water in the existence of surfactant and displayed a positive response with increased resistance behavior in detecting dichloromethane and various alcohols, negative response detection behavior to Cl_2 and NO_2. Furthermore, different carbon nanomaterials-based hybrid composites are employed to enhance the sensitivity and selectivity of these gas sensors. Chung et al. fabricated a flexible device using CVD graphene which is decked with palladium nanoparticles to further improve the sensitivity to 30 for 1% of H2 gas detection [143]. Wu et al. used a nanocomposite thin-film device which was prepared using graphene extracted using exfoliation and PANI for detecting methane gas in ambient temperature conditions [144]. This device has shown a considerable amount of 10% response to 10 ppm of gas. Using hydrothermal method Deng et al. synthesized rGO/Cu_2O nanowire mesocrystals which are integrated to detect NO_2 gas [145] (Fig. 15). Furthermore, there is a lot of scope for various hybrid carbon-based nanomaterials that are used for detecting various gases by fabricating suitable gas sensors. The sensing performance of the various carbon-based gas sensors is tabulated in Table 2.

4.4.2 Carbon-Based Nanomaterials for Humidity Sensors

The water vapor existing in the air is known as humidity which is measured as relative humidity (RH). The relative humidity is defined as the fraction between the fractional pressure of water and the vapor pressure at equilibrium at a given temperature. These sensors are mostly used in agricultural sectors and industries for keeping a track of the quality of food in food processing industries, medical equipment, weather prediction, etc. Polymers [150], metal oxides [151, 152], porous silicon [153, 154] are mostly used for commercial humidity sensors. Exceptional qualities such as larger sensing areas and high chemical inertness made researchers use carbon nanomaterials for humidity sensing applications. Zhang et al. synthesized carbon nanosheets and honeycombs in the nano range on Si substrates using a hot filament PVD (physical vapor deposition) technique and observed that carbon nanosheets have shown remarkable sensing response 11–95% humidity range under ambient conditions [155]. Luo et al. manufactured and characterized quartz crystal micro-balance (QCM) humidity sensor using a sensitive layer made by graphene oxide [156]. This QCM humidity sensor has shown a 10–60% sensing range, 1371(Q)/1% relative humidity, a recovery time of 3 s, and a response time of 20 s. Yadav et al. synthesized multiwalled CNTs using the direct liquid injection CVD method on cobalt substrate and utilized for designing humidity sensors to sense moisture with RH in the range of 10–90% for various applications [157]. Very recently, Borini et al. fabricated graphene oxide (GO) based humidity sensors using a spray coating method on silver electrodes (GO thin-films typically in the range of ~15 nm) [158]. Figure 16a depicts that GO-based thin film has shown characteristics almost similar to that of commercially available RH sensors. In Fig. 16b, GO-based thin film is almost invisible (i.e., it is in the order of few nm).

5 Carbon Nanomaterials-Based Electrochemical Sensors

Various nanomaterials such as carbon-based [8, 159], transition metal dichalco-genides [160], noble metals [161, 162], etc., are used for electrochemical detection. Among them, nanomaterials based on carbon are extensively used for elec-trochemical (electroanalytical and electrocatalytic) sensing due to their improved capability to transfer electrons in electrochemical reactions. Mostly reported carbon nanomaterials for electrochemical sensing are graphene [163, 164], CNTs [165, 166] fullerenes [167], carbon nanofibers [168], and carbon quantum dots [169], etc. These all nanomaterials when utilized for electrochemical sensing are either used in their pristine form/mixed with new materials/functionalizing their pristine form during device fabrication. Electrochemical sensing of the above-mentioned carbon-based nanomaterials is reviewed below.

5.1 CNTs-Based Nanomaterials

In the study of electrochemistry of carbon nanomaterials, electron transfer is surface-dependent. Electron transfer is improved by adding functional groups on the surface of these materials. Britto et al. first applied carbon nanotubes in electrochemistry as electrodes [170]. They concluded that the immobilization of activated CNTs on the surface of the electrode is the major drawback for electrochemical sensing applications because of their insolubility property in almost all solvents which would result in nonuniform and unstable films. To avoid this problem, different physical/chemical methods are used for dissolving CNTs first in various solutions and immobilizing on different electrode surfaces. Immobilization of CNTs on electrode surfaces is done using various methods that are used for electrochemical sensing applications which are explained in detail below.

(i)   **Solvent dispersion and immobilization by casting**: In this method, initially CNTs are cleaned/purified and activated by pretreatments then dispersed in solvents such as DMF [171, 172], toluene [173, 174], etc., using ultrasonication process and later this mixture is dropped on to the surface of electrode followed by drying it (i.e., casting) for electrochemical detection of analytes.

(ii)   **Additive assisted and immobilization**: In this method, additives such as surfactants, polymers are added into the solvents where CNTs are dispersed to enhance the stability and solubility of the solution (i.e., solution doesn't get settled easily) and then subjected to immobilization for applying in electrochemical sensing applications.

(iii)   **Carbon nanotube paste electrodes**: In this method, paste electrodes are fabricated using binders in liquid form and CNTs. These paste electrodes made from CNTs hold the properties of the old-style carbon paste electrodes (CPEs), for example, the attainability to consolidate different substances, the low currents in the background, the wide range of potential, and the composite nature. These electrodes also reserve exceptional electrochemical properties that enhance the electron transfer mechanism which is useful for increasing the efficiency for electrochemical detection of various analytes especially biomolecules. The carbon paste electrode fabrication doesn't require any pretreatment for CNTs (SWCNTs and MWCNTs) which are used.

Properties such as easily modifiable surface, solid electrocatalytic activity, the tube-like structure of CNTs in nano range (which helps to increase the surface activity), etc., have made CNTs unique materials for electrochemical sensing applications.

5.2 *Graphene-Based Nanomaterials*

Graphene is a two-dimensional material that has a single layer of sp^2 hybridized carbon atoms extracted from three-dimensional graphite and closely packed into a hexagonal lattice structure.

5.2.1   Graphene Oxides-Based Electrodes

Initially, GO aqueous electrolyte solution was prepared by using conventional methods of chemistry such as modified hummers methods which are used in the preparation of electrode materials using two-dimensional graphene for electrochemical sensing applications. In the next step, as prepared GO solution is dropped onto the surface of electrodes such as glassy carbon electrode [175, 176], gold electrode [177], disk shape platinum electrode [178]. For activating the carboxylic acid groups present in GO, the GO-coated electrode is immersed in buffer solutions such as EDC, NHS which helps in the formation of a covalent bond between electrode and enzyme [178]. This prepared electrode is used for the recognition of various target analytes based on its electrochemistry. The above mentioned prepared graphene-based sensors are used as enzyme biosensors, non-enzymatic biosensors that are applicable for simple biomolecules, immunosensors, etc. The performance of various graphene-based electrodes for electrochemical sensing is tabulated in Table 3.

6   Carbon Nanomaterials-Based Resistive Memories

Technology based on semiconductor devices has been improving rapidly by a decrease in their size. Advanced semiconductor device fabrication uses nanotechnology, where critical dimensions of these devices are found to be around 20 nm. These devices are categorized into display devices, memory devices, logic devices, etc. Among them, memory devices with large density, high-performance, and lower consumption of power are in great demand [183]. Portable electronic devices have unique non-volatility (where memory is not lost even when the device is off). These non-volatile memory property-based devices are further classified into flash memory and resistive switching memory devices [184]. A three-dimensional structure consisting of an insulating layer between two electrodes with a simple configuration is the resistive switching memory cell in the ReRAM device. Resistive exchanging memory gadgets use the contrast in opposition to the high and low resistance memory states and have a straightforward gadget structure (metal/separator/MIM metal structure) relevant to high-thickness memory gadgets with a minimal creation cost [184–187].

From Fig. 17 it is clear that the resistance of a resistive memory device is switched from high to low state and vice versa by applying an appropriate voltage to the device [188–190]. Recently, researchers around the world widely studied and found that nanomaterials like metal chalcogenides, metal oxides, carbon-based materials are suitable candidates for resistive-based memory devices [188–192]. Among them, carbon-based nanomaterials are found as efficient for non-volatile resistive-based memory devices [193].

Table 3 Carbon-based nanomaterials performance for electrochemical glucose sensing

Graphene-based materials	Linear range (mM)	Detection limit (μM)	Sensitivity ($\mu AmM^{-1}\ cm^{-2}$)	Reproducibility: RSD for (No. of electrodes)	Stability, % of initial response after (days)	References
CrGO	0.1–10	10 ± 2	110 ± 3	2.5% (6)	No obvious decrease (2 days)	[179]
CrGO-AuNPs	0.1–10	35	–	0.74% (3)	80% (4 months)	[180]
Sulfonated rGo-CdS	2–16	700	1.76	4.2% (5)	93% (30 days)	[181]
CrGO-Ni (II)-quercetin	0.003–0.9	0.5	187	5.1% (7)	92.1% (20 days)	[182]

Resistor-based memory

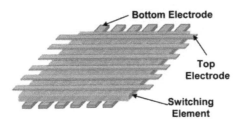

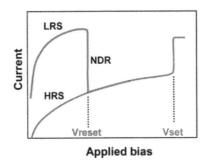

Applied bias

Fig. 17 Schematic of the resistive-based memory device and its operation (low resistance state, high resistance state, and negative differential resistance are LRS, HRS, and NDR, respectively). Reprinted with permission from [183]. Copyright © 2011, Royal Society of Chemistry

Graphene-based nanomaterials such as GO, rGO, hybrid GO, and hybrid rGO are used as dielectric materials in nonvolatile resistive-based memory devices which are depicted in Fig. 18. Rani et al. experimentally found that graphene-based non-volatile memory devices (pristine and hybrid GO, rGO materials) are having a broad range of scope for the advancement of new flexible memory applications by evaluating device characteristics [191]. Tsai et al. have demonstrated that $CNT/AlO_x/CNT$ material-based crossbar electrodes for RRAM device with current programmed around 1 nano ampere and ON/OFF ratio up to 5×10^5 have capable characteristics for non-volatile memory devices [194]. By embedding a graphene single-layer sheet into the interface between a transparent top cathode (formed of Indium-doped tin oxide substrate) and a zinc oxide resistive-exchanging layer, transparent RRAM innovation can be upgraded [195]. The resulting resistive RRAM (random access memory) device displays better-exchanging conduct with elevated exchanging yield and consistency than those of the gadget absence of graphene.

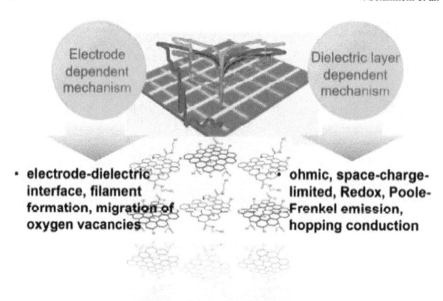

Fig.18 Mechanism of ReRAM device using graphene-based materials. Reprinted with permission from [191]. Copyright © 2016, Royal Society of Chemistry

7 Carbon Nanomaterials-Based Photodetectors

The photodetector is a semiconductor device that converts illuminated light into a precise electrical signal, photovoltage, or photocurrent [51, 196–199]. Photodetectors are considered as an interface among electrical circuitries and optical information. Therefore, photodetectors have several applications in digital imaging, security, environmental monitoring, digital communication [200]. Carbon nanomaterials have excellent optoelectronic properties with the potential of substituting inorganic semiconductors and metals. GQD, carbon allotropes (CQD), buckyballs, CNTs, graphene-based materials are extensively used in the fabrication of photodetectors because of their larger values of absorption coefficients and carrier mobility, tunable bandgap [201–206].

7.1 Fullerene Based Photodetectors

Fullerenes show a strong absorption below 400 nm, which could be utilized for UV light detection. Ma et al. showed Buckminster fullerenes (C_{60}) electronic potential in electronic devices, by forming a C_{60}-diode of relatively modest architecture [207]. C_{60} was inserted in between aluminum (anode) contact) and an ohmic copper (cathode), forming a diode that can withstand a very high current density of 363 A/cm^2. Szendrei et al. demonstrated hybrid thin-film photodetectors based on fullerene and PbS nanocrystals with a spectral range of visible and NIR regions up to 1300 nm [208]. The maximum responsivity at 514 nm is 1.6 A/W value which is recorded. Furthermore, sturdy gate dependence and electron mobility values up to 3×10^{-4} cm^2/Vs are displayed by the device. Guo et al. demonstrated fullerene-based organic photodetector (OPD) on ITO glass with very little noise current and at 3470 nm, a high detectivity of 3.6×10^{11} Jones, 90 dB of broad linear dynamic range (LDR), and a response speed greater than 20 kHz [209].

7.2 Graphene Quantum Dots Photodetectors

In GQD, the optical bandgap is strongly influenced by quantum confinement and edge effects. In GQD, bandgap tuning has been investigated and is considered to be an important advantage in optoelectronics over plane graphene and bandgap decreases with an increase in quantum size. A key absorption in the ultraviolet region is shown by graphene-based quantum dots. Zhang et al. fabricated a superior performance deep UV photodetector using GQD. Using a facile hydrothermal method, GQDs were synthesized with an average particle diameter of 4.5 nm and displayed a bandgap of 3.8 eV due to the quantum confinement effect which leads to an absorption peak found around the wavelength of 320 nm (Fig. 19) [210]. The built-in electric field was created due to the difference between work function of GQDs and two metal electrodes which is the major reason for extraction efficient photocurrent. The responsivity and detectivity of the photodetector were 2.1 A/W, 9.59×10^{11} Jones, respectively. Tang et al. synthesized n-doped GQDs using a microwave-assisted synthesis method. N-doped GQDs shown a wide absorption range covering UV, vis, and NIR range (300 to >1000 nm) [211]. The UV absorption was ascribed to the transitions in $C = N, C = O$, and $C = C$ bonds. The visible absorption is caused by extended partial conjugated π-electrons in single layers of paper. Absorption in the NIR region is due to the absorption is ascribed to the delocalization of π-electrons due to the layered structure. The responsivity at various excitation wavelengths was calculated to be at 365 nm, 405 nm, 808 nm, 980 nm wavelengths are 1.14 V/W, 325 V/W, 10.91 V/W, 10.91 V/W, respectively.

7.3 CNTs-Based Photodetectors

CNTs display opportunities for novel photodetectors because of their remarkable optical properties. Free electron–hole pair excitations and also strongly bound electron–hole pair states are called excitons present in semiconducting CNTs which have a direct bandgap. The exciton binding energy of carbon nanotubes depends on the diameter and dielectric constant of the surrounding environment. In recent years,

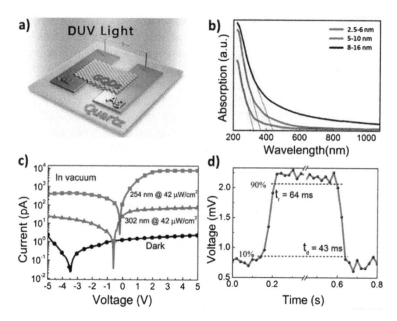

Fig. 19 **a** GQDs used photodiode structures illustration **b** Using particle size, the absorption performance of GQDs can be tailored **c** Current–voltage characteristics of the Au/GQDs/Ag diode. The photocurrent and rectification behavior is increased by a decrease in wavelength **d** The photovoltage had a response time and fall time of 64 ms, in response to a light pulse at the wavelength of 254 nm (43 ms) Reprinted with permission from [210]. Copyright © 2015, American Chemical Society

several research groups have reported that photocurrent could be produced in a CNT when illuminated, which is in contact with metal electrodes. The charges are separated due to the Schottky barrier on the metal–CNT interface that provides an internal electrical field. Rao et al. studied the photoconductivity of the pure CNT film. They developed SWCNTs on SiO_2/Si substrates using CVD across a channel in the layer of SiO_2 and 8–20 CNTs have connected the channel [212]. Alternative methods for charge separation are establishing p–n junction, applying asymmetrical metal contacts. Liu and co-authors demonstrated a photodiode using purified SWCNTs deposited on n-Si/SiO_2 by liquid phase, wherein asymmetrical metal contacts were used [213]. The I–V characteristics were well matched with the conventional diode equation which was ascribed to reduced defects in SWCNTs. Peak responsivity attained was 1.5×10^8 V W^{-1} and above 107 V W^{-1} for the wavelengths ranging between 1200 and 2100 nm. In a similar spectral range, detectivity stayed above 10^{10} Jones and achieved a peak with 1.25×10^{11} Jones at 1800 nm.

He et al. developed a photodetector by developing highly aligned SWCNTs between two Au contacts, where to establish p–n junction, half the intrinsically P-type CNTs were n-doped (Fig. 20) [214]. The highly aligned SWCNTs provide an extended absorption spectrum range from 3 μm to 3 mm. The responsivity was

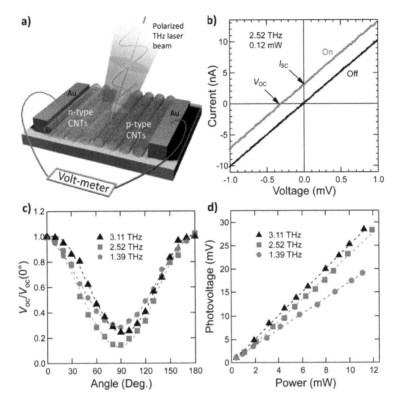

Fig. 20 Carbon nanotube Terra-Hertz detector characteristics. **a** Experimental geometry-based diagram schematic illustration. Linearly polarized Terra-Hertz beam is illuminated on the device at room temperature and I–V characteristics are measured **b** Graph showing the I–V characteristics without illumination and with the illumination of 2.THz beam **c** Polarization dependence of open-circuit voltage (V_{OC}) for frequencies of 3.11 THz, 2.52 THz, and 1.39 THz which was normalized by its value for parallel polarization, **d** Power dependence of V_{OC} for frequencies of 3.11 THz, 2.52 THz, and 1.39 THz, showing responsivities of 1.7 V/W, 2.4 V/W, and 2.5 V/W, respectively. Reprinted with permission from [214]. Copyright © 2014, American Chemical Society

observed to be 2.5 V/W at 96.5 μm (3.11 Tera-Hertz), 2.4 V/W at 19 μm (2.52 Tera-Hertz), and 1.7 V/W at 215 μm (1.39 Tera-Hertz). Liu et al. demonstrated hybrid broadband photodetector (across visible to NIR range, 400–1550 nm) using atomically thin SWCNTs and graphene with a high photoconductive gain ~10^5 [215]. The fabricated photodetector achieved a high responsivity of >100 A/W and a response time of ~100 sμ (electrical bandwidth of ~10^4 Hz). The high built-in potential at the 1D and 2D interface provides efficient separation of photogenerated electron–hole pairs and decreases the recombination. Further, a fast response rate was achieved due to a trap-free interface. Lu and co-workers explored the implementation of a novel mechanism for exciton dissociation using graphene/MWCNT nanohybrids [216]. The fabricated photodetector was used for infrared detection and achieved high responsivity of ~3065 V/W and detectivity of 1.57×10^7 cm. Hz$^{1/2}$/W.

7.4 Graphene-Based Photodetectors

Single-layer graphene has a wide-ranging but minimal absorption, due to this few-layer graphene was used to enhance the absorption. The other possibility to enhance the absorption in monolayer graphene was reported by Kang and co-workers for the first time [217]. They have used crumbled graphene which was intentionally deformed to get the continuous undulating 3D surface, to induce an increase in areal density that yields high optical absorption, hence improving the photoresponsivity. Mueller et al. reported graphene-based photodetector. In this asymmetric metal contacts and an interdigitated metal–graphene–metal was used on a Si/SiO$_2$ substrate [218]. The maximum responsivity of 6.1 mA/W was observed at 1550 mm. Liu et al. fabricated a photodetector using graphene double-layer heterostructure for ultra-broadband and high sensitivity of 4 A/W at room temperature. The device consists of two graphene layers sandwiching a thin SiO2 barrier. The device consists of two graphene layers sandwiching a thin SiO2 barrier.

7.5 Hybrid Photodetectors

Kim et al. reported a hybrid broadband photodetector based on carbon–carbon composite materials ranging from 300 nm (UV) to 1000 nm (NIR) wavelength, wherein GQDs were sandwiched between two graphene sheets (Fig. 21) [219]. SiO$_2$/Si was used as a substrate and two metal (silver) contacts were evaporated on top and bottom graphene layers. Although the design was symmetric, and asymmetric I–V characteristics were observed, which was ascribed to the charging or doping effects in the bottom layer of graphene. The fabricated device achieved the photoresponsivity of 0.5 A/W at 800 nm and the detectivity of 2.4 × 10^{11} Jones.

Sahatiya and co-workers reported a graphene/MoS$_2$ based visible light photodetector [220]. The internal electricity created due to graphene/MoS$_2$ heterostructure enhances the charge carriers separation which improves the responsivity of the photodetector. The same group demonstrated a flexible graphene-based infrared photodetector on PI substrate with responsivity and external quantum efficiency (EQE) of 0.4 A/W, 16.53%, respectively [120]. This project incorporates the synergistic advantages of the substrate (which acts as dielectric also) and sensing material. The same group optimized various parameters such as calcination temperature, time-dependent electrospinning of graphene/ZnO-based composite nanofiber across gold electrodes for UV detection which was fabricated by one-step in situ synthesis method [221]. The fabricated photodetector showed superior performance for UV sensing with an 1892 time increase in the conductance.

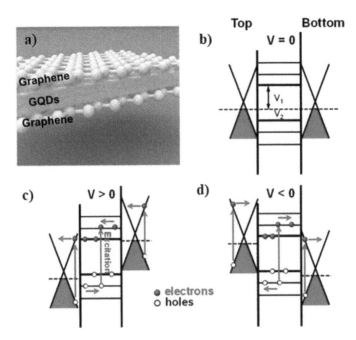

Fig. 21 Schematic illustration of band structure describing the photodetector **a** A typical GQD photodetector device and its band diagrams under **b** no, **c** positive (forward), and **d** negative (reverse) biases. Electrons and holes are represented by red and blank spots, respectively, and lateral arrows in the transport directions of holes and electrons contributing to dark- and photo-currents. Reprinted with permission from [219]. Copyright © 2014, Springer Nature

8 Conclusion and Outlook

It is apparent from above all sections in this chapter, carbon nanomaterials due to their unique thermal, mechanical, electronic properties have been proven to have the noteworthy potential to be used in various evolving electronic and sensing applications. Carbon nanomaterials which are available in different forms such as zero-dimensional (carbon quantum dots), one-dimensional (carbon nanotubes), two-dimensional (graphene), and their hybrids have shown significant impact in the fabrication of electronic and sensing devices. These fabricated devices helped in detecting various target analytes, dangerous gases, change in relative humidity/temperature/pressure, etc., and also increasing the non-volatile memory storage capacity by increasing the performance of fabricated electronic devices. Before realizing commercial carbon-based electronic and sensing devices for various applications, challenges must be addressed to avoid further complexities. Even though propels in development and post-manufactured partition strategies have significantly enhanced the carbon nanomaterials monodispersity, gadget-to-gadget inconstancy stays an issue. For instance, the threshold voltages of a variety of FETs created from semiconducting carbon nanotubes regularly differ by a few volts, which is, at any rate, a significant degree higher than satisfactory in contemporary ICs. These dissimilarities can be mostly tended to by further upgrades in the faultlessness and monodispersity of the semiconducting carbon nanotubes bandgap, yet a significant part of the gadget inhomogeneity can likely be ascribed to extraneous factors, for example, the impact of metal contacts, dielectric layers, fundamental substrate, and general condition. Therefore, we predict that carbon nanomaterials will keep on being the subject of serious and productive exploration even as the first applications arrive at the commercial center.

Acknowledgments The authors acknowledge support from the Research Initiation Grant (RIG and ACRG), Birla Institute of Technology Pilani, Hyderabad Campus.

References

1. Kroto HW, McKay K (1988) The formation of quasi-icosahedral spiral shell carbon particles. Nature 331:328–331
2. Ugarte D (1992) Curling and closure of graphitic networks under electron-beam irradiation. Nature 359:707–709
3. Wang X, Hofmann O, Das R, Barrett EM, DeMello AJ, DeMello JC, Bradley DDC (2007) Integrated thin-film polymer/fullerene photodetectors for on-chip microfluidic chemiluminescence detection. Lab Chip 7:58–63
4. Deibel C, Dyakonov V (2010) Polymer–fullerene bulk heterojunction solar cells. Mater Today 73:462–470
5. Haddon RC, Perel AS, Morris RC, Palstra TTM, Hebard AF, Fleming RM (2012) C_{60} thin film transistors. 121:1–4
6. Monthioux M, Kuznetsov VL (2006) Who should be given the credit for the discovery of carbon nanotubes? Carbon N Y 44:1621–1623
7. Radushkevich LV, Lukyanovich VM (1952) The structure of carbon forming in thermal decomposition of carbon monoxide on an iron catalyst. Russ J Phys Chem 26:88–95
8. Iijima S (1991) Helical microtubules of graphitic carbon. Nature 354:56–58
9. Inami N, Mohamed MA, Shikoh E, Fujiwara A (2007) Synthesis-condition dependence of carbon nanotube growth by alcohol catalytic chemical vapor deposition method. Sci Technol Adv Mater 8:292–295. https://doi.org/10.1016/j.stam.2007.02.009
10. Eftekhari A, Jafarkhani P, Moztarzadeh F (2006) High-yield synthesis of carbon nanotubes using a water-soluble catalyst support in catalytic chemical vapor deposition. Carbon N Y 44:1343–1345
11. Guo T, Nikolaev P, Rinzler AG, Tomanek D, Colbert DT, Smalley RE (1995) Self-assembly of tubular fullerenes. J Phys Chem 99:10694–10697
12. Ebbesen TW, Ajayan PM (1992) Large-scale synthesis of carbon nanotubes. Nature 358:220–222
13. Dai H (2002) Carbon nanotubes: opportunities and challenges. Surf Sci 500:218–241
14. Zhao X, Liu Y, Inoue S, Suzuki T, Jones RO, Ando Y (2004) Smallest carbon nanotube is 3Å in diameter. Phys Rev Lett 92:125502
15. Wen Q, Zhang R, Qian W, Wang Y, Tan P, Nie J, Wei F (2010) Growing 20 cm long DWNTs/TWNTs at a rapid growth rate of 80–90 μm/s. Chem Mater 22:1294–1296
16. Dresselhaus EMS, Dresselhaus G, Avouris P (2003) Carbon nanotubes:synthesis, structure, properties, and applications. vol 80. Springer Science & Business Media
17. Reich S, Christian Thomsen JM (2004) Carbon nanotubes: basic concepts and physical properties

18. Ivchenko EL, Spivak B (2002) Chirality effects in carbon nanotubes. Phys Rev B 66:155404
19. Nakayama Y, Akita S (2001) Field-emission device with carbon nanotubes for a flat panel display. Synth Met 117:207–210
20. Varghese OK, Kichambre PD, Gong D, Ong KG, Dickey EC, Grimes CA (2001) Gas sensing characteristics of multi-wall carbon nanotubes. Sensors Actuators B Chem 81:32–41
21. Jiang W, Xiao S, Zhang H, Dong Y, Li X (2007) Capacitive humidity sensing properties of carbon nanotubes grown on silicon nanoporous pillar array. Sci China Ser E Technol Sci 50:510–515
22. Gao B, Kleinhammes A, Tang XP, Bower C, Fleming L, Wu Y, Zhou O (1999) Electrochemical intercalation of single-walled carbon nanotubes with lithium. Chem Phys Lett 307:153–157
23. Saha MS, Li R, Sun X, Ye S (2009) 3-D composite electrodes for high performance PEM fuel cells composed of Pt supported on nitrogen-doped carbon nanotubes grown on carbon paper. Electrochem Commun 11:438–441
24. Sun X, Li R, Villers D, Dodelet JP, Désilets S (2003) Composite electrodes made of Pt nanoparticles deposited on carbon nanotubes grown on fuel cell backings. Chem Phys Lett 379:99–104
25. Ghasempour R, Narei H (2018) 1 - CNT Basics and characteristics. In: Rafiee RBT-CN-RP (ed) Micro and Nano Technologies. Elsevier, pp 1–24
26. Mashkoor F, Nasar A, Inamuddin, (2020) Carbon nanotube-based adsorbents for the removal of dyes from waters: a review. Environ Chem Lett 18:605–629
27. Messina G SS (2006) Carbon: the future material for advanced technology applications, p 530. Springer-Verlag Berlin Heidelberg
28. Morgan P (2005) Carbon fibers and their composites. Taylor & Francis Group, LLC, p P1200
29. Cheng H-Y, Zhu Y-A, Sui Z-J, Zhou X-G, Chen D (2012) Modeling of fishbone-type carbon nanofibers with cone-helix structures. Carbon N Y 50:4359–4372
30. Geim AK, Kim P (2008) Carbon wonderland. Sci Am 298:90–97
31. Novoselov KS, Geim AK, Morozov S V, Jiang D, Zhang Y, Dubonos S V, Grigorieva I V, Firsov AA (2004) Electric field effect in atomically thin carbon films. Science (80-) 306:666–669
32. Reina A, Jia X, Ho J, Nezich D, Son H, Bulovic V, Dresselhaus MS, Kong J (2009) Large area, few-layer graphene films on arbitrary substrates by chemical vapor deposition. Nano Lett 9:30–35
33. Dikin DA, Stankovich S, Zimney EJ, Piner RD, Dommett GHB, Evmenenko G, Nguyen ST, Ruoff RS (2007) Preparation and characterization of graphene oxide paper. Nature 448:457–460
34. Lotya M, Hernandez Y, King PJ, Smith RJ, Nicolosi V, Karlsson LS, Blighe FM, De S, Wang Z, McGovern IT, Duesberg GS, Coleman JN (2009) Liquid phase production of graphene by exfoliation of graphite in surfactant/water solutions. J Am Chem Soc 131:3611–3620
35. Geim AK, Novoselov KS (2007) The rise of graphene. Nat Mater 6:183–191
36. Lee C, Wei X, Kysar JW, Hone J (2008) Measurement of the elastic properties and intrinsic strength of monolayer graphene. Science (80-)321:385–388
37. Kuzmenko AB, van Heumen E, Carbone F, van der Marel D (2008) Universal optical conductance of graphite. Phys Rev Lett 100:117401
38. Balandin AA, Ghosh S, Bao W, Calizo I, Teweldebrhan D, Miao F, Lau CN (2008) Superior thermal conductivity of single-layer graphene. Nano Lett 8:902–907
39. Liu S, Chevali VS, Xu Z, Hui D, Wang H (2017) A review of extending performance of epoxy resins using carbon nanomaterials. Compos Part B Eng 136:197–214
40. Rizvi SB, Ghaderi S, Keshtgar M, Seifalian AM (2010) Semiconductor quantum dots as fluorescent probes for in vitro and in vivo bio-molecular and cellular imaging. Nano Rev 1:5161
41. Smith AM, Duan H, Mohs AM, Nie S (2008) Bioconjugated quantum dots for in vivo molecular and cellular imaging. Adv Drug Deliv Rev 60:1226–1240
42. Arya H, Kaul Z, Wadhwa R, Taira K, Hirano T, Kaul SC (2005) Quantum dots in bio-imaging: revolution by the small. Biochem Biophys Res Commun 329:1173–1177

43. Rizvi SB, Ghaderi S, Keshtgar M (2010) Semiconductor quantum dots as fluorescent probes for in vitro and in vivo bio-molecular and cellular imaging. Nano Rev 1(10):3402
44. Baker SN, Baker GA (2010) Luminescent carbon nanodots : emergent nanolights angewandte. Angew Chem Int Ed 49:6726–6744
45. Hu S-L, Niu K-Y, Sun J, Yang J, Zhao N-Q, Du X-W (2009) One-step synthesis of fluorescent carbon nanoparticles by laser irradiation. J Mater Chem 19:484–488
46. Ray SC, Saha A, Jana NR, Sarkar R (2009) Fluorescent carbon nanoparticles: synthesis, characterization, and bioimaging application. J Phys Chem C 113:18546–18551
47. Qiao Z-A, Wang Y, Gao Y, Li H, Dai T, Liu Y, Huo Q (2010) Commercially activated carbon as the source for producing multicolor photoluminescent carbon dots by chemical oxidation. Chem Commun 46:8812–8814
48. Zhou J, Booker C, Li R, Zhou X, Sham T-K, Sun X, Ding Z (2007) An electrochemical avenue to blue luminescent nanocrystals from multiwalled carbon nanotubes (MWCNTs). J Am Chem Soc 129:744–745
49. Bao L, Zhang Z-L, Tian Z-Q, Zhang L, Liu C, Lin Y, Qi B, Pang D-W (2011) Electrochemical tuning of luminescent carbon nanodots: from preparation to luminescence mechanism. Adv Mater 23:5801–5806
50. Sahatiya P, Jones SS, Badhulika S (2018) 2D MoS2–carbon quantum dot hybrid based large area, flexible UV–vis–NIR photodetector on paper substrate. Appl Mater Today 10:106–114
51. Koduvayur Ganeshan S, Selamneni V, Sahatiya P (2020) Water dissolvable MoS$_2$ quantum dots/PVA film as an active material for destructible memristors. New J Chem. https://doi.org/10.1039/D0NJ02053B
52. Jariwala D, Sangwan VK, Lauhon LJ, Marks TJ, Hersam MC (2013) Carbon nanomaterials for electronics, optoelectronics, photovoltaics, and sensing. Chem Soc Rev 42:2824–2860
53. Avouris P, Chen Z, Perebeinos V (2007) Carbon-Based Electronics. Nat Nanotech 2:605–615
54. Ando T, Nakanishi T (1998) Impurity scattering in carbon nanotubes–absence of back scattering. J Phys Soc Japan 67:1704–1713
55. Zhou X, Park J-Y, Huang S, Liu J, McEuen PL (2005) Band structure, phonon scattering, and the performance limit of single-walled carbon nanotube transistors. Phys Rev Lett 95:146805
56. Collins PG, Hersam M, Arnold M, Martel R, Avouris P (2001) Current saturation and electrical breakdown in multiwalled carbon nanotubes. Phys Rev Lett 86:3128–3131
57. Dürkop T, Getty SA, Cobas E, Fuhrer MS (2004) Extraordinary mobility in semiconducting carbon nanotubes. Nano Lett 4:35–39
58. Léonard F, Tersoff J (2000) Role of fermi-level pinning in nanotube schottky diodes. Phys Rev Lett 84:4693–4696
59. Chen Z, Appenzeller J, Knoch J, Lin Y (2005) The role of metal–nanotube contact in the performance of carbon nanotube field-effect transistors. Nano Lett 5:1–6
60. Sangwan VK, Ballarotto VW, Fuhrer MS, Williams ED, Sangwan VK, Ballarotto VW, Fuhrer MS, Williams ED (2014) Facile Fabrication of Suspended As-Grown Carbon Nanotube Devices. 93:113112
61. Franklin AD, Luisier M, Han S-J, Tulevski G, Breslin CM, Gignac L, Lundstrom MS, Haensch W (2012) Sub-10 nm carbon nanotube transistor. Nano Lett 12:758–762
62. Hickey BM, Oceanogr P, Emery WJ, Hamilton K, Res JG, Simpson JJ, Lett GR, Ely LL, Enzel Y, Cayan DR, Clim J, Prahl FG, Muehlhausen LA, Zahnle DL, Postma HWC, Teepen T, Yao Z, Grifoni M (2001) Carbon nanotube single-electron transistors at room temperature. Science (80-) 293:76–79
63. Javey A, Guo J, Wang Q, Lundstrom M, Dai H (2003) Ballistic carbon nanotube field-effect transistors. Nature 424:654–657
64. Bockrath M, Cobden DH, Lu J (1999) Luttinger-liquid behaviour in carbon nanotubes. Nature 397:598–601
65. Martel R, Derycke V, Lavoie C, Appenzeller J, Chan KK, Tersoff J, Avouris P (2001) Ambipolar electrical transport in semiconducting single-wall carbon nanotubes. Phys Rev Lett 87:256805

66. Snow ES, Novak JP, Campbell PM, Park D, Snow ES, Novak JP, Campbell PM, Park D (2003) Random networks of carbon nanotubes as an electronic material. Appl Phys Lett 82:2145

67. Topinka MA, Rowell MW, Goldhaber-gordon D, Mcgehee MD, Hecht DS, Gruner G (2009) Charge transport in interpenetrating networks of semiconducting and metallic carbon nanotubes. Nano Lett 9:1866–1871

68. Collins PG, Arnold MS, Avouris P (2001) Engineering Carbon Nanotubes and Nanotube Circuits Using Electrical Breakdown. Science (80-) 292:706–709

69. Separated UH, Wang C, Zhang J, Zhou C (2010) Macroelectronic integrated circuits using high-performance separated carbon nanotube thin-film transistors. ACS Nano 4:7123–7132

70. Thin-film CN, Sangwan VK, Ortiz RP, Alaboson JMP, Emery JD, Bedzyk MJ, Lauhon LJ, Marks TJ, Hersam MC (2012) Fundamental performance limits of transistors achieved using hybrid molecular dielectrics. ACS Nano 6:7480–7488

71. Roberts ME, Lemieux MC, Sokolov AN, Bao Z (2009) Self-sorted nanotube networks on polymer dielectrics for low-voltage thin-film transistors. Nano Lett 9:2526–2531

72. Tans SJ, Verschueren ARM, Dekker C (1998) Room-temperature transistor based on a single carbon nanotube. Nature 393:49–52

73. Yao Z, Kane CL, Dekker C (2000) High-field electrical transport in single-wall carbon nanotubes. Phys Rev Lett 84:2941–2944

74. Steiner M, Engel M, Lin Y, Wu Y, Jenkins K, Farmer DB, Humes JJ, Yoder NL, Seo JT, Green AA, Hersam MC, Krupke R, Avouris P, Steiner M, Engel M, Lin Y, Wu Y, Jenkins K, Green AA, Hersam MC, Krupke R, Avouris P (2014) High-frequency performance of scaled carbon nanotube array field-effect transistors. Appl Phys Lett 101:053123

75. Cao Y, Brady GJ, Gui H, Rutherglen C, Arnold MS, Zhou C (2016) Radio frequency transistors using aligned semiconducting carbon nanotubes with current-gain cutoff frequency and maximum oscillation frequency simultaneously greater than 70 GHz. ACS Nano 10:6782–6790

76. Zhong D, Shi H, Ding L, Zhao C, Liu J, Zhou J, Zhang Z, Peng L (2019) Carbon nanotube film-based radio frequency transistors with maximum oscillation frequency above 100 GHz. ACS Nano 11:42496–42503

77. Raimond JM, Brune M, Computation Q, Martini F De, Monroe C (2004) Electric field effect in atomically thin carbon films. Science (80-) 306:666–669

78. Schwierz F (2010) Graphene transistors. Nat Publ Gr 5:487–496

79. Neto AHC (2009) The electronic properties of graphene. RevModPhys 81:109–162

80. Lin Y, Dimitrakopoulos C, Jenkins KA, Farmer DB, Chiu H, Grill A, Avouris P (2010) 100-GHz transistors from wafer-scale epitaxial graphene. Science (80-) 327:662

81. Badmaev A, Che Y, Li Z, Wang C, Zhou C (2012) Self-aligned fabrication of graphene RF transistors with T-shaped gate. ACS Nano 6:3371–3376

82. Guo Z, Dong R, Chakraborty PS, Lourenco N, Palmer J, Hu Y, Ruan M, Hankinson J, Kunc J, Cressler JD, Berger C, De HWA (2013) Record maximum oscillation frequency in C-face epitaxial graphene transistors. Nano Lett 13:942–947

83. Sur UK (2012) Graphene: a rising star on the horizon of materials science. Int J Electrochem 2012:237689

84. Khan K, Tareen AK, Aslam M, Wang R, Zhang Y, Mahmood A, Ouyang Z, Zhang H, Guo Z (2020) Recent developments in emerging two-dimensional materials and their applications. J Mater Chem C 8:387–440

85. Kalavakunda V, Hosmane NS (2016) Mini review graphene and its analogues. Nanotechnol Rev 5:369–376

86. Kim K, Choi J, Kim T, Cho S, Chung H (2011) A role for graphene in silicon-based semiconductor devices. Nature 479:338–344

87. Feng X, Zhao X, Yang L, Li M, Qie F, Guo J, Zhang Y, Li T (2018) All carbon materials pn diode. Nat Commun 9:3750

88. Jariwala D, Sangwan VK, Wu C, Prabhumirashi PL, Geier ML (2013) Gate-tunable carbon nanotube–MoS_2 heterojunction p-n diode. PNAS 110:18076–18080

89. Yang Y, Zhao Q, Feng W, Li F (2013) Luminescent chemodosimeters for bioimaging. ChemRev 113:192–270
90. Fan Q, Li J, Zhu Y, Yang Z, Shen T, Guo Y, Wang L, Mei T, Wang J, Wang X (2020) Functional carbon quantum dots for highly sensitive graphene transistors for Cu^{2+} ion detection. ACS Appl Mater Interfaces 12:4797–4803
91. Liu W, Song M, Kong B, Cui Y (2017) Flexible and stretchable energy storage: recent advances and future perspectives. Adv Mater 29:1603436
92. Khang D-Y, Jiang H, Huang Y, Rogers JA (2006) A stretchable form of single-crystal silicon for high-performance electronics on rubber substrates. Science (80-) 311:208 LP–212
93. Nanotube GC, Huang J, Fang J, Liu C, Chu C (2011) Effective work function modulation of graphene/carbon nanotube composite films as transparent cathodes for organic optoelectronics. ACS Nano 5:6262–6271
94. Cao Q, Hur S-H, Zhu Z-T, Sun YG, Wang C-J, Meitl MA, Shim M, Rogers JA (2006) Highly bendable, transparent thin-film transistors that use carbon-nanotube-based conductors and semiconductors with elastomeric dielectrics. Adv Mater 18:304–309
95. Aikawa S Transparent all-carbon-nanotube transistors D transparent all-carbon-nanotube transistors, Einarsson E, Thurakitseree T, Chiashi S, Nishikawa E (2012) Deformable transparent all-carbon-nanotube transistors. Appl Phys Lett 100:063502
96. Sun D, Timmermans MY, Tian Y, Nasibulin AG, Kauppinen EI, Kishimoto S, Mizutani T, Ohno Y (2011) Flexible high-performance carbon nanotube integrated circuits. Nat Nanotech 6:156–161
97. Sun D, Timmermans MY, Kaskela A, Nasibulin AG, Kishimoto S, Mizutani T, Kauppinen EI, Ohno Y (2013) Mouldable all-carbon integrated circuits. Nat Commun 4:2302
98. Lu R, Christianson C, Weintrub B, Wu JZ (2013) High photoresponse in hybrid graphene–carbon nanotube infrared detectors. ACS Appl Mater Interfaces 5:11703–11707
99. Kim SH, Song W, Jung MW, Kang M, Kim K (2014) Carbon nanotube and graphene hybrid thin film for transparent electrodes and field effect transistors. Adv Mater 26:4247–4252
100. Tung VC, Chen L, Allen MJ, Wassei JK, Nelson K, Kaner RB, Yang Y (2009) Low-temperature solution processing of graphene-carbon nanotube hybrid materials for high-performance transparent conductors. Nano Lett 9:1949–1955
101. Yakobson BI, Brabec CJ, Bernholc J (1996) Nanomechanics of carbon tubes: instabilities beyond linear response. Phys Rev Lett 76:2511–2514
102. Wong EW, Sheehan PE, Lieber CM (1997) Nanobeam mechanics: elasticity, strength, and toughness of nanorods and nanotubes. Science (80-) 277:1971–1975
103. Selamneni V, Barya P, Deshpande N, Sahatiya P (2019) Low-cost, disposable, flexible, and smartphone enabled pressure sensor for monitoring drug dosage in smart medicine applications. IEEE Sens J 19:11255–11261
104. Selamneni V, Dave A, Mihailovic P, Mondal S, Sahatiya P (2020) Large area pressure sensor for smart floor sensor applications—an occupancy limiting technology to combat social distancing. IEEE Consum Electron Mag. https://doi.org/10.1109/MCE.2020.3033932
105. Selamneni V, Dave A, Mihailovic P , Mondal S and Sahatiya P (2020) Large area pressure sensor for smart floor sensor applications—an occupancy limiting technology to combat social distancing. IEEE Consum Electron Mag. https://doi.org/10.1109/MCE.2020.3033932.
106. Selamneni V, B S A, Sahatiya P(2020) Highly air-stabilized black phosphorus on disposable paper substrate as a tunnelling effect-based highly sensitive piezoresistive strain sensor. Med Dev Sensors 3:e10099
107. Zhan Z, Lin R, Tran V, An J, Wei Y, Du H, Tran T, Lu W (2017) Paper/carbon nanotube-based wearable pressure sensor for physiological signal acquisition and soft robotic skin. ACS Appl Mater Interfaces 9:37921–37928
108. Sahatiya P, Badhulika S (2016) Solvent-free fabrication of multi-walled carbon nanotube based flexible pressure sensors for ultra-sensitive touch pad and electronic skin applications. RSC Adv 6:95836–95845
109. Park J, Kim M, Lee Y, Lee HS, Ko H (2015) Fingertip skin–inspired microstructured ferroelectric skins discriminate static / dynamic pressure and temperature stimuli. Sci Adv 1:e1500661

110. Zhu S, Ghatkesar MK, Zhang C, Janssen GCAM, Zhu S, Ghatkesar K, Zhang C, Janssen GCAM (2013) Graphene based piezoresistive pressure sensor. Appl Phys Lett 102:161904
111. Yao H, Ge J, Wang C, Wang X, Hu W, Zheng Z (2013) A flexible and highly pressure-sensitive graphene-polyurethane sponge based on fractured microstructure design. Adv Mater 25:6692–6698
112. Jian M, Xia K, Wang Q, Yin Z, Wang H, Wang C (2017) Flexible and highly sensitive pressure sensors based on bionic hierarchical structures. RSC Adv 9:22740–22748
113. Tian H, Shu Y, Wang X, Mohammad MA, Bie Z, Xie Q, Li C, Mi W, Yang Y, Ren T (2015) A graphene-based resistive pressure sensor with record-high sensitivity in a wide pressure range. Sci Rep 5:8603
114. Lee MJ, Hong HP, Min NK, Lee D (2012) A fully-microfabricated SWCNT film strain sensor. J Korean Phys Soc 61:1656–1659
115. Dharap P, Li Z, Nagarajaiah S (2004) Nanotube film based on single-wall carbon nanotubes for strain sensing. Nanotechnology 15:379–382
116. Zhang S, Zhang H, Yao G, Liao F, Gao M, Huang Z, Li K, Lin Y (2015) Highly stretchable, sensitive, and flexible strain sensors based on silver nanoparticles/carbon nanotubes composites. J Alloys Compd 652:48–54
117. Zhao J, He C, Yang R, Shi Z, Cheng M, Yang W, Xie G, Wang D, Shi D, Zhang G (2012) Ultra-sensitive strain sensors based on piezoresistive nanographene films. Appl Phys Lett 101:63112
118. Li X, Zhang R, Yu W, Wang K, Wei J, Wu D, Cao A, Li Z, Cheng Y, Zheng Q, Ruoff RS, Zhu H (2012) Stretchable and highly sensitive graphene-on-polymer strain sensors. Sci Rep 2:870
119. Davaji B, Cho HD, Malakoutian M, Lee J, Panin G, Kang TW, Lee CH (2017) A patterned single layer graphene resistance temperature sensor. Sci Rep 7:8811
120. Sahatiya P, Puttapati SK, Srikanth VVSS, Badhulika S (2016) Graphene-based wearable temperature sensor and infrared photodetector on a flexible polyimide substrate. Flex Print Electron 1:25006
121. Compagnone D, Di Francia G, Di Natale C, Neri G, Seeber R, Tajani A (2017) Chemical sensors and biosensors in Italy: a review of the 2015 literature. Sensors (Switzerland) 17:1–22
122. Veeralingam S, Sahatiya P, Badhulika S (2019) Low cost, flexible and disposable SnSe$_2$ based photoresponsive ammonia sensor for detection of ammonia in urine samples. Sens Actuators B Chem 297:126725
123. Bokka N, Selamneni V, Sahatiya P (2020) A water destructible SnS$_2$ QD/PVA film based transient multifunctional sensor and machine learning assisted stimulus identification for non-invasive personal care diagnostics. Mater Adv. https://doi.org/10.1039/d0ma00573h
124. Selamneni V, Gohel K, Bokka N, Sharma S, Sahatiya P (2020) MoS$_2$ based Multifunctional sensor for both chemical and physical stimuli and their classification using machine learning algorithms. IEEE Sens J. https://doi.org/10.1109/JSEN.2020.3023309
125. Leelasree T, Selamneni V, Akshaya T, Sahatiya P, Aggarwal H (2020) MOF based flexible, low-cost chemiresistive device as a respiration sensor for sleep apnea diagnosis. J Mater Chem B. https://doi.org/10.1039/D0TB01748E
126. Sahatiya P, Badhulika S (2016) Graphene hybrid architectures for chemical sensors. Springer, Cham, Switzerland, pp 259–285
127. Yang S, Jiang C, Wei SH (2017) Gas sensing in 2D materials. Appl Phys Rev 4:021304
128. Fine GF, Cavanagh LM, Afonja A, Binions R (2010) Metal oxide semi-conductor gas sensors in environmental monitoring. Sensors 10:5469–5502
129. Zhang J, Liu X, Neri G, Pinna N (2016) Nanostructured materials for room-temperature gas sensors. Adv Mater 28:795–831
130. Choi S-J, Jang B-H, Lee S-J, Min BK, Rothschild A, Kim I-D (2014) Selective detection of acetone and hydrogen sulfide for the diagnosis of diabetes and halitosis using SnO$_2$ nanofibers functionalized with reduced graphene oxide nanosheets. ACS Appl Mater Interfaces 6:2588–2597

131. Li W, Geng X, Guo Y, Rong J, Gong Y, Wu L, Zhang X, Li P, Xu J, Cheng G, Sun M, Liu L (2011) Reduced graphene oxide electrically contacted graphene sensor for highly sensitive nitric oxide detection. ACS Nano 5:6955–6961
132. Schedin F, Geim AK, Morozov SV, Hill EW, Blake P, Katsnelson MI, Novoselov KS (2007) Detection of individual gas molecules adsorbed on graphene. Nat Mater 6:652–655
133. Dan Y, Lu Y, Kybert NJ, Luo Z, Johnson ATC (2009) Intrinsic response of graphene vapor sensors. Nano Lett 9:1472–1475
134. Rumyantsev S, Liu G, Shur MS, Potyrailo RA, Balandin AA (2012) Selective gas sensing with a single pristine graphene transistor. Nano Lett 12:2294–2298
135. Gautam M, Jayatissa AH (2012) Detection of organic vapors by graphene films functionalized with metallic nanoparticles. J Appl Phys 112:114326
136. Yavari F, Chen Z, Thomas AV, Ren W, Cheng H-M, Koratkar N (2011) High sensitivity gas detection using a macroscopic three-dimensional graphene foam network. Sci Rep 1:166
137. Kumar S, Kaushik S, Pratap R, Raghavan S (2015) Graphene on paper: a simple, low-cost chemical sensing platform. ACS Appl Mater Interfaces 7:2189–2194
138. Park S, Park M, Kim S, Yi S, Kim M, Son J, Cha J, Hong J, Park S, Park M, Kim S, Yi S, Kim M, Son J (2017) NO_2 gas sensor based on hydrogenated graphene. Appl Phys Lett 111:213102
139. Kim YH, Kim SJ, Kim Y-J, Shim Y-S, Kim SY, Hong BH, Jang HW (2015) Self-activated transparent all-graphene gas sensor with endurance to humidity and mechanical bending. ACS Nano 9:10453–10460
140. Choi H, Choi JS, Kim J-S, Choe J-H, Chung KH, Shin J-W, Kim JT, Youn D-H, Kim K-C, Lee J-I, Choi S-Y, Kim P, Choi C-G, Yu Y-J (2014) Flexible and transparent gas molecule sensor integrated with sensing and heating graphene layers. Small 10:3685–3691
141. Robinson JT, Perkins FK, Snow ES, Wei Z, Sheehan PE (2008) Reduced graphene oxide molecular sensors. Nano Lett 8:3137–3140
142. Dua V, Surwade SP, Ammu S, Agnihotra SR, Jain S, Roberts KE, Park S, Ruoff RS, Manohar SK (2010) All-organic vapor sensor using inkjet-printed reduced graphene oxide. Angew Chemie Int Ed 49:2154–2157
143. Chung MG, Kim DH, Seo DK, Kim T, Im HU, Lee HM, Yoo JB, Hong SH, Kang TJ, Kim YH (2012) Flexible hydrogen sensors using graphene with palladium nanoparticle decoration. Sens Actuators B Chem 169:387–392
144. Khalaf AL, Mohamad FS, Rahman NA, Lim HN, Paiman S, Yusof NA, Mahdi MA, Yaacob MH (2017) Room temperature ammonia sensor using side-polished optical fiber coated with graphene/polyaniline nanocomposite. Opt Mater Express 7:1858–1870
145. Deng S, Tjoa V, Fan HM, Tan HR, Sayle DC, Olivo M, Mhaisalkar S, Wei J, Sow CH (2012) Reduced Graphene oxide conjugated Cu_2O nanowire mesocrystals for high-performance NO_2 gas sensor. J Am Chem Soc 134:4905–4917
146. Al-Mashat L, Shin K, Kalantar-zadeh K, Plessis JD, Han SH, Kojima RW, Kaner RB, Li D, Gou X, Ippolito SJ, Wlodarski W (2010) Graphene/polyaniline nanocomposite for hydrogen sensing. J Phys Chem C 114:16168–16173
147. Jeong HY, Lee DS, Choi HK, Lee DH, Kim JE, Lee JY, Lee WJ, Kim SO, Choi SY (2010) Flexible room-temperature NO_2 gas sensors based on carbon nanotubes/reduced graphene hybrid films. Appl Phys Lett 96:2010–2013
148. Yang W, Wan P, Zhou X, Hu J, Guan Y, Feng L (2014) Additive-Free Synthesis Of In_2O_3 cubes embedded into graphene sheets and their enhanced NO_2 sensing performance at room temperature. ACS Appl Mater Interfaces 6:21093–21100
149. Zhang D, Chang H, Li P, Liu R (2016) Characterization of nickel oxide decorated-reduced graphene oxide nanocomposite and its sensing properties toward methane gas detection. J Mater Sci Mater Electron 27:3723–3730
150. Kraus F, Cruz S, Müller J (2003) Plasmapolymerized silicon organic thin films from HMDSN for capacitive humidity sensors. Sens Actuators B Chem 88:300–311
151. Rittersma ZM (2002) Recent achievements in miniaturised humidity sensors-a review of transduction techniques. Sensors Actuators a 96:196–210

152. Varghese OK, Grimes CA (2003) Metal oxide nanoarchitectures for environmental sensing. J Nanosci Nanotechnol 3:277–293

153. Rittersma ZM, Splinter A, Bodecker A, Benecke W (2000) A novel surface-micromachined capacitive porous silicon humidity sensor. Sens Actuators B 68:210–217

154. Björkqvist M, Salonen J, Paski J, Laine E (2004) Characterization of thermally carbonized porous silicon humidity sensor. Sens Actuators a Phys 112:244–247

155. Chu J, Peng X, Feng P, Sheng Y, Zhang J (2013) Sensors and actuators B: chemical Study of humidity sensors based on nanostructured carbon films produced by physical vapor deposition. Sens Actuators B Chem 178:508–513

156. Jin H, Tao X, Feng B, Yu L, Wang D, Dong S (2017) A humidity sensor based on quartz crystal microbalance using graphene oxide as a sensitive layer. Vaccum 140:101–105

157. Kumar U, Yadav BC (2019) Development of humidity sensor using modified curved MWCNT based thin film with DFT calculations. Sens Actuators B Chem 288:399–407

158. Borini S, White R, Wei D, Astley M, Haque S, Spigone E, Harris N (2013) Ultrafast graphene oxide humidity sensors. ACS Nano 7:11166–11173

159. Dresselhaus BMS, Terrones M (2013) Carbon-based nanomaterials from a historical perspective. Proc IEEE 101:1522–1535

160. Wang Y, Huang K, Wu X (2017) Recent advances in transition-metal dichalcogenides based electrochemical biosensors: a review. Biosens Bioelectron 97:305–316

161. Arvizo RR, Bhattacharyya S, Kudgus RA, Giri K, Bhattacharya RMP (2012) Intrinsic therapeutic applications of noble metal nanoparticles: past, present and future. Chem Soc Rev 41:2943–2970

162. Kwon SJ, Bard AJ (2012) DNA analysis by application of Pt nanoparticle electrochemical amplification with single label response. J Am Chem Soc 134:10777–10779

163. Wu S, He Q, Tan C, Wang Y, Zhang H (2013) Graphene-**based electrochemical sensors**. Small 9:1160–1172

164. Kochmann S, Hirsch T, Wolfbeis OS (2012) Graphenes in chemical sensors and biosensors. TrAC Trends Anal Chem 39:87–113

165. Balasubramanian RK, Burghard M (2006) Biosensors based on carbon nanotubes. Anal Bioanal Chem 385:452–468

166. Wang J, Lin Y (2008) Functionalized carbon nanotubes and nanofibers for biosensing applications. Trends Analyt Chem 27:619–626

167. Griese S, Kampouris DK, Kadara RO, Banks CE (2008) A critical review of the electrocatalysis reported at C_{60} modified electrodes. Electroanalysis 20:1507–1512

168. Tang X, Liu Y, Hou H, You T (2010) Electrochemical determination of L -Tryptophan, L -Tyrosine and L-Cysteine using electrospun carbon nanofibers modified electrode. Talanta 80:2182–2186

169. Nguyen HV, Richtera L, Moulick A, Xhaxhiu K, Kudr J, Cernei N, Polanska H, Heger Z, Masarik M, Kopel P, Stiborova M, Eckschlager T, Adam V, Kizek R (2016) Electrochemical sensing of etoposide using carbon quantum dot modified glassy carbon electrode. Analyst 141:2665–2675

170. Santhanam KSV, Ajayan PM (1996) Carbon nanotube electrode for oxidation of dopamine. Bioelectrochemistry Bioenerg 41:121–125

171. Luo H, Shi Z, Li N, Gu Z, Zhuang Q (2001) Investigation of the electrochemical and electrocatalytic behavior of single-wall carbon nanotube film on a glassy carbon electrode. Anal Chem 73:915–920

172. Joshi KA, Tang J, Haddon R, Wang J, Chen W, Mulchandani A (2005) A disposable biosensor for organophosphorus nerve agents based on carbon nanotubes modified thick film strip electrode. Electroanalysis 17:54–58

173. Lefrant S, Baibarac M, Baltog I, Mevellec JY, Mihut L, Chauvet O (2004) SERS spectroscopy studies on the electrochemical oxidation of single-walled carbon nanotubes in sulfuric acid solutions. Synth Met 144:133–142

174. Rakhi RB, Sethupathi K, Ramaprabhu S (2009) A Glucose biosensor based on deposition of glucose oxidase onto crystalline gold nanoparticle modified carbon nanotube electrode. J Phys Chem B 113:3190–3194

175. Qiu J-D, Huang J, Liang R-P (2011) Nanocomposite film based on graphene oxide for high performance flexible glucose biosensor. Sens Actuators B Chem 160:287–294

176. Lu W, Luo Y, Chang G, Sun X (2011) Synthesis of functional SiO_2-coated graphene oxide nanosheets decorated with Ag nanoparticles for H_2O_2 and glucose detection. Biosens Bioelectron 26:4791–4797

177. Shan C, Yang H, Han D, Zhang Q, Ivaska A, Niu L (2010) Graphene/AuNPs/chitosan nanocomposites film for glucose biosensing. Biosens Bioelectron 25:1070–1074

178. Liu Y, Yu D, Zeng C, Miao Z, Dai L (2010) Biocompatible graphene oxide-based glucose biosensors. Langmuir 26:6158–6160

179. Wu P, Shao Q, Hu Y, Jin J, Yin Y, Zhang H, Cai C (2010) Direct electrochemistry of glucose oxidase assembled on graphene and application to glucose detection. Electrochim Acta 55:8606–8614

180. Chen Y, Li Y, Sun D, Tian D, Zhang J, Zhu J-J (2011) Fabrication of gold nanoparticles on bilayer graphene for glucose electrochemical biosensing. J Mater Chem 21:7604–7611

181. Wang K, Liu Q, Guan Q-M, Wu J, Li H-N, Yan J-J (2011) Enhanced direct electrochemistry of glucose oxidase and biosensing for glucose via synergy effect of graphene and CdS nanocrystals. Biosens Bioelectron 26:2252–2257

182. Sun J-Y, Huang K-J, Fan Y, Wu Z-W, Li D-D (2011) Glassy carbon electrode modified with a film composed of Ni(II), quercetin and graphene for enzyme-less sensing of glucose. Microchim Acta 174:289

183. Lee J-S (2011) Progress in non-volatile memory devices based on nanostructured materials and nanofabrication. J Mater Chem 21:14097–14112

184. Sawa A (2008) Resistive switching in transition metal oxides. Mater Today 11:28–36

185. Waser R, Aono M (2007) Nanoionics-based resistive switching memories. Nat Mater 6:833–840

186. Lee M, Han S, Jeon SH, Park BH, Kang BS, Ahn S, Kim KH, Lee CB, Kim CJ, Yoo I, Seo DH, Li X, Park J, Lee J, Park Y (2009) Electrical manipulation of nanofilaments in transition-metal oxides for resistance-based memory. Nano Lett 9:1476–1481

187. Waser R, Dittmann R, Staikov G, Szot K (2009) Redox-based resistive switching memories–nanoionic mechanisms, prospects, and challenges. Adv Mater 21:2632–2663

188. Watanabe Y, Bednorz JG, Bietsch A, Gerber C, Widmer D, Beck A, Wind SJ, Watanabe Y, Bednorz JG, Bietsch A, Gerber C, Widmer D, Beck A (2001) Current-driven insulator–conductor transition and nonvolatile memory in chromium-doped $SrTiO_3$ single crystals Current-driven insulator–conductor transition and nonvolatile memory in chromium-doped $SrTiO_3$ single crystals. Appl Phys Lett 78:3738

189. Seo S, Lee MJ, Seo DH, Jeoung EJ, Suh D, Seo S, Lee MJ, Seo DH, Jeoung EJ, Suh D, Joung YS, Yoo IK (2004) Reproducible resistance switching in polycrystalline NiO films. Appl Phys Lett 85:5655

190. Beck A, Bednorz JG, Gerber C, Rossel C, Widmer D, Beck A, Bednorz JG, Gerber C, Rossel C, Widmer D (2000) Reproducible switching effect in thin oxide films for memory applications. Appl Phys Lett 77:139

191. Rani.A and Kim D.H. (2016) A mechanistic study on graphene-based nonvolatile ReRAM devices. J Mater Chem C 4:11007–11031

192. Hlee Ã, Hen PC, Ang CW, Aikap SM (2007) Low-Power Switching of nonvolatile resistive memory using hafnium oxide low-power switching of nonvolatile resistive memory using hafnium oxide. Jpn J Appl Phys 46:2175–2179

193. Yalagala B, Sahatiya P, Mattela V, Badhulika S (2019) Ultra-low cost, large area graphene/MoS_2-based piezotronic memristor on paper: a systematic study for both direct current and alternating current inputs. ACS Appl Electron Mater 1:883–891

194. Tsai C, Xiong F, Pop E, Shim M, Science M, Seitz F, Engineering C, States U (2013) Resistive random access memory enabled by carbon nanotube. ACS Nano 7:5360–5366

195. Yang P, Chang W, Teng P, Jeng S, Lin S, Chiu P, He J (2013) Fully transparent resistive memory employing graphene electrodes for eliminating undesired surface effects. Proc IEEE 101:1732–1739

196. Selamneni V, Nerurkar N, Sahatiya P (2020) Large area deposition of $MoSe_2$ on paper as a flexible near-infrared photodetector. IEEE Sens Lett 4:1–4
197. Sahatiya P, Solomon Jones S, Thanga Gomathi P, Badhulika S (2017) Flexible substrate based 2D ZnO (n)/graphene (p) rectifying junction as enhanced broadband photodetector using strain modulation. 2D Mater 4:25053
198. Veerla RS, Sahatiya P, Badhulika S (2017) Fabrication of a flexible UV photodetector and disposable photoresponsive uric acid sensor by direct writing of ZnO pencil on paper. J Mater Chem C 5:10231–10240
199. Selamneni V, Koduvayur Ganeshan S, Sahatiya P (2020) All MoS_2 based 2D/0D localized unipolar heterojunctions as a flexible broadband (UV-Vis-NIR) photodetector. J Mater Chem C. https://doi.org/10.1039/D0TC02651D
200. Yang D, Ma D (2019) Development of organic semiconductor photodetectors: from mechanism to applications. Adv Opt Mater 7:1800522
201. Selamneni V, Sahatiya P (2020) Bolometric effect enhanced ultrafast graphene based do-it-yourself wearable respiration sensor for personal healthcare monitoring. IEEE Sens J 20:3452–3459
202. Sahatiya P, Badhulika S (2017) Strain-modulation-assisted enhanced broadband photodetector based on large-area, flexible, few-layered Gr/MoS_2 on cellulose paper. Nanotechnology 28:455204
203. Joshna P, Gollu SR, Raj PMP, Rao BVVSNP, Sahatiya P, Kundu S (2019) Plasmonic Ag nanoparticles arbitrated enhanced photodetection in p-NiO/n-rGO heterojunction for future self-powered UV photodetectors. Nanotechnology 30:365201
204. Sahatiya P, Shinde A, Badhulika S (2018) Pyro-phototronic nanogenerator based on flexible 2D ZnO/graphene heterojunction and its application in self-powered near infrared photodetector and active analog frequency modulation. Nanotechnology 29:325205
205. Sahatiya P, Badhulika S (2016) UV/ozone assisted local graphene (p)/ZnO(n) heterojunctions as a nanodiode rectifier. J Phys D Appl Phys 49:265101
206. Sahatiya P, Gopalakrishnan A, Badhulika S (2017) Paper based large area Graphene/MoS_2 visible light photodetector. In: 2017 IEEE 17th International Conference on Nanotechnology (IEEE-NANO), pp 728–730
207. Ma L, Ouyang J, Yang Y (2004) High-speed and high-current density C60 diodes. Appl Phys Lett 84:4786–4788
208. Szendrei K, Cordella F, Kovalenko MV, Böberl M, Hesser G, Yarema M, Jarzab D, Mikhnenko OV, Gocalinska A, Saba M, Quochi F, Mura A, Bongiovanni G, Blom PWM, Heiss W, Loi MA (2009) Solution-Processable near-IR photodetectors based on electron transfer from PbS nanocrystals to fullerene derivatives. Adv Mater 21:683–687
209. Guo F, Xiao Z, Huang J (2013) Photodetectors: fullerene photodetectors with a linear dynamic range of 90 dB enabled by a cross-linkable buffer layer (Advanced Optical Materials 4/2013). Adv Opt Mater 1:275
210. Zhang Q, Jie J, Diao S, Shao Z, Zhang Q, Wang L, Deng W, Hu W, Xia H, Yuan X, Lee S-T (2015) Solution-processed graphene quantum dot deep-UV photodetectors. ACS Nano 9:1561–1570
211. Tang L, Ji R, Li X, Bai G, Liu CP, Hao J, Lin J, Jiang H, Teng KS, Yang Z, Lau SP (2014) Deep ultraviolet to near-infrared emission and photoresponse in layered N-doped graphene quantum dots. ACS Nano 8:6312–6320
212. Rao F, Liu X, Li T, Zhou Y, Wang Y (2009) The synthesis and fabrication of horizontally aligned single-walled carbon nanotubes suspended across wide trenches for infrared detecting application. Nanotechnology 20:55501
213. Liu Y, Wei N, Zeng Q, Han J, Huang H, Zhong D, Wang F, Ding L, Xia J, Xu H, Ma Z, Qiu S, Li Q, Liang X, Zhang Z, Wang S, Peng L-M (2016) Room temperature broadband infrared carbon nanotube photodetector with high detectivity and stability. Adv Opt Mater 4:238–245
214. He X, Fujimura N, Lloyd JM, Erickson KJ, Talin AA, Zhang Q, Gao W, Jiang Q, Kawano Y, Hauge RH, Léonard F, Kono J (2014) Carbon nanotube terahertz detector. Nano Lett 14:3953–3958

215. Liu Y, Wang F, Wang X, Wang X, Flahaut E, Liu X, Li Y, Wang X, Xu Y, Shi Y, Zhang R (2015) Planar carbon nanotube–graphene hybrid films for high-performance broadband photodetectors. Nat Commun 6:8589
216. Lu R, Christianson C, Weintrub B, Wu JZ (2013) High photoresponse in hybrid graphene-carbon nanotube infrared detectors. ACS Appl Mater Interfaces 5:11703–11707
217. Kang P, Wang MC, Knapp PM, Nam S (2016) Crumpled graphene photodetector with enhanced, strain-tunable, and wavelength-selective photoresponsivity. Adv Mater 28:4639–4645
218. Mueller T, Xia F, Avouris P (2010) Graphene photodetectors for high-speed optical communications. Nat Photonics 4:297–301
219. Kim CO, Hwang SW, Kim S, Shin DH, Kang SS, Kim JM, Jang CW, Kim JH, Lee KW, Choi S-H, Hwang E (2014) High-performance graphene-quantum-dot photodetectors. Sci Rep 4:5603
220. Gomathi PT, Sahatiya P, Badhulika S (2017) Large-area, flexible broadband photodetector based on ZnS–MoS$_2$ hybrid on paper substrate. Adv Funct Mater 27:1701611
221. Sahatiya P, Badhulika S (2015) One-step in situ synthesis of single aligned graphene–ZnO nanofiber for UV sensing. RSC Adv 5:82481–82487

Devices and Applications

Applications of Carbon-Based Nanomaterials in Health and Environment: Biosensors, Medicine and Water Treatment

Velpula Divya, Sai Kumar Pavar, Chidurala Shilpa Chakra, Thida Rakesh Kumar, Konda Shireesha, and Sakaray Madhuri

Abstract The carbon nanomaterials have been receiving great interest in department of nanoscience and technology an account of their extraordinary physical, chemical and electronic properties. Carbon nanomaterials have found emerging applications in various fields, viz. drug delivery, energy conversion and storage devices, field emission electronics, biosensors and water treatment. This book chapter is focused on health and environmental applications of fullerenes, carbon nanotubes and graphene. The closed cage structure, various redox states, stability, functionalization ability and light-induced switching behaviour of fullerenes trend in development of supercapacitors, sensors, optical and other electronic devices. The large surface area and fast charge transfer ability of carbon nanotubes enable their sensing ability for the detection of catechol, para-cresol and para-nitrophenol, hydroquinone, etc. that are widely located in aqueous and diverse biological systems. The adsorption and conjugating ability of carbon nanotubes with therapeutic and diagnostic agents signifies their importance in pharmaceutical and medical applications. Carbon nanotubes are also important in regeneration of tissues, diagnosis of biomolecules, extraction or enantiomer separation of chiral drug molecules and analysis of various drug molecules. The high surface-to-volume ratio and hydrophobic nature of carbon nanotubes facilitate strong affinities towards adsorption and removal of wide range of aliphatic and aromatic contaminants which include pathogenic organisms, and cyanobacterial toxins in water samples. The anti-microbial activity of carbon nanotubes helps in killing of pathogen present in water treatment plants. The unique morphological and structural features of graphene make them suitable for emerging energy and environmental applications, ranging from energy conversion and storage to green corona discharges for pollution control.

Keywords Carbon nanotubes · Biosensors · Fullerenes · Water treatment · Carbon nanomaterials · Graphene

V. Divya · S. K. Pavar · C. Shilpa Chakra (✉) · T. Rakesh Kumar · K. Shireesha · S. Madhuri
Center for Nano Science and Technology, Institute of Science and Technology, JNTU,
500085 Hyderabad, India
e-mail: shilpachakra.nano@jntuh.ac.in

A. Hazra and R. Goswami (eds.), *Carbon Nanomaterial Electronics: Devices and Applications*, Advances in Sustainability Science and Technology,
https://doi.org/10.1007/978-981-16-1052-3_11

1   Introduction

The idiosyncratic and controllable physico-chemical, electrical properties of carbon-based nanomaterials empower new technologies for recognizing and addressing various health and environmental challenges.

The significant optical, mechanical, electronic, chemical and thermal properties attracted researchers to work in the field of health and environmental applications. These applications include the removal of harmful contaminants, manufacture of nanomembrane structures for water purification, carbon-based biosensors, targeted cancerous cells, transport of drugs through denser tissues, etc. which mainly depend on structure, size and hybridization of carbon nanomaterials. Variable hybridization states classified the carbon nanomaterials into different configurations like nanodiamond, fullerene C_{60}, C_{80},....C_n, carbon onion, carbon nanotubes (single-walled and multi-walled), graphene, etc. These configurations have inherent features that can be easily exploited in the enhancement of advanced technologies for various applications.

Fullerene is curved structure molecule formed by fused pentagons and hexagons. Among various fullerenes, C_{60} (Buckminsterfullerene) is the more abundant and stable due its electronic bonding mechanism, followed by C_{70} and then C_{74}, C_{76}, C_{78}, C_{80}, C_{82}, C_{84}, etc. [1]. Buckminsterfullerene was given higher priority to researchers due to its significant stability and extraordinary properties. In Buckminsterfullerene, each carbon atom is undergoing sp^2 hybridization and is surrounded with three other carbon atoms. Buckminsterfullerene is an electron-deficient alkene and readily reacts with electron-rich species. Electron affinity (2.7 eV) and ionization potential (7.8 eV) of C_{60} indicate high electron transfer capability for various electroanalytical applications [2–4]. Irradiating laser beam, laser ablation, arc discharge, non-equilibrium plasma, etc. are some of the methods for the synthesis of C60 fullerene [5–9]; the synthesized fullerene can be functionalized by nucleophilic, radical addition, as well as cyclo addition reactions [10]. The high symmetric nature, low cost, stability and mild toxic nature with the additional unique properties such as stable in multiple redox states, easy functionalization abilities and light-induced switching behaviour make fullerenes as promising materials in various health and environment applications [11, 12].

Carbon Nanotubes (CNTs) are provided with unique strength since their chemical bonding is completely composed of sp^2-hybridized bonds and are stronger than sp^3-hybridized bonds [11]. CNTs are allotropes of carbon (rolled cylindrical sheets and tube-like structures) which are composed of graphite [12]. Depending on the physical structure CNTs are classified into two types (i) single graphene sheet rolled /single-walled CNT (SWCNT) and (ii) multilayer graphene sheets/multi-walled CNTs (MWCNT) [13]. Laser ablation, arc discharge and chemical vapour deposition are the most used synthesizing routes for the production of CNTs [14]. The extraordinary electrochemical and physico-chemical properties of CNTs provide potential applications in sustainable environment, green technologies, medicine and biosensors [15–17].

Graphite is composed of several layers of single sheet called graphene, which is an eco-friendly material with lower toxic levels in comparison to other inorganic materials. Discovery of graphene was honoured with a Nobel Prize in physics in year 2010 which reflects their importance. Graphene possesses a honey comb lattice with arrangement of single-layer carbon atoms in a hexagonal manner and carbon atoms in sp^2-hybridized state. The extraordinary characteristics (higher surface-to-volume ratio, high electrical conductivity, strong mechanical property, high thermal conductivity, functionalization capability, density, chemical stability, optical transmittance and high hydrophobicity [18, 19, 22]) made them immense interest in various fields (nanoelectronics, nanocomposites, opto-electronic devices, electrochemical supercapacitors, solar cells, pH sensors, gas sensors and other biosensors [20–23]).

Hence, carbon nanomaterials have attracted researchers in various applications in health and environmental applications including biomedical sciences, biosensors, targeted drug delivery, waste water treatments, electrochemical applications, energy storage devices, supercapacitors, etc. Numerous applications of carbon nanomaterials have found an account of excellence in physico-chemical properties. In this chapter, we focus on the importance of carbon nanomaterials, especially fullerenes, CNTs, graphene in the biosensors, medicine and wastewater treatment applications.

2 Applications of Carbon-Based Nanomaterials in Biosensors

A biosensor is an analytical device used for the detection of biological elements in direct spatial contact with a transduction element that converts biological events into quantifiable signals [24]. The reorganization of biomolecules is important in many areas such as medicines, drug and clinical domain, food analysis, genetic analysis, protein engineering, biomolecular sequencing, environmental monitoring, toxicity measurement and in the detection and identification of diseases and new drug molecules. Thus, an ideal biosensor should be high selective, high sensitive, high service life, good reproducibility, simplicity, low cost, scalability, direct edible (real-time monitoring) and rapid analysis of biomolecules in wide measurement range and should have a positive impact on human health and environment [25].

The basic working principle of a sensor is as follows (Fig. 1). It consists of an active sensing component; a signal converter and originates an electrical, optical, thermal or magnetic output signal. The sensor component detects analyte which causes generation of signal and is amplified if needed to generate required information [26, 27].

The sensitivity and selectivity of biosensors depend on the physico-chemical properties of the material used for sensing, for transducer, for enzyme immobilization and other stabilizers, mediators used in this process. The drawback of biosensor for large-scale utilization is an account of their low reproducibility and signal strength

Fig. 1 The working
principle of biosensors [25]

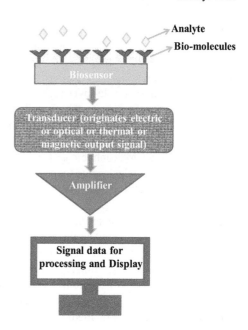

which depends mainly on the materials used in the construction of biosensors and transducers, while it can be overcome by using various nanomaterials. The carbon (1D, 2D and 3D) nanomaterial biosensors have good biocompatibility, high sensitivity, good selectivity, excellent transduction for the biorecognition, decrease in response times and lower limits of detection for detecting a wide range of chemical and biomolecules [28].

2.1 Fullerene-Based Biosensors

Fullerene exhibits unique and favourable characteristics such as good conductivity, wide absorption of light in the UV–Vis spectral region, easy of chemical modification, combination of nucleophilic and electrophilic property, angle strain produce from its structure, production of singlet oxygen, long life triple state, etc.[29]. These properties enable the potentials of fullerenes in the design of novel biosensor along with enhanced sensitivity, selectivity, real-time responsiveness and rapid signal transmission towards various biomolecules [30].

It is important to note that fullerenes are not harmful to biomolecules and are sufficiently small to detect the analyte in a closed manner. Fullerenes are having capability to exchange electrons with the surrounding biomolecules and to a transducer. During the identification of a biosensor, utilization of biomolecules takes place and fullerenes act as a mediator between the target and the electrode of a biosensor.

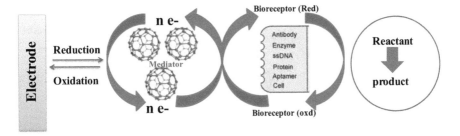

Fig. 2 Mechanism of sensing an analyte and the role of fullerene in a biosensor [29]

Further fullerenes can speed up the e^- transfer which enhance the identification ability and sensitivity than biological element without fullerene (Fig. 2) [29].

Fullerenes and functionalized fullerenes have potentials in detection and diagnosis of glucose with enhanced sensitivity and selectivity. Fullerene-based materials are also showing their importance in trace-level detection of urea with high sensitivity, repeatability and relatively good selectivity [25]. Thus, fullerenes act as an important component in immunosensors for the detection of diverse organic and inorganic bio-species in the blood (cysteine, ascorbic acid, tyrosine, urea, Na^+, K^+ and Ca^{2+}) that can be produced due to metabolism [31].

2.2 CNT-Based Biosensors

The extraordinary properties such as electrical conductivity, electron mobility and field effect mobility, small dimensions, high surface area and surface modification capability make CNTs as the perfect candidate for biosensor application. Even though CNT acts as biosensor, its solubility is the biggest problem. To overcome this problem, functionalization of CNT is required [32] and is carried out in three ways: (i) the skeleton of CNTs is linked with chemical groups by covalent or non-covalent linkage, (ii) adsorption of different functional groups to CNTs, and (iii) the internal cavities of CNTs are filled with Endohedrals [33]. Thus, increase in research on CNT conjugates has been developed for detection of DNA biomarkers, cell-surface sugars, protein receptors, etc., with high success rate [34].

Functionalized CNTs develop the sensitivity, selectivity and response time than conventional carbon biosensor and act as mediators between working electrode and enzymatic molecules. The CNT biosensors are further divided into electronic transducers (cancer biomarkers or aptamers and sensing of antibodies, peptides, proteins, enzymes, etc.,), electrochemical CNT biosensors (e.g. glucose, urea, dopamine, nitric oxide, epinephrine sensing), immunosensors (α-fetoprotein (AFP) in human serum, glucose, urea, Cardiac Troponin I (cTnI)) and optical CNT-based biosensors (detection of carbon dioxide, oxygen, alcohol, fluorescent molecules, ATP in living cells,

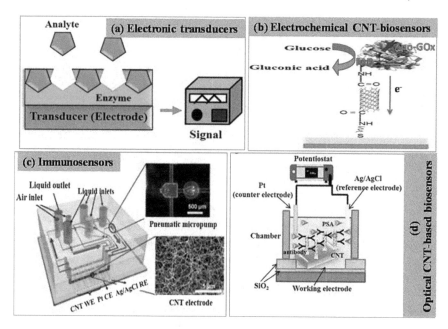

Fig. 3 a A Typical enzyme-based biosensor. **b** SWNT electrochemical glucose biosensor. **c** Amperometric immunobiosensor. **d** Microfluidic chip-based optical CNT biosensor [35] (Received copyright permission)

single-stranded DNA and cyclin A) on the bases of their mechanism towards target recognition and transduction while some of the examples are predicted in Fig. 3.

2.3 Graphene-Based Biosensors

The outstanding electrostatic π–π stacking interaction, high surface area, immobilization of different molecules, and electrochemical and optical properties make graphene as an ideal material for constructing biosensors and loading drug molecules. Graphene-based nanomaterials are used as transducers in a biosensor, which convert the interactions between the receptor (the organic/inorganic material) and the target molecules (can be organic/inorganic/whole cells) into detectable measurements, while the conjugated structure of graphene can facilitate high signal sensitivity to the biosensor. Furthermore, graphene and functionalized graphene can quench the absorbance in fluorescent biosensors. The sensitivity, selectivity and detection limit of graphene-based biosensors depends on synthetic methods, orientations and number of layers between the graphene, functional groups and oxidation states of graphene and number of functional groups. Graphene-based biosensors are of various types, viz. immunosensors (the measure of specific conjugation reaction between antibody and antigen), electrochemical sensors (measures any electrochemical changes at the interface of an electrode–electrolyte), optical/fluorescence biosensors (devices that derive an analytical signal from a photoluminescent process) and enzyme-based biosensor (enzyme and the transducer combination produces a signal that can be proportional to an analyte concentration).

The sp^2-bonded carbon atoms' utilization of large surface area makes graphenes as ideal substrates for selective detection of aromatic molecules and ssDNA by π–π stacking interactions [36]. In electrochemical DNA biosensors, the electron loss and conductivity changes that are caused due to change in voltage, current or impedance reflect the hybridization of DNA. The electrochemical signals are generated by this process and are detected in the biosensors either by cyclic voltammetry or differential pulse voltammetry or electrochemical impedance spectroscopy (Fig. 4a).

The fluorescent DNA nanosensors are developed using hybridization of two single-stranded DNA (ssDNA) and among them one is labelled with a fluorescent dye and the other corresponds to the target. Another widely used graphene-based DNA biosensor is carried out using the fluorescence-sensing approach. In this approach, the DNA probe is quenched to the surface of graphene-based nanomaterials through fluorescence resonance energy transfer (FRET) reaction that causes fluorescent signal to turn off (Fig. 4b). Upon hybridization of the probe with the target DNA, graphene surface releases the fluorescent molecule along with the dsDNA which causes fluorescent signal to turn on and can be detected.

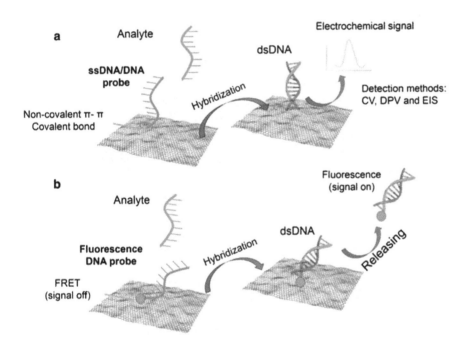

Fig. 4 Schematic illustration of graphene-based biosensors **a** electrochemical and **b** fluorescence detection [38]. (Received copyright permission)

Graphene is also used as a biosensor in detection of fluorescein [37], DNA, Celiac disease, HCV, HIV, catechol's, dopamine [38], hydrogen, lead, cadmium, silver, mercury, arsenic, acetaminophen, catecholamines, ascorbic acid, uric acid, hydroquinone, nicotinamide adenine dinucleotide hydrogen, carcinoembryonic antigen, haemoglobin, myoglobin, proteins [39] and cancer biomarkers [40].

3 Applications of Carbon-Based Nanomaterials in Medicine

3.1 Fullerenes in Medicine

The closed spherical C_{60} cage structure, and other exceptional properties such as multiple redox states and their stability, easy functionalization and light-induced switching behaviour make them promising in diverse applications. Fullerenes are used in various applications such as development of capacitors, biosensors, catalysis, optical, electronic devices, etc. [41, 42], while one of the most important applications of fullerene is medicine (Fig. 5) [43].

1. Fullerene molecule can be used as an antioxidant, antiviral agents due to their high affinity of the electron causes rapid reaction with radicals. Fullerenes are also used as anti-ageing and anti-damage and anti-bacterial agent in the cosmetic sector.

Fig. 5 The medical applications of fullerene [41–43]

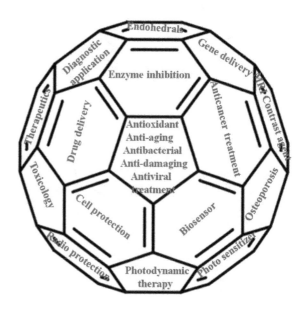

2. Fullerenes are able to conjugate with proteins, DNA and different functional groups which possess biological affinity to nucleic acids or proteins. This property assists the scientist to develop anti-cancer treatments.

3. Fullerene contributes a large extent in designing of various biosensors such as glucose, urea, microorganisms, etc.

4. Fullerenes can be used in drug delivery for finding cancer cell using surface-covered chemotherapeutic agents [25].

5. Fullerenes are world's strongest free-radical collectors and attract electrons due to their more available $C = C$ and lowest unoccupied molecular orbital. During the diseased state, fullerene helps successfully to protect cells from toxins by the collecting free radicals from the local body in mitochondria [44].

6. The easy functionalized capacity and mild toxicity of fullerenes make them an ideal to serve as 3D scaffolds for carrying the drug molecules. They can be found to go to tissues and make them respond to carry drug molecules to hydrophobic tissues in liver. The high collecting capacity and reactivity make them conjugate to various drugs in drug delivery [45].

7. Fullerene structure is highly complementary, both sterically and chemically, to the HIV-P active site, and hence binds effectively and inhibits the HIV protease enzyme [46].

8. Fullerenes with additional atoms, ions or clusters enclosed within their inner spheres are called Endohedral fullerenes, which are very useful for biomedical applications [47] such as MRI, X-ray imaging, radiopharmaceutical applications, etc. [48].

9. Fullerenes are used for osteoporosis treatment due to its preferential localization.

10. Fullerenes and their derivatives will have a neuroprotective effect in the central nervous system due to its therapeutic action [44].

Further, Fullerenes and their derivatives are widely employed in photosensitization, photodynamic therapy, radio protection, toxicology and diagnostic applications.

3.2 CNTs in Medicine

CNTs have been employed successfully in medicine since from the beginning of the twenty-first century on account of their unique properties, such as ultra-small size, large surface area, high aspect ratio, high reactivity, distinct optical properties, conjugating ability with a variety of therapeutic molecules (drugs, proteins, antibodies, DNA, enzymes, etc.) and diagnostic nature (vaccines, antibodies, biosensors, etc.). CNTs are also proved as excellent vehicles for drug delivery which penetrates into the cells directly and keeping the drug intact without metabolism during transport in the body. Figure 6 depicts various applications of CNTs and other functionalized CNTs in different fields of medicine.

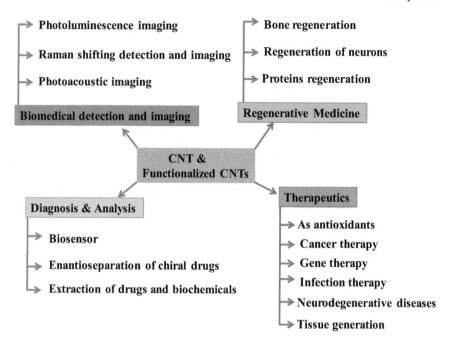

Fig. 6 Schematic representation of CNTs and other functionalized CNT applications in medicine [48]

1. Regenerative medicine

   The easy chemical functionalization of CNTs provides a platform which customizes to a range of functions related to regenerative medicine. The bone regeneration is being developed using CNTs that uses negatively charged functional groups with calcium bonded to them. CNTs act as an ideal alternative to the titanium or ceramic bone scaffolds due to their strength, stiffness and flexibility. In tissue regeneration process, regulation of the electroactive behaviour of cardiac or nervous cells is important, where CNT provides an excellent platform due to their improved mechanical properties and electrical conductivity. The enhancement of physical properties with the combination of CNTs is particularly promising for cardiac reinforcement applications. Nerve conduits composed of PEG functionalized CNT/oligo (poly (ethylene glycol) fumarate) (OPF) nanocomposite were generated using injection moulding technique and are filled in the gaps resulting in spinal cord injury, (Fig. 7) while the hydrogel enhances the cell attachment, and proliferation that conforms axon recovery [48, 49].

2. Therapeutics

   CNTs and their functional derivatives are having antioxidant nature which helps in various biomedical applications and prevents chronic ailments, ageing and food preservations. The –COOH groups present in carboxylated SWCNTs

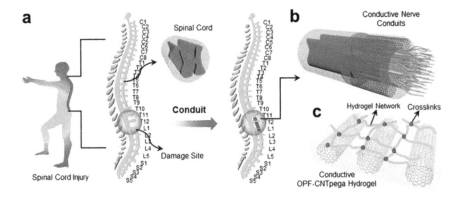

Fig. 7 (a) Schematic depiction of the spinal cord. (b) Conductive nerve conduits for spinal cord injury treatment. (c) Structure of the conductive OPF-CNT pega hydrogel [49]. (Received copyright permission)

increase the free-radical scavenging activity and improve antioxidant property. The antioxidant property is very important in anti-ageing cosmetics and sunscreen creams which protect skin against free radicals produced by the human body or by sunlight. Many successful anti-cancer drugs (epirubicin, doxorubicin, cisplatin, paclitaxel, etc.) have been conjugated with functionalized CNTs. Functionalized CNTs are used for the treatment of infectious diseases such as TB, cancer and also protect from anti-microbial and anti-fungal infections. The tiny dimensions and accessible external modifications property of CNTs helps them in neurosciences for the treatment of neurodegenerative diseases, while it is found that conjugation of CNTs with therapeutic molecules has positive effects on neuronal growth than the single drug. CNTs and their composites (synthetic biocompatible polymers) are evaluated as scaffolds for tissue regeneration.

3. Diagnosis and Analysis

The length scale, specific structure and unique physico-chemical properties of CNTs make them popular tool in cancer diagnosis and therapy [50]. Addition of CNTs provides more accuracy and simpler manipulation than biosensors alone in detection of glucose-oxidase biosensors for blood sugar control in diabetic patient [51–54]. The sensitivity of the assay using SWCNT-DNA sensor was much higher than traditional fluorescent and hybridization assays in detection of antigen. CNTs provide a fast and simple solution for molecular (DNA or protein) diagnosis in pathologies. [53, 55]. The organophosphoric pesticides are also detected by immobilization of acetylcholine esterase over CNT surface [52–54].

4.   Biomedical detection and imaging

The metallic or semi-conductive nature of CNTs depends on the different ways of rolling a graphene sheet into a cylinder. The semi-conductive SWNTs with appropriate chirality will generate a small bandgap fluorescence of 1 eV, which corresponds to NIR region (900–1600 nm) and is important for bio-imaging. The inherent graphene structure provides SWNTs with specific Raman scattering signature [56], which is strong enough for imaging of live cells and small animal models [57]. Thus, SWNTs act as a contrast agents for near-infrared (NIR) photoluminescence imaging [58–61], Raman imaging and optical absorption agent for photoacoustic imaging [56, 62, 63]. The strong light absorption characteristic [64] of CNTs helps as photoacoustic contrast agents which are used for imaging of tumors in mice [65–67].

3.3   Graphene in Medicine

The extraordinary properties, such as extreme hardness, wear resistant, high elasticity, transparent nature, lightweight, excellent thermal conductivity, chemically reactivity, anti-bacterial activity, good electrical conductivity, high density and withstanding ability towards ionizing radiation, make graphene as a novel material for the synthesis of new composites as well as for endless applications. Graphene is one of the most important materials that can be used for various medical applications. Graphene can be used frequently in the treatment of Cancer disease due to their improved property towards fighting and destroying the diseased cells. Graphene can also be used for developing muscle and in bone implants. Some of the medical applications of graphene are provided in the Fig. 8.

1.   Graphene as an Anticancer agent

By the injection of graphene particles into the patient they can be chemically modified and attached to cancer cells. The capability of absorption of IR radiation tends to treat the tumour directly on the damaged cell in the radiological treatment, while the rest of the body is unaffected. Thus, graphene is used as a diagnostic tool against cancer

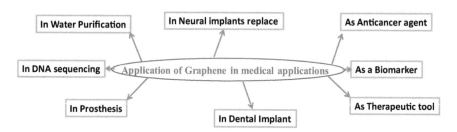

Fig. 8 Various applications of graphene in medical field [68–71]

cell [68]. The functionalized graphene is used as a drug carrier for delivery of anti-cancer chemotherapeutic drugs. In addition to these, graphene oxide microfluidic chips are developed to arrest tumour cells from blood and to support their growth for further analysis [69].

2. Graphene in Neural Implants

As 3D-graphene foams are predominant to traditional inert biomaterials, it is used as neural implants for energy conversion, storage system, tissue engineering, construction of a scaffold of neural stem cell, etc. [70].

3. Graphene in Dental Implants

The application of graphene in dental implant was initiated at the University of Alicante and is found that the durability of material increases with the incorporating graphene. The graphene can also be used to create a layer of vertical splinters, which can manage to form a protective surface to prevent bacteria in dental implants. Graphene composed with the prostheses (placed on dental implants) has higher resistance and structural stability, flexibility and is lighter. The use of graphene in this field helps the patients to protect from infections and reduces the use of antibiotics.

4. Graphene as a Therapeutic tool

Graphene nanomaterials have capability to combine with medications and other molecules lead to improve the function of the drug towards target material. The graphene nanomaterials injected into the blood can be attached to different types of cells of our immune system without defences attacking the intruder.

5. Graphene in Prosthesis

Rubber composed graphene is an ideal material for an efficient bionic muscle. The electrical stimulation on this compound makes them ideal to control tension and relaxation. The dental prosthesis is the only quick fix to restore missing natural teeth which might improve oral functions and comfort. Graphene is a new material that can meet the desirable features to be used for this purpose. Graphene is also useful for making splints and that combination makes them hard and highly strengthens materials.

6. Graphene in Water Purification

Graphene nanosheets are highly efficient and cost-effective materials for the purification water. The graphene nanosheets are impermeable for passage of impurities through it due to the small pore size and interactions, but allow water and provide clear and pure water as filtrate. This shows the importance of graphene in protection of humans from various diseases.

7. Graphene as a Biomarker

Graphene can be employed in biosensors for detection of blood pressure, blood sugar levels, nitric oxide in oxygen, etc. Graphene is used as a biomarker in detection of bacteria and pathology that causes sick. Graphene also acts as an immunosensor used for small-scale detection of growth hormone. Graphene plays a crucial role in fighting against various diseases, especially cancer. Graphene-based biosensors have capability to detect E. coli bacteria. As the surface of graphene is attached by a number of drug molecules, it can be used in drug delivery to target the diseases present over the cell surfaces. Graphene is used in birth control and also in anti-microbial applications [69].

Thus, graphene is an important material that enhances the potentials in various medical applications such as cancer therapies, disease diagnostic tools, tissue engineering, implants, DNA sequencing, biomarker and transfer of genetic material, biomedical imaging and neuroscience, etc. [71].

4 Applications of Carbon-Based Nanomaterials in Water Treatment

Water is a natural gift on the earth and is essential for human, animal and plants. Human being uses water for all domestic and industrial purposes. Without food humans can survive for a number of days, but water is such an essential commodity that without it, one cannot survive. In nature, water is present in 75% of the earth's surface; however, much of it is not suitable for the needs due to the presence of impurities from various sources. The process of removing all types of impurities from water and making it fit for domestic or industrial purposes is called water technology or water treatment.

Even though there are many varieties of materials developed for the treatment of water such as zeolites, polymers, metal–organic frameworks, etc., carbon-based adsorbents like carbon nanotubes, fullerenes and graphene are given higher importance since they possess high adsorption capacity due to their higher surface area and large pore volume which are the primary requirements in achieving success in any wastewater treatments. They are also considered to be superior in treating organic and inorganic pollutants [72]. Thus, carbon nanomaterials play a key role in purification of contaminated water with low cost and high efficiency. The importance of fullerenes, CNT and graphene in water treatment is discussed in the below section.

4.1 Fullerene in Water Treatment

The high electron affinity, reactivity and capability of producing reactive oxygen species (ROS) make fullerenes as efficient nanomaterials in water treatment.

1. By photosensitization

Fullerenes are used in engineered systems to photoactively oxidize organic contaminants or to inhibit microbe activities. The ability of fullerene in tailoring the surfaces

Fig. 9 Potential photosensitization mechanism of C_{60} [74]. (Received Copyright permission)

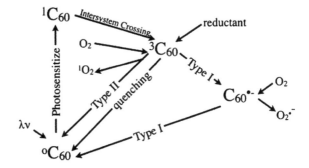

makes them specific to increase adsorbing capacities or to recognize specific contaminants [73]. C_{60} acts as a photosensitizer (photoactive material) in which the electrons are excited within their molecular orbital leading to produce ROS in solution. Type 1 and Type 2 are the two mechanisms followed by sensitized electrons, i.e. electrons transfer to a donor molecule, energy transfer to ground state oxygen, respectively. In both the processes, light energy is converted into chemical energy which is utilized in oxidizing the contaminants and microorganisms from wastewater. Figure 9 depicts the C60 molecule sensitizer that follows Type 1 and Type 2 mechanisms.

The unique geometry of C_{60} molecule is responsible for efficient intersystem crossing. Intersystem crossing occurs very efficiently due to the unique geometry of the molecule. The addition of C_{60} to the aqueous system reduces the efficiency of some of these processes. C_{60} molecule forms clusters in the aqueous solution resulting in reduced ROS formation due to decrease in lifetime of triplet state [75]. Though proper addition of functional groups can reduce the clustering and increase ROS production, addition of these groups can lower the triplet state which is responsible for ROS formation [76].

Fullerene derivatives (fullerol) are also important in production of ROS which helps in disinfecting wastewater or destroy the organic compounds by avoiding the undesirable oxidation by-products. Fullerol (C_{60} (OH) 2224) was prepared [77] and a comparative study was made with known photosensitizer Rose Bengal dye (RB) [78] on destruction of organic compound 2-chlorophenol (2-CP) irradiating under UV light. For the same conditions maintained for both results shown that fullerol was useful in 17% of the 2-CP destruction while RB was 28% in destruction.

2.   By Anti-microbial property

Medical literature survey on role of C_{60}, C_{70} fullerene membranes in wastewater treatment reflects the deactivation of bacteria and other water-borne viruses [77] and their anti-fouling property [79]. Fullerene derivatives (fullerol) act as an effective disinfectant which can decrease the activity of MS2 bacteriophage (virus) in wastewater treatments [80]. The ceramic membranes modified with depositing a layer of C_{60} molecules are prepared to study their impact on bacterial (E. coli K12) attachment and metabolic activities and it has been observed that the CTC/DAPI (metabolically active/total number of bacteria present) ratio has significantly decreased on increasing C_{60} molecules over the membrane. Long-term efficiency of C_{60} membranes was expected in anti-biofouling activities [77].

4.2 Carbon Nanotubes in Water Treatment

The pore size and adsorbent capability have made CNTs as extensive materials in determination of various aqueous pollutants. They have also attained significant considerations for innovative membrane developments for wastewater treatments [81].

The classification of CNT members is as follows:

1. Free standing CNT membranes:

    (a) Vertically Aligned (VA) CNT membranes.
    (b) Bucky-paper membranes.

2. Mixed CNT membranes

Free standing CNT membranes are majority membranes used in desalination of water applications. CNTs are typically lined up as cylindrical pores by which the fluid can be passing within the hollow interior of CNT or in between the bundles of CNT, in the case of VA-CNT membranes Fig. 10a. In the case of Bucky-paper CNT membranes [82] (Fig. 10b), CNTs were randomly arranged which are highly porous with higher specific area. Mixed CNT membranes are somehow similar to thin-film RO membranes, where a top layer is composed of both CNT and a polymer. Figure 10b depicts the respective images of VA-CNT and mixed CNT membranes. CNT membranes have the potential of higher performance such that

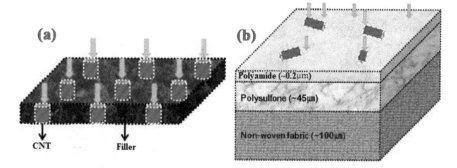

Fig. 10 Representation of **a** VA-CNT membranes, and **b** mixed (nanocomposite) CNT membranes with the top layer of mixed CNT membranes [83]. (Received Copyright permission)

they can replace/improve the conventional membranes (RO, NF, UF, MF, etc.), and hydrophobic property of CNT promotes the faster flow of water molecules since water is a polar molecule.

1. Applications of VA-CNT membranes in water treatment

VA-CNT membranes not only exhibit higher flow rate but also possess anti-bacterial activity that shows appreciable performance in removing various unwanted salts [84]. Li et al. manufactured a membrane which is useful in ultra-filtration, while the high flux flow was made with the help of polyether sulfone (PES) and pre-aligned MWCNT array was made by simple phase inversion, drop-casting process [85] and the results were compared with the pure PES. It is observed that the prepared membranes were so impressive that water speed through the prepared membranes was 10 times higher than the pure PES membranes and also 3 times higher than the mixed CNT/PES membranes.

Monolithic freestanding uniform macroscopic hollow cylindrical membrane filters were fabricated using spray pyrolysis of ferrocene/benzene [86]. The prepared filters were effective and reliable in separating heavy hydrocarbon and also in removal of bacteria.

The VA-CNTs have higher fluid transport efficiency but it is worthy to mention that their alignment control in the membrane matrix and control of agglomeration is still challenging.

2. Applications of Bucky-paper membranes in water treatment

    a. The high thermal conductivity, good porous structure and hydrophobic nature of Bucky-paper CNT membranes made them promising in direct membrane distillation process. Dumee et al. reported that 99% of salt was rejected by using Bucky-paper CNT membranes in direct membrane distillation method [87]. Bucky-paper membranes were also reported in filtration of fine particles (100–500 nm diameter) and are highly efficient in air filtration [88].

    b. Anti-bacterial properties were also strengthened on using Bucky-paper CNT membranes. CNT Bucky-paper membrane exhibits higher removal rate of virus MS2 bacteriophage of 27 nm through depth filtration and was also effective in retaining E. coli cells (2 μm), damaging cell membranes and inactivating the E. coli cells [84].

3. Applications of Mixed CNT membranes in water treatment

CNT-based nanocomposite membranes show higher resistance to BSA, protein and bacteria fouling in various reports [89]. Comparison study was made on bovine serum albumin (BSA) fouling mechanism over pristine membranes and GO/MWCNTs/PVDF membranes by Zhang et al. [90]. Results were likely dependent over the adhesion force of the membrane foulant. The CNT membrane decreases the adhesion force of membrane, enhances the pure water flux and maintains good anti-fouling performance.

4.3 Graphene in Water Treatment

Monolayer membranes as well as stacked multilayer membranes are extensively used in desalination membrane preparations for water purification.

1. Monolayer graphene desalination membrane

Pristine graphene membranes are impermeable to water, pollutants and also to the smallest mono-atomic molecule of 1.3 Å (Helum) [91]. By controlling the pore size, density and functionality, current desalination membranes were showing magnitude of higher permeability and selectivity [92, 93].

2. Multilayer membrane for desalination purpose

Monolayer membranes are effective in water permeability, while its production at industrial scale is still challenging on account of leak-proof fabrication and there is a difficulty in manufacturing large area monolayer graphene with controlled pore size and density (Fig. 11a) [94]. The multilayer graphene membranes consist of assembled graphene oxide sheets and are promising for desalination purposes. The multilayer graphene membranes have high attention to the industries as the fabrication of highly stackable, single atom thick layer of graphene oxide can be easily obtained by a simple chemical oxidation and ultrasonic exfoliation of graphite, while is cost-effective. The

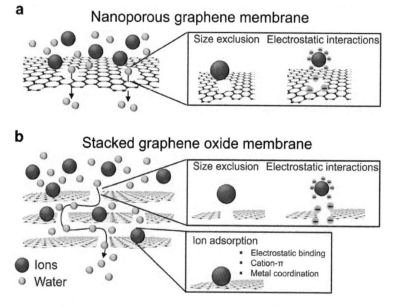

Fig. 11 Schematic representation for the separation mechanism of **a** a monolayer graphene membrane with nanopores of controlled size and **b** a multilayer graphene membrane composed of stacked GO sheets [96]. (Received Copyright permission)

structure channels in GO layers facilitate the permeation of water rejecting undesired matter, while functionalization enables charge-based interactions in treating water pollutants with lower cost. These promising reliable features of multilayer graphene make the possibility of producing advanced ionic and molecular membranes for effective desalination (Fig. 11b) [95, 96].

5 Conclusion and Outlook

In this book chapter, the basic concepts, importance and recent trends of carbon nanomaterials in various applications have been reported. In specific, the role of carbon nanomaterials in biosensors, medicine and water treatment has been described with various examples. The extraordinary physio-chemical property, electro-optical nature, high surface area and easy functionalization make fullerene, CNTs and graphene as ideal candidates for various applications. Wide research has been carried out on carbon-based nanomaterials which is widely used classes of nanomaterials. The research of carbon-based nanomaterials for health and environment applications has attracted great attention due to their unique physical and chemical properties including thermal, electrical, mechanical and structural diversity. In addition, owing to their versatile surface properties, size and shape over the past decade, carbon-based nanomaterials have considerable attention.

Acknowledgements This work has been supported in part by DST Project Central Government Fundamental Research Funds (no. SERB/F/7867/2019-2020).

References

1. Sanaz P, Karolien DW (2015) Recent Advances in Electrochemical Biosensors Based on Fullerene-C_{60} Nano-Structured Platforms. Biosensors 5:712–735. https://doi.org/10.3390/bios5040712
2. Jariwala D, Sangwan VK, Lauhon LJ, Marks TJ, Hersam MC (2013) Carbon nanomaterials for electronics, optoelectronics, photovoltaics, and sensing. Chem. Soc. Rev 42:2824–2860. https://doi.org/10.1039/C2CS35335K
3. Langa F, Nierengarten JF (2007) Fullerenes: Principles and Applications. The Royal Society of Chemistry, Cambridge, UK
4. Shinar J, Vardeny ZV, Kafafi ZH (2000) Optical and Electronic Properties of Fullerene and Fullerene-Based Materials. Marcel Dekker, New York, NY, USA
5. Lieber CM, Chen CC (1994) Preparation of fullerenes and fullerene-based materials. In Solid State Physics. Henry, E., Frans, S., Eds.; Academic Press: 48: 109–148
6. Scott LT (2004) Methods for the chemical synthesis of fullerenes. Angew. Chem. Int. Ed 43:4994–5007. https://doi.org/10.1002/anie.200400661
7. Dai L, Mau AWH (2001) Controlled synthesis and modification of carbon nanotubes and C_{60}: Carbon nanostructures for advanced polymeric composite materials. Adv. Mater 13:899–913. https://doi.org/10.1002/1521-4095(200107)13:12/13%3c899::AID-ADMA899%3e3.0.CO;2-G

8. Kharlamov AI, Bondarenko ME, Kirillova NV (2012) New method for synthesis of fullerenes and fullerene hydrides from benzene. Russ. J. Appl. Chem 85:233–238. https://doi.org/10.1134/S1070427212020127

9. Inomata K, Aoki N, Koinuma H (1994) Production of fullerenes by low temperature plasma chemical vaper deposition under atmospheric pressure. Jpn. J. Appl. Phys 33:197–199. https://doi.org/10.1143/JJAP.33.L197

10. Hirsch A (2006) Functionalization of fullerenes and carbon nanotubes. Phys. Status Solidi B 243:3209–3212. https://doi.org/10.1002/pssb.200669191

11. Wang X, Li Q, Xie J, Jin Z, Wang J, Li Y, Jiang K, Fan S (2009) Fabrication of ultralong and electrically uniform single-walled carbon nanotubes on clean substrates. Nano Lett 3137–3141. https://doi.org/https://doi.org/10.1021/nl901260b

12. Dai H (2002) Carbon nanotubes: Opportunities and challenges. Surf. Sci 500:218–241. https://doi.org/10.1016/S0039-6028(01)01558-8

13. Zhao YL, Stoddart JF (2009) Noncovalent functionalization of single-walled carbon nanotubes. Acc. Chem. Res 42:1161–1171. https://doi.org/10.1021/ar900056z

14. Manawi YM, Samara IA, Al-Ansari T, Atieh MA (2018) A review of carbon nanomaterials' synthesis via the chemical vapor deposition (CVD) method. Materials (Basel). 11:822. https://doi.org/10.3390/ma11050822

15. Asmaly HA, Abussaud B, Ihsanullah TAS, Alaadin A, Laoui T, Shemsi AM, Gupta VK, Atieh MA, Asmaly HA, Abussaud B, Saleh TA, Alaadin A (2015) Evaluation of micro- and nano-carbon-based adsorbents for the removal of phenol from aqueous solutions. Toxicol. Environ. Chem 97:1164–1179. https://doi.org/10.1080/02772248.2015.1092543

16. Lalia BS, Ahmed FE, Shah T, Hilal N, Hashaikeh R (2015) Electrically conductive membranes based on carbon nanostructures for self-cleaning of biofouling. Desalination 360:8–12. https://doi.org/10.1016/j.desal.2015.01.006

17. Ranveer A, Attar S (2015) Carbon Nanotubes and Its Environmental Applications. Journal of Environmental Science 4(2):304–311

18. Chen J, Jang C, Xiao S et al (2008) Intrinsic and extrinsic performance limits of graphene devices on SiO_2. Nature Nanotech 3:206–209. https://doi.org/10.1038/nnano.2008.58

19. Wonbong C, Indranil L, Raghunandan S, Yong SK (2010) Synthesis of Graphene and Its Applications: A Review. Crit Rev Solid State Mater Sci 35(1):52–71. https://doi.org/10.1080/10408430903505036

20. Gurunathan S, Han JW, Dayem AA, Eppakayala V, Kim JH (2012) Oxidative stress-mediated antibacterial activity of graphene oxide and reduced graphene oxide in Pseudomonas aeruginosa. Int J Nanomed 7:5901–5914. https://doi.org/10.2147/IJN.S37397

21. Gurunathan S, Han JW, Eppakayala V, Kim JH (2013) Green synthesis of graphene and its cytotoxic effects in human breast cancer cells. Int J Nanomed 8:1015–1027. https://doi.org/10.2147/IJN.S42047

22. Gurunathan S, Han JW, Kim JH (2013) Green chemistry approach for the synthesis of biocompatible graphene. Int J Nanomed 8:2719–2732. https://doi.org/10.2147/IJN.S45174

23. Singh J, Rathi A, Rawat M et al (2018) Graphene: from synthesis to engineering to biosensor applications. Front. Mater. Sci 12:1–20. https://doi.org/10.1007/s11706-018-0409-0

24. Jianrong C, Yuqing M, Nongyue H, Xiaohua W, Sijiao L (2004) Nanotechnology and biosensors. Biotechnol. Adv 22:505–518. https://doi.org/10.1016/j.biotechadv.2004.03.004

25. Hazal G, Serdar Y, Mehmet FE (2020) Nano-carbons in biosensor applications: an overview of carbon nanotubes (CNTs) and fullerenes (C_{60}). SN Applied Sciences 2:603. https://doi.org/10.1007/s42452-020-2404-1

26. Solanki PR, Kaushik A, Agrawal VV, Malhotra BD (2011) Nanostructured metal oxide-based biosensors. NPG Asia Mater 3:17. https://doi.org/10.1038/asiamat.2010.137

27. Balasubramanian K, Burghard M (2006) Biosensors based on carbon nanotubes. Anal BioanalChem 385(3):452–468. https://doi.org/10.1007/s00216-006-0314-8

28. Gogotsi Y, Presser V (2014) Carbon Nanomaterial. Taylor & Francis Group, LLC: New York, USA.

29. Afreen S, Muthoosamy K, Manickam S, Hashim U (2015) Functionalized fullerene (C_{60}) as a potential nanomediator in the fabrication of highly sensitive biosensors. BiosensBioelectron 63:354–364. https://doi.org/10.1016/j.bios.2014.07.044

30. Langa F, Nierengarten JF (2007) Fullerenes: Principles and Applications. the Royal Society of Chemistry. https://doi.org/10.1039/9781847557711

31. Pan NY, Shih JS (2004) Piezoelectric crystal immunosensors based on immobilized fullerene C_{60}-antibodies. Sens Actuators B Chem 98:180–187

32. Same S, Samee G (2018) Carbon nanotube biosensor for diabetes disease. Crescent J Med Biol Sci 5:1–6

33. Wu HC, Chang X, Liu L, Zhao F, Zhao Y (2010) Chemistry of carbon nanotubes in biomedical applications. J Mater Chem 20:1036–1052. https://doi.org/10.1039/B911099M

34. Tîlmaciu CM, Morris MC (2015) Carbon Nanotube Biosensors. Front. Chem 3:59. https://doi.org/10.3389/fchem.2015.00059

35. Yuichi T, Kenzo M, Kazuhiko M, Miyuki C, Yuzuru T, Eiichi T (2009) Microfluidic and label-free multi-immunosensors based on carbon nanotube microelectrodes. Jpn. J. Appl. Phys 48:06FJ02. https://doi.org/https://doi.org/10.1143/JJAP.48.06FJ02

36. Yang Y, Asiri AM, Tang Z, Du D, Lin Y (2013) Graphene based materials for biomedical applications. Mater Today 16:365–373. https://doi.org/10.1016/j.mattod.2013.09.004

37. Usikov AS, Lebedev SP, Roenkov AD, Barash IS, Novikov SV, Puzyk MV, Zubov AV, Makarov YN, Lebedev AA (2020) Studying the Sensitivity of Graphene for Biosensor Applications. Tech Phys Lett 46:462–465

38. Bahamonde JP, Nguyen HN, Fanourakis SK, Rodrigues DF (2018) Recent advances in graphene-based biosensor technology with applications in life sciences. J Nanobiotechnol 16:75. https://doi.org/10.1186/s12951-018-0400-z

39. Atta NF, Galal A, El-Ads EH (2015) Graphene A Platform for Sensor and Biosensor Applications. ToonikaRinken, Intech Open, Biosensors - Micro and Nanoscale Applications. https://doi.org/10.5772/60676

40. Zhou L, Wang K, Sun H et al (2019) Novel Graphene Biosensor Based on the Functionalization of Multifunctional Nano-bovine Serum Albumin for the Highly Sensitive Detection of Cancer Biomarkers. Nano-Micro Lett 11:20. https://doi.org/10.1007/s40820-019-0250-8

41. Bosi S, da Ros T, Spalluto G, Prato M (2003) Fullerene derivatives: An attractive tool for biological applications. Eur. J. Med. Chem 38:913–923. https://doi.org/10.1016/j.ejmech.2003.09.005

42. Dresselhaus MS, Dresselhaus G, Eklund PC (1996) Science of Fullerenes and Carbon Nanotubes: Their Properties and Applications. Elsevier Science, London, UK

43. Da RT (2008) Twenty years of promises: fullerene in medicinal chemistry. Medicinal chemistry and pharmacological potential of fullerenes and carbon nanotubes. Springer, Netherlands 1–21. https://doi.org/https://doi.org/10.1007/978-1-4020-6845-4_1

44. Shan-hui H, Pei-wen L (2019) Nano architectonics to Tissue Architectonics: Nanomaterials for Tissue Engineering. Elsevier, Advanced Supramolecular Nanoarchitectonics, pp 277–288

45. Anilkumar P, Lu F, Cao L, Luo PG, Liu JH, Sahu S, Tackett KN, Wang Y, Sun YP (2011) Fullerenes for applications in biology and medicine. Curr Med Chem 18(14):2045–2059. https://doi.org/10.2174/092986711795656225

46. Wilson SR (2002) Nanomedicine: Fullerene and Carbon Nanotube Biology. In: Ōsawa E. (ed) Perspectives of Fullerene Nanotechnology. Springer, Dordrecht 155–163. https://doi.org/10.1007/978-94-010-9598-3_14

47. Bakry R, Vallant RM, Najam-ul-Haq M, Rainer M, Szabo Z, Huck CW, Bonn GK (2007) Medicinal applications of fullerenes. Int J Nanomed 2(4):639–649

48. Juliette S, Emmanuel F, Muriel G (2019) Overview of Carbon Nanotubes for Biomedical Applications. Materials (Basel) 12(4):624. https://doi.org/10.3390/ma12040624

49. Liu X, Kim JC, Miller AL, Waletzki BE, Lu L (2018) Electrically Conductive Nanocomposite Hydrogels Embedded with Functionalized Carbon Nanotubes for Spinal Cord Injury. New J. Chem 42:17671–17681. https://doi.org/10.1039/C8NJ03038C

50. Ji SR, Liu C, Zhang B, Yang F, Xu J, Long J, Jin C, Fu DL, Ni QX, Yu XJ (2010) Carbon nanotubes in cancer diagnosis and therapy. Biochimicaetbiophysicaacta 1806(1):29–35. https://doi.org/10.1016/j.bbcan.2010.02.004
51. Usui Y, Haniu H, Tsuruoka S, Saito N (2012) Carbon nanotubes innovate on medical technology. Med Chem 2(1):1–6. https://doi.org/10.4172/2161-0444.1000105
52. Digge MS, Moon RS, Gattani SG (2012) Applications of carbon nanotubes in drug delivery: a review. International Journal of PharmTech Research 4(2):839–847
53. Wang J (2005) Carbon-nanotube based electrochemical biosensors: a review. Electroanalysis 17(1):7–14. https://doi.org/10.1002/elan.200403113
54. Zhu Y, Wang L, Xu C (2011) Carbon nanotubes in biomedicine and biosensing in Carbon Nanotubes-Growth and Applications. M. Naraghi, Ed 135–162. https://doi.org/https://doi.org/10.5772/16558
55. Liao H, Paratala B, Sitharaman B, Wang Y (2011) Applications of carbon nanotubes in biomedical studies. Methods Mol Biol 726:223–241. https://doi.org/10.1155/2013/578290
56. Tasis D et al (2008) Diameter-selective solubilization of carbon nanotubes by lipid micelles. J NanosciNanotechnol 8(1):420–423. https://doi.org/10.1166/jnn.2008.104
57. Karmakar A et al (2011) Raman spectroscopy as a detection and analysis tool for in vitro specific targeting of pancreatic cancer cells by EGF-conjugated, single-walled carbon nanotubes. J ApplToxicol 32(5):365–375. https://doi.org/10.1002/jat.1742
58. Jin H, Heller D, Strano M (2008) Single-Particle Tracking of Endocytosis and Exocytosis of Single-Walled Carbon Nanotubes in NIH-3T3 Cells. Nano Lett 8:1577–1585. https://doi.org/10.1021/nl072969s
59. Strano MS, Jin H (2008) Where is it heading? Single-particle tracking of single-walled carbon nanotubes. ACS Nano 2(9):1749–1752. https://doi.org/10.1021/nn800550u
60. Liu Z et al (2010) Multiplexed Five-Color Molecular Imaging of Cancer Cells and Tumor Tissues with Carbon Nanotube Raman Tags in the Near-Infrared. Nano Res 3(3):222–233. https://doi.org/10.1007/s12274-010-1025-1
61. Welsher K et al (2008) Selective probing and imaging of cells with single walled carbon nanotubes as near-infrared fluorescent molecules. Nano Lett 8(2):586–590. https://doi.org/10.1021/nl072949q
62. Keren S et al (2008) Non-invasive molecular imaging of small living subjects using Raman spectroscopy. Proc Natl Acad Sci USA 105(15):5844–5849. https://doi.org/10.1073/pnas.0710575105
63. Welsher K et al (2009) A route to brightly fluorescent carbon nanotubes for near-infrared imaging in mice. Nat Nanotechnol 4(11):773–780. https://doi.org/10.1038/nnano.2009.294
64. Berber S, Kwon YK, Tomanek D (2000) Unusually high thermal conductivity of carbon nanotubes. Phys Rev Lett 84(20):4613–4616. https://doi.org/10.1103/PhysRevLett.84.4613
65. De la Zerda A et al (2008) Carbon nanotubes as photoacoustic molecular imaging agents in living mice. Nat Nanotechnol 3(9):557–562. https://doi.org/10.1038/nnano.2008.231
66. Xiang L et al (2009) Photoacoustic molecular imaging with antibody-functionalized single-walled carbon nanotubes for early diagnosis of tumor. J Biomed Opt 14(2): 021008. https://doi.org/https://doi.org/10.1117/1.3078809
67. Wei S, Paul A, Mai Y, Laetitia R, Satya P (2013). Carbon Nanotubes for Use in Medicine: Potentials and Limitations. https://doi.org/10.5772/51785
68. Subhashree P, Swaraj M, Sumit M, Srirupa B, Monalisa M (2018) Graphene and graphene oxide as nanomaterials for medicine and biology application. Journal of Nanostructure in Chemistry 8:123–137. https://doi.org/10.1007/s40097-018-0265-6
69. Pirolini, Alessandro. "Applications of Graphene in Medicine". AZoNano. https://www.azonano.com/article.aspx?ArticleID=3723.
70. Tyagi M, Albert A, Tyagi V, Hema R (2013) Graphene nanomaterials and applications in biomedical sciences. World J. Pharm. Pharmacol. Sci 3(1):339–345. https://doi.org/10.1007/s40097-018-0265-6
71. Priyadarsini S, Mohanty S, Mukherjee S et al (2018) Graphene and graphene oxide as nanomaterials for medicine and biology application. J Nanostruct Chem 8:123–137. https://doi.org/10.1007/s40097-018-0265-6

72. Yang K, Zhu L, Xing B (2006) Adsorption of polycyclic aromatic hydrocarbons by carbon nanomaterials. Environ. Sci. Technol 40:1855–1861. https://doi.org/10.1021/es052208w
73. Bottero JY, Rose J, Wiesner MR (2006) Nanotechnologies: tools for sustainability in a new wave of water treatment processes. Integr Environ Assess Manag 2(4):391–395
74. So-Ryong C, Ernest MH, Mark RW (2014) Chapter 21 - Possible Applications of Fullerene Nanomaterials in Water Treatment and Reuse. Micro and Nano Technologies p 329–338. https://doi.org/https://doi.org/10.1016/B978-1-4557-3116-9.00021-4
75. Anderson T, Nilsson K, Sundahl M, Westman G, Wennerstrom O (1992) C_{60} Embedded in γ-cyclodextrin: a water-soluble fullerene. J. Chem. Soc., Chem. Commun 604–606. https://doi.org/https://doi.org/10.1039/C39920000604
76. Hotze EM, Labille J, Alvarez P, Wiesner MR (2008) Mechanisms of photochemistry and reactive oxygen production by fullerene suspensions in water. Environ. Sci. Technol 42(11):4175–4180. https://doi.org/10.1021/es702172w
77. Badireddy AR, Hotze EM, Chellam S, Alvarez P, Wiesner MR (2007) Inactivation of bacteriophages via photosensitization of fullerol nanoparticles. Environ. Sci. Technol 41(18):6627–6632. https://doi.org/10.1021/es0708215
78. Pickering KD, Wiesner MR (2005) Fullerol-sensitized production of reactive oxygen species in aqueous solution. Environ. Sci. Technol 39(5):1359–1365. https://doi.org/10.1021/es048940x
79. Mallevialle J, Odendaal PE, Wiesner MR (1996) Water treatment membrane processes. NY, McGraw-Hill, New York
80. Batch LE, Schulz CR, Linden KG (2004) Evaluating water qualify effects on UV disinfection of MS2 coliphage. J. Am. Water Works Assoc 96(7):75–87. https://doi.org/10.1002/j.1551-8833.2004.tb10651.x
81. Ihsanullah, (2019) Carbon nanotube membranes for water purification: Developments, challenges, and prospects for the future. Sep Purif Technol 209:307–337. https://doi.org/10.1016/j.seppur.2018.07.043
82. Sears K, Dumée L, Schütz J, She M, Huynh C (2010) Recent developments in carbon nanotube membranes for water purification and gas separation. Materials (Basel) 127–149. https://doi.org/https://doi.org/10.3390/ma3010127
83. Hoon C, Baek Y, Lee C, Ouk S, Kim S, Lee S, Kim S, Seek S, Park J, Yoon J (2012) Carbon nanotube-based membranes: Fabrication and application to desalination. J. Ind. Eng. Chem 18:1551–1559. https://doi.org/10.1016/j.jiec.2012.04.005
84. Brady-Estevez AS, Kang S, Elimelech M (2008) A single-walled-carbon-nanotube filter for removal of viral and bacterial pathogens. Small 4:481–484. https://doi.org/10.1002/smll.200700863
85. Li S, Liao G, Liu Z, Pan Y, Wu Q, Weng Y, Zhang X, Yang Z, Tsui OKC (2014) Enhanced water flux in vertically aligned carbon nanotube arrays and polyethersulfone composite membranes. J. Mater. Chem. A 2:12171–12176. https://doi.org/10.1039/C4TA02119C
86. Srivastava A, Srivastava ON, Talapatra S, Vajtai R, Ajayan PM (2004) Carbon nanotube filters. Nat. Mater 3:610–614. https://doi.org/10.1038/nmat1192
87. Dumée LF, Sears K, Schütz J, Finn N, Huynh C, Hawkins S, Duke M, Gray S (2010) Characterization and evaluation of carbon nanotube bucky-paper membranes for direct contact membrane distillation. J. Memb. Sci 351:36–43. https://doi.org/10.1016/j.memsci.2010.01.025
88. Viswanathan BG, Kane DB, Lipowicz PJ (2004) High efficiency fine particulate filtration using carbon nanotube coatings. Adv. Mater 16:2045–2049. https://doi.org/10.1002/adma.200400463
89. Daraei P, Siavash S, Ghaemi N, Ali M, Astinchap B, Moradian R (2013) Enhancing antifouling capability of PES membrane via mixing with various types of polymer modified multi-walled carbon nanotube. J. Memb. Sci 444:184–191. https://doi.org/10.1016/j.memsci.2013.05.020
90. Zhang J, Xu Z, Shan M, Zhou B, Li Y, Li B (2013) Synergetic effects of oxidized carbon nanotubes and graphene oxide on fouling control and anti-fouling mechanism of polyvinylidene fluoride ultrafiltration membranes. J. Memb. Sci 448:81–92. https://doi.org/10.1016/j.carbon.2010.10.014

91. Bunch JS, Verbridge SS, Alden JS, van der Zande AM, Parpia JM, Craighead HG, McEuen PL (2008) Impermeable atomic membranes from graphene sheets. Nano Lett 8:2458–2462. https://doi.org/10.1021/nl801457b

92. O'Hern SC, Stewart CA, Boutilier MS, Idrobo JC, Bhaviripudi S, Das SK, Kong J, Laoui T, Atieh M, Kornik R (2012) Selective molecular transport through intrinsic defects in a single layer of CVD graphene. ACS Nano 6:10130–10138. https://doi.org/10.1021/nn303869m

93. Wang EN, Karnik R (2012) Water desalination: Graphene cleans up water. Nat. Nanotechnol 7:552–554. https://doi.org/10.1038/nnano.2012.153

94. Shahin H, Mady E (2017) Graphene membranes for water desalination. NPG Asia Materials 9:e427. https://doi.org/https://doi.org/10.1038/am.2017.135

95. An D, Yang L, Wang TJ, Liu B (2016) Separation performance of graphene oxide membrane in aqueous solution. Ind. Eng. Chem. Res 55:4803–4810. https://doi.org/10.1021/acs.iecr.6b00620

96. Perreault F, Fonseca de Faria A, Elimelech M (2015) Environmental applications of graphene-based nanomaterials. Chem. Soc. Rev 44:5861–5896. https://doi.org/10.1039/C5CS00021A

Large Area Graphene and Their Use as Flexible Touchscreens

Surender P. Gaur, Sk Riyajuddin, Sushil Kumar, and Kaushik Ghosh

Abstract An enormous demand for advance touchscreen gadgets has drawn significant scientific attention for last few years due to the substantial developments in the field of flexible and portable electronics. Thin coating of tin doped indium oxide (ITO) material offers good electrical conductivity and high optical transparency and thus, is widely adopted as the transparent conductive material for touchscreens and optoelectronic display devices. However, limited availability of indium (In), and brittleness of ITO thin coatings limits their use in next-generation flexible displays. Graphene is an emerging material in this aspect due to good electrical conductivity, high optical transparency and mechanical stretchability makes graphene a better choice for flexible electronics and display devices. Graphene possesses sufficient robustness to be used in the harsh environment. Though, defect free-high quality, volume production and limited fabrication compatibility are the major challenges in commercialization of graphene-based flexible devices. In this chapter we have briefly reviewed the basic understanding of touchscreen technology and the importance of graphene in flexible touchscreens as well as the remaining challenges for commercialization of graphene-based touchscreens.

Keywords Graphene · Flexible electronics · Transparent electrodes · Touchscreens · Chemical vapor deposition

Surender P. Gaur and Sk Riyajuddin—Both authors have an equal contribution.

S. P. Gaur · Sk Riyajuddin · S. Kumar · K. Ghosh (✉)
Institute of Nano Science and Technology, Knowledge city, Sector 81, Mohali, Manauli P.O-140306, Punjab, India
e-mail: kaushik@inst.ac.in

1 Introduction

Advancement in graphene-based ultrathin, flexible and conductive films has widened the scopes of transparent and flexible device applications such as electronic circuits, flexible displays, touchscreens, optical sensors, charge storage, energy harvesting and the wearable biosensors. Thin oxide coating of tin doped indium oxide (ITO) offers good electrical conductivity and high optical transparency. As a result, ITO coatings are the most popular conducting transparent oxide which is used for touchscreen applications. Apart from this, ITO has a huge demand in various technological aspects like plasma screens, LEDs, OLEDs, electronic networks, solar cells, smart heating windscreens, etc. [1, 2]. ITO thin films are very easy to handle, inert and can be developed with available deposition methods like sputtering, chemical vapor deposition, physical vapor deposition, etc. However, scarcity of indium (In) makes it an expensive choice for the use in coatings technology. Simultaneously the brittleness of ITO coating imposes the major limitation for its use in flexible touchscreen gadgets [3].

Graphene is an emerging material in this context, having high optical transparency, mechanical flexibility and electrical conductivity that makes it a preferred choice for the next-generation flexible touchscreens and optoelectronic devices. Owing to its robustness, graphene offers physical stability in harsh environment. It has Young's modulus of 0.5–1.0 T Pa, tensile strength up to 130 GPa, spring constant 1–5 N/m and physical stretchability of up to 20%. These properties promote graphene as a preferred choice for flexible optoelectronic and wearable biomedical devices. Furthermore, graphene offers larger 2D surface that allows to make large area seamless interfacing with other organic materials and the substrate itself. Thermal and chemical stability, optical transparency, thermo-resistivity and piezoelectric responsivity make this material a multifunctional agent [4–7]. In contrast to commercially available indium tin oxide, graphene has various additive properties like ultrahigh electronic mobility, variable resistivity, and high conductivity required for better performance of the electronic devices.

The present technology markets demand advanced electronic devices with flexible circuits and foldable displays. Graphene has ideal 2D structure that makes it compatible with the modern top down fabrication approaches. Graphene is being explored to use in electronic devices like transistor, logic circuits, energy harvesting devices, touchscreens displays, etc. As graphene has low sheet resistance, tunable electronic mobility, and thermo-resistivity. However, some major and critical concerns are yet to be overcome like presence of multiple pinholes defects, micro-cracks, and overlapped grain boundaries in the graphene films [8]. With the help of this chapter, we are discussing the processing and application of graphene in flexible touchscreens including the development reported so far and the major challenges to be overcome.

1.1 What is Touchscreen?

Touchscreen is a graphical user interface in computing display devices. Touchscreens are input–output devices mostly integrated with optical displays. Touchscreen is sensitive to physical touch of finger or stylus by means of either of the pressure, electrostatic charge and optical sensing to prove an input command to the computer [9]. The touchscreen enables users to communicate with a computer directly through display rather than the conventional methods of a mouse, keypad or a joystick. This has been widely adopted by the electronic gadgets like mobile phones, laptops, digital game consoles, voting machines, ticket vending machines, ATMs, etc. The idea of touchscreen was first suggested by Johnson of the Royal Radar Establishment in 1965 [10]. The first transparent touchscreen is being developed by Frank Beck and Bent Stumpe at CERN in 1970 [11] and later a prototype device has been successfully demonstrated in 1973.

1.2 Types of Touchscreen

1.2.1 Resistive Touchscreen

The first resistive touchscreen has been developed by George Samuel Hurst in 1975 but not much attention has been paid till 1982 [12]. The resistive touchscreen, which is one of the simplest types of touch sensor technology, comprises a few stacks of transparent layers; mainly the substrate layer, bottom side resistive layer, intermediate patterned insulating layer, top side resistive layer and the top protective layer, respectively. The important part of this assembly is the isolated arrangement of top and bottom resistive layers. These two layers facing each other are isolated by an intermediate patterned insulating layer to avoid any unwanted contact and maintain an air gap at possible contact regions. The bottom side resistive layer has two parallel conductive tracks alongside and two parallel conductive layers at top and bottom sides of top resistive layer. A voltage is applied at either side of two layers and measured from the other end to estimate the sensitivity of the touch [13]. When external pressure is applied on the outer surface by stylus or finger the two isolated layers make a contact at that point and this arrangement behaves as voltage divider circuit. The x-y coordinates can be identified by measuring the voltage between electrodes of each layer by rapid switching. The resistive touchscreens are used in multiuser harsh environments like factories, restaurants, hospitals, ticket wending machines, ATMs, etc. due to its high tolerance of liquid and dust contaminants. Resistive touchscreens are relatively low-cost technology and can be operated with gloved fingers and non-conducting rigid objects like pointer, pen, stylus, etc. The main limitations of resistive touchscreens are the facile degradation of the top surface with sharp objects, limited resolution and manual pressing operation [14].

1.2.2 Capacitive Touchscreens

Working of the capacitive touchscreens is based on the change in the capacitance at the location of touch over the touchscreen when contacted with an electrically conductive object such as human finger. The capacitive touchscreens are fabricated on a transparent substrate like glass coated with the transparent conductive film like ITO and thereafter passivating dielectric film such as PET. Touching this arrangement by a conductive object creates distortion in the screen's parasitic capacitance or the electrostatic field that can be measured as the change in the effective capacitance at the sensitive area. Different approaches are used to detect the location of touch. The electrostatic turbulence created by the touch is processed by the controller circuit. Some capacitive touchscreens are unable to locate the touch of an insulating object like gloved finger or stylus. This is due to the fact that touching the screen with finger adds conductive area that add some capacitance (C_F, finger capacitance) to the parasitic capacitance of (C_P) of the touchscreen. This drawback of capacitive touchscreen limits the widespread use of these touchscreens in consumer applications like tablet PCs, smartphones, digital books and cold weather regions where people wear gloves [13]. However, special capacitive stylus or conductive thread gloves can be used to overcome this issue, but this added an additional accessory to the product. The capacitive touchscreens are in development with reduced film thicknesses that minimize the gap between the finger and the display. In parallel plate capacitive touchscreen network, most of the electrostatic charge accumulates between the dielectric layer and the transparent electrodes. This accumulated charge creates an electrostatic field over the sensitive area by projecting the electrostatic field lines. Any turbulence in these field lines due to the physical touch on the touchscreen causes change in the capacitance of that region. Since the capacitance change is generating due the electrostatic field turbulence, this type of capacitive touchscreens is also sensitive to nonconductive objects. For uniform field coverage, parallel plate capacitor network is identically printed all over the touchscreen. The capacitance touchscreens are of two types (a) surface capacitance touchscreen and (b) projected capacitance touchscreen [14].

1.2.3 Surface Capacitance Touchscreen

This is the basic capacitive touchscreen technology. In this type of touchscreens one side of the panel is coated with conductive transparent thin film like ITO and the other side is protected with a transparent polymeric layer such as PET. The conductive coating is biased with small DC voltage at each corner. When the panel is touched by a finger, a dynamic capacitor is formed at that region due to the charge accumulation. The current driven from each electrode of the panel will be as per the distance of the touch from electrode. This technique allows the controller to sense the location of touch by calculating the ratio of current driven from all electrodes. The surface capacitance touchscreens offer good durability due to its simple architecture, however, suffer with some serious infirmities like low resolution, false capacitive coupling signal and excessive sensitivity to temperature that result in a change in the capacitance due to temperature fluctuations. This sensitivity to temperature creates lots of background noise; therefore, a large area capacitive coupling is prime requirement. Surface capacitance touchscreens require frequent calibration and their applications are limited to industrial controls, vendor machines, electronic machine control panels [13, 14].

1.2.4 Projected Capacitance Touchscreens

The projected capacitance technology provides a significant improvement in sensitivity, resolution, accuracy and response time over surface capacitance touch technology. Projected touchscreens are made of rows and columns of transparent conductive electrodes on glass substrates. Conductive transparent thin film like ITO is used to deposit and pattern using microelectronic processing to generate grid line electrodes, see Fig. 8a. Voltage applied on these electrodes produces electrostatic field lines. When a conductive object such as finger comes in contact with these grid lines, leads to turbulence in electrostatic field and, therefore, changes the capacitance of that region. Projected capacitance touch sensing allows multiple touch and scrolling of the touch panel as the capacitance is simultaneously measured at every point of the touchscreen. Unlike surface capacitance, projected capacitance touchscreen facilitates passive touch sensing like stylus and gloved finger. However, excessive humidity, moisture droplet on the surface and dust particles can reduce the performance of touchscreen. These issues, however, can be overcome using fine wire track pattern on the touch panel as the fine wire tracks touchscreens have very low parasitic capacitance and larger distance between two electrodes [12, 13]. Projected capacitance touchscreens are of the following two types; mutual capacitance and self-conductive layers close to each other. An inherent capacitance is generated at the intersection of two grid lines, i.e., column and row when biased electrically. When bringing a finger or stylus near to the point of mutual capacitance or the intersection changes the electrostatic field that reduces mutual capacitance. Capacitance change to these points is measured by voltage of neighboring axis. The self-capacitance touchscreen is also based on column and rows-based architecture; however, each grid line, i.e., row or column is operated individualistically. The capacitive load or the touch is measured as current or frequency change. The touch can be detected anywhere on the row, if the touch has also been sensed by a column at the same time, it is assumed that the touch is at the intersection of the row and column. The self-capacitance allows us for faster and precise detection of the touch but created obscurity for more than one touch. This issue is a de-sensitizing signal is applied to all columns except one that leaves some limited area sensitive to the touch. This sequential scanning of short regions enables precise detection on the touch panel and the multiple touches.

1.2.5 Surface Acoustic Wave (SAW) Touchscreen

Surface acoustic wave touchscreens use an array of saw generators and detectors at opposite sides, X-Y plane on the top of touchscreen. Touching the panel with finger will absorb some portion of the saw wave resulting in the loss in the response of the respective detector. The SAW touchscreen can detect multiple touches and the scrawling as well; however, contamination of the panel can interfere with the function of the touchscreen [9].

1.2.6 Infrared Touchscreen

The infrared touchscreen uses infrared LED and detector array on X-Y plane of the top layer of touchscreen. The IR-LED and detectors create horizontal and vertical pattern on the touchscreen. The touch of an object on the screen results as an obstacle to the IR optical beam and reduces the response of detector. The IR detector does not need conductive material to be sense and enable all types of passive sensing. Moreover, these do not require any patterning or the electrically conductive layer to realize the touchscreen. The touchscreens are sensitive to dust and any kind of IR opaque material that obstacle the infrared beam and also the parallax in curved surfaces that generally create accidentally when user hovers a finger on the screen while searching an option [9].

2 Synthesis and Transfer Methods of Large Area Graphene Films

Since its discovery, various efforts have been made for growth and synthesis of graphene such as liquid phase chemical exfoliation, mechanical exfoliation, epitaxial growth on SiC substrate, and chemical vapor deposition. From various experimental studies reported in this field, we concluded that CVD is the promising, reliable and feasible method for commercial scale production of large area graphene. The liquid phase chemical exfoliation and mechanical exfoliation methods have been widely adopted for growing graphene films or flakes in laboratories for research purpose. Synthesizing the graphene flakes using chemical exfoliation method consists of three following steps: (a) oxidation of graphite, (b) exfoliation of graphite layers and (c) purification. The metal salts are inexpensive exfoliation agent and produce high-quality graphene flakes, however, produces serious defects in graphene. Production of graphene by the mechanical exfoliation is limited due to small flake size and reproducibility.

Epitaxial growth of graphene on SiC substrate is a relatively costly process as this requires special synthesis facility and the semiconductor grade substrates. Harvesting the graphene from its growth substrate is also a tedious process. Since the size of epitaxial grown graphene is limited by the size of the substrate itself and, therefore, the large-scale volume production is restricted. Thus, developing techniques for economical large volume high-quality graphene production is highly required. Bae et al. [14] have demonstrated the role-to-role (R2R) method production of large area graphene film on 30 cm wide Cu foil role using CVD process. The graphene film developed was transferred on polymeric substrate using heat and press process and has been demonstrated as the resistive touchscreen on commercial gadgets. The chemical vapor deposition of graphene allows mass production of graphene to satisfy consumer needs. Moreover, the growth of graphene is carried out in an inert environment at an elevated temperature of nearly 1000 °C so the films are of high crystallinity and low defects.

The number of layers, however, can be controlled with the flow of hydrocarbon precursor to the metallic catalyst, i.e., Cu or Ni. The process is based on the thermal decomposition of hydrocarbon gases at elevated temperature. Several metal substrates such as Ni, Fe, Cu has been used as a catalyst to get the film deposited. Complete synthesis process of graphene using CVD procedure is involved in the following subsequent steps; (a) diffusion of carbon atoms into metal film, (b) precipitation of carbon atoms during cooling, (c) formation of multiple layers on metal surface. For the process to be carried out a chamber with optimized growth temperature and hydrocarbon precursor and preplaced thin metal such as Cu or Ni is required. A hydrocarbon molecule is made to decompose due to high temperature and thereafter the carbon atom is diffused into the substrate. The temperature is reducing gradually such that the graphene layer is developed [15].

Solubility of carbon atoms on Cu is lower than the Ni that leads to the lower rate of decomposition and thus diffusion of carbon atoms on Cu. This helps in monolayer graphene growth on the Cu substrate. The Ni substrate is preferably used for poly-layer-graphene synthesis. However, the growth of layers depends on the reaction time and the cooling process. For multilayer graphene synthesis faster controlled cooling process is required. Due to the self-limiting effect of Cu, the entire process including adsorption, decomposition and diffusion occurs on surface only. Consequently, CVD technique in which Cu used as metal catalyst is the preferred method for monolayer growth of graphene [14, 15].

The domain size and crystallinity of graphene are investigated on variable synthesis parameters for optimal quality graphene with low defects and cracks [19]. The major limitation of the process so far is the high-temperature growth and, therefore, do not suit the polymeric substrates. The low-temperature synthesis of the large area graphene over polymeric substrate has been demonstrated by plasma enhanced chemical vapor deposition technique to realize the flexible-transparent electrodes. We have demonstrated the detailed optimization of the synthesis process of few-layer and multilayer graphene film that meets the touchscreen standards in this work. Effect of various growth parameters like temperature, rate of gas flow, growth time, annealing time, etc. formation of the layer of graphene has been verified with detailed characterization of developed films.

Surface mediated reaction and precipitation are the two mechanisms of graphene growth by a thermal CVD process in which hydrocarbon gas molecules acting as a precursor were introduced in the thermally reactive chamber and decompose to active carbon species which defuse on the active catalyst surface [14, 15]. The desired layer of graphene grows on the surface by out-diffusion while cooling the substrate. Since carbon atoms possess lesser solubility on Cu surface than Ni, the entire growth phenomenon, i.e., adsorption, decomposition and diffusion take place at the surface only (self-limiting effect) that preferentially leads to a monolayer or few-layer of formation of graphene on Cu. We have synthesized few-layers of graphene film on Cu substrate using acetylene as the hydrocarbon precursor. The schematic synthesis process is illustrated in Fig. 1a.

Graphene growth was carried out on a 25 ± 1 μm thick, 99.99% pure Cu foil of dimensions 6 cm × 12 cm. For the removal of the native oxide layer from the Cu foil, it was initially dipped in 0.1 M Hydrochloric (HCL) solution for 10–15 s and then rinsed in deionized (DI) water thrice. The Cu foil substrate was then dried using N_2 gas flow. The quartz tube used for the synthesis process was subsequently cleaned by acetone, isopropanol and deionized water to remove inorganic/organic impurities from its inner wall.

Fig. 1 a Schematic representation of the thermal CVD process used for the synthesis of large area graphene, **b** Cu foils containing few-layer and multilayer graphene films grown with different flow time of C_2H_2 gas

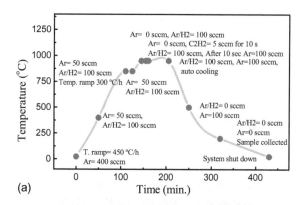

(a)

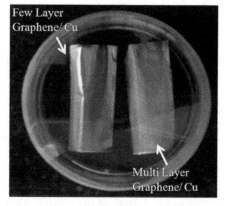

(b)

Continuous flow of Ar gas (200 sccm) was introduced to exhaust the atmospheric impurities from the quartz tube and to create an inert environment before the growth. Furthermore, if any oxide content remains on the Cu foil then it was removed by a continuous flow of H_2 gas through mixed Ar/H_2 (4:1) gases. Presence of H_2 gas in the synthesis helps in surface activation via removal of native oxides from the larger grain of catalyst surface required for high crystalline growth of graphene. The Cu foil was kept in an inclined position at an angle of 30 °C in the quartz tube to get uniform distribution leading to identical diffusion of carbon atoms on its exposed surface. Initially, quartz tube was heated with a ramp rate of 450 °C/h till it reaches 850 °C. After getting this temperature an intermediate halt of 15 min was invoked in the furnace with the help of a temperature controller to make the rising temperature to be stable.

The thermal CVD furnace was further heated to 950 °C by lowering the heating rate to 300 °C/hr for better and uniform growth. After the attainment of the desired temperature, the excess supply of Ar to the quartz tube was reduced from 200 to 80 sccm. However, the supply of H_2 gas to the chamber was flowing continuously to avoid oxidization of Cu foil at high temperature. Depending on the thickness of graphene, i.e., number of layers of graphene, acetylene (C_2H_2) gas which is used as a precursor was precisely controlled to flow 5sccm. In order to have uniform dispersion and dissociation of C_2H_2 molecules on the surface of Cu foil, Ar gas with a flow rate of 200 sccm was again flowed immediately after switching off the acetylene gas. Figure 1b represents two Cu foils containing the different number of graphitic layers being synthesized by the varying time of flow of C_2H_2 also known as the effective growth time.

3 Large Area Transparent Graphene Transfer

Qualitative transfer of large area graphene on flexible, transparent and insulating substrate is required to avail its inherent electrical and optical properties. However, transferring the large area graphene from substrate of growth to secondary substrates is a difficult process that can induce cracks, crumpling and opacity in the film. Since after its initial physical discovery, numerous transfer processes of graphene like PMMA coating, PDMS Curing and role-to-role transfer have been demonstrated. The widely used method is the polymethyl methacrylate (PMMA) thin layer casting; a uniform thin layer of PMMA is spin coated on the graphene containing metal (Cu) surface after which the carrier metal foil is used to etched away to get the free-floating PMMA/graphene stack in Cu etchant solution like ammonium persulfate. This free-floating PMMA/graphene stack is rinsed in highly resistive deionized water to remove the ammonium persulfate content which was used to etch the Cu foil. The PMMA/graphene stack possesses low density and high surface area to volume ratio by which it can float on the surface of water.

Acetone was drop casted in the water containing the floating PMMA/graphene solution to dissolve the PMMA thin layer. The quality transfer of graphene in this process demands complete removal of the supporting PMMA film from the transferred graphene. However, complete removal of the polymer film from the graphene layer is a tedious process and many times the PMMA traces are still left even after long treatment with acetone [15, 16]. The only disadvantage of using the wet transfer method by PMMA is it reintroduces the structural discontinuities. As the scooping of floating graphene from DI water to the secondary substrate introduces cracks in the graphene and the multiple folding on the edges of the transferred graphene. Schematics of the graphene scooping transfer process has been shown in Fig. 2a. The other methods of transferring the large area graphene film from primary substrate to secondary polymeric substrate are the direct substrate development method in which the polymer substrate such as PDMS is directly molded on the graphene/Cu stack. The detailed schematic process of PDMS molding method is shown in Fig. 2b. The graphene/Cu sample is first mounted on a flat surface in such a way that liquid PDMS can't penetrate or leak inside the substrate. For this purpose, the edges of the copper foil can be attached to the surface with a tape. Thereafter, the liquid PDMS and curing agent is drop cast on graphene coated copper sample to get molded. After molding and curing the PDMS on graphene samples, the samples are used to etch out the Cu foil leaving behind the graphene/PDMS stack. Detailed process schematic of transferring the graphene on PDMS substrate is shown in Fig. 2b [7].

However, an efficient method is still requiring for the qualitative transfer of graphene film. The above-discussed methods have some major limitations, such as; formation of cracks, lattice defects and flipping of graphene on edges area in free-floating graphene scooping process. Similarly, transferring the graphene on PDMS substrate using PDMS molding and curing method degrades the electronic properties of graphene. To overcome these issues another method called role-to-role transfer

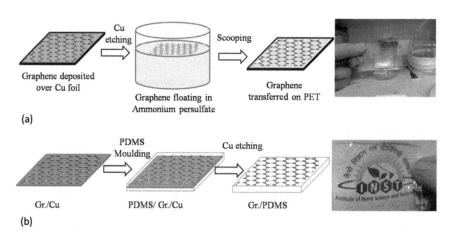

Fig. 2 **a** Schematic of few-layer graphene film transfer on a secondary substrate by traditional free-floating film scooping process, and **b** Schematic representation of transfer process of the large area few-layers graphene film from the metal catalyst (Cu foil) to PDMS substrate by PDMS molding technique

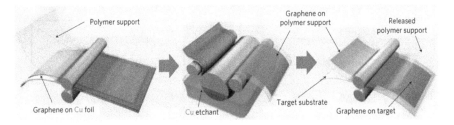

Fig. 3 Schematics of role-to-role transfer process of graphene film grown on Cu foil to polymeric substrate by heating and pressing. *Image source* Bae et al. [14]

of graphene from Cu foil to PET substrate was demonstrated by Bae et al. [14]. This method is suitable for commercial scale graphene production and transfer. In their work Bae et al. have demonstrated the growth and transferring of monolayer graphene of size 30-inch diagonal by subsequent etching of Cu foil in liquid etchant [14]. A polymer support, namely, the thermal release tape (TRT) was first attached to the Cu foil containing grown graphene by thermal compression method, after attaching the TRT to the Gr/Cu stack the Cu foil was wet etched to get the graphene film attached on the TRT. The graphene film from TRT was again transferred to the polymer substrate (PET) by hot rolling compression. Schematic process of R2R process is shown in Fig. 3. Growth and transfer process of graphene promise good possibilities of integration of various electronic applications.

4 Characterization of Large Area Graphene Used in Flexible Touchscreens

It has been known for over 40 years that CVD of hydrocarbons on reactive nickel/copper or transition-metal-carbide surfaces can produce thin graphitic layers [16]. However, the large amount of carbon source which is absorbed on copper foils usually forms thick graphite crystals rather than graphene films. In order to overcome this issue, a controlled amount of hydrocarbon precursor is allowed to pass through the reactive chamber that contains metal catalysts (Cu) substrate which was heated up to 950 °C inside the quartz tube consisting of an argon atmosphere. After flowing reaction gas mixtures, i.e., C_2H_4 and H_2/Ar (20:80), 10 and 100 standard cubic centimeters per minute, respectively. The system was rapidly cooled to room temperature, i.e., 25 °C at the rate of 10 °C/s by flowing H_2/Ar. We believe that this fast cooling rate is critical in suppressing the formation of multiple layers and for separating graphene layers efficiently from the substrate in the later process [7, 16].

For the macroscopic transport electrode applications, the optical and electrical properties of graphene films were, respectively, measured by ultraviolet-visible spectrophotometer and two-probe current-voltage ($I-V$) measurement methods. Figure 4a and b presents the optical transparency and electrical resistance of the films. The

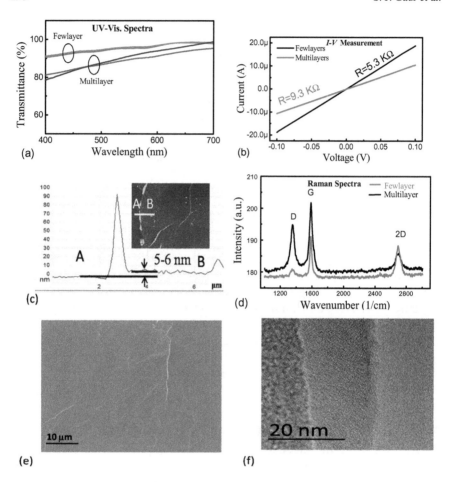

Fig. 4 **a** UV–Vis. Transmittance spectra of few-layer and multilayer graphene film. **b** Current–voltage graph of few-layers and multi-layers graphene film obtained by 2-probe measurement method. **c** AFM micrograph of multilayer graphene film on Si substrate. **d** Raman shift obtained for the multilayer and the few-layer graphene film on Si substrate. **e** SEM of few-layer graphene film on Silicon substrate and **f** TEM of multilayer graphene film flake synthesized by thermal CVD method

optical transmittance of the developed film was measured on Shimadzu UV-3600 UV-Vis. spectrometer using two different floating graphene films of a different number of layers, i.e., few-layer and multilayer, respectively, on a glass plate. In the visible range of spectra, the transmittance of the film grown on a 25 μm thick copper foil for 30 s was found to be 84% over wavelength 550 nm, a value similar to those found for previously studied assembled films [7]. As the transmittance of an individual graphene layer is 2.3%, the transmittance value indicates that the average number of graphene layers is 8–10. The transmittance can be increased up to 93% by further reducing the growth time and flows of acetylene, resulting in a thinner graphene film, i.e., few-layers of graphene. The few-layers graphene synthesized in this work show optical transparency of 92% at 550 nm which corresponds to the average number of layers as 3–5.

The Cr/Au electrodes were deposited on the Gr/Si samples to minimize the contact resistance in current–voltage measurement. The minimum resistance obtained for the few-layers graphene samples is found to be 5.3 kΩ between the electrical probes placed 500 μm apart. However, the resistance value for multilayer graphene samples is found to be 9.3 kΩ at similar spacing. The value of sheet resistance, however, gets an increase with increasing the number of graphene layers. Figure 4c shows the atomic force microscopy (AFM) image of the transferred graphene layer on Si substrates. The effective step height of nearly 6 nm has been measured by the study, this height is consistent with that of 8–10 layers of graphene. Based on the data collected on various isolated graphene flakes using AFM, it is estimated that $\sim$80% of the graphene flakes comprise 8–10 layers of graphene. Effective control of the layer thickness of graphene flakes can be achieved by varying the growth conditions, such as growth temperature and methane concentration [7, 8].

Raman analysis of 2D/G and D/G band intensity ratios for the graphene layers grown at different temperatures and acetylene concentrations is shown in Fig. 4d. Notably, the intensity ratio of D/G peaks decreases with either the increase of growth temperature from 850 to 950 °C at constant acetylene flow of 10 sccm or the increase of acetylene concentration at a constant growth temperature of 950 °C, while the ratio of 2D/G peaks does not change much. This suggests that moderate increases in temperature and methane concentration are reliable for decreasing the defect density and improving the crystallinity of graphene. However, with an enhanced growth temperature above 950 °C, the D/G ratio improves along with the lowering of the 2D/G ratio. The improved D/G ratio is well understandable. Higher growth temperature can accelerate the pyrolysis of carbon species and increase the nucleation density of graphene, leading to reduced domain size and increased sp3 defects. Besides this, weakening and broadening of 2D peaks should indicate the variation of layer thicknesses of a graphene flakes from few-layers to multi-layers [7, 8].

This has been further confirmed by the corresponding AFM analysis of multilayer graphene/Si sample. A combined scan of Si and multilayer graphene is shown in Fig. 4c, depicting a clear step height of 5–6 nm between two regions that confirms the total graphitic layers of nearly 8–10. On increasing growth temperature from 850 to 950 °C, and the additional annealing time has improved the crystallinity of the synthesized graphene film significantly as is evident from the detailed Raman spectroscopy analysis, Fig. 4d. In brief, it seems that suitable growth temperature and acetylene flow rate should play an important role in the growth of high-quality graphene. A scanning electron microscope image of graphene films on thick silicon substrate shows a clear contrast between areas with different numbers of graphene layers as shown in Fig. 4e. The transmission electron microscope (TEM) image of multilayer graphene is shown in Fig. 4f. The TEM analysis indicates that the film mostly consists of 8–10 layers of graphene.

5   Fabrication Techniques of Graphene-Based Flexible Touchscreens

Broadly used touchscreens in electronic gadgets are the analog resistive and the capacitive types touchscreens. All types of touchscreens, however, predominantly require transparent conductive film in their construction. The resistance of conductive transparent films plays a major role in the resistive touchscreens, while in capacitive touchscreens the sheet resistance of the transparent film should be nominal. Therefore, an additional conductive material can be doped in graphene to improve its electrical conductivity for this purpose [14, 17, 18].

Bae et al. [14] have demonstrated the role-to-role synthesis and transfer of ultra large graphene film that meets the commercial requirements. The film was multiple transferred and chemically doped to improve the electrical and optical properties of the graphene. The process demonstrated has scalability and process abilities to transfer the ultra large graphene production and their transformation in flexible touchscreen panels and are seen as the possible replacement of ITO coating. Complete process demonstration of the work has been given in Fig. 5.

Synthesis of graphene using thermal CVD is a lengthy process due to low heating and cooling requirements of synthesis process and the large size process setup which limits the production of material. Ryu et al. [18] have presented faster method growing graphene using rapid heating of the Cu foils with halogen-based heating lamps. This process is named as rapid thermal-chemical vapor deposition (RT-CVD). This process supports the H_2 free growth, wet etching of Cu and role-to-role transfer that enable faster mass production of large area graphene. Schematic of the synthesis process setup and the process graph has been shown in Fig. 6. The RT-CVD graphene growth system allows the synthesis of high-quality graphene at lower temperature, hydrogen free faster growth process and high uniformity. The setup consists of halogen lamp as heating source, graphitic susceptors between lamp and the Cu foil samples to convert NIR radiation into heat, CH_4 and N_2 gas inlets, and Cu foil for the graphene growth. The relative graphical illustration of graphene growth using thermal CVD and rapid thermal CVD process indicates that the process takes 40 min in graphene growth in RT-CVD than the 280 min for thermal CVD that makes a significant improvement in the synthesis time.

The high-quality graphene developed by RT-CVD is demonstrated in practical application of multi-touch capacitive touchscreen. Developed graphene film was transferred to transparent PET substrate using thermal release tape (TRT) and the wet etching of Cu foil in H_2O_2 and H_2SO_4 solution. The graphene attached on TRT was released on PET using lamination process. The transferred graphene was thereafter patterned using O_2 plasma method for developing grid line electrode of capacitive touchscreens. Ag metallic electrodes on the panel were deposited using printing thereafter for integrating the panel to controller circuit. Photograph of completely assembled capacitive touchscreen is given in Fig. 7f.

The applicability of graphene as a transparent and flexible electrode has also been demonstrated by Kang et al. [19] by developing graphene-based flexible touch

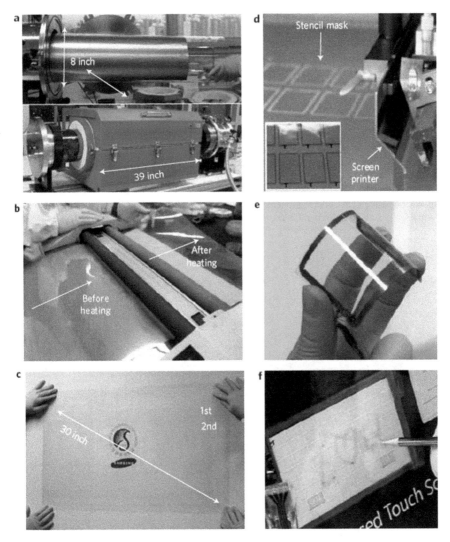

Fig. 5 Photographs of fabrication process of graphene-based resistive touchscreen, **a** copper foil wrapping to be used for the graphene growth, **b** thermal releasing of graphene from thermal release tape to PET, **c** large area transparent graphene transferred to PET substrate, **d** printing the silver paste electrode on the patterned graphene film, **e** flexible graphene/PET touchscreen panel, and **f** fabricated graphene-based resistive touchscreen integrated on a computer. *Image source* Bae et al. [14]

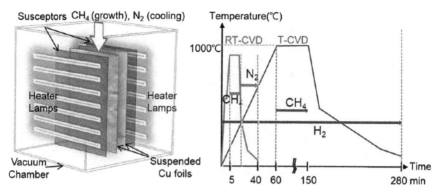

Fig. 6 Schematic of RT-CVD graphene synthesis setup and comparative graphical illustration of thermal CVD and rapid thermal CVD processes of graphene growth. *Image source* Ryu et al. [18]

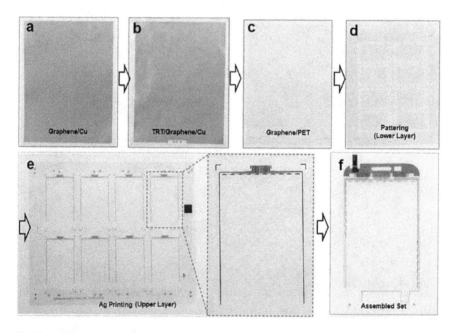

Fig. 7 **a** Photograph of RT-CVD grown graphene on Cu foil, **b** TRT attached on graphene/Cu sample, **c** graphene transferred on PET after TRT attachment and etching out the Cu foil, **d** patterning of graphene using selective O_2 plasma exposure, **e** printing of the Ag electrodes on the graphene/PET panel, and **f** completely assembled touchscreen. *Image source* Ryu et al. [18]

sensor array for wearable applications. The array works in both contact as well non-contact mode that has been enabled using graphene and thin geometry architecture. The device provides high deformation sensitivity and, therefore, can be mounted on different parts of the human body like forearm and legs of different deformation levels. The touch sensor allows multiple touch sensing and recognition of distance

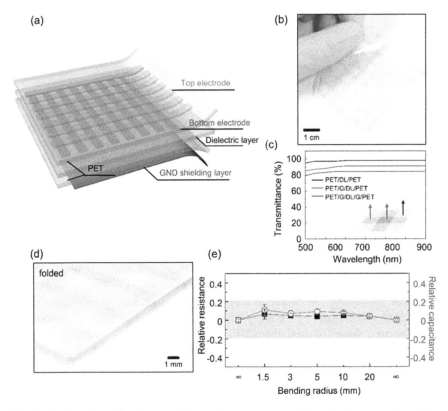

Fig. 8 Graphene-based flexible capacitive touch sensor array. **a** Schematic diagram of capacitive sensor array showing the arrangement of top graphene electrode array, bottom graphene electrode array, transparent dielectric layer and supporting PET layers. **b** Photograph of fabricated graphene-based flexible 3D capacitive touch sensor array. **c** Optical transmittance characteristics of the sensor array. **d** Photograph of foldable, ultrathin, capacitive sensor. **e** Resistive and capacitive response of the 3D sensor array under different bending radius. *Image source* Kang et al. [19]

and shape of the approaching object like finger before the touch. The technology offers advance machine–human interface and other utilities for multiple sensing applications. The device was initially analyzed using FEA analysis on ABAQUS. The unit representative volume element was modeled with periodic boundary conditions and nonlinear deformation under tensile stress in X-Y plane.

For the implementation of the touch sensor array, the large area graphene was synthesized using thermal CVD method on Cu foil. The graphene was transferred to PET substrate after spin casting and curing the PET on Cu foil. The Cu foil was subsequently etched away in ammonium persulfate solution. Transferred graphene was patterned using photolithography and O_2 plasma etching. A significant improvement of 30% in electrical conductivity of PET transferred graphene was achieved after chemical doping with trifluoromethane sulfonamide. Capacitive response of this sensor array was measured on a CV analyzer under dynamic touching action

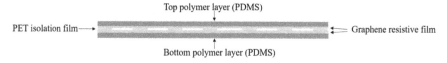

PET isolation film ⟶   ⟵ Graphene resistive film
Bottom polymer layer (PDMS)

Fig. 9 Schematic layout of a resistive touchscreen

with finger and a touch pen. The study evolved the complete resettling of the capacitance after removal of the touch. All measurements were executed in controlled ambient. The schematic diagram and real-time image of graphene-based capacitive touch sensor array with response characteristics have been shown in Fig. 8.

Thermal CVD grown large area graphene films have gained high significance in recent years in resistive touchscreen applications. Resistive touchscreen is simply an arrangement of two resistive and transparent films separated by tiny insulating patterns, when one side of the arrangement is pressed by a finger, stylus or any passive component by which it gets in touch with other side resistive film and makes a closed-loop for current conduction. Rate of current flow is always in proportion with the distance of the touchpoint from the current source [17].

We are discussing our work on fabrication of large area graphene-based 16 elements resistive touch sensor array. Graphene film used to fabricate the sensor array had been deposited using thermal CVD synthesis process by acetylene (C_2H_2) gas as hydrocarbon precursor. The graphene film had been patterned using fine tip laser beam over PDMS substrate. As discussed previously, the resistive touchscreen is an arrangement of two resistive films facing each other with intermediate patterned isolation layer. The schematic of a graphene (Gr) film-based resistive touchscreen is shown in Fig. 9.

We patterned the top and bottom graphene using laser engraving, to make each element operates individually without electronic control. The layout used to pattern the bottom electrode, Fig. 10a, top electrode, Fig. 10c and the intermediate PET isolation layer, Fig. 10c. Electrode 1 is used as the reference electrode to provide electrical current to the sensor array. Electrode 2 is the set of sixteen individual electrodes that makes closed-loop for switching arrangement for current flow upon the physical touch by finger or stylus. The final arrangement of both electrodes and the intermediate PET isolation layer is shown in Fig. 10d. The current flow within each loop can be measured by connecting an ammeter in series with top electrode and the ground. Patterning of graphene film was carried out using laser engraving of the large area graphene film on flexible PDMS substrate. Electrical connections were drown using silver pasting, Cu wires of thickness 100 μm were connected on each pad.

The characterization setup of graphene-based touch sensor array uses a dual source power supply of voltage up to 30 V and current up to 5 A. Source and measurement unit, Keithley 2450 SMU was used to detect the output current from the sensor under touch condition. The sensor array was biased through electrode 1 at 15 V, whereas the negative terminal of the power supply was kept at ground. The discrete electrodes of all devices from the top electrode panel were kept at ground potential through an

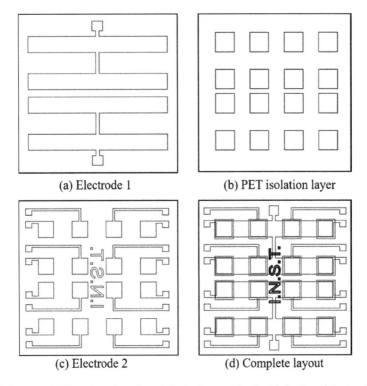

(a) Electrode 1   (b) PET isolation layer

(c) Electrode 2   (d) Complete layout

Fig. 10 Layout of different graphene-based electrodes used for the fabrication of the touch sensor array

ammeter in series connection to the power supply. The resistive sensor fabricated in this work offers maximum current of 7.8 μA at biasing voltage of 15 V. The graphene film used in this work and the current response characteristic of the touch sensor are shown in Fig. 11. Current responsivity of the sensor, however, need to

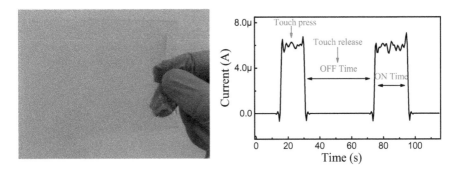

Fig. 11 a Graphene transferred on PDMS substrate, **b** ON–OFF characteristics of the fabricated graphene touch sensor array

be improved by reducing the polymer residue effects and the electronic structure-controlled processing [20, 21]. Scaling down the dimension of the sensor array can also significantly increase the current conductivity as the larger area array causes larger electrical resistance.

6 Conclusion

Graphene can be seen as the possible replacement material of ITO in commercial displays and devices due to its ideal electrical, optical, mechanical properties and economical production approaches. Graphene is, therefore, emerging for low cost, flexible-transparent applications. The demand of an alternative material for ITO coating is reported due to the high cost of indium and its adverse effect on environment and human health. Moreover, the brittleness of ITO coatings restricts their use in next-generation flexible devices. Graphene possesses sufficient robustness to be used in harsh environment with excellent stretchability, but defect free-high quality, large volume production and limited fabrication compatibility are some major challenges in commercialization of large area graphene applications. In this chapter we have discussed briefly the use of graphene in flexible touchscreens. For this purpose, brief overview of touchscreen technology and its types is included. The large area graphene synthesis and transfer and characterization procedure reported by different research groups have been discussed. The fabrication techniques of graphene-based resistive and capacitive touchscreens are also presented. In the overall study of the chapter it can be concluded that most of the work reported on the graphene-based touchscreens is limited to the laboratory label only, and yet several challenges are there to be get over to bring the technology to the market level. Major of these challenges are the production of highly conductive graphene films in large volumes, production of defect free-high uniformity graphene films, long term stability and durability of the graphene films and device process feasibility of these films. However, the issues and challenges are still existing, but some technically feasible approaches are also emerging. Resolving these issues will bring the advent of graphene-based touchscreens to the advanced level.

Acknowledgments The authors are thankful to the research grant of SR/NM/NS-91/2016, Department of Science and Technology (DST), Government of India.

References

1. Geim AK, Novoselov KS (2007) The rise of graphene. Nat Mat 6(3):183–191
2. Geim AK (2009) Graphene: status and prospects. Sci 324(5934):1530–1534
3. Castro AH, Neto GF, Novoselov KS et al (2009) The electronic properties of graphene. Rev Mod Phys 81(1):109–162

4. Stylianakis MM, Konios D, Petridis K et al (2017) Solution-processed graphene-based transparent conductive electrodes as ideal ITO alternatives for organic solar cells. In: George ZK, Mitropoulos AC (eds) graphene materials—advanced applications. Int OP. https://doi.org/10.5772/67919
5. Akinwande D, Brennan CJ, Bunch S et al (2017) A review on mechanics and mechanical properties of 2D materials—graphene and beyond. Ext Mech Lett 13:42–77
6. Lee C, Wei X, Kysar JW et al (2008) Measurement of the elastic properties and intrinsic strength of monolayer graphene. Sci 321(5887):385–388
7. Riyajuddin SK, Kumar S, Gaur SP et al (2020) Linear piezoresistive strain sensor based on graphene/g-C_3N_4/PDMS heterostructure. Nanotech 31(29). https://doi.org/10.1088/1361-6528/ab7b88
8. Riyajuddin SK, Kumar S, Gaur SP et al (2019) Study of Field Emission property of pure graphene-CNT heterostructure connected via seamless interface. Nanotech 30(38). https://doi.org/10.1088/1361-6528/ab1774
9. Walker G (2012) A review of technologies for sensing contact location on the surface of a display. J Soc Info Disp 20(8):413–440
10. Johnson EA (1965) Touch display—a novel input/output device for computers. Elect Lett 1(8):219–220
11. Lowe JF (1974) Computer creates custom control panel. Design News 54–55, November 18, 1974
12. Hurst GS (2010) 'Tom Edison' of ORNL, December 14 2010. Archived from the original on April 10, 2020. https://www.oakridger.com
13. Lee D et al (2012) Capacitive vs. Resistive touchscreens. R-Tools Tech Inc. Retrieved 9 September 2012. https://www.makeuseof.com/tag/differences-capacitive-resistive-touchscreens-si/
14. Bae S, Kim H, Lee Y et al (2010) Roll-to-roll production of 30-inch graphene films for transparent electrodes. Nat Nanotech 5(8):574–578
15. Li X, Cai W, An J et al (2009) Large-area synthesis of high-quality and uniform graphene films on copper foils. Sci 324(5932):1312–1320
16. Suk JW, Kitt A, Magnuson CW et al (2011) Transfer of CVD grown monolayer graphene onto arbitrary substrates. ACS Nano 5(9):6916–6925
17. Tian H, Shu Y, Wang XF et al (2015) A graphene-based resistive pressure sensor with record-high sensitivity in a wide pressure range. Sci Rep 5(1):8603–8610
18. Ryu J, Kim Y, Won D et al (2014) Fast synthesis of high-performance graphene films by hydrogen-free rapid thermal chemical vapor deposition. ACS Nano 8(1):950–956
19. Kang M, Kim J, Jang B et al (2017) Graphene-based three-dimensional capacitive touch sensor for wearable electronics. ACS Nano 11(8):7950
20. Ohta T, Bostwick A, Seyller T et al (2006) Controlling the electronic structure of bilayer graphene. Sci 313(5789):951
21. Suk JW, Lee WH, Lee J et al (2013) Enhancement of the electrical properties of graphene grown by chemical vapor deposition via controlling the effects of polymer residue. Nano Lett 13(4):1462

Carbon Nanotube Alignment Techniques and Their Sensing Applications

Pankaj B. Agarwal⬭, **Sk. Masiul Islam**⬭, **Ravi Agarwal**⬭, **Nitin Kumar, and Avshish Kumar**⬭

Abstract Recent progress on the synthesis and scalable manufacturing of carbon nanotubes (CNTs) remain critical to exploit various commercial applications. Here we review breakthroughs in the alignment of CNTs, and highlight related major ongoing research domain along with their challenges. Some promising applications capitalizing the synthesis techniques along with the characteristics of CNTs are also explained in context to the recent developments of CNT alignment. The prime objective of this chapter is to provide an up-to-date scientific framework of this niche emerging research area as well as on the growth of CNTs either by in-situ or ex-situ synthesis techniques followed by its alignment during growth or post-growth processing. This chapter deals with various mechanism of CNTs alignment, its process parameters, and the critical challenges associated with the individual technique. Numerous novel applications utilizing the characteristics of aligned CNTs are also discussed.

P. B. Agarwal (✉)
Nano Bio Sensors, Smart Sensors Area, Council of Scientific and Industrial Research-Central Electronics Engineering Research Institute (CSIR-CEERI), Pilani 333031, Rajasthan, India

P. B. Agarwal · Sk. M. Islam
Academy of Scientific and Innovative Research (AcSIR), CSIR-CEERI Campus, Pilani 333031, Rajasthan, India

Sk. M. Islam
Optoelectronics and MOEMS, Smart Sensors Area, Council of Scientific and Industrial Research-Central Electronics Engineering Research Institute (CSIR-CEERI), Pilani 333031, Rajasthan, India

R. Agarwal
Centre for Converging Technologies, University of Rajasthan, Jaipur 302004, India

N. Kumar
Department of Physics, National Institute of Technology Mizoram, Aizawl 796012, India

A. Kumar
Amity Institute for Advanced Research and Studies (Materials and Devices), Amity University, Noida 201313, India

© The Author(s), under exclusive license to Springer Nature Singapore Pte Ltd. 2021   307
A. Hazra and R. Goswami (eds.), *Carbon Nanomaterial Electronics: Devices and Applications*, Advances in Sustainability Science and Technology,
https://doi.org/10.1007/978-981-16-1052-3_13

1 Introduction

Currently, carbon nanotubes (CNTs) and its quasi-one-dimensional (1D) nanostructures are found to be the most technologically significant material system due to its superior mechanical, optical, and electrical behaviour, which enable its potential to fabricate various electronic devices such as wearable sensors, biochemical sensors, water purification systems, energy storage, and harvesting devices [1–3]. Unique intrinsic electrical, thermal, mechanical, and optical properties of CNTs are attributed to its very high aspect ratio (length-to-diameter) along with the presence of strong covalent bond followed by the electronic configuration [4–6]. However, reports are available in the literature where electron mobility of aligned CNTs matrix is observed to be 43 times larger than the random configuration [7]. Moreover, well-aligned CNTs composite film exhibits 6 times higher strain than its random counterpart [8]. Furthermore, Thotenson et al. [9] reported the mechanical property of highly ordered CNTs. The storage modulus of highly ordered CNTs is determined to be 40% as compared to its randomly distributed network. It may be mentioned here that proper alignment and uniform distribution of CNTs are responsible for reproducibility [10]. Also, random network CNTs-based devices usually consist of broad size (length and diameter) distribution of CNTs and different inter-tube junctions. These factors lead to the change of CNT electrical resistance, metal/CNT contact resistance, and inter-tube junction resistance, which are most likely the reasons for inadequate sensing responses. It is worth mentioning that selective alignment of CNTs on silicon substrates may help in solving such issues in the present sensor and device fabrication technologies. Selectivity and sensitivity of CNTs-based sensors can be increased with the precise positioning of CNTs. There are many techniques available to align such nanostructures, which include dielectrophoresis, dip-pen nanolithography (DPN), fluidic manipulation, self-assembly, nanorobotic manipulation, microcontact printing, etc. [11–15]. Controlled position management and alignment of CNTs have always been the bottleneck for their applications [16]. The techniques to align CNTs can be broadly categorized as (1) in-situ, where alignment is done at the time of CNT growth, and (2) ex-situ, where CNTs are originally synthesized followed by their alignment during the particular application such as device fabrication and gas/biosensors. Chemical vapour deposition (CVD) technique is found to be used most frequently during in-situ alignment method, in which the growth of the nanotubes can be controlled over specifically positioned catalyst nanoparticles on device structures [17].

Through the above discussion, it is evident that the in-depth study about the realization of individual technique, exploring the underlying science and technology of each process is still needed. This motivates further research to explore the fundamental mechanisms for various recent CNTs alignment techniques and CNTs-based devices. Herein, we represent a comprehensive review focusing on state-of-the-art progress made on CNTs alignment mechanisms via physical and chemical routes, discussing the underlying principle, and envisaging the potential of the individual method for various CNTs networks.

2   In-Situ Alignment Techniques

In-situ alignment method defines directly growing the CNTs on a substrate. This method does not require any sonication or oxidative purification. The invention of CNTs has drawn considerable attention in many research activities as physical (electrical, thermal, electronic, and magnetic) and mechanical features. These key features make CNTs ideal, not only for extensive areas of applications but also for the fundamental scientific and engineering studies [18]. The exact magnitudes of these properties are observed at the nanoscale range that mainly depends on the chirality and diameter of nanotubes along with its single or multi-walled structures. The CNTs having strong anisotropy properties are essential to replicate the alignment of its specific orientation for their further scientific applications. The anisotropic trend of specific CNTs is better conceived and preserved when all CNTs are associated in the equal direction as an array.

Few sets of techniques are also available to align CNTs array in the programmed orientation. These techniques depend on several mechanisms which are also valid for other conditions. These techniques are mainly classified into two sets, depending on when the alignment is done. In the in-situ method, the alignment is attained through the growth of CNTs, whereas in the case of ex-situ methods, the CNTs are developed in random alignment and orientations are completed through the device combination subsequently. These classifications are not perfect, however, the changes between systems are of great significance for extensive manufacturing. On the other hand, the in-situ methods directly bear the advantages of simplicity and straightforwardness. Moreover, the ex-situ methods are free from the restrictions of growth on substrate material and temperature. Furthermore, simplicity and efficient processing always favours a manufacturer. The in-situ system allows direct alignment of the CNTs.

Therefore, the majority of in-situ alignment systems/techniques discovered until now are summarized below. In contrast, in-situ methods of CNTs synthesis such as arc-discharge, laser ablation, flame, and hydrothermal are discussed [19, 20]. Herein, efforts have been made on in-situ CNT orientation/alignment through chemical vapour deposition (CVD) and plasma-enhanced chemical vapour deposition (PECVD) techniques using catalytic nanoparticles in different orientations.

2.1   Chemical Vapour Deposition (CVD)

In this method, a thin film of the catalytic metal is deposited using sputtering or evaporation methods followed by annealing to form nanoparticles of size 10–100 nm. The decomposition of a carbon-based gaseous precursor is catalyzed by deposited nanoparticles, which act as the nucleation sites for the growth initiation of CNTs. Mostly used catalysts nanoparticles include transition metal oxides (Fe, Co, and Ni) and typically the carbon precursors are alcohols, carbon monoxide (CO), and methane (CH_4). CVD method is considered suitable for scaling up to commercial level synthesis [21]. Locations of catalyst nanoparticles would also define the aligned growth of CNTs. The schematic of a typical in-situ CVD-based alignment method is shown in Fig. 1 [22].

The growth process of CNTs can be controlled by suitable carbon precursors and varying the reaction conditions, viz., catalysts, substrate temperature, heating ramp, precursor flow rate, or an external field. Further, the synthesis process of catalyst nanoparticles decides their size, which is responsible for the aligned growth of single and multiple CNTs. CVD is a chemical technique to yield the high performance and high-purity of solid specimen materials. This technique is often utilized in semiconductor thin films manufacturing. Large-scale of aligned CNTs were produced through the thermal CVD using iron nanoparticles as a catalyst in mesoporous silica [23]. During CVD growth of CNTs, the substrate is polished with a nano-layer of particles (metal acts as a catalyst), as in the form of commonly iron, nickel, cobalt, gold, stainless steel, platinum, or a suitable combination of each other. The CNTs have been grown-up vertically on the substrates utilizing thermal CVD when the density of the catalyst is much higher. The catalytic CVD growth procedure employs crowding influence. CNTs are grown-up from the catalyst nanoparticles deposited on the substrate. The aligned CNTs are nearly perpendicular to the silica surface and generate an aligned array of isolated tubes through an inter-tube spacing (nearly 100 nm). The nanotubes growth rate might be controlled through the orientation of pores. Aligned nanotube packets were also grown-up (over thin films) via CVD technique using a catalyst (cobalt) on a silica substrate [24]. During nanotube synthesis, CVD method paves the best potential for industrial growth due to lower expenses. The advantages of CVD techniques are simple arrangement setup, easy operation, and further scale-up at the lower cost. It is also capable to grow nanotubes directly on a preferred substrate. Tons of the CNTs are commercially produced using CVD technique. Thus, the CVD technique is found to be used extensively and significant development has been made in the past decades.

On the other hand, the CVD takes advantage of lower preparation/synthesis temperatures than other techniques (laser ablation and arc-discharge). It still needs

Fig. 1 Schematic representation of the conventional procedure to prepare catalyst nanoparticles or nanoislands followed by CVD growth (Copyright (2007) [22])

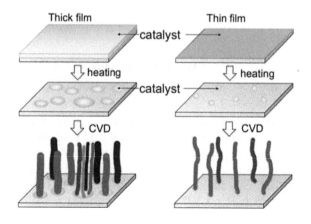

a growth temperature in the range of 600–950 °C. To grow the CNTs at lower temperatures, a plasma-based CVD technique is employed. The CNTs array could be grown on the glass specimen using plasma at 660 °C [25]. According to literature, the industrial-scale/large-scale of aligned CNTs array have already been developed through CVD and PECVD techniques [26–30].

2.1.1 Vertical Alignment

The CNTs could grow vertically on the substrates using the CVD method when the catalyst density is higher enough. For the process of catalytic CVD growth, the crowding effect is employed. As a result, the CNTs are grown by the catalyst nanoparticles and deposited on the substrate. The nanoparticles are made of the solid-state dewetting method [20, 31, 32]. The proper polycrystalline thin film was deposited through the physical vapour deposition, thermal evaporation, and magnetron sputtering techniques. Once the thin film of a catalytic specimen is excited in a vacuum, voids appear at the grain boundary triple junctions and grow larger through the surface energy minimization. During CNT growth through the CVD process, catalytic nanoparticle sizes are generally in the range of 0.5–10 nm. It involves precise thin films with nominal particle spacing.

Initially, the CNTs grow from these closely spread catalyst elements. In general, it is considered that the Van der Waals force among densely CNTs causes to produce vertically to the substrate. The assured growth conditions facilitate a reduction in the growth or density of CNTs [33]. The crowding effect also influences the alignment process for closely packed CNT arrays.

Several reports are available on the enhanced growth of CNTs from the catalyst. Higher growth in the proper oxidative agents (oxygen, water) into the gas ambient to catalyst particle surfaces is attained through the balancing between carbon growth and the sp^2 graphitic construction on the catalyst elements. The oxidants could not eliminate from the carbon growth; however, it may also scratch the graphite layers when it is used at high concentrations. It also eliminates unnecessary water to mitigate negative effects (NE) on the CNTs growth in the incubation phase [20, 33, 34]. Controlling of oxidants stage in the growth situation mainly subjects to the configuration of deposition apparatus. When an appropriate alignment is performed, it is expected to grow long-aligned arrays of CNTs in the millimeter range.

The CNTs can develop into vertical groups along the pre-designed coordinate and shape. Moreover, these aligned growths are independent of the substrate morphology (macroscopic). Figure 2 display a scanning electron microscopy (SEM)) and atomic force microscopy (AFM) images of single-walled CNT (SWNT) arrays with several sets of characters. Alignment of these SWNT arrays is encompassing the crowding influence the growth.

Moreover, the thin-film dewetting method utilizes a higher density of catalyst particles. Besides, there are also different ways to deposit as-synthesized catalyst nanoparticles at higher density on growth substrates using solution-based, sublimation, deposition methods, etc. [35–37]. Thus, the CNTs are grown at higher areal

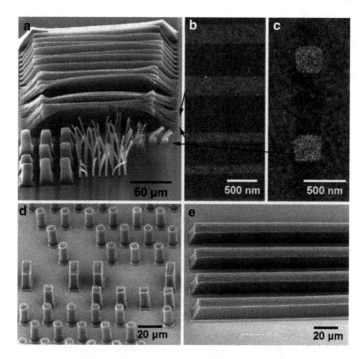

Fig. 2 Molecular O_2-assisted growth of vertical SWNT towers and sheets, **a** SEM image of vertical SWNT sheets (upper part, top to bottom thickness 20, 5, 1 μm, 500, 300, and 100 nm) and SWNT towers (lower part, left to right width 20, 5, 1 μm, 500 and 300 nm) after 30 min of growth, **b** AFM image of the patterned catalyst strips (bright 300 and 100 nm-wide regions) composed of densely packed Fe nanoparticles used for the growth of the 300 and 100 nm-thick vertical SWNT sheets (pointed to by arrows) in **a**, **c** AFM image of two 300 nm width patterned catalyst squares used for the growth of the smallest towers (tilted due to high aspect ratio and pointed by an arrow) in **a**, **d** SEM image of square and circular towers of vertical SWNTs and **e** SEM image of lines of vertical SWNTs; [33], Copyright (2005) National Academy of Sciences, USA

density having close spacing. Crowding influence plays a pivotal role to vertically align all separate CNTs.

2.1.2 Horizontal Alignment

Dai's group reported the growth of SWNTs using CH_4 as a precursor in the CVD method [35]. Rogers et al. used CVD for large-scale synthesis of horizontally aligned CNTs using the patterned iron catalyst and CH_4 [38]. In-situ growth of horizontally aligned CNTs could be achieved through external guidance/forces such as electrical field, magnetic field, and gravity. These forces utilize the polarization property of SWNTs. Dai et al. demonstrated the use of an electric field for the growth of aligned SWNTs [39].

The CNTs can grow horizontally on the substrates through thermal CVD along with epitaxy approach. Epitaxy method is a very appropriate technique to align CNTs horizontally. There are several epitaxial CNTs growth techniques [40]. (i) the lattice-directed growth epitaxy through the atomic rows, working on atomically flat planes; (ii) the ledge-directed growth epitaxy through the atomic steps, working on vicinal planes; and (iii) the graphoepitaxy through nanofacets, working on the nanostructured planes.

Lattice-Directed Growth

As-prepared CNTs are generally shorter in size. Long SWNTs are horizontally aligned in parallel planes (a-plane and R-plane) sapphire. For lattice-directed progress/growth, the sapphire without miscut is generally used as a substrate. The initial surface layer of the O_2 particles of sapphire substrate might be depleted in the CVD method and surface is terminated by aluminium atoms. The CNTs grow along with the surface atoms [20, 41–43]. The better growth of SWNTs along lattice directions is ascribed to higher density of charges of atomic rows caused by Van der Waals and electrostatic forces.

Ledge-Directed Growth

The development of highly dense and aligned arrays of long SWNTs was initially achieved on the c-plane sapphires using CVD technique [44]. The c-plane sapphire wafers produce a discontinuity in the c-plane to yield an atomic step of vicinal α-Al_2O_3 (0001) surfaces (Fig. 3). Ledge-directed growth causes CNTs alignment along the edge direction. The uncompensated surface elements at the step edges make the electrostatic interactions and it increases Van der Waals forces at the step edge, causing the CNTs to align [44].

Graphoepitaxy

Graphoepitaxy usually results in incommensurate orientation/alignment of the crystals [45]. The ledge-directed growth, wafers, c-plane sapphire is miscut and automatically polished to yield an atomic step near particular directions and height. In this growth process, annealing of the sapphire substrates is undertaken at a higher temperature. Further annealing facilitates thermodynamically unstable atomic movement by reducing the surface energy followed by bunching of atomic movement into faceted nanosteps spaced through the flat c-plane terraces (Fig. 4a). Figure 4b–d displays the AFM images of graphoepitaxial growth of the CNTs at nanosteps surface of annealed miscut c-plane sapphire [45].

Atomic steps direction along $[10\bar{1}0]$ result in the growth of graphoepitaxial nanotube along $[10\bar{1}0]$ direction (Fig. 4b). The nanosteps height is three times greater than the unit cell (1.3–3.8 nm).

Fig. 3 a AFM image of the SWNTs (scale bar: 1 μm), **b** Asymmetric double-exposure back-reflection XRD indicating relevant low-index directions and the resulting step vector(s). The green and red triangles indicate the reference reflections arising from the first exposure (2 h) and second exposure (1 h, after 1808 sample rotation), respectively, and (c) AFM micrograph of the α-Al₂O₃ substrate surface (annealed at 1100 °C, scale bar: 100 nm). The darker blue region shows low terraces that edges correspond to the micro-steps. Inset indicates a segment study along the red line [44].

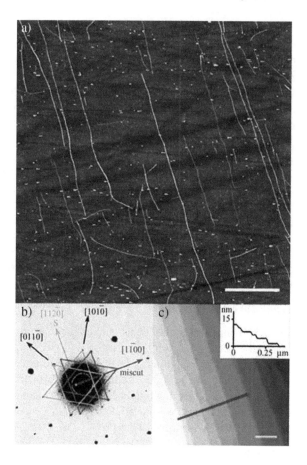

2.2 *Plasma-Enhanced Chemical Vapour Deposition (PECVD)*

PECVD technique utilizes plasma to improve the chemical response rates of a precursor and deposit thin film in the form of vapour (solid form) on the substrate. The PECVD processing permits the deposition at lower temperatures compared to the CVD technique, the growth temperature of CNTs is slightly reduced through the plasma during the PECVD process [46, 47].

The chemical reactions involve the deposition process after the formation of plasmas. The plasma is usually made through direct current or radio-frequency (RF) release between both electrodes, where space is filled by reacting gases. The physical characteristics of the developed CNTs are simply affected by the PECVD than that of CVD deposition method. Additional process factors/parameters could vary in the PECVD process. Also, the hydrogen content, etching rate, step coverage, etching ability along with stoichiometry could be tuned throughout PECVD growth. There

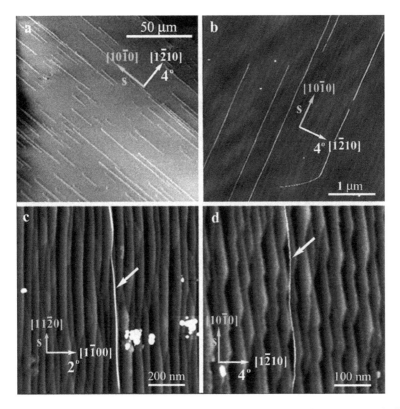

Fig. 4 Graphoepitaxial growth at several annealed miscut c-plane on sapphire substrates, **a** the SEM micrograph of straight nanosteps along the [10$\bar{1}$0], **b** the AFM micrograph of the CNTs grown on nanosteps along [10$\bar{1}$0], **c** the nanosteps along the [11$\bar{2}$0] and **d** much faceted sawtooth-shaped nanosteps along the [10$\bar{1}$0]. Arrows symbol in **c** and **d** represent grown CNTs. Reprinted (adapted) with permission from [45]. Copyright (2005) American Chemical Society

are several ways to synthesize the CNTs for PECVD [47, 48]. PECVD is classified as RF-PECVD, DC-PECVD, microwave-PECVD, etc. PECVD scheme employs many combinations of the plasma heating source. During DC-PECVD process, the gas precursors are decomposed using DC plasma (without heating source) and formed radicals are simply deposited on the substrate surface. Here, the growth temperature of the substrate is higher compared to room temperature. A substrate is heated by plasma depending on the energy and intensity of plasma. Through literature, it is evident that during PECVD processes, the substrate is generally heated by external heating. In general, the infra-red (IR) lamp/electric furnace is used as an external heating source. In the case of back-heated PECVD process, the substrate is heated through the resistance heater under the substrate-holding stage to manipulate the growth temperature precisely [49, 50]. For the hot-filament PECVD process, the hot filament is hanging above the substrate acting as a heating source [20, 25, 51–53].

2.2.1 Vertical Alignment

The CNT arrays having in-situ alignment are mostly grown by PECVD technique. Initially, CNT arrays were aligned in-situ along vertical direction on the glass substrates by DC-PECVD technique [25, 51, 54, 55]. Several other kinds of PECVD techniques, like hot-filament PECVD, RF-PECVD, electron cyclotron PECVD, microwave-PECVD, etc., are the mature techniques to grow the aligned CNTs. In this section, we will briefly discuss DC-PECVD as well as the deposition conditions and experimental setup. Additional PECVD techniques are also quite similar to the DC-PECVD technique. In these PECVD procedures, the DC electric fields, and radio-frequency/microwaves fields yield plasma to primarily lower the growth temperature of CNTs.

The schematic experimental scheme of DC-PECVD instrument is depicted in Fig. 5 [16, 56]. Inside vacuum chamber (denoted as 1), a rod-form of molybdenum anode (denoted as 2) suspends vertically around 1.5–2 cm above horizontal molybdenum cathode plate (denoted as 3) that also assists as a model stage. The diameters of the plate and rod are 4 and 0.25 mm, respectively. Below the cathode plate, a resistive plate heater (denoted as 4) is powered through the external AC voltage source (denoted as 5). Both electrodes are associated with the outer power supply (denoted as 6), the MDX-1 K Magnetron Drive, that runs on continuous power mode or current and voltage. Feedstock gases are passed to the chamber through a single gas pipe (denoted as 7) containing flow controllers (denoted as 8, 9). Chamber evacuation is done through the mechanical pump (denoted as 10).

The evacuation speed and pressure are physically organized through a gate valve (denoted as 11) that throttles the evacuation channel. Additionally, a pressure sensor (shown in Fig. 5) is connected in the chamber along with a thermocouple in the quartz tube. A cathode is attached for temperature monitoring.

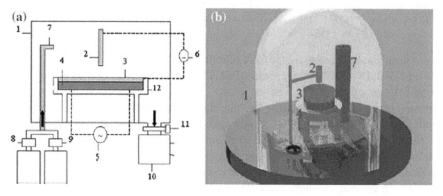

Fig. 5 **a** The DC-PECVD setup, and **b** three-dimensional representation of the PECVD growth chamber to align CNTs [16] (reprinted by permission of the publisher Taylor &Francis Ltd, https://www.tandfonline.com)

The DC-PECVD method for the vertically aligned CNTs array contains four simple stages, such as evacuation, heating, plasma formation, and cooling. In this procedure, the specimen samples are loaded in the sample chamber. The chamber stage is evacuated near the base pressure of 10^{-2} Torr. Then the NH_3 is subjected to the chamber to reach desired pressure (about 8 Torr) and then the heater is turned to slowly raise stage temperature. The characteristic growth temperatures are 450–600 °C and usually extended in 15–30 min.

The pressure is optimized through the movable gate valve. DC voltage (about 400–700 V) is used in the gap between electrodes to ignite electrical discharge (plasma) over the specimen. The plasma current density may vary in the range of 0.1–0.5 A. The C_2H_2 is then subjected to grow the CNTs. The flow rate of NH_3 and C_2H_2 are taken as 4:1 that makes the lowest quantity of amorphous carbon formation. As discussed, apparatus and the process having significant straightforwardness are adopted by several researchers around the globe. It also assists to push forward the research and development of aligned CNTs in a significant manner.

2.2.2 Horizontal Alignment

PECVD process successfully synthesizes the horizontally aligned CNTs. This technique is complicated and accurate horizontal alignment with closely packed densities is required to achieve precise functionalities. The CNTs array could be aligned through post-synthesis alignment, in which an external force orientates the nanotubes in liquid medium [57–60]. Minor manipulation has been described by the processing that involved an ultimate order and needs time-consuming electron microscopy monitoring system [61]. However, despite of in-plane alignment of CNTs, they have minor or nonlinear directions.

The catalyst activity was confirmed through the samples grown by thermal CVD, horizontal PECVD, and in-situ alignment techniques [20, 62]. The grown CNTs having a length of 3 μm along with the diameter of ~75 nm (by the PECVD) and ~77 nm (by the thermal CVD). To examine the interaction between as-prepared CNTs (by thermal CVD) and the local electric field. DC electric field (~6 V/μm) was applied during post-growth. In a few cases, the CNTs intensely interacted with the electric field and became progressively dispersed up to 20 μm, by the minor alignment. Variations in the alignment during electrode production, the alignment could be attributed to the electric field and gives linear growth. On the other side, the low-linear densities reveal a major limitation of the local electric field alignment, and it is the essential condition for microelectrodes patterned on the substrate. However, new techniques to align CNTs horizontally through the PECVD reactor were briefly explained.

Figure 6 shows an optical image of the apparatus during characteristic deposition. The specimen samples were electrically isolated from the heating stage (graphite spacers) to minimize parasitic charging of degenerately doped Si. The device contains double-polished stainless steel electrodes of dimension 1.5 mm × 70 mm × 70 mm. The gas/cathode inlet is electrically grounded, and gas/anode outlet is associated

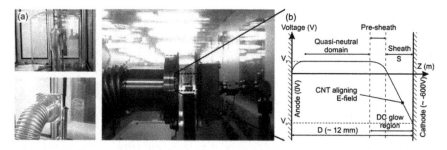

Fig. 6 **a** The horizontal aligned PECVD process; **b** Typical potential conditions (700 V/50 W, 700 °C, 200:50 sccm, 3 mbar) and variation in a DC glow gas discharge plasma. Setup collected and reproduced from [59]

with a vacuum system attached with a computer-controlled variable high-voltage setup (DC supply; 0–1000 V). The separation between the electrode could be done by adjusting the spacers (ceramic). Heating of the specimen sample is done through the graphite (0.13 mm × 10 mm × 100 mm) stage and electrically isolated by cover-sheet.

It is assumed that the CNT alignment is accomplished through anisotropic torque induced in a narrow plasma sheath [63]. For collision-less, quasi-neutral, there is the Child's relation between the ion sheath of glow potential (Vp) and electron temperature (Te).

The relation between sheath width (S), cathode potential (V_0), and charge carrier density (n) can be expressed as Eq. (1) [64, 65]

$$S = \frac{\sqrt{2}}{3} \lambda_D \left(\frac{2e(V_0 + V_p)}{k\,Te} \right)^{3/4} \tag{1}$$

Debye length (λ_D) is expressed as Eq. (2)

$$\lambda_D = \sqrt{\frac{\varepsilon_0 k Te}{n_e e^2}} \tag{2}$$

Herein, ε_0 represents the permittivity of free space, k the Boltzmann's constant, and the term e represents an electronic charge. Debye length was observed to be ~22 μm, which is less than the dimensions of the apparatus (~1 mm).

3 Ex-Situ Alignment Techniques

The complex process for growing quality CNTs at a lower temperature compatible with working devices is expensive and challenging and remains a subject for intense research. Ex-situ techniques are preferred over in-situ ones due to the dependency on the substrate material, catalytic nanoparticles sizes, growth temperature, etc. in case of the latter alignment methods. In the ex-situ alignment method, different electric fields, magnetic fields, and force fields are used to assemble the CNTs at predefined locations.

3.1 Dielectrophoresis

The most commonly used ex-situ technique is dielectrophoresis [66]. In this method, CNTs are polarized in a solvent by the application of a suitable field (between two electrodes) and due to the dielectric variation between CNTs and the solvent, the polarization field placed the CNTs in the direction of electric field leading to their alignment in a common direction [12, 67–70] (Fig. 7).

Banerjee and co-workers have optimized the parameters such as electrode geometry, applied bias, frequency, load resistance, and type of the nanotube sample for aligning the SWNTs in device architectures by alternating current (AC) dielectrophoresis. The alignment of the SWNTs was achieved using pre-patterned microelectrodes up to a certain extent. The floating potential was used to guide the nanotubes across devices along predefined and predictable direction [67] (Fig. 8). Stokes et al. reported a facile low-temperature growth approach to assemble local-gated carbon nanotube field-effect transistors (CNTFETs) devices from solvent [68]. This method finds its compatibility with the current process lines of the microfabrication manufacturing technology. In a report, controlled deposition of CNTs from solution onto microfabricated electrodes using dielectrophoresis technique is achieved [12]. Reports are available in the literature where the self-limiting dielectrophoresis technique is employed to capture single or multiple of SWNTs and multi-walled CNTs (MWNTs). This method is found to be superior compared to its traditional

Fig. 7 The principle of dielectrophoresis deposition and alignment of a carbon nanotube (© 2011 Xue W, Li P. Published in [short citation] under CC BY 3.0 license. Available from: https://dx.doi.org/10.5772/16487, [66])

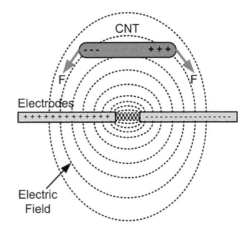

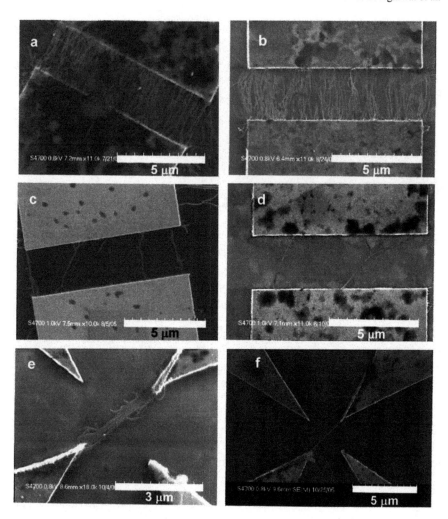

Fig. 8 Aligned CNTs after applying an AC voltage between the opposite electrodes (Reproduced from [67], with the permission of the American Vacuum Society)

counterpart without using any resistor [69]. Development of SWNTs-based FETs using semiconducting materials such as Sb_2Te_3 or Bi_2Te_2Se as the source/drain contact materials is reported by Xiao and Camino [70]. In the literature, efforts had been made to align metallic CNTs having paramagnetism behaviour in the direction of long axes [71].

3.2 Self-assembled Monolayers

In the ex-situ alignment method, the end-functionalized CNTs are arranged in the form of ordered structures on a chemically modified substrate. Liu et al. prepared the gold surface with thiol group followed by attachment of carboxylic (–COOH) functionalized SWNTs [72]. Similarly, patterns of polar group-based self-assembled monolayers (SAM) for the positioning of CNTs and non-polar SAM layers for passivation have been successfully used by Rao et al. [73] and Wang et al. [11], respectively. The direct assembly of CNTs was demonstrated over –COOH self-assembled monolayer followed by patterning via a dip-pen nanowriting (DPN) process over gold substrates using 16-MHA (Fig. 9). Functionalized patterns are initially created on a substrate followed by the alignment of CNTs over them.

In literature, it is reported that adsorption kinetics about rotational and sliding motions of SWNTs play a pivotal role to define the final alignment. Based on diffusion and free energy, the authors proposed a design to describe the motion of SWNTs sliding above SAM [74]. Lee et al. reported a novel method for the fabrication of 100 nm low-noise sensors using aligned SWNT architectures. Superior device performance in terms of significantly high sensitivity along with low signal-to-noise ratio(SNR) is reported. The authors found increasing SNR with decreasing channel

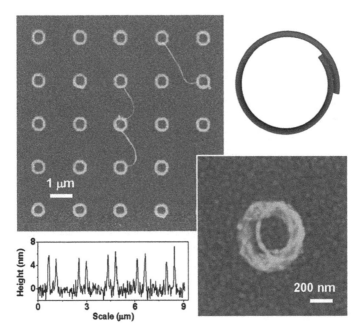

Fig. 9 SWNTs assembled into rings and nano letters. (Left) AFM tapping mode topographic images (upper) and height profiles (lower) of SWNT rings in a 5 X 5 array are shown. (Right) A zoomed-in view of one SWNT ring (lower) and a molecular model of a coiled SWNT (upper) (Copyright (2006) National Academy of Sciences, USA [11])

Fig. 10 A schematic diagram depicting the procedure to align CNTs over polar SAM Reproduced from [74], with the permission of AIP Publishing

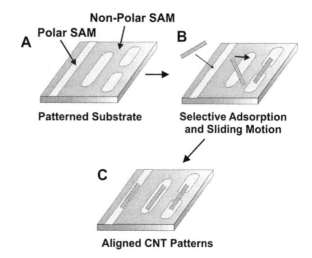

width which differs from the characteristics of sensors fabricated by bulk-film or random structures [75]. The procedure of CNT alignment using molecular patterning is shown in Fig. 10.

3.3 Langmuir–Blodgett (LB) and Langmuir–Schaefer (LS) Assembly

Langmuir–Blodgett (LB) technique has also been used to fabricate the monolayers of aligned SWNTs [76] and SWNTs-based electrochemical sensors [77]. In a report, the LB method is used to grow densely packed monolayers of aligned SWNTs having non-covalent functionalization from organic compounds. The monolayer SWNTs allow microfabrication during device integration and facilitates high currents (~3 mA) SWNT devices having narrow channel width. This method is found to be a generic technique to grow various bulk materials with different diameters [76]. Yu et al. reported a highly sensitive and selective electrochemical technique of protein cancer biomarker detection using SWNT immunosensors [77]. Prostate-specific antigen measurement was carried out in complex biomedical samples, viz. human serum and tissue lysates. The limit of detection was determined to be 4 pg/mL. The authors claimed that SWNT immunosensors have the potential to detect other relevant biomarkers leading to multiplexed detection, point-of-care detection of cancer and other diseases, pathological testing, proteomics, and biological systems.

Functionalization of SWNTs using ODA after purification is reported by Jia et al. [78]. Low density and highly aligned SWNTs were grown by the LB technique. The authors proposed a model to discuss the alignment of the SWNTs. SWNTs having high aspect ratio tend to orient in dipping direction on the application of compression. Good quality multiple layers of SWNT LB films were also grown

[78]. Reports are available in the literature where CNTFETs having highly dense and well-fashioned CNT channel were grown at room temperature using LB method. The current, transconductance, and subthreshold swing of CNTs devices were determined to be 1.8×10^{-6} A, 3.19×10^{-6} S, and 164 mV/dec, respectively. The carrier mobility was found to be 30.81 cm^2/(V-s). The authors compared the performance between CNTFETs and CNTs-based thin-film transistors device and observed a significant enhancement in mobility, drive current, and switching speed [79]. Several reports are also available on the recent progress for the synthesis of aligned SWNTs), particularly in the horizontal direction [80–82].

Multiple Alzheimer's disease core biomarkers detections in human plasma using densely aligned CNTs are reported by Kim et al. [83]. The close-packed aligned CNT-based sensor module showed 2.29 times higher sensitivity compared to random-architecture CNT device array caused by the low density of nanotube junctions and the uniform number of CNTs [83].

Thin-film deposition by Langmuir–Schaefer (LS) method has drawn interest due to its precise control over molecular packing along with the bidimensional arrangement of the nanostructures. In the literature, this technique has been adopted to deposit thin films for applications in organic electronics, sensors, transistors, and biotechnological devices. Sgobba et al. [84] reported the functionalization of aligned CNTs with organic conjugated polymers. The authors transferred the SWNT blend onto different solid substrates using Langmuir–Schaefer deposition technique. In another report, the LB technique is used to achieve lower density and highly oriented SWNTs. Functionalization of SWNTs is done by octadecylamine after purification. A model is proposed to explain its compression-induced alignment [78].

3.4 Microfluidic Techniques

Lee et al. [85] adopted an alignment technique using thermally enhanced microfluidic template and fabricated suspended bridges for nanoelectromechanical switching application. This alignment scheme paves a way to fabricate dense films and alignment method realization, which is scalable and processable to produce CNT nanostructures. The resistivity of aligned CNTs film is determined as 2.2×10^{-3} Ω-cm, high Young's modulus of 635 GPa along with yield strength of 2.4 GPa. In this work, the lithographic process was employed for large production of CNT-based devices as shown in Fig. 11.

Ye et al. reported thermally enhanced fluidic alignment of SWNTs for a very high degree of alignment in large scale [86]. The authors reported a good degree of alignment having the G/D band ratio up to 30.

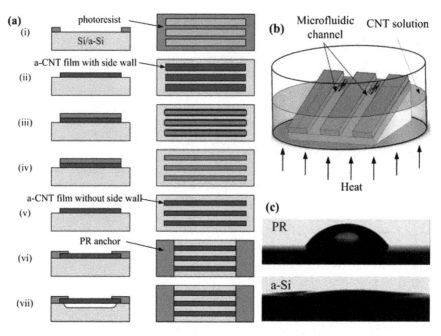

Fig. 11 Microfluidic alignment and device fabrication technique **a** flow chart for device fabrication (i) fabrication of microfluidic channel using photoresist (ii) alignment process of CNT followed by lift-off (iii)–(v) oxygen plasma etching technique to remove sidewall (vi) fabrication of anchor using photoresist (vii) etching of sacrificial layer **b** schematic of aligning technique using microfluidic template and **c** measurement of contact angle before the alignment to ensure hydrophobicity of the photoresist channel (top) and superhydrophilic a-Si surface (bottom) that are suitable for CNT alignment. In **a**, the left column represents a cross-sectional image, whereas the top view is on the right. Evaporation of CNT solution is carried out at 80 °C to align CNTs towards fluidic channels (Copyright with permission from [85])

3.5 Mechanical Techniques

There are few other alignment methods including robotic arm-based alignment, controlled flow of nitrogen-based alignment, spin-based alignment, etc. [14, 87] Nanoassembly can suitably be achieved by nanomanipulation. Several methods of nanoassembly include the fabrication of nanoscale building blocks, its characterizations, proper positioning, alignment of building blocks, nanometer-scale resolution, and connection processes. Nanorobotic manipulations characterized by the degree of freedom, better positioning and alignment controls, multiprobes actuation, and a real-time observation system are found suitable to assemble nanotube-based devices in 3D space [14]. The report is available in the literature for the fabrication of SWNT architecture-based FETs using effective solution possessing method and the on/off ratio is reported as high as 900,000. In this work, tuning of the spin-assisted orientation and SWNTs density are controlled by different surfaces, which changes the interaction between surface functionalities into the device channel. As a result, separation in nanotube chirality, manipulation of density, and alignment occur into self-sorted SWNT platform during device fabrication [87].

3.6 Selective Removal

In the category of ex-situ alignment, there are few reports on the fabrication of aligned CNTs devices using a top-down approach by selectively removing the CNTs from the substrates, except those aligned between the electrodes. It is a kind of facile approach to effectively differentiate the semiconducting SWNT arrays onto the surface by cleaning metallic SWNTs using sodium dodecyl sulphate (SDS) solution. Moreover, this method eliminates the metallic SWNTs from horizontally aligned nanotube arrays using SDS solution. The authors found a very long and good quality 90% semiconducting horizontally aligned SWNT arrays [88]. In a recent report silicon shadow mask-based chemical-free technology is used for sorting of semiconducting SWNTs as well as for alignment between electrodes as shown in Fig. 12 [89, 90]. The cost-effectively prepared aligned semiconducting SWNT-based devices are useful for different biochemical sensors, logics, and electronic circuits. It may be mentioned here that the technique is simple, cost-effective, and easy to scale for mass production of aligned CNT-based devices.

4 Applications of Aligned CNTs

The extraordinary properties of CNTs include high electrical and thermal conductivity, high mechanical strength, and optical polarizability [91]. Due to such exceptional properties, CNTs are potentially suitable for diverse realistic applications in physical, chemical, biological sciences including electronic, magnetic, optical, and optoelectronic applications [92–95]. The characteristics of CNTs are very sensitive and depend on the positioning, alignment, chirality, and length of CNTs. Most importantly, the alignment of CNTs is exclusively significant for various applications and the maximum output of the device would not be possible without alignment. Aligned CNT arrays have certain properties in comparison with a random network of CNTs and have applications in field emission displays (FEDs), nanoelectrode arrays, supercapacitors, FETs, fuel cells, solar cells, and biochemical sensors [18, 96–101]. In this section, detail description of the applications of the well-aligned CNT arrays is done.

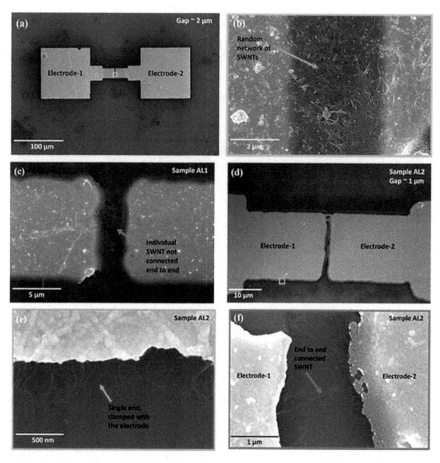

Fig. 12 a FESEM image of the spray-coated random dense network of SWNTs between two electrodes; **b** zoomed-in image of the green circle in **a**; **c** FESEM image of sample AL1 after bath sonication in 4 wt % SDS solution for 15 min, which shows that the SWNTs are not end-to-end connected with the electrodes due to large gap size; **d** FESEM image of AL2 indicating the minimum gap of ~1 μm between the electrodes; **e** the almost complete absence of SWNTs beyond the electrode area as well as SWNTs with their single end connected/clamped with the electrodes; **f** SWNTs end-to-end connected with the two electrodes (Copyright with permission from [90])

4.1 Field Emission Display

CNTs-based FEDs are the next-generation display devices having a high current density at low turn-on field with low power consumption and quick response time [102–105]. The high aspect ratio of CNTs and its atomically sharp tip curvature increases the local field and reduces the turn-on field for effective electron emission. Also, the extensive electrical, thermal conductivity, and long-term chemical and temperature stability are some of the significant factors responsible for electron field emission. The superior electrical and thermal conductivities of CNTs lead to rapid charge transfer between the substrate and CNTs.

The field emission behaviour of the CNTs could be measured in vacuum using a diode configuration where CNTs are placed as a cathode and metallic material is used as an anode as shown in Fig. 13.

The field emission (FE) property of a field emitting material (CNTs) could be explained using Fowler Nordheim (FN) mechanism. FN mechanism was proposed in the year 1928 with a complete explanation of electron field emission on the basis of quantum tunnelling phenomena [106–108]. The characteristic properties of any CNTs-based FEDs depend on various FE parameters which are responsible to influence the behaviour of the device. These parameters include current density (J), effective emitting area (α), field-enhancement factor (β), and material's work function (Φ). The effective emitting area is directly related to the current density as the maximum area provides maximum current density and indicates the size of the FE device [109]. Another promising parameter, i.e., β is an enhancement in the local electric field at the emission sites and can be estimated by the geometrical shape (ratio of length and radius) of electron emitters. Therefore, it shows the microscopic or localized electric field at the tip of the emitters and hence, electric field lines concentrated at a small radius of curvature results in considerable higher localized field potentials. In case of the work function of CNTs field emitters, the maximum emission of the electron occurs at the tip of CNTs and therefore, it is known as the local work function which is responsible for the properties of CNTs field emitters.

The field-enhancement factor can be determined by using the FN plot, where the relation between $ln\left(\frac{J}{V^2}\right)$ versus $\frac{1}{V}$ gives the value of field-enhancement factor (β) as Eq. (3)

$$\beta = \frac{B\varnothing^{\frac{3}{2}}}{m} \qquad (3)$$

Fig. 13 Schematic of field emission setup

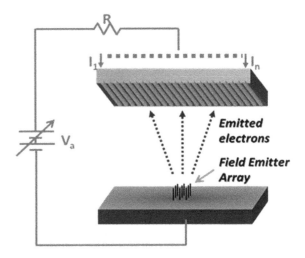

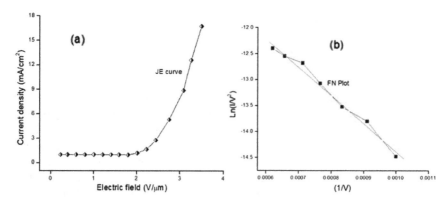

Fig. 14 **a** JE curve and **b** FN plot

Here, B is a constant, ∅ is known as material's work function whereas m (slope) can be calculated from FN plot (as shown in Fig. 14b).

The value of β is extremely important to understand that how the field could be enhanced at the tip of CNTs field emitters and the value may vary approximately between 10 and 10^4.

The field emission properties of CNTs can also be affected due to the proximity of neighbouring CNTs where electric field over one CNT is screened by the neighbouring nanotube. This phenomenon is known as a screening effect [109–111]. Therefore, densely packed CNTs arrays do not reveal the enhanced field emission properties because of the screening effect. Lin et al. worked on vertically aligned CNTs bundles and optimized the square and hexagonal pattern of these bundles to obtain high current density with a low turn-on the field [112].

Due to the screening effect, the whole CNT film does not take part in the field emission and therefore, an effective emitting area takes part in the emission of the electrons [109, 112] which can be derived from Eq. (4)

$$ J = \frac{i}{\alpha} = \frac{A}{\varnothing} E^2 \exp\left(-\frac{B\varnothing^{\frac{3}{2}}}{E}\right) \qquad (4) $$

Here, J is the current density, α is the effective emitting area, A and B are constant, E is called the electric field whereas ∅ is the material's work function. If a graph is plotted between current density (J) and electric field (E) known as JE curve (Fig. 14a), then exponential increments in current density with respect to the applying electric field can be shown. Therefore, the turn-on the electric field and corresponding current density could be observed using JE curve as shown in Fig. 14a whereas Fig. 14b shows the FN plot to calculate the field-enhancement factor (β).

Vertically aligned CNTs have been proved to be an ideal emitter which emits electron at low turn-on voltage and generates a highly saturated current density along with a high field-enhancement factor. Gupta et al. reported that pillar-shaped

Fig. 15 Schematic of
dual-layer deposition of
CNTs

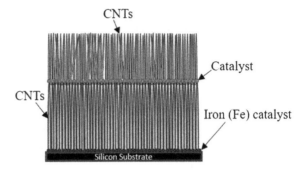

vertically aligned CNTs exhibit a current density of $\sim$4 mA/cm^2 at low turn-on
field $\sim$1.2 V/μm suitable for futuristic thin film-based display applications [113].
Chhowalla et al. showed that CNTs with short length (0.7 μm) and stubby type structure with diameters 200 nm are best for field emission characteristics with emission
current density of 10 mA/cm^2 and the threshold voltage of 2 V/μm. In this work,
they have explained that the best field emission characteristics can be obtained from
vertically aligned CNTs forest with a shorter length and intermediate diameters
[114]. Our group [115] has also worked on field emission studies of CNTs where it
has been observed that vertically alignment of CNTs plays an important role in the
enhancement of field emission properties. It has also been shown that the field emission measurement of as-grown vertically aligned CNTs forest was observed to be
extremely superior with turn-on field of 1.91 V/m at the current density 10 mA/cm^2
wherein field-enhancement factor is calculated to be 7.82 $\times$ 10^3 [115]. In another
work, it has been shown that the current density of vertically aligned CNTs can further
be enhanced by depositing a double layer of vertically aligned CNTs as shown in
Fig. 15 [116]. Since the aspect ratio of CNTs is the ratio of length-to-diameter, the
length of CNTs should be large and thereby the aspect ratio would be high. Hence,
the field emission current density was found to be increased from 1.4 to 3.0 mA/cm^2
[116]. Field emission properties of aligned CNT arrays can also be enhanced using
some effective techniques such as plasma treatments in H$_2$ [117], in O$_2$, and N$_2$
environment [118].

4.2 Gas Sensors

CNT-based gas sensor has drawn intense research interest due to its effective electronic transport and thermopower properties [99]. CNT gas sensors have several
advantages over conventional metal-oxide-based sensors such as room temperature
operation, high sensitivity, selectivity, low power consumption, most reliable with
the miniaturized device and have revealed trace level detection (ppm and ppb) of
various toxic gases such as NH$_3$, CO, NO$_2$, CH$_4$, H$_2$, SO$_2$, and H$_2$S. Most of these
CNTs sensors have been fabricated using a random and aligned network of CNTs

whereas some researchers also used individual CNTs in FETs configuration [99, 119, 120]. However, the aligned network of CNTs has shown more promising results in terms of high sensitivity (2) [90], selectivity, quick response, and recovery time including stability of the device for a longer period (Fig. 16).

Fabrication of CNTs-Based Gas Sensor and Sensing Mechanism

The gas sensing setup consists of a gas chamber with a heater, small fan, needle for the gas injection at trace level inside the gas chamber, and the whole setup is associated with data acquisition system along with a computer. The schematic of gas sensing setup is shown in Fig. 17.

The fabrication of the gas sensor is a very much sensitive work in terms of synthesis of materials, electrode preparation, and proper connection with the source meter, etc. The fabrication of CNT sensor consists of silicon (Si) as the desired substrate, an oxide layer (SiO_2), electrode preparation, and CNTs network. The schematic of the fabricated CNT-based sensor has been shown in Fig. 18. The schematic diagram (Fig. 18) also reveals the interaction of gas molecules with the surface of CNTs and then change in the charge carriers corresponding to nature (reducing/oxidizing) of the gas. The gas sensing phenomenon is defined as the adsorption of the gas molecules on the surface of CNTs and therefore, change in the electrical resistance takes place. During the interaction of the gas molecules, chemisorption /physisorption, or even both the processes may occur depending on the nature of target gas. The reducing gas species such as NH_3 usually donate electrons to the CNT surface, on the other hand, the oxidizing gas species like NO_2 accept electrons from CNTs surface during the interaction process and therefore accordingly change in resistance takes place [121, 122].

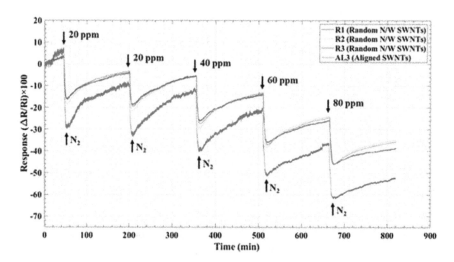

Fig. 16 NO_2 gas sensing response comparing the performances of aligned (AL3) and random network (R1, R2, and R3) SWNT-based chemiresistors (Copyright with permission from [90])

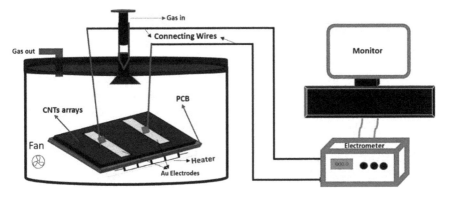

Fig. 17 Schematic of the gas sensing set up

Fig. 18 Schematic of the interaction of the analytes with the surface of CNTs [121] (Copyright Elsevier 2019)

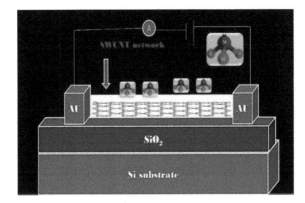

There are some basic parameters for an ideal and efficient gas sensor which include huge response/sensitivity, stability, quick response & recovery time, low analyst consumption, selectivity, and room temperature operation.

(I)   **Sensitivity and response**: The sensitivity and response are the initial and most important parameters to evaluate the performance of any gas sensor. The sensitivity of the sensor can be defined as the corresponding change in electrical resistance (ΔR) for variation in analyte concentration (Δc). Hence, the sensitivity can be measured with the help of the slope of a calibration graph as $S = \Delta R/\Delta c$. The sensor response, on the other hand, can be defined as the ratio of change in resistance before and after gas exposure, i.e. (R_{gas}/R_{air}), where R_{gas} represent the resistance of the sensor in target gas atmosphere and R_{air} denotes the resistance in the air [123].

(II) **Response time and recovery time**: CNTs-based gas sensors are always known for quick response and recovery time after exposure with toxic gases. The response time (R_{res}) of a sensor can be defined as the time in which approximately ~90% resistance change (increase/decrease) is observed after exposure with target gas. On the other hand, the recovery time (R_{recv}) is the time in which resistance of the sensor recovered (~90%) to its original value after the removal of the target gas.

(III)   **Stability of the sensor**: It can be defined as the ability of a gas sensor to show consistency in sensitivity, selectivity, response, and recovery for a certain time, i.e., the results of the gas sensor must be reproducible for some particular period.

(IV)   **Selectivity**: It is another important characteristic of a sensor that determines the ability of a sensor to distinguish among the group of analytes or specifically can detect a single analyte.

(V)   **Limit of detection (LOD)**: Limit of detection of a gas sensor can be defined as least concentration (at ppb and ppm level) of the analyte (gas) that can be sensed under specific conditions such as temperature.

(VI)   **Operating temperature of the sensor**: The working temperature of a gas sensor corresponds to its maximum sensitivity. Usually, the sensor response increases with increasing temperature because the adsorption or diffusion of the target gas is usually temperature dependent. CNTs-based gas sensors are generally operated at room temperature.

Hence, aligned CNTs gas sensors have been observed to be very promising and fulfil all the characteristics of an ideal sensor. Because they possess very high surface to volume ratio along with excellent properties. Also, a large number of target gas molecules are adsorbed on the surface of CNT structures due to the defects sites present and therefore characteristic properties of the sensor are enhanced multiple times. Although all characteristic properties of a sensor are observed from pristine CNTs still, more efforts are being employed to further improve the efficiency of a gas sensor. These efforts include functionalization of CNTs surface with binding legends like polymers. Snow et al. showed the detection of toxic gases using PEI-functionalized SWNT array [99]. In another work, Snow et al. showed a novel SWNT chemi-capacitor for gas sensing [124]. Nguyet et al. worked on hybrid heterojunctions of MWNTs and on-chip grown SnO_2 nanowires to detect NO_2 gas [125].

4.3   FET Devices

Due to 1D structure, remarkable electronic properties, excellent mobility, and ballistic transport CNTs are considered to be very promising material and a substitute of silicon-based FETs. Moreover, the bandgap of CNTs is directly related to its chiral angle and diameter and therefore it can be varied with controlling these parameters. That's why CNTs are being considered as a promising material for future devices and the fabrication of CNTFET devices would be an excellent alternative for all transistor devices [126] (Fig. 19).

The current technology of FET devices with CNTs has shown extraordinary performance and outperformed the Si-based metal oxide semiconductor field-effect

Fig. 19 Schematic of the CNTFET

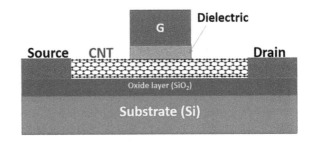

transistors (MOSFETs) [127, 128]. It has been revealed by various groups [129, 130] that CNTFETs behave like conventional MOSFETs. Although there are some important differences like due to the 1D structure of CNT, the scattering probability is less, and therefore, the CNTFETs which are having channel length < the mean free path of the charge carrier, probably would activate in the ballistic regime, whereas it does not occur in MOSFETs based on Si technology even if the length is in the nanometer range. Another difference between CNTFET and conventional MOSFET is that the CNTs structure are chemically satisfied carbon atoms and there is no dangling bond present at the surface. Thus, there is no requirement for vigilant passivation on the interface between the CNTs channel and the gate dielectric. Hence, these properties make CNTFET devices very promising, unique, and interesting [130].

Aligned CNTs have been used to fabricate FET devices and observed to be ultra-sensitive for adsorption events on their surface. The aligned arrays of CNTs enhance the device performance in comparison to the random CNTs network having same density and can provide high current outputs and large active areas as suggested by Kang et al. [38, 131]. Hence, the aligned CNTFETs are expected to work with excellent speed thereby enhancement in device mobilities and fabrication of the devices would remarkably be very small. Besteman et al. showed the first SWNT-based FET device to demonstrate a change in conductance after exposure with glucose [132]. In 1998, Trans et al. [133] demonstrated a room temperature FET device using SWNTs. Just after this, Martel et al. (IBM group) also showed an SWNTs-based FET device and indicated that channel conductance can be improved by an externally applied gate voltage in a similar way as in MOSFET [130]. Rutherglen et al. fabricated an aligned CNTFET device which operates at gigahertz frequencies and revealed its performance close to GaAs technology [134]. Cao et al. reported aligned CNTs-based RF transistor, which shows a remarkable direct current performance, high transconductance, and higher current saturation along with superior linearity performance [135].

4.4 Biosensors

Biosensors are the type of devices, modules, or machines which can be used for the detection of changes in its environment like the detection of different gas molecules and tracking of chemical signals in biological cells. A sensor consists of two main components, namely, an active sensing element and a signal transducer. The sensing element detects the changes in its environment or detect the selective analyte and passes the signal to the transducer, then the transducer converts those chemical signals into an output signal generates which can be electrical, optical, thermal, or magnetic. These outputs can be used by the user to determine the concentration of the detected analyte in a given sample or environment [136]. A biosensor is a type of sensors which generally monitor the biological processes or recognize various biomolecules and modified them into electrical signals depending on the concentration of target analyte [137]. Figure 20 shows a schematic diagram of a liquid-gated CNTFET-based biosensing platform. The CNT channel between source and drain acts as a sensing element. According to the required detection of antigen, the corresponding antibodies are immobilized on the functionalized CNTs via suitable coupling chemistry.

The first biosensor was reported in 1956 by Clark et al. who is also known as the father of biosensors [138]. In biosensors, the sensing element is a biological material like proteins such as antibodies, enzymes and cell receptors, oligo-nucleotides, polynucleotides, microorganisms, or even whole biological tissues [136]. Biosensors could be used for both in vitro as well as in vivo biosensing. For the electrochemical detection of DNA, the first sensor was fabricated by Cai et al. [139].

Although there are several classical techniques used for biosensing, they cannot be miniaturized. Aligned CNTs are nanomaterials and have unique electrical and electrochemical properties; therefore, they can be used as electrodes and transducer elements in biosensors and can also be miniaturizing the whole biosensor. Aligned CNTs can overcome the dispersion problems of CNTs, prevent the degradation of the biological compound on electrodes like enzymes, reduces tube-to-tube resistance, easily be functionalized, immobilized and offers good electrical contact without the use of any binders and without losing their stability which makes them an appropriate

Fig. 20 Schamatic of liquid-gated CNTFET-based biosensor

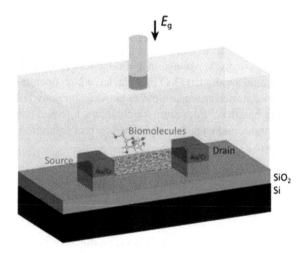

material for highly stable, cost-effective, flexible and miniaturized biosensors with very good performance [136, 140–143].

Based on the process of transduction and the mechanism of the recognition of the analytes, CNT-based biosensors can be broadly classified into the following classes [144]:

- CNT-based electronic and electrochemical biosensors
- CNT immunosensors
- Optical CNT biosensors

Electrochemical biosensors become popular because they are cost-effective, user-friendly, small in size and give fast response as compared to other sensors. The electrochemical biosensor consists of three types of electrodes, namely, counter, working, and reference electrode [144].

Immunosensors are the devices those work based on the affinity of antigens and antibodies. Various recombinant antibodies and antibody fragments are used as sensing elements and antigens, and biomarkers are the target analytes that are recognized by the antibodies because of their affinity. In recent years, aligned CNTs immunosensors have been fabricated by immobilizing or functionalizing recombinant antibodies or antigens onto it which increases the binding capacity and sensitivity as compared to the traditional biosensors [144].

Optical biosensor, as the name itself, suggests, detects the target molecules or analytes based on the change in the emission of light. There are different readout approaches by which optical biosensors can be used to detect the analytes such as absorbance, reflectance, surface plasmon resonance, luminescence, fluorescence, phosphorescence, lifetime, quenching, and fluorescence energy transfer [144]. Colorimetric biosensors and photometric light intensity biosensors are also a part of optical biosensors. Aligned CNTs and functionalized aligned CNTs have enhanced properties due to which they can be used for CNT-based optical biosensors.

Aligned CNTs have enhanced performance than randomly oriented CNTs. Golizhadeh et al. used vertically aligned carbon nanotube nanoelectrode array (CNT-NEA) and high density vertically aligned CNT as electrochemical glutamate biosensor. The recorded detection limit was 10 nM for glutamate [142]. Chen et al. proposed a novel FET-based biosensor using horizontally aligned CNTs for the detection of PSA (Prostate-Specific Antigen) proteins. The limit of detection proposed by them was 84 pM, which was lower than the clinically relevant level of PSA in blood, i.e., 133 pM [140]. Gao et al. used aligned MWNT as electrodes platform for inherently conducting polymer (i.e., polypyrrole-based glucose oxidase system) based biosensors to detect glucose. This biosensor proves to be a highly selective and sensitive biosensor for glucose [145]. Golizhadeh et al. prepared a glutamate biosensor which was based on the GLDH/vertically aligned CNTs without any electron mediator [142]. Fayazfar et al. developed an impedance-based biosensor using gold nanoparticles/ aligned CNTs modified as electrodes for the detection of DNA of the TP53 gene mutation which is one of the most common genes in cancer research [146]. Zhu et al. reviewed work on a novel biosensor for the detection of glucose in several beverages by using aligned CNTs and a conjugated biosensor [147]. Zhao et al. combined platinum nanoparticles with aligned CNTs electrodes having high electrocatalytic activity and develop a sensitive amperometric glucose biosensor with a very fast response time of 5 s [148]. Yang et al. developed herbicide atrazine biosensor by using glucose oxidase assembled on aligned CNTs on the surface of copper electrodes and determined atrazine in environmental water samples [149].

4.5 Energy Applications

Recently, the demand for small, lightweight, and long-lasting portable energy storage devices has been increased due to their promising use in mobiles, camera flashes, power electronic toys, security alarm systems, Uninterruptible Power Supplies (UPS) systems, solar power, hybrid electric vehicles to wearable are other electronic applications [150–152]. Therefore, the need to develop more advanced and efficient energy storage devices has triggered the pace of research to meet the above future energy demands. The latest research shows that CNTs-based energy generation and storage devices specifically in solar cells [153], fuel cells [154], Li-ion batteries [155], and electrochemical capacitors (ECs) [152] exhibit great improvement in processability, scalability, and efficiency of these devices. Due to remarkable, electrical conductivity, mechanical properties, chemical stability, and high specific surface area, CNTs are considered to be the best materials particularly as electrode materials or conductive fillers for future energy applications [153–155]. Although CNTs as excellent electrode materials in supercapacitors can be extensively used as compatible electrochemical active materials. But, particularly, a vertically aligned forest of CNTs is significantly suited as they permit ions to pass freely through the electrolyte between two electrodes. Wang et al. [156] showed that highly aligned CNTs are capable to use them as current collectors in Li-Ion batteries. It has also been demonstrated by Park et al. [157] that vertically aligned CNTs doped with CO_3O_4 nanoparticles as an anode material reveals a superior reversible capacity (627 mAh/g). Supercapacitors are another energy storage device where CNTs could be extensively used as electrode material. Supercapacitors have the capabilities to store more energy and deliver it very quickly and efficiently for a long period in terms of its life cycles. Dogru et al. reported that vertically aligned CNTs can be used as electrodes material wherein internal resistance of the electrodes as well as charge transfer resistance between electrolyte and electrodes is negligible due to synthesis of CNTs on aluminium foil [158]. Also, CNTs as electrode materials can be restricted to an extremely lesser area and thus increase the electrode–electrolyte contact and decrease the overall weight of the device. Hence the gravimetric performance of the energy storage devices is enhanced. Niu et al. [159] in 1997, first suggested that CNTs could be the future of supercapacitors. Later in 2000 [160], Barisci et al. used SWNTs-based electrodes observed a redox response from cyclic voltammetric (CV) plot and revealed that due to the functional groups and impurities, the pseudocapacitance has occurred. Pan et al. reviewed the effects of physical and chemical properties of CNTs including purity, size, shape, defect, annealing, and functionalization on the super capacitance [161].

CNTs have also proven a crucial role in solar cell devices because they have high mobility, superior strength, chemical inertness, and most importantly CNTs are photoactive [162]. There has been a lot of work to accomplish a mature technology based on CNTs solar cell devices. Li et al. reported an improved efficiency of the solar cells fabricated using aligned CNTs/Si hybrid nanostructure and observed a high fill factor with low ideality factor [163].

4.6 Nano Electromechanical System (NEMS)

The further miniaturization of MEMS devices to nanoscale comes under the regime of nanotechnology and is known as nanoelectromechanical systems (NEMS). Currently, NEMS devices have been identified as the topic of innovative research and generated a huge interest among researchers because of its wide range of scientific and technological applications at commercial level. These are the devices which effectively integrate electronic as well as mechanical functionality on the nanoscale, for example, integration of CNTFET with actuators, motors or pump. More specifically, NEMS devices have been extensively used in signal processing [164], mass detection [165], and actuators [166].

Among various nanomaterials, CNTs have shown very promising applications in NEMS devices because of their extraordinary properties which include low mass, high Young's modulus, high thermal conductivity, and high surface to volume ratio. These applications include resonators, FET devices, NEMS switches including high-frequency oscillators, nanomotors, etc. The ultrasensitive mass can be detected by CNT-based NEMS resonator or also known as a mass detector. Since in CNTs, the mass of a single nanotube is ultra-low, hardly in the range of few atto-grams. Therefore, a very tiny number of atoms deposited on the surface of CNT make up a notable fraction of the total mass. The highly rigid nature of CNTs makes them superior for high resonance frequency and hence, due to ultra-low mass and high resonance frequency, CNTs are proposed to show an extraordinary mass responsivity. Lassagne et al. reported a CNT mechanical resonator for ultrasensitive mass sensing and demonstrated a high mass responsivity and a good mass resolution at room temperature [167].

In NEMS devices, it is also important to consider that alignment of CNTs is crucial while fabricating the device because the high surface area and low contact resistance make CNTs a perfect electrode material in electrochemical systems. Eom et al. reviewed in their work about the development of NEMS devices and discussed in detail that how these devices have superior capabilities to detect not only bio or chemical molecules but also used for the detection of physical quantities like elastic stiffness, surface stress, molecular weight. The developed devices also can detect the surface elastic stiffness for the molecules which are adsorbed on the surface of CNTs [168]. In another work, this group has described the CNTs resonance behaviour on the adsorption of DNA. They developed a new process based on multiscale modelling to understand the characteristic of DNA adsorption on the CNTs surface [169].

Laocharoensu et al. demonstrated an efficient and controllable CNT incorporated Au/Pt nanomotor which enhanced the oxidation of the hydrogen peroxide fuel [170]. Zettl's group also reported a CNT-supported shafts for rotational actuator [171]. Hence, it is expected that these next-generation NEMS devices could also open the door for powerful nano-vehicle systems in the near future.

5   Challenges and Future Prospects

Despite rapid development in CNT alignment technique, further research and development need to be undertaken to circumvent the challenges associated with CNT-based technology. However, some critical issues are as follows [1]:

1. Controlled growth and fabrication of aligned CNTs having a uniform diameter and chiral nature using the CVD technique.
2. Removal of metallic residue from CNTs with high packing density.
3. Prone to aggregate and misalign due to strong randomized inter-nanotube interactions.
4. There is a gradual upsurge to optimize each nanotube separately.
5. Ensure strong bonding between nanotubes and subsequent materials for sustaining sufficient loads.
6. The loads must be distributed uniformly to avoid misalignment followed by its fragmentation.
7. Because of insolubility in organic solvents, the addition of polymers such as epoxy with nanotubes results in increased viscosity, facilitating difficulty in subsequent device processing.

Moreover, the challenges about the smooth interface between nanotubes and polymer composites as well as transferring loads from the matrix to nanotubes must be intriguingly taken care. Furthermore, the non-uniform distribution of loads due to concentric MWNTs along with bundle-like arrangements in SWNTs can be addressed by direct formation of 3D composite structures of aligned interlaminar CNT arrays onto the fibre stacks of SiC. This structure favours enhancing the interlaminar strength and fracture toughness to a great extent. It may be mentioned here that amid very high aspect ratio, CNTs are not considered as a suitable substitute as reinforcing material [172].

Future roadmap on aligned CNTs should be focused on scalable, developing newer approaches and mechanisms to integrate with the existing process lines and thereby to circumvent the challenges mentioned above. Owing to its lightweight and superior characteristics, nanotubes may be considered as a suitable material to replace carbon fibres. Simple design, lightweight, non-flammable nature coupled with profound improvement in the structural and thermal barrier, it has great potential for automobile application. Thus, it is evident that CNTs are the extraordinary material having a plethora of applications including nanotube capacitors, hydrogen storage, biochemical sensors, FET, NEMS, computer, instruments, etc.

6 Conclusion

In conclusion, recent progress on possible approaches to align CNTs is summarized in this chapter. Important concepts along with the underlying mechanism of fabricating aligned CNTs devices from in-situ and ex-situ methods were elucidated. Numerous modified approaches have been adopted for large-scale production of CNTs for commercial applications. It is evident that controlled in-situ growth of aligned CNT arrays having desired orientation, diameter, length, location, and site density along with reasonably low cost make PECVD and CVD methods, a promising growth technique compared to the other available methods. CNTs may align into 1D or 3D orderings based on catalytic nanoparticles distribution on substrates. The dimension, composition, surface characteristics, phase purity, and crystallinity play a crucial role in the functionality of these nanostructures. Potential applications towards a large number of electronic devices such as gas sensors, nanobiosensors, FETs, FEDs, optoelectronics, electromechanical, and electrochemical devices were discussed. It is pertinent to further study the reliability and performance dependence on the number of CNT layers of above-mentioned devices. Rapid progress on CNT technology makes a feasible solution for the development of numerous CNTs-based devices, which needs to be commercialized in the long run. However, this material may play a pivotal role in transforming chemical/biomolecular species into electronic/digital signals to bridge the gap between nanoelectronic and biological information processing systems in future. Although there have been widespread research reports on the alignment of CNTs, underlying physics of growth and properties, still substantial efforts need to be carried out to develop more mechanism and mature technology to minimize the effect of matrix interference from actual samples, non-specific bindings, precise morphology manipulation along with device homogeneity. Since research interest in this domain is rapidly growing, hence this review may open a new avenue to further explore this niche area to enrich scientific knowledge, research excellence and innovative technology having a high societal, industrial and strategic impact.

Acknowledgments The authors wish to thank Director, CSIR-CEERI, for his keen interest and encouragement. One of the authors (P. B. Agarwal) gratefully acknowledges the members of Smart Sensors Area, CSIR-CEERI, Pilani, Rajasthan, India for their valuable suggestions and continuous cooperation. Support of CSIR project OLP5101 and HCP0012 is also gratefully acknowledged.

References

1. Rao R, Pint CL, Islam AE et al (2018) Carbon nanotubes and related nanomaterials: critical advances and challenges for synthesis toward mainstream commercial applications. ACS Nano 12:11756–11784. https://doi.org/10.1021/acsnano.8b06511
2. Mudimela PR, Scardamaglia M, González-León O et al (2014) Gas sensing with gold-decorated vertically aligned carbon nanotubes. Beilstein J Nanotechnol 5:910–918. https://doi.org/10.3762/bjnano.5.104
3. Dai L, Patil A, Gong X et al (2003) Aligned nanotubes. ChemPhysChem 4:1150–1169. https://doi.org/10.1002/cphc.200300770
4. Fam DWHWH, Palaniappan A, Tok AIYIY et al (2011) A review on technological aspects influencing commercialization of carbon nanotube sensors. Sens Actuat, B Chem 157:1–7. https://doi.org/10.1016/j.snb.2011.03.040
5. Saifuddin N, Raziah AZ, Junizah AR (2013) Carbon nanotubes: a review on structure and their interaction with proteins. J Chem 2013:1–18. https://doi.org/10.1155/2013/676815
6. Cao Q, Rogers JA (2008) Random networks and aligned arrays of single-walled carbon nanotubes for electronic device applications. Nano Res 1:259–272. https://doi.org/10.1007/s12274-008-8033-4
7. Ishikawa FN, Chang H, Ryu K et al (2009) Transparent electronics based on transfer printed aligned carbon nanotubes on rigid and flexible substrates. ACS Nano 3:73–79. https://doi.org/10.1021/nn800434d
8. Rahman R, Servati P (2012) Effects of inter-tube distance and alignment on tunnelling resistance and strain sensitivity of nanotube/polymer composite films. Nanotechnology 23:055703. https://doi.org/10.1088/0957-4484/23/5/055703
9. Thostenson ET, Chou T-W (2002) Aligned multi-walled carbon nanotube-reinforced composites: processing and mechanical characterization. J Phys D Appl Phys 35:L77–L80. https://doi.org/10.1088/0022-3727/35/16/103
10. Goh GL, Agarwala S, Yeong WY (2019) Directed and on-demand alignment of carbon nanotube: a review toward 3D printing of electronics. Adv Mater Interf 6:1801318. https://doi.org/10.1002/admi.201801318
11. Wang Y, Maspoch D, Zou S et al (2006) Controlling the shape, orientation, and linkage of carbon nanotube features with nano affinity templates. Proc Natl Acad Sci 103:2026–2031. https://doi.org/10.1073/pnas.0511022103
12. Duchamp M, Lee K, Dwir B et al (2010) Controlled positioning of carbon nanotubes by dielectrophoresis: insights into the solvent and substrate role. ACS Nano 4:279–284. https://doi.org/10.1021/nn901559q
13. Huang Y (2001) Directed assembly of one-dimensional nanostructures into functional networks. Science 291:630–633. https://doi.org/10.1126/science.291.5504.630
14. Fukuda T, Arai F, Dong L (2003) Assembly of nanodevices with carbon nanotubes through nanorobotic manipulations. Proc IEEE 9:1803–1818. https://doi.org/10.1109/JPROC.2003.818334
15. Ko H, Peleshanko S, Tsukruk VV (2004) Combing and bending of carbon nanotube arrays with confined microfluidic flow on patterned surfaces. J Phys Chem B 108:4385–4393. https://doi.org/10.1021/jp031229e
16. Lan Y, Wang Y, Ren ZF (2011) Physics and applications of aligned carbon nanotubes. Adv Phys 60:553–678. https://doi.org/10.1080/00018732.2011.599963
17. Oh BS, Min Y-S, Bae EJ et al (2006) Fabrication of suspended single-walled carbon nanotubes via a direct lithographic route. J Mater Chem 16:174–178. https://doi.org/10.1039/B510742C
18. Baughman RH (2002) Carbon nanotubes—the route toward applications. Science (80-) 297:787–792. https://doi.org/10.1126/science.1060928
19. Dresselhaus MS (1997) Future directions in carbon science. Ann Rev Mater Sci 27:1–34. https://doi.org/10.1146/annurev.matsci.27.1.1
20. Ren Z, Lan Y, Wang Y (2013) Aligned carbon nanotubes. Springer, Berlin, Heidelberg

21. Dai H (2002) Carbon nanotubes: opportunities and challenges. Surf Sci 500:218–241. https://doi.org/10.1016/S0039-6028(01)01558-8

22. Hiramatsu M, Deguchi T, Nagao H, Hori M (2007) Aligned growth of single-walled and double-walled carbon nanotube films by control of catalyst preparation. Jpn J Appl Phys 46:L303–L306. https://doi.org/10.1143/JJAP.46.L303

23. Li WZ, Xie SS, Qian LX et al (1996) Large-scale synthesis of aligned carbon nanotubes. Science (80–)274:1701–1703. https://doi.org/10.1126/science.274.5293.1701

24. Terrones M, Grobert N, Olivares J et al (1997) Controlled production of aligned-nanotube bundles. Nature 388:52–55. https://doi.org/10.1038/40369

25. Ren ZF (1998) Synthesis of large arrays of well-aligned carbon nanotubes on glass. Science (80–) 282:1105–1107. https://doi.org/10.1126/science.282.5391.1105

26. Pan ZW, Xie SS, Chang BH et al (1998) Very long carbon nanotubes. Nature 394:631–632. https://doi.org/10.1038/29206

27. Fan S (1999) self-oriented regular arrays of carbon nanotubes and their field emission properties. Science (80–) 283:512–514. https://doi.org/10.1126/science.283.5401.512

28. Choi YC, Shin YM, Lee YH et al (2000) Controlling the diameter, growth rate, and density of vertically aligned carbon nanotubes synthesized by microwave plasma-enhanced chemical vapor deposition. Appl Phys Lett 76:2367–2369. https://doi.org/10.1063/1.126348

29. Bower C, Zhu W, Jin S, Zhou O (2000) Plasma-induced alignment of carbon nanotubes. Appl Phys Lett 77:830–832. https://doi.org/10.1063/1.1306658

30. Bower C, Zhou O, Zhu W et al (2000) Nucleation and growth of carbon nanotubes by microwave plasma chemical vapor deposition. Appl Phys Lett 77:2767–2769. https://doi.org/10.1063/1.1319529

31. Jiran E, Thompson CV (1990) Capillary instabilities in thin films. J Electron Mater 19:1153–1160. https://doi.org/10.1007/BF02673327

32. Jiran E, Thompson CV (1992) Capillary instabilities in thin, continuous films. Thin Solid Films 208:23–28. https://doi.org/10.1016/0040-6090(92)90941-4

33. Zhang G, Mann D, Zhang L et al (2005) Ultra-high-yield growth of vertical single-walled carbon nanotubes: hhidden roles of hydrogen and oxygen. Proc Natl Acad Sci 102:16141–16145. https://doi.org/10.1073/pnas.0507064102

34. Hata K (2004) Water-assisted highly efficient synthesis of impurity-free single-walled carbon nanotubes. Science (80–) 306:1362–1364. https://doi.org/10.1126/science.1104962

35. Kong J, Soh HT, Cassell AM et al (1998) Synthesis of individual single-walled carbon nanotubes on patterned silicon wafers. Nature 395:878–881. https://doi.org/10.1038/27632

36. Sato S, Kawabata A, Nihei M, Awano Y (2003) Growth of diameter-controlled carbon nanotubes using monodisperse nickel nanoparticles obtained with a differential mobility analyzer. Chem Phys Lett 382:361–366. https://doi.org/10.1016/j.cplett.2003.10.076

37. Andrews R, Jacques D, Rao AM et al (1999) Continuous production of aligned carbon nanotubes: a step closer to commercial realization. Chem Phys Lett 303:467–474. https://doi.org/10.1016/S0009-2614(99)00282-1

38. Kang SJ, Kocabas C, Ozel T et al (2007) High-performance electronics using dense, perfectly aligned arrays of single-walled carbon nanotubes. Nat Nanotechnol 2:230–236. https://doi.org/10.1038/nnano.2007.77

39. Zhang Y, Chang A, Cao J et al (2001) Electric-field-directed growth of aligned single-walled carbon nanotubes. Appl Phys Lett 79:3155–3157. https://doi.org/10.1063/1.1415412

40. Jorio A, Dresselhaus G, Dresselhaus MS (2008) Carbon nanotubes advanced topics in the synthesis, structure. Properties and applications. Springer, , Berlin, Heidelberg

41. Han S, Liu X, Zhou C (2005) Template-free directional growth of single-walled carbon nanotubes on a- and r-Plane Sapphire. J Am Chem Soc 127:5294–5295. https://doi.org/10.1021/ja042544x

42. Ago H, Nakamura K, Ikeda K et al (2005) Aligned growth of isolated single-walled carbon nanotubes programmed by atomic arrangement of substrate surface. Chem Phys Lett 408:433–438. https://doi.org/10.1016/j.cplett.2005.04.054

43. Rutkowska A, Walker D, Gorfman S et al (2009) Horizontal alignment of chemical vapor-deposited swnts on single-crystal quartz surfaces: further evidence for epitaxial alignment. J Phys Chem C 113:17087–17096. https://doi.org/10.1021/jp9048555
44. Ismach A, Segev L, Wachtel E, Joselevich E (2004) Atomic-step-templated formation of single wall carbon nanotube patterns. Angew Chemie Int Ed 43:6140–6143. https://doi.org/10.1002/anie.200460356
45. Ismach A, Kantorovich D, Joselevich E (2005) Carbon nanotube graphoepitaxy: highly oriented growth by faceted nanosteps. J Am Chem Soc 127:11554–11555. https://doi.org/10.1021/ja052759m
46. Hofmann S, Ducati C, Kleinsorge B, Robertson J (2003) Direct growth of aligned carbon nanotube field emitter arrays onto plastic substrates. Appl Phys Lett 83:4661–4663. https://doi.org/10.1063/1.1630167
47. Meyyappan M (2009) A review of plasma enhanced chemical vapour deposition of carbon nanotubes. J Phys D Appl Phys 42:213001. https://doi.org/10.1088/0022-3727/42/21/213001
48. Meyyappan M, Delzeit L, Cassell A, Hash D (2003) Carbon nanotube growth by PECVD: a review. Plasma Sources Sci Technol 12:205–216. https://doi.org/10.1088/0963-0252/12/2/312
49. Merkulov VI, Melechko AV, Guillorn MA et al (2002) Growth rate of plasma-synthesized vertically aligned carbon nanofibers. Chem Phys Lett 361:492–498. https://doi.org/10.1016/S0009-2614(02)01016-3
50. Teo KBK, Chhowalla M, Amaratunga GAJ et al (2001) Uniform patterned growth of carbon nanotubes without surface carbon. Appl Phys Lett 79:1534–1536. https://doi.org/10.1063/1.1400085
51. Teo KBK, Hash DB, Lacerda RG et al (2004) The significance of plasma heating in carbon nanotube and nanofiber growth. Nano Lett 4:921–926. https://doi.org/10.1021/nl049629g
52. Cruden BA, Cassell AM, Ye Q, Meyyappan M (2003) Reactor design considerations in the hot filament/direct current plasma synthesis of carbon nanofibers. J Appl Phys 94:4070–4078. https://doi.org/10.1063/1.1601293
53. Han J, Yang W-S, Yoo J-B, Park C-Y (2000) Growth and emission characteristics of vertically well-aligned carbon nanotubes grown on glass substrate by hot filament plasma-enhanced chemical vapor deposition. J Appl Phys 88:7363–7365. https://doi.org/10.1063/1.1322378
54. Wang Y, Rybczynski J, Wang DZ et al (2004) Periodicity and alignment of large-scale carbon nanotubes arrays. Appl Phys Lett 85:4741–4743. https://doi.org/10.1063/1.1819992
55. Tu Y, Lin Y, Ren ZF (2003) Nanoelectrode arrays based on low site density aligned carbon nanotubes. Nano Lett 3:107–109. https://doi.org/10.1021/nl025879q
56. Wang Y (2006) Nanophotonics of vertically aligned carbon nanotubes: two-dimensional photonic crystals and optical dipole antenna. Boston College, Chestnut Hill
57. Cao Q, Rogers JA (2009) Ultrathin films of single-walled carbon nanotubes for electronics and sensors: a review of fundamental and applied Aspects. Adv Mater 21:29–53. https://doi.org/10.1002/adma.200801995
58. Smith PA, Nordquist CD, Jackson TN et al (2000) Electric-field assisted assembly and alignment of metallic nanowires. Appl Phys Lett 77:1399–1401. https://doi.org/10.1063/1.1290272
59. Cole M, Milne W (2013) Plasma enhanced chemical vapour deposition of horizontally aligned carbon nanotubes. Materials (Basel) 6:2262–2273. https://doi.org/10.3390/ma6062262
60. Cole M, Hiralal P, Ying K et al (2012) Dry-transfer of aligned multiwalled carbon nanotubes for flexible transparent thin films. J Nanomater 2012:1–8. https://doi.org/10.1155/2012/272960
61. de Jonge N, Allioux M, Doytcheva M et al (2004) Characterization of the field emission properties of individual thin carbon nanotubes. Appl Phys Lett 85:1607–1609. https://doi.org/10.1063/1.1786634
62. Teo KBK, Lee S-B, Chhowalla M et al (2003) Plasma enhanced chemical vapour deposition carbon nanotubes/nanofibres how uniform do they grow? Nanotechnology 14:204–211. https://doi.org/10.1088/0957-4484/14/2/321

63. Chhowalla M, Teo KBK, Ducati C et al (2001) Growth process conditions of vertically aligned carbon nanotubes using plasma enhanced chemical vapor deposition. J Appl Phys 90:5308–5317. https://doi.org/10.1063/1.1410322

64. Lieberman MA, Lichtenberg AJ (2005) Principles of plasma discharges and materials processing. Wiley, Hoboken, NJ, USA

65. Bell MS, Teo KBK, Lacerda RG et al (2006) Carbon nanotubes by plasma-enhanced chemical vapor deposition. Pure Appl Chem 78:1117–1125. https://doi.org/10.1351/pac200678061117

66. Xue W, Li P (2011) Dielectrophoretic deposition and alignment of carbon nanotubes. In: Carbon nanotubes—synthesis, characterization, applications. InTech, pp 171–190

67. Banerjee S, White BE, Huang L et al (2006) Precise positioning of single-walled carbon nanotubes by AC dielectrophoresis. J Vac Sci Technol B Microelectron Nanom Struct 24:3173. https://doi.org/10.1116/1.2387155

68. Stokes P, Khondaker SI (2008) Local-gated single-walled carbon nanotube field effect transistors assembled by AC dielectrophoresis. Nanotechnology 19:175202. https://doi.org/10.1088/0957-4484/19/17/175202

69. Arun A, Salet P, Ionescu AM (2009) A study of deterministic positioning of carbon nanotubes by dielectrophoresis. J Electron Mater 38:742–749. https://doi.org/10.1007/s11664-009-0797-0

70. Xiao Z, Camino FE (2009) The fabrication of carbon nanotube field-effect transistors with semiconductors as the source and drain contact materials. Nanotechnology 20:135205. https://doi.org/10.1088/0957-4484/20/13/135205

71. Heremans J, Olk CH, Morelli DT (1994) Magnetic susceptibility of carbon structures. Phys Rev B 49:15122–15125. https://doi.org/10.1103/PhysRevB.49.15122

72. Wu B, Zhang J, Wei Z et al (2001) Chemical alignment of oxidatively shortened single-walled carbon nanotubes on silver surface. J Phys Chem B 105:5075–5078. https://doi.org/10.1021/jp0101256

73. Rao SG, Huang L, Setyawan W, Hong S (2003) Large-scale assembly of carbon nanotubes. Nature 425:36–37. https://doi.org/10.1038/425036a

74. Im J, Huang L, Kang J et al (2006) "Sliding kinetics" of single-walled carbon nanotubes on self-assembled monolayer patterns: beyond random adsorption. J Chem Phys 124:224707. https://doi.org/10.1063/1.2206590

75. Lee M, Lee J, Kim TH, et al (2010) 100 nm scale low-noise sensors based on aligned carbon nanotube networks: overcoming the fundamental limitation of network-based sensors. Nanotechnology 21:055504. https://doi.org/10.1088/0957-4484/21/5/055504

76. Li X, Zhang L, Wang X et al (2007) Langmuir-Blodgett assembly of densely aligned single-walled carbon nanotubes from bulk materials. J Am Chem Soc 129:4890–4891. https://doi.org/10.1021/ja071114e

77. Yu X, Munge B, Patel V et al (2006) Carbon nanotube amplification strategies for highly sensitive immunodetection of cancer biomarkers. J Am Chem Soc 128:11199–11205. https://doi.org/10.1021/ja062117e

78. Jia L, Zhang Y, Li J et al (2008) Aligned single-walled carbon nanotubes by Langmuir-Blodgett technique. J Appl Phys 104:074318. https://doi.org/10.1063/1.2996033

79. Gao Y, Deng Y, Liao Z, Zhang M (2017) Aligned carbon nanotube field effect transistors by repeated compression-expansion cycles in Langmuir-Blodgett. In: 2017 IEEE 17th international conference on nanotechnology (IEEE-NANO). IEEE, pp 731–734

80. Giancane G, Ruland A, Sgobba V et al (2010) Aligning single-walled carbon nanotubes by means of langmuir-blodgett film deposition: optical, morphological, and photo-electrochemical studies. Adv Funct Mater 20:2481–2488. https://doi.org/10.1002/adfm.201000290

81. Ma Y, Wang B, Wu Y et al (2011) The production of horizontally aligned single-walled carbon nanotubes. Carbon 49:4098–4110. https://doi.org/10.1016/j.carbon.2011.06.068

82. Sadovoy A, Dubovik Y, Nazvanov V (2007) Carbon nanotubes aligning by Langmuir-Blodgett technique and visualizing by nematic liquid crystals. In: Zimnyakov DA, Khlebtsov NG (eds) Saratov pp 653609–653609–5. https://doi.org/10.1117/12.753440

83. Kim K, Kim MJ, Kim DW et al (2020) Clinically accurate diagnosis of Alzheimer's disease via multiplexed sensing of core biomarkers in human plasma. Nat Commun 11:1–9. https://doi.org/10.1038/s41467-019-13901-z

84. Sgobba V, Giancane G, Cannoletta D et al (2014) Langmuir-Schaefer films for aligned carbon nanotubes functionalized with a conjugate polymer and photoelectrochemical response enhancement. ACS Appl Mater Interfaces 6:153–158. https://doi.org/10.1021/am403656k

85. Lee D, Ye Z, Campbell SA, Cui T (2012) Suspended and highly aligned carbon nanotube thin-film structures using open microfluidic channel template. Sensors Actuat A Phys 188:434–441. https://doi.org/10.1016/j.sna.2012.06.013

86. Ye Z, Lee D, Campbell SA, Cui T (2011) Thermally enhanced single-walled carbon nanotube microfluidic alignment. Microelectron Eng 88:2919–2923. https://doi.org/10.1016/j.mee.2011.03.158

87. LeMieux MC, Roberts M, Barman S et al (2008) Self-sorted, aligned nanotube networks for thin-film transistors. Science (80–) 321:101–104. https://doi.org/10.1126/science.1156588

88. Hu Y, Chen Y, Li P, Zhang J (2013) Sorting out semiconducting single-walled carbon nanotube arrays by washing off metallic tubes using SDS aqueous solution. Small 9:1306–1311. https://doi.org/10.1002/smll.201202940

89. Agarwal PB, Pawar S, Reddy SM et al (2016) Reusable silicon shadow mask with sub-5 μm gap for low cost patterning. Sens Actuat A Phys 242:67–72. https://doi.org/10.1016/j.sna.2016.02.040

90. Agarwal PB, Sharma R, Mishra D et al (2020) Silicon shadow mask technology for aligning and in situ sorting of semiconducting SWNTs for sensitivity enhancement: a case study of NO_2 gas sensor. ACS Appl Mater Interf 12:40901–40909. https://doi.org/10.1021/acsami.0c10189

91. Reich S, Thomsen C, Maultzsch J (2004) Carbon nanotubes: basic concepts and physical properties. Wiley-VCH Verlag GmbH & Co, KGaA

92. de Heer WA, Ch telain A, Ugarte D (1995) A carbon nanotube field-emission electron source. Science (80–) 270:1179–1180. https://doi.org/10.1126/science.270.5239.1179

93. Hu Y, Chiang S-W, Chu X et al (2020) Vertically aligned carbon nanotubes grown on reduced graphene oxide as high-performance thermal interface materials. J Mater Sci. https://doi.org/10.1007/s10853-020-04681-9

94. de Jonge N, Lamy Y, Schoots K, Oosterkamp TH (2002) High brightness electron beam from a multi-walled carbon nanotube. Nature 420:393–395. https://doi.org/10.1038/nature01233

95. Modi A, Koratkar N, Lass E et al (2003) Miniaturized gas ionization sensors using carbon nanotubes. Nature 424:171–174. https://doi.org/10.1038/nature01777

96. Popov VN (2004) Carbon nanotubes: properties and application. Mater Sci Eng R Reports 43:61–102. https://doi.org/10.1016/j.mser.2003.10.001

97. Schnorr JM, Swager TM (2011) Emerging applications of carbon nanotubes. Chem Mater 23:646–657. https://doi.org/10.1021/cm102406h

98. Dai H, Hafner JH, Rinzler AG et al (1996) Nanotubes as nanoprobes in scanning probe microscopy. Nature 384:147–150. https://doi.org/10.1038/384147a0

99. Kong J (2000) Nanotube molecular wires as chemical sensors. Science (80–) 287:622–625. https://doi.org/10.1126/science.287.5453.622

100. Yao Z, Postma HWC, Balents L, Dekker C (1999) Carbon nanotube intramolecular junctions. Nature 402:273–276. https://doi.org/10.1038/46241

101. Wei J, Zhu H, Wu D, Wei B (2004) Carbon nanotube filaments in household light bulbs. Appl Phys Lett 84:4869–4871. https://doi.org/10.1063/1.1762697

102. Wang QH, Setlur AA, Lauerhaas JM et al (1998) A nanotube-based field-emission flat panel display. Appl Phys Lett 72:2912–2913. https://doi.org/10.1063/1.121493

103. Nakayama Y, Akita S (2001) Field-emission device with carbon nanotubes for a flat panel display. Synth Met 117:207–210. https://doi.org/10.1016/S0379-6779(00)00365-9

104. Thapa A, Jungjohann KL, Wang X, Li W (2020) Improving field emission properties of vertically aligned carbon nanotube arrays through a structure modification. J Mater Sci 55:2101–2117. https://doi.org/10.1007/s10853-019-04156-6

105. Milne WI, Teo KBK, Amaratunga GAJ et al (2004) Carbon nanotubes as field emission sources. J Mater Chem 14:933. https://doi.org/10.1039/b314155c

106. Fowler RH, Nordheim L (1928) Electron emission in intense electric fields. Proc R Soc London Ser A, Contain Pap a Math Phys Character 119:173–181. https://doi.org/10.1098/rspa.1928.0091

107. Mayer A, Vigneron J-P (1998) Quantum-mechanical theory of field electron emission under axially symmetric forces. J Phys Condens Matter 10:869–881. https://doi.org/10.1088/0953-8984/10/4/015

108. Jensen KL (2003) Electron emission theory and its application: Fowler-Nordheim equation and beyond. J Vac Sci Technol B Microelectron Nanom Struct 21:1528. https://doi.org/10.1116/1.1573664

109. Parveen S, Kumar A, Husain S, Husain M (2017) Fowler Nordheim theory of carbon nanotube based field emitters. Phys B Condens Matter 505:1–8. https://doi.org/10.1016/j.physb.2016.10.031

110. Li Y, Sun Y, Yeow JTW (2015) Nanotube field electron emission: principles, development, and applications. Nanotechnology 26:242001. https://doi.org/10.1088/0957-4484/26/24/242001

111. Chouhan V, Noguchi T, Kato S (2016) Field emission from optimized structure of carbon nanotube field emitter array. J Appl Phys 119:134303. https://doi.org/10.1063/1.4945581

112. Lin P-H, Sie C-L, Chen C-A et al (2015) Field emission characteristics of the structure of vertically aligned carbon nanotube bundles. Nanoscale Res Lett 10:297. https://doi.org/10.1186/s11671-015-1005-1

113. Gupta BK, Kedawat G, Gangwar AK et al (2018) High-performance field emission device utilizing vertically aligned carbon nanotubes-based pillar architectures. AIP Adv 8:015117. https://doi.org/10.1063/1.5004769

114. Chhowalla M, Ducati C, Rupesinghe NL et al (2001) Field emission from short and stubby vertically aligned carbon nanotubes. Appl Phys Lett 79:2079–2081. https://doi.org/10.1063/1.1406557

115. Kumar A, Husain S, Ali J et al (2012) Field emission study of carbon nanotubes forest and array grown on si using fe as catalyst deposited by electro-chemical method. J Nanosci Nanotechnol 12:2829–2832. https://doi.org/10.1166/jnn.2012.5806

116. Parveen S, Husain S, Kumar A et al (2015) Improved field emission properties of carbon nanotubes by dual layer deposition. J Exp Nanosci 10:499–510. https://doi.org/10.1080/17458080.2013.845914

117. Cheng TC, Shieh J, Huang WJ et al (2006) Hydrogen plasma dry etching method for field emission application. Appl Phys Lett 88:263118. https://doi.org/10.1063/1.2218824

118. Kumar A, Parveen S, Husain S et al (2014) Effect of oxygen plasma on field emission characteristics of single-wall carbon nanotubes grown by plasma enhanced chemical vapour deposition system. J Appl Phys 115:084308. https://doi.org/10.1063/1.4866995

119. Lone MY, Kumar A, Husain S et al (2017) Growth of single wall carbon nanotubes using PECVD technique: an efficient chemiresistor gas sensor. Phys E Low-Dimensional Syst Nanostructures 87:261–265. https://doi.org/10.1016/j.physe.2016.10.049

120. Lone MY, Kumar A, Ansari N et al (2019) Structural effect of SWCNTs grown by PECVD towards NH_3 gas sensing and field emission properties. Mater Res Bull 119:110532. https://doi.org/10.1016/j.materresbull.2019.110532

121. Lone MY, Kumar A, Husain S et al (2019) Fabrication of sensitive SWCNT sensor for trace level detection of reducing and oxidizing gases (NH_3 and NO_2) at room temperature. Phys E Low-Dimensional Syst Nanostruct 108:206–214. https://doi.org/10.1016/j.physe.2018.11.020

122. Song H, Li K, Wang C (2018) Selective detection of NO and NO_2 with CNTs-based ionization sensor array. Micromachines 9:354. https://doi.org/10.3390/mi9070354

123. Mirzaei A, Lee J-H, Majhi SM et al (2019) Resistive gas sensors based on metal-oxide nanowires. J Appl Phys 126:241102. https://doi.org/10.1063/1.5118805

124. Snow ES, Perkins FK, Houser EJ et al (2005) Chemical detection with a single-walled carbon nanotube capacitor. Science 307:1942–1945. https://doi.org/10.1126/science.1109128

125. Nguyet QTM, Van Duy N, Manh Hung C et al (2018) Ultrasensitive NO_2 gas sensors using hybrid heterojunctions of multi-walled carbon nanotubes and on-chip grown SnO_2 nanowires. Appl Phys Lett 112:153110. https://doi.org/10.1063/1.5023851

126. Obite F, Ijeomah G, Bassi JS (2019) Carbon nanotube field effect transistors: toward future nanoscale electronics. Int J Comput Appl 41:149–164. https://doi.org/10.1080/1206212X.2017.1415111

127. Wind SJ, Appenzeller J, Martel R et al (2002) Vertical scaling of carbon nanotube field-effect transistors using top gate electrodes. Appl Phys Lett 80:3817–3819. https://doi.org/10.1063/1.1480877

128. Javey A, Kim H, Brink M et al (2002) High-κ dielectrics for advanced carbon-nanotube transistors and logic gates. Nat Mater 1:241–246. https://doi.org/10.1038/nmat769

129. Javey A, Guo J, Wang Q et al (2003) Ballistic carbon nanotube field-effect transistors. Nature 424:654–657. https://doi.org/10.1038/nature01797

130. Martel R, Schmidt T, Shea HR et al (1998) Single- and multi-wall carbon nanotube field-effect transistors. Appl Phys Lett 73:2447. https://doi.org/10.1063/1.122477

131. Kocabas C, Hur S-H, Gaur A et al (2005) Guided growth of large-scale, horizontally aligned arrays of single-walled carbon nanotubes and their use in thin-film transistors. Small 1:1110–1116. https://doi.org/10.1002/smll.200500120

132. Besteman K, Lee J-O, Wiertz FGM et al (2003) Enzyme-coated carbon nanotubes as single-molecule biosensors. Nano Lett 3:727–730. https://doi.org/10.1021/nl034139u

133. Tans SJ, Verschueren ARM, Dekker C (1998) Room-temperature transistor based on a single carbon nanotube. Nature 393:49–52. https://doi.org/10.1038/29954

134. Rutherglen C, Kane AA, Marsh PF et al (2019) Wafer-scalable, aligned carbon nanotube transistors operating at frequencies of over 100 GHz. Nat Electron 2:530–539. https://doi.org/10.1038/s41928-019-0326-y

135. Cao Y, Brady GJ, Gui H et al (2016) Radio frequency transistors using aligned semiconducting carbon nanotubes with current-gain cutoff frequency and maximum oscillation frequency simultaneously greater than 70 GHz. ACS Nano 10:6782–6790. https://doi.org/10.1021/acsnano.6b02395

136. Balasubramanian K, Burghard M (2006) Biosensors based on carbon nanotubes. Anal Bioanal Chem 385:452–468. https://doi.org/10.1007/s00216-006-0314-8

137. Kumar MA, Jung S, Ji T (2011) Protein biosensors based on polymer nanowires, carbon nanotubes and zinc oxide nanorods. Sensors 11:5087–5111. https://doi.org/10.3390/s110505087

138. Sireesha M, Jagadeesh Babu V, Kranthi Kiran AS, Ramakrishna S (2018) A review on carbon nanotubes in biosensor devices and their applications in medicine. Nanocomposites 4:36–57. https://doi.org/10.1080/20550324.2018.1478765

139. Cai H, Cao X, Jiang Y et al (2003) Carbon nanotube-enhanced electrochemical DNA biosensor for DNA hybridization detection. Anal Bioanal Chem 375:287–293. https://doi.org/10.1007/s00216-002-1652-9

140. Chen H, Huang J, Fam D, Tok A (2016) Horizontally aligned carbon nanotube based biosensors for protein detection. Bioengineering 3:23. https://doi.org/10.3390/bioengineering3040023

141. Gholizadeh A, Shahrokhian S, Iraji zad A, et al (2012) Fabrication of sensitive glutamate biosensor based on vertically aligned CNT nanoelectrode array and investigating the effect of CNTs density on the electrode performance. Anal Chem 84:5932–5938. https://doi.org/10.1021/ac300463x

142. Gholizadeh A, Shahrokhian S, Iraji zad A et al (2012) Mediator-less highly sensitive voltammetric detection of glutamate using glutamate dehydrogenase/vertically aligned CNTs grown on silicon substrate. Biosens Bioelectron 31:110–115. https://doi.org/10.1016/j.bios.2011.10.002

143. Kim W-S, Lee G-J, Ryu J-H et al (2014) A flexible, nonenzymatic glucose biosensor based on Ni-coordinated, vertically aligned carbon nanotube arrays. RSC Adv 4:48310–48316. https://doi.org/10.1039/C4RA07615J

144. Tîlmaciu C-M, Morris MC (2015) Carbon nanotube biosensors. Front Chem 3. https://doi. org/10.3389/fchem.2015.00059

145. Gao M, Dai L, Wallace GG (2003) Biosensors based on aligned carbon nanotubes coated with inherently conducting polymers. Electroanalysis 15:1089–1094. https://doi.org/10.1002/elan. 200390131

146. Fayazfar H, Afshar A, Dolati M, Dolati A (2014) DNA impedance biosensor for detection of cancer, TP53 gene mutation, based on gold nanoparticles/aligned carbon nanotubes modified electrode. Anal Chim Acta 836:34–44. https://doi.org/10.1016/j.aca.2014.05.029

147. Zhu Z, Garcia-Gancedo L, Flewitt AJ et al (2012) A critical review of glucose biosensors based on carbon nanomaterials: carbon nanotubes and graphene. Sensors 12:5996–6022. https://doi. org/10.3390/s120505996

148. Zhao K, Zhuang S, Chang Z et al (2007) Amperometric glucose biosensor based on platinum nanoparticles combined aligned carbon nanotubes electrode. Electroanalysis 19:1069–1074. https://doi.org/10.1002/elan.200603823

149. Yang Q, Qu Y, Bo Y et al (2010) Biosensor for atrazin based on aligned carbon nanotubes modified with glucose oxidase. Microchim Acta 168:197–203. https://doi.org/10.1007/s00 604-009-0272-x

150. Zilli D, Bonelli PR, Cukierman AL (2006) Effect of alignment on adsorption characteristics of self-oriented multi-walled carbon nanotube arrays. Nanotechnology 17:5136–5141. https:// doi.org/10.1088/0957-4484/17/20/016

151. Lu Z, Raad R, Safaei F et al (2019) Carbon nanotube based fiber supercapacitor as wearable energy storage. Front Mater 6. https://doi.org/10.3389/fmats.2019.00138

152. Kim YJ, Kim YA, Chino T et al (2006) Chemically modified multiwalled carbon nanotubes as an additive for supercapacitors. Small 2:339–345. https://doi.org/10.1002/smll.200500327

153. Landi BJ, Castro SL, Ruf HJ et al (2005) CdSe quantum dot-single wall carbon nanotube complexes for polymeric solar cells. Sol Energy Mater Sol Cells 87:733–746. https://doi.org/ 10.1016/j.solmat.2004.07.047

154. Kim C, Kim YJ, Kim YA et al (2004) High performance of cup-stacked-type carbon nanotubes as a Pt–Ru catalyst support for fuel cell applications. J Appl Phys 96:5903–5905. https://doi. org/10.1063/1.1804242

155. Shimoda H, Gao B, Tang XP et al (2001) Lithium intercalation into opened single-wall carbon nanotubes: storage capacity and electronic properties. Phys Rev Lett 88:015502. https://doi. org/10.1103/PhysRevLett.88.015502

156. Wang K, Luo S, Wu Y et al (2013) Super-aligned carbon nanotube films as current collectors for lightweight and flexible lithium ion batteries. Adv Funct Mater 23:846–853. https://doi. org/10.1002/adfm.201202412

157. Park J, Moon WG, Kim G-P et al (2013) Three-dimensional aligned mesoporous carbon nanotubes filled with CO_3O_4 nanoparticles for Li-ion battery anode applications. Electrochim Acta 105:110–114. https://doi.org/10.1016/j.electacta.2013.04.170

158. Dogru IB, Durukan MB, Turel O, Unalan HE (2016) Flexible supercapacitor electrodes with vertically aligned carbon nanotubes grown on aluminum foils. Prog Nat Sci Mater Int 26:232–236. https://doi.org/10.1016/j.pnsc.2016.05.011

159. Niu C, Sichel EK, Hoch R et al (1997) High power electrochemical capacitors based on carbon nanotube electrodes. Appl Phys Lett 70:1480–1482. https://doi.org/10.1063/1.118568

160. Barisci JN, Wallace GG, Baughman RH (2000) Electrochemical characterization of single-walled carbon nanotube electrodes. J Electrochem Soc 147:4580. https://doi.org/10.1149/1. 1394104

161. Pan H, Li J, Feng YP (2010) Carbon nanotubes for supercapacitor. Nanoscale Res Lett 5:654–668. https://doi.org/10.1007/s11671-009-9508-2

162. Klinger C, Patel Y, Postma HWC (2012) Carbon nanotube solar cells. PLoS ONE 7:e37806. https://doi.org/10.1371/journal.pone.0037806

163. Li X, Jung Y, Sakimoto K et al (2013) Improved efficiency of smooth and aligned single walled carbon nanotube/silicon hybrid solar cells. Energy Environ Sci 6:879. https://doi.org/ 10.1039/c2ee23716d

164. Nguyen CT-C (1999) Frequency-selective MEMS for miniaturized low-power communication devices. IEEE Trans Microw Theory Tech 47:1486–1503. https://doi.org/10.1109/22.780400

165. Roukes M (2001) Nanoelectromechanical systems face the future. Phys World 14:25–32. https://doi.org/10.1088/2058-7058/14/2/29

166. LaHaye MD (2004) Approaching the quantum limit of a nanomechanical resonator. Science (80–) 304:74–77. https://doi.org/10.1126/science.1094419

167. Lassagne B, Garcia-Sanchez D, Aguasca A, Bachtold A (2008) Ultrasensitive mass sensing with a nanotube electromechanical resonator. Nano Lett 8:3735–3738. https://doi.org/10.1021/nl801982v

168. Eom K, Park HS, Yoon DS, Kwon T (2011) Nanomechanical resonators and their applications in biological/chemical detection: nanomechanics principles. Phys Rep 503:115–163. https://doi.org/10.1016/j.physrep.2011.03.002

169. Zheng M, Eom K, Ke C (2009) Calculations of the resonant response of carbon nanotubes to binding of DNA. J Phys D Appl Phys 42:145408. https://doi.org/10.1088/0022-3727/42/14/145408

170. Laocharoensuk R, Burdick J, Wang J (2008) Carbon-nanotube-induced acceleration of catalytic nanomotors. ACS Nano 2:1069–1075. https://doi.org/10.1021/nn800154g

171. Fennimore AM, Yuzvinsky TD, Han W-Q et al (2003) Rotational actuators based on carbon nanotubes. Nature 424:408–410. https://doi.org/10.1038/nature01823

172. Narang J, Pundir CS (2018) Current and future developments in nanomaterials and carbon nanotubes introduction to carbon nanomaterials. Bentham Science Publishers

Reduced Graphene Oxide Photodetector Devices for Infra-Red Sensing

Vinayak Kamble, Soumya Biswas, V. R. Appu, and Arun Kumar

Abstract Infra-Red (IR) radiation is the thermal radiation which is characterized by the temperature of the emitting source. Hence, IR photodetectors could be used for a number of applications such as surveillance in defence, non-contact thermometry, non-contact human access control, bolometers and terahertz, etc. Infra-Red spans over a vast range of wavelengths, i.e. from 1 μm to several tens of μm. While there are several materials used for sensing the IR radiation, broadly the underlying physical principles of IR detections could be classified into three distinct categories i.e. photothermoelectrics, photovoltaics and photogating. The heating effect of IR radiation brings about the thermopower considerations in case of non-uniform illumination or inhomogeneity of the sample leading to photothermoelectric effect. While, in case photovoltaics, the generation of photocurrent as a result of exitonic contribution modulates the conductivity of the device. However, in photogating, the Fermi energy of the sensor material is controlled through optical illuminations. In this chapter, we restrict ourselves to photovoltaic and photothermoelectrics in case of reduced graphene oxide and its composites. Graphene and its derivatives, such as graphene oxide, graphene nanoribbons, graphene quantum dots, etc., have revealed a wide range of novel physical properties and led to a spectrum of functional devices. Because of its small yet tunable bandgap through controlled reduction, graphene oxide is a potential choice for IR detection devices. Here in this chapter, we discuss its physical attributes which could be utilized for IR detection.

Keywords Reduced graphene oxide · Photodetector · Infra-Red · Heterojunction · Photothermoelctrics

V. Kamble (✉) · S. Biswas · A. Kumar
School of Physics, Indian Institute of Science Education and Research, Vithura,
Thiruvananthapuram 695551, Kerala, India
e-mail: kbvinayak@iisertvm.ac.in

V. R. Appu
Sustainable Energy and Materials Lab, Korea Maritime and Ocean University, Yeongdo-GU,
Busan, South Korea

© The Author(s), under exclusive license to Springer Nature Singapore Pte Ltd. 2021   349
A. Hazra and R. Goswami (eds.), *Carbon Nanomaterial Electronics: Devices
and Applications*, Advances in Sustainability Science and Technology,
https://doi.org/10.1007/978-981-16-1052-3_14

1 Introduction

Photodetectors are the devices which respond to the light energy incident on it. The response is usually measured in electrical signals, i.e. either current or voltage [1] . In a typical photosensor device, the material is deposited with two metal electrodes for electrical measurements. A constant bias voltage is applied to these metal electrodes and the resulting current is measured in the absence as well as the presence of any light, called dark and illuminated current, respectively. Often, the bias is also swept to measure the I–V characteristics in dark and under illumination. In the transistor configurations, there exists a third terminal, i.e. gate, which is realized either through the bottom SiO_2 layer (200–300 nm) on the Si substrate or a metal contact deposited on top (called top gate). The idea of a gate is to tune the channel conductivity through the application of electric field (positive or negative bias). For instance, the Dirac point (lowest conductivity) for p-type semiconductors is obtained for negative bias and vice versa for n-type semiconductors [2, 3].

In an ideal setting, the current is continuously monitored as a function of time under a constantly applied bias and the the change in current is monitored upon illumination with desired wavelength(s) of light. The responsivity or response is calculated as change in current (ΔI_{Ph}) upon illumination to that of power (P) and area of the device (A), as shown in Eq. (1) [1, 3].

$$R = \frac{\Delta I_{Ph}}{P.A} \tag{1}$$

Thus, the unit of response or responsivity is ampere/watt.

Such response can be measured as a function of the wavelength of light to estimate the bandwidth (span of frequencies) of the response of the photodetector. Besides, the noise of the detector is quantified in terms of the spectral power density as a function of the frequency bandwidth of the noise. For practical applications, it is ideal that the device shows a high signal-to-noise ratio with large bandwidth. However, this quantity is seldom reported in many of the research papers.

Normally, it is observed in most materials that the change in photocurrent is positive upon illumination (i.e. rise in current). This indicated that the rise in current of the device is due to photogenerated carriers in semiconductors. Here, the electron-hole pairs are generated upon illumination with light of photon energy higher than that of the bandgap [1]. However, when Infra-Red (IR) light is incident on a material because of the thermal energy density of the radiation, there may exist some local heating in the sample. This rise in temperature may also contribute towards the increase in carrier density (in semiconductors) through thermally generated charge carriers in addition to the photogenerated ones. Thus, one also has to consider the Temperature Coefficient of Resistance (TCR) of the material. Such devices are called bolometer which are sensitive to temperature changes, for example, vanadium dioxide [4, 5]. Besides, the heating effect due to the IR exposure introduces a thermoelectric contribution to the current if the exposure is partial or non-uniform. Thus, all these processes collectively participate in the photodetection using semiconductor devices.

While the photodetector term refers to the detection of light of any wavelength, we restrict ourselves to the detection of IR wavelengths throughout the scope of the chapter. The IR radiation is the thermal radiation as emitted from any given body in equilibrium at a finite temperature (T > 0 K). The governing equation of the relation between the wavelength of thermal radiation and its temperature is given by Wein displacement law. The emission spectrum wavelength maxima (λ_{max}) of any given body is related inversely with the temperature of the body, i.e. $\lambda_{max} \propto 1/T$. Thus, for bodies with temperature in the range 10–100 emitt thermal radiation in the IR region of the electromagnetic spectrum.

These photodetector devices are highly useful for multiple technological, domestic and strategic applications such as contactless access doors, smart devices for Internet of Things (IoT), UV detectors, etc. Particularly, IR-sensitive devices are useful for non-contact access as they can sense human radiations which happens to be in mid-IR. These are also of prime importance for defence applications for surveillance [3].

2 Structure of Graphene and Its Derivatives

Although it was known for years that graphite consists of a stack of two-dimensional (2D) layers, there was no experimental evidence of someone isolating such sheets individually. In fact, Landau, the famous condensed matter physicist, in one of his papers said that the isolation of such 2D sheets would not be possible as they would be thermodynamically unstable, despite their presence as a part of bulk three-dimensional solid. However, in 1991, a Japanese physicist, Prof. Ilijima [6], could isolate what is now called multiwalled carbon nanotubes (MWCNTs) and subsequently single-walled carbon nanotubes (SWCNTs). These were ultra-small (a few nanometres thick), hollow tubes of only carbon atoms [7, 8]. About a decade people celebrated this discovery and used CNTs in pretty much everything from electronic devices to high-strength fibre composites [9–13].

In the year 2004, Prof Andre Geim and Prof. Konstantin Novoselov of the University of Manchester published a paper which discussed the effect of electric field (FET) on single-atom-thick sheets of carbon, which showed ultra-high conductivity and tunability in Fermi level with the application of electric field [14]. For this, the duo won the Nobel prize in Physics in the year 2010. The first decade of the century unequivocally celebrated the new allotrope of carbon which was discovered in the form of 2D sheet of carbon atoms. This discovery of Prof A. K. Geim and Prof Sir K. S. Novoselov which they called 'Graphene', was arranged in a hexagonal lattice which attracted enormous attention since then [15–17]. This is due to its remarkable physical, chemical, and mechanical properties such as high electrical conductivity, high thermal conductivity, high carrier motilies and ultra-high mechanical hardness along with flexibility. Because of these unusual combinations of properties along with ultra-thin nature, graphene has turned out to be a potential novel electronic material in a vast range of applications [18–23].

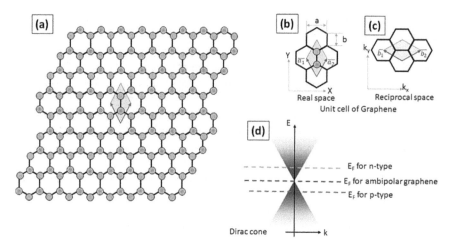

Fig. 1 The schematic diagram of showing **a** the hexagonal structure of graphene, the unit cell of graphene in **b** real and **c** reciprocal space and the band structure of graphene showing Dirac cone with position of Fermi energy for ambipolar, p-doped and n-doped graphene

We introduce ourselves to the graphene layers in terms of their structure and morphology. As shown in Fig. 1, the graphene sheet is a single-atom-thick layer of sp²-hybridized carbon atoms which is essentially building block of most of the carbon allotropes. These carbon atoms are arranged in a regular hexagonal arrangement called a honeycomb lattice [24]. Wherein each hexagon is naturally composed of six carbon atoms (like benzene) and the interatomic distance between the two neighbouring carbon atoms is, $b = 1.42$ Å, while the linear distance between the alternate carbon atoms is, $a = \sqrt{3}b = 2.45$ Å [25], as shown in Fig. 1. The unit cell of graphene layer has been shown in Fig. 1b using the yellow parallelogram which is marked by the two unit cell vectors $\overline{a_1}$ and $\overline{a_2}$; whose magnitudes are $|\overline{a_1}| = |\overline{a_2}| = a$. These unit vectors make an angle of 60° with respect to each other. The unit cell contains two carbon atoms each.

The reciprocal lattice (k-space) of graphene is also a honeycomb lattice. The reciprocal lattice parameters are $\overline{b_1}$ and $\overline{b_2}$; are given by

$$\overline{a_i} . \overline{b_j} = 2\pi \delta_{ij} \tag{2}$$

where, $\overline{a_i}$ and $\overline{b_j}$ are real and reciprocal space vectors, respectively, and the $i, j = 1$, 2. In other words, the $2\pi\delta_{ij}$ is equal to 1 when indices are same and 0 when indices are mixed.

The k-space unit vectors are having a magnitude of $|\overline{b_1}| = |\overline{b_2}| = b = 4\pi\sqrt{3}a$; making an angle of 120° with each other as shown in Fig. 1c.

Thus, by virtue of the symmetry of the honeycomb lattice, the electronic bands of graphene layers exhibit linear dispersion in the vicinity of the Fermi level (E_F) i.e. $E \propto k$. This characteristic band structure is called a "Dirac cone", and results in unusual characteristics for the charge carriers such as relativistic nature, ultra-high mobility, etc. [24–26]. At ground state, i.e. 0 K the electrons are completely occupying the valance band states making a null electrochemical potential for graphene. However, as the temperature is raised the carrier gets thermally activated to the available continuous levels and thus results in an *n*-type nature.

It has been demonstrated that graphene has zero electronic bandgap. However, the presence of some function groups can change the dominant type of charge carriers as well as may open up a finite bandgap close to the Dirac point. The calculations performed by several researchers indicated this band opening can result from oxygen functionalization or patterning the graphene sheets into lower dimensional structures such as nanoribbons or quantum dots. This patterning might involve a stringent lithographic process and hence are little tedious. On the other hand, solution-based chemical methods offer a controlled synthesis of different graphene nanostructures as well as controlled oxidation/reduction.

The graphene can be obtained from graphite by simple mechanical exfoliation or chemical exfoliation methods. The chemical exfoliation involves oxidation of graphite into graphite oxide (GO) followed by reduction to give reduced graphene oxide as shown in Fig. 2.

Graphite oxide is a brown-coloured solid and is an insulator. GO is known as a non-conductor due to the disconnected network of its conjugated carbon domains by oxygenated groups including alcohols, epoxies and carboxylic groups [27]. The bandgap of GO is estimated to be more than 2 eV [26, 28]. Because of the presence of intercalated oxygen-containing functional groups, the interlayer spacing of GO is

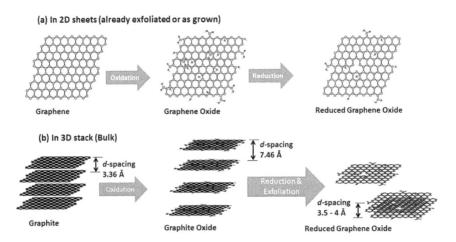

Fig. 2 The schematic showing the process flow for obtaining reduced graphene oxide sheets from as grown 2D sheets or bulk graphite material

much larger than that of the graphite. GO has a d-spacing of typically 7.46 Å and the corresponding diffraction peak is observed in XRD at a much lower Bragg angle of about 10–12° [29].

When most of the oxygen-containing functional groups of GO are removed either via heat treatment or via chemical reduction, the electrical properties are largely restored similar to prior to the oxidation, however not completely restored as some residual functional groups may remain. The extent of reduction of GO is quantified in terms of various characterization techniques such as X-ray diffraction (the peak at 10° disappears), X-ray photoelectron spectroscopy (the photoemissions observed in C 1 s spectrum are diminished except 284.5 eV), and Raman as well as Infra-Red (IR) spectroscopy [28, 30]. Other techniques are not mentioned in this chapter as it is out of scope, while IR spectroscopy is mentioned in detail in subsequent sections.

The reduced graphene oxide (RGO) which is obtained by chemical exfoliation has a small electronic bandgap of 10–200 meV [28]. This bandgap corresponds to a near band edge absorption in IR range with a wavelength of 5–25 μm. This could be further extended by additional reduction reducing the bandgap to approach to zero like pristine graphene. This narrow electronic bandgap in RGO is thus very suitable for the detection of electromagnetic radiation of these near, mid- and far IR extending to even Tera Hz waves [31–35].

3 Synthesis of Reduced Graphene Oxide

Several techniques for GO reduction have been reported, including thermal annealing in vacuum or inert atmosphere, electrical reduction, chemical reduction by hydrazine or acids, and reduction by IR or even UV light. The most commonly used methods of RGO synthesis are by solution-based method resulting from exfoliation of graphite powders with strong oxidizing reagents using modified Hummers' method [36]. Finally, graphene oxide is reduced into graphene-like material. Previous works have reported the reduction by using reducing agents such as sodium borohydride, hydrazine, etc. However, these chemicals might be mild to highly toxic to human and environment [3, 4]. This has, therefore, stimulated a number of researches in recent years to develop simpler and more benign methods for the reduction of graphene oxide. The details of strategies employed to reduce graphene oxide to restore its conductivity are out of scope of this chapter and interested readers are suggested to look at other texts available [24, 30, 37]. However, we would like to emphasize here that graphene oxide reduced using chemical routes show diverse degrees of reduction and types of residual functional groups depending on the reducing agent used and temperature of the reaction. On the other hand, the other strategy to reduce GO is heating the GO to high temperature which gives RGO [38]. Similar heating-induced reductions have been reported using different heating mechanisms such as microwave heating or localized heating using intense IR lights [39, 40]. There has been consistency in the type of residual functional groups with thermally exfoliated graphene oxide and it is discussed in detail in the next section.

4   Infra-Red Absorption of Graphene Oxide and RGO

Depending on the method is employed in reduction process, the nature of residual functional groups in RGO differs and hence the IR spectra would have some qualitative change. Both the scenarios are discussed in this section. The typical spectra of IR absorption 400–4000 cm^{-1} correspond to wavelength absorption of 2.5–25 μm in mid to far IR.

4.1   Thermally Reduced Graphene Oxide

Graphene Oxide when heated at high temperatures, not only exfoliates but also loses the oxygen-containing functional groups. Thus, thermal oxidation of GO is one of the simple and popular **strategies** to obtain the RGO. The degree of reduction is a strong function of heat treatment temperature. Spray pyrolysis is one such popular technique which involves thermal reduction of GO.

Acik et.al. [41] have studied the Infra-Red absorption spectrum of graphene oxide as a function of thermal reduction temperature in vacuum. As seen in Fig. 3, the single-layer graphene oxide shows absorption bands ascribed to carbon bonded to various oxygen-containing functional groups summarized in Table 1.

As shown in Fig. 3, Acik et.al. [41] reported that upon annealing the RGO at high temperature, as the temperature increases, the oxygen-containing functional groups are slowly desorbed and the corresponding IR absorptions (mentioned in Table 1) peaks disappear. It is clear from Fig. 3c that by 300 °C, most of the functional groups are desorbed except for the strongly bonded oxygen present at the edges which shows a strong peak at 850 cm^{-1}. Further, it was also deduced from the simulations that isolated $C-O-C$ asymmetric stretching mode observed at about 1200 to 1300 cm^{-1}, gradually decreases in wave number as the oxygen number increases. This mode further asymptotically reduces to about 850 cm^{-1}, which corresponds to the edge oxygen species in an infinite series similar to edge oxidized graphene nanoribbons.

4.2   Chemically Exfoliated RGO

Similar to thermally exfoliated GO, the chemical methods of reduction of GO show that most of the IR absorption peaks observed in case of GO are disappeared in case of RGO. However, as mentioned earlier presence of individual peak is sensitive to the reducing agent employed to carry out this reduction process. Nonetheless, additional peaks of C-N are bond absorption are seen in the reports where nitrogen containing compounds are used for reduction such as pyrroles, hydrazine, etc. [28]. Rest energies of particular IR absorption are similar to those of RGO obtained by thermal route.

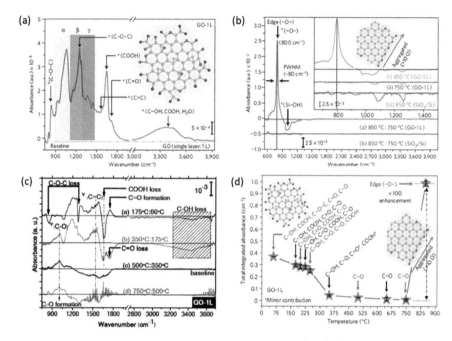

Fig. 3 The Infra-Red absorption spectra of (**a**) single-layer of graphite oxide and (**b**) that of the same after reducing it in vacuum at 850 °C showing strong absorption band at 800 cm^{-1}. (**c**) The IR absorption spectra of GO annealed at 175, 350, 500, 750 °C showing diminishing absorption bands. (**d**) The total integrated absorbance (cm^{-1}) at 60 °C, plotted in samples at various annealing temperatures showing the gradually decreasing trend marking the removal of said functional group from GO sample (Reproduced from [41] LICENSE #: 4901770786349 Order Date: 09/04/2020)

Table 1 Summary of the Infra-Red absorption bands observed in spectrum of thermally annealed graphite oxide with their corresponding wave numbers (from [41])

Sr no	Functional group name	Functional group structure	Wave number (cm^{-1})
1	Hydroxyl (including all C−OH vibrations from COOH and H$_2$O)	−OH	3050–3800 and ~1070
2	Ketonic groups	C=O	1600–1650, 1750–1850
3	Carboxyl	−COOH	1650–1750
4	Epoxide	C−O−C (ring)	1230–1320 (asymmetric stretching) ~850 (bending)
5	sp^2-hybridize	C=C	1500–1600 (in plane)
6	Etheric groups	C−O−C	900–1100

5 Electrical Response of RGO and Its Heterostructures

It was soon after the discovery of graphene, various electronic devices based on graphene and RGO were reported, for instance, transistors, gas sensors and photodetectors. The early reports on photodetection using graphene try various material and device configurations for the photodetectors. Some of the representative device designs and their responses are summarized in Table 2. Predominantly the devices can be classified into two categories, i.e. RGO as single-active material and heterostructures based on graphene.

5.1 RGO as a Single-Active Material

In a usual setup, two metal electrodes (Au, Pt, Ag, etc.) are deposited onto the Graphene flakes using lithography or RGO suspension is drop casted in to the predeposited electrodes. RGO shows a large change in its conductivity when illuminated

Table 2 Summary and comparison of RGO-based IR Sensors from the literature

RGO as single-active material

Sr no	Graphene form	Configuration	Response (mA/W) at given wavelength	Year	References
1	Graphene Nanoribbon (GNR)	Transistor (top gate) modelling	50–250 A/W	2008	[42]
2	Reduced Graphene Oxide	Thin film (large area, two probe)	0.01 mA/W at 808 nm	2010	[43]
3	GNR	Transistor Modelling		2009	[44]
4	Semi reduced RGO	Film with two probe	0.18 A/W at 1,550 nm	2018	[45]
5	RGO film	Film with two probe	0.04 mA/W for 1064 nm	2018	[46]
6	RGO sheet Thermally oxidized	Transistor	0.7 A/W at 895 nm	2013	[47]
7	RGO film	Two probe on flexible substrate	1.5 mA/W at 940 nm	2020	[48]
8	RGO film	Two probe on flexible substrate	800 mA/W at λ > 780 nm	2016	[49]
9	RGO film	Free standing RGO film with two probes	96 mA/W at 1064 nm	2017	[50]
10	RGO film	Film with two probe	0.3 mA/W at 808 nm	2018	[51]
11	RGO	Two probe	4 mA/W at 1550 nm	2011	[52]

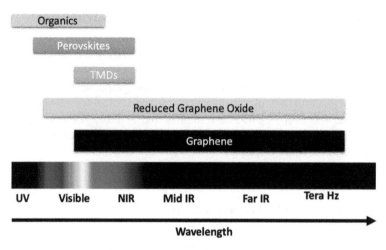

Fig. 4 The comparison of bandwidth of RGO and Graphene-based photodetectors with that of others (Adopted from [35])

with IR light.[1] Joung and Khondar [28] showed as the oxygen-containing functional groups reduce on graphene oxide (i.e. reduction in **sp²** hybridized carbon), the conductivity increases. Viz a viz the direct point shifts from negative gate bias to positive gate bias signifying a change over from p-type conduction in GO to ambipolar conduction in heavily reduced graphene oxide. Further, they also demonstrated that the conduction in RGO is not band like rather it happens with variable range hopping of Efros-Shklovskii type through localized states [28] (Fig. 4).

Kurra et al. [53] have shown that the dominant type of mechanism being either photoconduction or bolometric is predominantly dependent on the synthesis method in pristine graphene. The graphene which is made by pyrolysis of hydrocarbons over nickel metal shows a large number of defects and photoconductive type of response. While that of the sufficiently thin graphitic film made by mechanical exfoliation and Chemical vapor deposition showed a rise in resistance upon exposure to IR radiation which is typical of a bolometric response.[2] Similar, bolometer like (positive TCR) response was shown by RGO annealed at 900 °C, in a study reported by Liang [54] This is very good agreement with the discussion above, as reducing at 900 °C would result in the removal of almost all the oxygen-containing functional group and finally end up making a graphite-like graphene flakes. Since, the graphite acts more like a dirty metal whose resistance increases upon heating, while that the graphene incorporated with hydrocarbon residues acts like a semiconductor where the carriers are

---

[1] Some of the researchers also found the response for shorter wavelengths of near IR and part of visible spectrum as shown in Fig. 5. However, we restrict ourselves to IR region only for the scope of this chapter.

[2] The bolometric response depends on the Temperature Coefficient of Resistance (TCR) and TCR is positive in metals. Hence the rise in temperature causes rise in resistance of bolometer.

excited under the illumination of suitable light wavelengths. However, their wavelengths reported by various researchers are quite diverse and suggest that these are highly processing condition-specific which govern the surface functionalization of graphene or RGO.

Thus, because of the heating effect of IR radiation the two mechanisms might be involved in the IR sensing using RGO are photothermoelectrics and Photoconductivity.

5.1.1 Photothermoelectricity (PTE)

When there exists a temperature gradient in a solid, an electrical potential is generated between the hot and cold ends of solid depending on its thermopower (S). The photo-induced temperature can be estimated using the input power and the specific heat of the material.

$$Q_{\text{heat}} - Q_{\text{loss}} = C.m.\Delta T = P_i \tag{3}$$

where $Q_{\text{heat}} - Q_{\text{loss}}$ is the heat supplied to heat lost to the surrounding, C is the specific heat of material, m is mass of material, P_i is the input radiation power and ΔT is the rise in temperature. Thus some part of the incident radiation power is utilized in the rise in temperature and a fraction is lost to surrounding depending on the emissivity of material. This rise in temperature causes the ΔV of potential gradient across the sample which is governed by Eq. (4).

$$\Delta V = -S.\Delta T \tag{4}$$

S is usually of the order of few μV/K and varies inversely with that of carrier concentration (n) of a material.

$$S = \frac{8\pi^2 k_B^2}{3eh^2} m^* T \left(\frac{\pi}{3n}\right)^{2/3} \tag{5}$$

Here, k_B is Boltzmann constant, m^* is the effective mass and e is the charge on electron, h is Planck's constant, T is the absolute temperature and n is the carrier concentration. It is discernible from Eq. (5) that the seebeck coefficient for RGO will be higher than that of pristine graphene due to relatively low electrical conductivity (and carriers concentration) [55]. Thus, a thermoemf generated will be higher for RGO when exposed to **non-uniform** heating such as local IR exposure.

Seebeck coefficient can also be obtained by Eq. (6) [35].

$$S = \frac{2\pi^2 k_B T_e}{3e T_F} \tag{6}$$

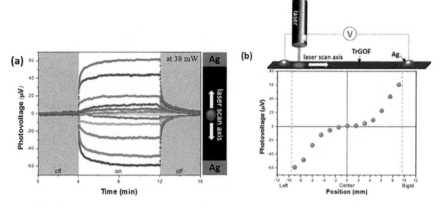

Fig. 5 a The thermoemf response measured along the sample length (s) showing time response and **b** The steady state photovoltaic responses' saturation voltages showing non-linear dependence on the laser spot positions symmetric about the center. From [57][3]

where T_F is the Fermi temperature i.e. E_F/k_B is the temperature of the electrons at energy of Fermi energy and T_e is the temperature of electrons at a given energy. This expression holds for large carrier density limit where the carrier scattering is independent of change in carrier density. Assuming that on the site of illumination, some of the carriers attain a temperature which is slightly higher than that of lattice temperature, then the thermoemf generated varies as $n^{-2/3}$ and thus, the resulting current varies as $P_i^{2/3}$ [35].

Xu et al. [56] studied the field-effect transistor geometry of graphene-based devices and found that the photocurrent generation is due to thermoelectric voltage arising at the interface of single and bi-layer graphene. This was confirmed by the spatially resolved photocurrent mapping. The photocurrent so obtained was expressed as

$$I_{ph} = \frac{S \times \Delta T}{R} \tag{7}$$

where, the numerator on RHS is actually given by Eq. (4).

The wavelength used was 635 nm and measurements were done under no bias condition between source to drain. Thus, the current measured upon illumination was solely due to the internal diffusion of carriers setup due to unbalanced thermal gradient. Moreover, it was observed that the photocurrent due to PTE effect increases as temperature decreases (largely a linear dependence). This is consistent with the fact that the thermopower decreasing with increasing temperature for mechanically exfoliated graphene which is assumed to have no or insignificant surface oxidation.

Moon et al. [57] showed that the photo response of RGO film showed systematic photo-induced voltage as shown in Fig. 5. This built in electric field resulted in a

---

[3]Reproduced under Creative Commons Attribution 4.0 International License.

photocurrent in the device with a responsivity of 0.27 mA/W for 633 nm. Similar study of PTE effect in RGO was reported by Ghosh et al. [43] In such cases the photovoltage produced and hence the photocurrent varies symmetrically about the centre of the electrodes and had maximum value towards the edges. (as seen in Fig. 5b) This was explained with the help of enhance charge carrier extraction efficiency near the electrodes and possible loos of one of the charge carriers due to recombination in due course of reaching to the other electrode. However, if the generation sit is equally far from both the electrodes, the photovoltage is rather low indicating loss of both types of charge carriers to recombination. The hypothesis of recombination-based loss has been claimed from the deviation of exponent of power from unity as it is ideally unity for efficient charge carrier extraction solely due to photoconductivity as discussed in next section. Moreover, in a device there may be two competing processes like photothemoelectrics and photoconductivity [58].

5.1.2 Photoconductivity

Photoconductivity is also often referred as photovoltaics which in a broader sense implies the generation of photoexcited carriers within a material. The applied electric field accelerates the photocarriers towards the opposite polarities of the bias where it is collected and measured. The observed photocurrent is expected to have a linear dependence with respect to the incident optical power i.e. $I_{ph} \propto P_i$. The exact expression for the photocurrent is given by Eq. (8);

$$I_{ph} = \left[\frac{\eta_i}{h\nu} \frac{W}{L} \tau_c q (\mu_e + \mu_h) V \right] P_i \qquad (8)$$

where, η_e is the internal quantum efficiency, $h\nu$ is the photon energy, w and L are the width and length of the exposed area, μ_e and μ_h are the mobilities of electron and hole, respectively. V is applied bias voltage. Thus, the photocurrent depends crucially on the average lifetime of the charge carrier τ_c and mobilities. Thus, the device geometry i.e. the spacing between the electrodes and recombination rate determines the efficiency of charge carrier extractions. There are two ways of estimating the charge carrier extraction efficiency; namely internal and external quantum efficiency given by Eqs. (8) and (9), respectively.

$$IQE, \eta_i = \frac{I_{Ph}}{q . \phi_{abs}} \qquad (9)$$

$$EQE, \eta_e = \frac{I_{Ph}}{q . \phi_{in}} \qquad (10)$$

The quantities ϕ_{in} and ϕ_{abs} are the incident photon flux and absorbed photon flux, respectively, where,

$$\phi_{abs} \text{ Absorbed photon flux} = \phi_{in} - (\text{transmitted} + \text{reflected}) \qquad (11)$$

For example, for IR light of wavelength (λ) 1500 nm.
Energy E = hc/ λ = 0.827 eV

$$\text{No of Photons/area} = \frac{\textbf{Incident power}}{\text{Energy of each photon}} \qquad (12)$$

$$\text{EQE} = \eta_e = \frac{I_{Ph}}{q \cdot \phi_{in}} = \frac{I_{Ph} \times E}{q \times P_{opt}} = \frac{E}{q} \times slope$$

Here, slope refers to the slope of the graph current versus IR power i.e. nothing but the responsivity of the device.

Chang et.al. [47] showed that the photoresponse of pristine (as prepared) RGO (~10 mA/W) is enhanced by about an order of magnitude (~700 mA/W) with increasing heat treatment time 20, 40, 90 and 260 min at 150 °C (see Fig. 6). This result essentially highlighted the as prepared GO sheets having larger O:C ratio showed little or smaller response. On the other hand, after thermal reduction at low temperature as some of the oxygen-containing functional groups are lost (supported by XPS data) the conductivity of RGO is partially restored. The response is measured in three terminals (i.e. transistor) fashion and the silicon oxide beneath graphene is used as bottom gate. The graphene flake here acts as a channel. The Dirac point for all the RGO sample however is found to be in negative gate voltages and thus marks all of them to be hole doped (p-type) systems. Moreover, no significant change is observed in this Dirac point with increased reduction as it is rather broad unlike conventional sharp profiles.

Effect of electrode metals Other than the primary photogeneration process, the nature of electrical contact metal also has significant effect on the overall device performance [59]. For instance, if the work function of the metal used is smaller

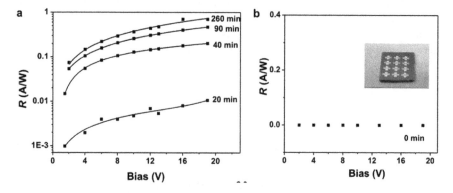

Fig. 6 a The photoresponse of RGO as a function of thermal reduction time at 150 °C for incident irradiation power is 14 mW/cm². **b** The response of as synthesized RGO flakes. (Reprinted with permission from [47] Copyright (YEAR) American Chemical Society)

or larger than that of the RGO flakes, decides whether there exists a barrier at the interface. The work function of RGO has been nominally take as 4.5 eV as reported in most studies [57, 60] and the data for the work functions of metals have been obtained from standard texts [2]. Assuming that the RGO has a p-type conductivity, i.e. hole conductor, the nominal values of contact potential difference (CPD) which is the difference in the Fermi energy of RGO and that of the metal, tells us whether the contact is Schottky or ohmic in nature. Thus, as can be seen in Figure (a) for some of the commonly used electrode metals, RGO is expected to form Schottky with silver, aluminium, titanium while it is ohmic for tungsten, gold, platinum and palladium. Thus, if it forms a Schottky barrier, in a double Schottky contact as shown in Fig. 7c, under the application of a bias, one of the side will have a forward bias and other will have a reverse bias for holes. While, the band bending this case is favourable for electron flow on either sides as seen in Fig. 7d. Thus, under illumination depending on the charge carrier generation site, it will be difficult of holes to move to the side having a reverse bias condition. Thus, the charge separation is not symmetric in this case. This could be one of the reasons for more pronounced photothermoelectric effect for RGO having Ag contacts predominantly [43, 57].

Effect of substrate Tian et al. [50] shown that the response is increased in suspended graphene films than that of the one on glass substrate. The RGO films which are free standing without any substrate showed 7 fold increase in the response upon removing the substrate as shown in Fig. 8.

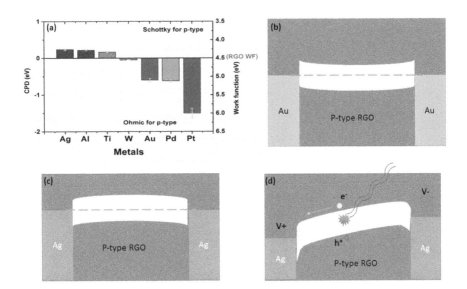

Fig. 7 **a** The variation of contact potential difference(CPD) for different metals and the nature of contact they form with RGO. The typical band alignment with **b** Au and **c** Ag metal electrodes showing ohmic and Schottky contact for RGO, respectively. **d** The formation and movement of the photogenerated carriers under applied bias (2 V)

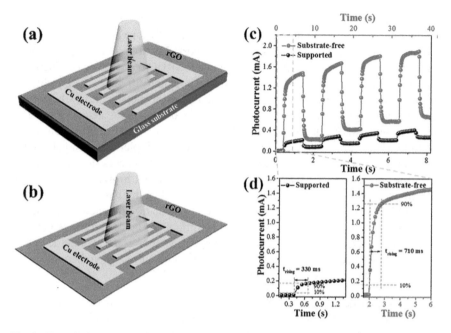

Fig. 8 The schematic showing **a** supported and **b** released or substrate free RGO film with Cu electrodes and **c** the effect of substrate on the photocurrent response of RGO film with **d** magnified view depicting the response times in either cases (From Ref [50] under Creative Commons Attribution-NonCommercial 3.0 Unported Licence)

The origin for the enhancement in the response for the free standing film was credited to the enhanced photothermal response of the film due to reduction in the thermal mass. While this is indeed true that low thermal mass shall increase the heating effect in the film and thereby shall have higher photothermal contribution in the film, given the smaller gap (200 μm) between of the interdigitated electrode spacing it is rather unlikely to have large temperature gradient within the electrodes. Thus, the contribution due to thermopower shall be insignificant. However, due to rise in temperature, there may be higher conductivity of RGO and subsequently upon cooling the resistance shall be back to baseline. In addition to this, in absence of the substrate, the carrier mobilities in graphene and other 2D materials increases due to reduced phononic interaction of substrate. This rise in mobility may have an added contribution in addition to those discussed above.

5.2 Heterostructures Based on RGO

There are numerous reports on the photodetector device designs and particularly their broadband response using heterojunction of graphene and other semiconducting or metallic materials. Table 3.

As seen from Table 3, the heterojunctions not only have high responsivity (photocurrent produced per unit power) but also show broad spectral range of response ranging from UV to NIR. Moreover, it can be particularly generalised that for making the heterojunction which shows a broadband response (not just IR),

Table 3 Some of the device architectures of RGO heterojunction with other materials and their response

RGO heterojunction with other materials

Sr no	Graphene form	Configuration	Response (mA/W) at given wavelength	Year	References
1	Reduced Graphene Oxide/n-Si	Vertical heterojunction	–	2016	[61]
2	RGO nanosheets on Si nanowires	Vertical heterojunction	–	2014	[62]
3	3D RGO–MoS$_2$/Pyramid Si Heterojunction	RGO sandwiched between MoS$_2$ sheets and placed on Si pyramids	10.6 mA/W at 1550 nm and 11.8 mA/W at 1310 nm	2018	[63]
4	Germanium-Graphene/Zinc-Oxide Heterostructure	Ge Quantum Dot Decorated Graphene/Zinc-Oxide Heterostructure	9.7 A/W, 1400 nm	2015	[64]
5	Monolayer Graphene (MLG)/germanium (Ge) heterojunction	Graphene layer on the copper foil attached to Ge wafer by Silver paste	51.8 mA/W, 1400 nm	2013	[59]
6	Layered Graphene-Black phosphorous heterostructure	Graphene put on BP layer and placed on a Si/SiO$_2$ wafer	3.3×10^3 A/W, 1550 nm	2017	[65]
7	CdS nanorod array/reduced graphene oxide film heterojunction	RGO layer is placed on the CdS nanorods array grown on the Fluorine doped Tin Oxide layer on glass substrate	0.58 mA/W, 1450 nm	2018	[66]
8	ZnO Nanowire Array/Reduced Graphene Oxide	RGO layer on ZnO nanowire array	0.55 mV/W, 1064 nm	2015	[67]

a material with suitable bandgap is used. Suitable bandgap refers to small bandgap of 0.7 to 1 eV for Si, Ge and MoS_2, black phosphorous (BP), etc., for NIR and long wavelength visible region, but wide bandgap oxides such as ZnO, TiO_2, etc., for the UV and short wavelength visible region.

The mechanism of detection can be represented as follows,

Under the illumination of IR (or visible light) the electron hole pairs are generated by breaking the excitons in the material other than RGO like Si, BP, TMDs, etc. these photogenerated carriers drift towards opposite polarities of the applied bias. Many reports use RGO as contact material or one of the heterojunction constituents for efficient recovery of electrons given its p-type nature with relatively low work function.

Certainly there is a large improvement in the responsivity in heterojunctions as compared to those of RGO as single active material exhibiting the change in current. For instance, the response reported Liu et.al. [65] the heterojunction made between RGO and black phosphorous (BP), showed a large response of the order of 10^3 A/W for very low incident photon power of a few nanowatts. Further, the response time is drastically lowered and spectral region on interest was tuned by varying the thickness of BP whose bandgap changes from bulk to few layers.

6 Summary and Outlook

To summarise, we have reviewed the structure of reduced graphene oxide (RGO) and the tunability of its electronic conductivity, carrier type and Fermi level through external stimulus such as temperature and electromagnetic radiation, particularly Infra-Red light. It is observed that when IR light is shined on RGO sheets depending on nature of device symmetry the device shows a combination of photothermoelectric and photoconductive response. For very small device sized and symmetric configurations, the device could be purely photoconductive where resistance decreases upon illumination due to photo-carrier generation. On the other hand, for highly reduced RGO (with minimal sp^2 fraction) shows more of photothermal effect, i.e. Bolometer type response which depends on the thermal mass and temperature coefficient of resistivity. In addition, the diverse spectrum of heterojunction formed between RGO and other semiconductors show much larger response with instantaneous timescales as a result of junction barrier formation at the interface which has built-in electric field.

Acknowledgements The authors are thankful to the inspire grant (Number DST/INSPIRE/04/2015/002111) of Dr. Vinayak Kamble for the funding support and supporting the fellowship of Dr. Appu V R.

References

1. Potter R, Eisenman W (1962) Infrared photodetectors: a review of operational detectors. Appl Opt 1(5):567–574
2. Grundmann M (2010) Physics of semiconductors. Springer
3. García de Arquer FP, Armin A, Meredith P, Sargent EH (2017) Solution-processed semiconductors for next-generation photodetectors. Nature Rev Mater 2(3):16100
4. Pirro S, Mauskopf P (2017) Advances in bolometer technology for fundamental physics. Ann Rev Nucl Particle Sci 67(1):161–181
5. Richards PL (1994) Bolometers for infrared and millimeter waves. J Appl Phys 76(1):1–24
6. Iijima S (1991) Helical microtubules of graphitic carbon. Nature 354(6348):56–58
7. Iijima S, Ichihashi T (1993) Single-shell carbon nanotubes of 1-nm diameter. Nature 363(6430):603–605
8. Qin L-C, Zhao X, Hirahara K, Miyamoto Y, Ando Y, Iijima S (2000) The smallest carbon nanotube. Nature 408(6808):50–50
9. Liu X-M, Dong Huang Z, Woon Oh S, Zhang B, Ma P-C, Yuen MM, Kim J-K (2012) Carbon nanotube (CNT)-based composites as electrode material for rechargeable Li-ion batteries: a review. Compos Sci Technol 72(2):121–144
10. Karthikeyan A, Mallick P (2017) Optimization techniques for CNT based VLSI interconnects—a review. J Circuit Syst Comput 26(03):1730002
11. Bakshi SR, Lahiri D, Agarwal A (2010) Carbon nanotube reinforced metal matrix composites-a review. Int Mater Rev 55(1):41–64
12. Trojanowicz M (2006) Analytical applications of carbon nanotubes: a review. TrAC Trends Analyt Chem 25(5):480–489
13. Goldoni A, Petaccia L, Lizzit S, Larciprete R (2009) Sensing gases with carbon nanotubes: a review of the actual situation. J Phys: Condensed Matter 22(1):013001
14. Novoselov KS, Geim AK, Morozov SV, Jiang D, Zhang Y, Dubonos SV, Grigorieva IV, Firsov AA (2004) Electric field effect in atomically thin carbon films. Science 306(5696):666–669
15. Geim AK, Novoselov KS (2007) The rise of graphene. Nature Mater 6(3):183–191
16. Geim AK (2009) Graphene: status and prospects. Science 324(5934):1530–1534
17. Novoselov KS, Fal'ko VI, Colombo L, Gellert PR, Schwab MG, Kim K (2012) A roadmap for graphene. Nature 490(7419):192–200
18. Schwierz F (2010) Graphene transistors. Nature Nanotechnol 5(7):487
19. Grigorenko A, Polini M, Novoselov K (2012) Graphene plasmonics. Nature Photon 6(11):749–758
20. Li D, Kaner RB (2008) Graphene-based materials. Nat Nanotechnol 3:101
21. Stoller MD, Park S, Zhu Y, An J, Ruoff RS (2008) Graphene-based ultracapacitors. Nano Lett 8(10):3498–3502
22. Stankovich S, Dikin DA, Dommett GH, Kohlhaas KM, Zimney EJ, Stach EA, Piner RD, Nguyen ST, Ruoff RS (2006) Graphene-based composite materials. Nature 442(7100):282–286
23. Xia F, Mueller T, Lin Y-M, Valdes –Garcia A, Avouris P, (2009) Ultrafast graphene photodetector. Nature Nanotechnol 4(12):839–843
24. Zhu Y, Murali S, Cai W, Li X, Suk JW, Potts JR, Ruoff RS, (2010) Graphene and Graphene oxide: synthesis, properties, and applications. Adv Mater 22(35):3906–3924
25. Biró LP, Nemes-Incze P, Lambin P (2012) Graphene: nanoscale processing and recent applications. Nanoscale 4(6):1824–1839
26. Son Y-W, Cohen ML, Louie SG (2006) Energy gaps in graphene nanoribbons. Phys Rev Lett 97(21):216803
27. Gao W, The chemistry of graphene oxide. Graphene oxide. Springer, pp 61–95
28. Joung D, Khondaker SI (2012) Efros-Shklovskii variable-range hopping in reduced graphene oxide sheets of varying carbon $s\{p\}^{[2]}$ fraction. Phys Rev B 86(23):235423
29. Pei S, Cheng H-M (2012) The reduction of graphene oxide. Carbon 50(9):3210–3228

30. Guex LG, Sacchi B, Peuvot KF, Andersson RL, Pourrahimi AM, Ström V, Farris S, Olsson RT (2017) Experimental review: chemical reduction of graphene oxide (GO) to reduced graphene oxide (rGO) by aqueous chemistry. Nanoscale 9(27):9562–9571
31. Gupta Chatterjee S, Chatterjee S, Ray AK, Chakraborty AK (2015) Graphene–metal oxide nanohybrids for toxic gas sensor: a review. Sens Actuat B: Chem 221:1170–1181
32. Son DI, Kwon BW, Park DH, Seo W-S, Yi Y, Angadi B, Lee C-L, Choi WK (2012) Emissive ZnO–graphene quantum dots for white-light-emitting diodes. Nat Nanotechnol 7(7):465–471
33. Zheng P, Liu T, Su Y, Zhang L, Guo S (2016) TiO_2 nanotubes wrapped with reduced graphene oxide as a high-performance anode material for lithium-ion batteries. Scientific Reports 6(1):36580
34. Kamble V (2019) Facile and in-situ spray deposition of SnO_2—reduced graphene oxide heterostructure sensor devices. J Phys Commun 3(1):
35. De Sanctis A, Mehew JD, Craciun MF, Russo S (2018) Graphene-based light sensing: fabrication, characterisation, physical properties and performance. Materials 11(9):1762
36. Hummers WS, Offeman RE (1958) Preparation of graphitic oxide. J Am Chem Soc 80(6):1339–1339
37. Ismail Z (2019) Green reduction of graphene oxide by plant extracts: a short review. Ceram Int 45(18):23857–23868
38. Acik M, Chabal YJ (2013) A review on thermal exfoliation of graphene oxide. J Mater Sci Res 2(1):101
39. Guo H, Peng M, Zhu Z, Sun L (2013) Preparation of reduced graphene oxide by infrared irradiation induced photothermal reduction. Nanoscale 5(19):9040–9048
40. Zhu Y, Murali S, Stoller MD, Velamakanni A, Piner RD, Ruoff RS (2010) Microwave assisted exfoliation and reduction of graphite oxide for ultracapacitors. Carbon 48(7):2118–2122
41. Acik M, Lee G, Mattevi C, Chhowalla M, Cho K, Chabal YJ (2010) Unusual infrared-absorption mechanism in thermally reduced graphene oxide. Nat Mater 9(10):840–845
42. Ryzhii V, Mitin V, Ryzhii M, Ryabova N, Otsuji T (2008) Device Model for Graphene Nanoribbon phototransistor. Appl Phys Express 1:063002
43. Ghosh S, Sarker BK, Chunder A, Zhai L, Khondaker SI (2010) Position dependent photodetector from large area reduced graphene oxide thin films. Appl Phys Lett 96(16):163109
44. Ryzhii V, Ryzhii M, Ryabova N, Mitin V, Otsuji T (2009) Graphene Nanoribbon phototransistor: proposal and analysis. Jpn J Appl Phys 48(4):04C144
45. Feng R, Zhang Y, Hu L, Chen J, Zaheer M, Qiu Z-J, Tian P, Cong C, Nie Q, Jin W, Liu R (2017) Laser-scribed highly responsive infrared detectors with semi-reduced graphene oxide. Appl Phys Express 11(1):015101
46. Abid P, Sehrawat SS, Islam P, Mishra S Ahmad (2018) Reduced graphene oxide (rGO) based wideband optical sensor and the role of Temperature. Defect States Quant Eff Scientific Reports 8(1):3537
47. Chang H, Sun Z, Saito M, Yuan Q, Zhang H, Li J, Wang Z, Fujita T, Ding F, Zheng Z, Yan F, Wu H, Chen M, Ikuhara Y (2013) Regulating infrared Photoresponses in reduced graphene oxide phototransistors by defect and atomic structure control. ACS Nano 7(7):6310–6320
48. Yang HY, Lee HJ, Jun Y, Yun YJ (2020) Broadband photoresponse of flexible textured reduced graphene oxide films. Thin Solid Films 697:137785
49. Sahatiya P, Puttapati SK, Srikanth VVSS, Badhulika S (2016) Graphene-based wearable temperature sensor and infrared photodetector on a flexible polyimide substrate. Flexible Printed Electron 1(2):025006
50. Tian H, Cao Y, Sun J, He J (2017) Enhanced broadband photoresponse of substrate-free reduced graphene oxide photodetectors. RSC Advances 7(74):46536–46544
51. Karimzadeh R, Assar M, Jahanbakhshian M (2018) Low cost and facile fabrication of broadband laser power meter based on reduced graphene oxide film. Mater Res Bull 100:42–48
52. Chitara B, Panchakarla LS, Krupanidhi SB, Rao CNR (2011) Infrared Photodetectors Based on Reduced Graphene Oxide and Graphene Nanoribbons. Adv Mater 23(45):5419–5424
53. Kurra N, Bhadram VS, Narayana C, Kulkarni GU (2013) Few layer graphene to graphitic films: infrared photoconductive versus bolometric response. Nanoscale 5(1):381–389

54. Liang H (2014) Mid-infrared response of reduced graphene oxide and its high-temperature coefficient of resistance. AIP Adv 4(10):

55. Choi J, Tu NDK, Lee S-S, Lee H, Kim JS, Kim H (2014) Controlled oxidation level of reduced graphene oxides and its effect on thermoelectric properties. Macromol Res 22(10):1104–1108

56. Xu X, Gabor NM, Alden JS, van der Zande AM, McEuen PL (2010) Photo-thermoelectric effect at a graphene interface junction. Nano Lett 10(2):562–566

57. Moon IK, Ki B, Yoon S, Choi J, Oh J (2016) Lateral photovoltaic effect in flexible free-standing reduced graphene oxide film for self-powered position-sensitive detection. Scientific Reports 6(1):33525

58. Zhang Y, Zheng H, Wang Q, Cong C, Hu L, Tian P, Liu R, Zhang S-L, Qiu Z-J (2018) Competing mechanisms for photocurrent induced at the monolayer-multilayer Graphene junction. Small 14(24):1800691

59. Zeng L-H, Wang M-Z, Hu H, Nie B, Yu Y-Q, Wu C-Y, Wang L, Hu J-G, Xie C, Liang F-X, Luo L-B (2013) Monolayer Graphene/germanium Schottky junction as high-performance self-driven infrared light photodetector. ACS Appl Mater Interfaces 5(19):9362–9366

60. Sygellou L, Paterakis G, Galiotis C, Tasis D (2016) Work function tuning of reduced graphene oxide thin films. J Phys Chem C 120(1):281–290

61. Li G, Liu L, Wu G, Chen W, Qin S, Wang Y, Zhang T (2016) Self-powered UV–near infrared photodetector based on reduced Graphene Oxide/n-Si vertical Heterojunction. Small 12(36):5019–5026

62. Cao Y, Zhu J, Xu J, He J, Sun J-L, Wang Y, Zhao Z (2014) Ultra-broadband photodetector for the visible to terahertz range by self-assembling reduced Graphene oxide-silicon nanowire array heterojunctions. Small 10(12):2345–2351

63. Xiao P, Mao J, Ding K, Luo W, Hu W, Zhang X, Zhang X, Jie J (2018) Solution-processed 3D RGO–MoS2/Pyramid Si Heterojunction for ultrahigh Detectivity and ultra-broadband photodetection. Adv Mater 30(31):1801729

64. Liu X, Ji X, Liu M, Liu N, Tao Z, Dai Q, Wei L, Li C, Zhang X, Wang B (2015) High-Performance ge quantum dot decorated Graphene/Zinc-oxide heterostructure infrared photodetector. ACS Appl Mater Interfaces 7(4):2452–2458

65. Liu Y, Shivananju BN, Wang Y, Zhang Y, Yu W, Xiao S, Sun T, Ma W, Mu H, Lin S, Zhang H, Lu Y, Qiu C-W, Li S, Bao Q (2017) Highly efficient and air-stable infrared photodetector based on 2D layered graphene-black phosphorus heterostructure. ACS Appl Mater Interfaces 9(41):36137–36145

66. Yu X-X, Yin H, Li H-X, Zhao H, Li C, Zhu M-Q (2018) A novel high-performance self-powered UV-vis-NIR photodetector based on a CdS nanorod array/reduced graphene oxide film heterojunction and its piezo-phototronic regulation. J Mater Chem C 6(3):630–636

67. Liu H, Sun Q, Xing J, Zheng Z, Zhang Z, Lü Z, Zhao K (2015) Fast and enhanced broadband Photoresponse of a ZnO Nanowire array/reduced graphene oxide film hybrid photodetector from the visible to the near-infrared range. ACS Appl Mater Interfaces 7(12):6645–6651

Characteristic Response Transition of Reduced Graphene Oxide as Hydrogen Gas Sensor-The Effect of Temperature and Doping Concentration

Anuradha Kashyap, Shikha Sinha, Sekhar Bhattacharya, Partha Bir Barman, Surajit Kumar Hazra, and Sukumar Basu

Abstract Catalytic palladium nanoparticle modified reduced graphene oxide (rGO) prepared in the laboratory showed a characteristic variation in hydrogen response when the operating temperature was raised from 30 °C to 125 °C. When exposed to a particular hydrogen concentration, the response comprising of initial increase in device resistance (response-1) followed by decrease (response-2) was observed up to 75 °C. Beyond 75 °C a transition in hydrogen response was observed, and only the response-1 prevailed from 100 °C. This transition temperature was reduced to 50 °C, when rGO was pretreated with ammonia solution at a temperature of 100 °C. On pretreatment with lower ammonia concentration, the response-2 was completely eliminated. Furthermore, without treatment with palladium nanoparticles, rGO films (both untreated and ammonia treated) showed negligible response toward various concentrations of hydrogen in the studied temperature range (30 °C–125 °C). The material characterization and sensing results were analyzed in detail and a suitable sensing model was proposed to explain the performance of these sensors.

Keywords rGO · Palladium · Dual response · Transition · Hydrogen sensor

1 Introduction

In the field of Condensed Matter Physics, one finds that after every interval of time there comes a material which becomes the paramount of that decade. The same happened in the case of Graphene, the mother of 2-dimensional materials. Attributing to its phenomenal properties, Graphene is known to be a real stunner in the field of research. It has a wide range of applicability, which is why researchers all over the globe are outworking in this field. The remarkable nanoscale properties of

A. Kashyap · S. Sinha · P. B. Barman · S. K. Hazra · S. Basu (✉)
Department of Physics and Materials Science, Jaypee University of Information Technology, Waknaghat, Solan, Himachal Pradesh 173234, India

S. Bhattacharya
Core Laboratories, King Abdullah University of Science and Technology, Thuwal, Saudi Arabia

© The Author(s), under exclusive license to Springer Nature Singapore Pte Ltd. 2021
A. Hazra and R. Goswami (eds.), *Carbon Nanomaterial Electronics: Devices and Applications*, Advances in Sustainability Science and Technology,
https://doi.org/10.1007/978-981-16-1052-3_15

graphene have been a breakthrough in the field of nanomaterials and nanotechnology. A plethora of research is enduring in the field of mass production of graphene and how this graphene could be capitalized in the field of technology. The flap in this field has also captivated the governments to invest in bulk. This commotion in the minds of researchers commenced ever since the isolation of graphene by Andre Geim and Konstantin Novoselov at the University of Manchester [1]. Graphene having aromatic nature can include functional groups in its structure. This can lead to band gap tuning that is beneficial for different applications. The reduced form of graphene oxide bears much resemblance with graphene rather than graphene oxide in respect that all the functional groups have been removed with some oxygen traces. Researchers all over the globe have continuously worked to find ways of synthesizing graphene on large scale by various methods [2–4].

The role of graphene and its derivatives in hydrogen related applications is gaining prominence. For instance, an alternative to fossil fuels is hydrogen, which is highly inflammable and explosive in nature. Therefore, safe use of hydrogen requires accurate gas sensors. Reduced graphene oxide (rGO) based resistive sensors are found to be very practical in this regard due to tremendous material properties of rGO such as large surface to volume ratio, low energy adsorption sites, good electrical and thermal properties, etc. Also it is economical to use rGO due to its large-scale low-cost synthesis. Moreover, the gas sensing properties like sensitivity and selectivity of rGO can be improved by functionalizing with metallic nanoparticles [5–8].

In recent years, a lot of work has been done in the field of hydrogen sensors based on rGO. A room temperature sensor based on thermally reduced (partially) rGO sheets shows excellent response to hydrogen gas. The temperature at which rGO sheets are reduced affects the density of states (DOS) distribution of band gap and band size which further affects sensitivity [9]. Pd doped rGO films were studied in the temperature range 30 °C–75 °C, and the sensor response was found to increase with the increase in temperature [10]. Room temperature response was observed by alternating current dielectrophoresis (DEP) method toward 200 ppm hydrogen with Pt decorated graphene oxide (GO) sensor [11].

Hydrogen sensing properties of rGO/ZnO nano-composite was reported, in which 1.2 wt% of rGO/ZnO nano-composite films deposited on alumina substrate showed highest sensitivity to hydrogen at 150 °C [12]. This high sensitivity is attributed to good conductivity of rGO and the p-n hetero contact between p-type rGO and n-type ZnO. Effect of nitrogen doping on rGO films decorated with Pd nanoparticles for hydrogen storage was reported by Ramaprabhu and group in 2012 [13]. They reported a 66% increase in hydrogen intake of nitrogen doped graphene. The hydrogen intake capability further increased up to 124% on decoration of nitrogen doped graphene with Pd nanoparticles [13].

In this work, we highlight the response difference between palladium loaded pure rGO and palladium loaded nitrogen doped rGO in hydrogen. Special emphasis is given to the nature of response and its characteristic variation with the change in operating temperature. The data is thoroughly analyzed and a suitable sensing scheme is proposed to support the experimental results.

2 Materials Preparation

Natural graphite flakes (>99%), sulfuric acid (H_2SO_4), potassium permanganate ($KMnO_4$), hydrogen peroxide (H_2O_2), distilled water, hydrazine hydrate (N_2H_4), hydrochloric acid (HCl), 0.04% ammonia solution (NH_3), ethylene glycol (99%), sodium tetra chloropalladate (II) hydrate (99.9999% metal basis) (Na_2PdCl_4), polyvinyl pyrrolidone (PVP Mw = 40,000), sodium hydroxide (NaOH Mw = 40 gm/L) were purchased from Sigma Aldrich and Alfa Aeser.

2.1 Synthesis of Reduced Graphene Oxide(RGO) by Chemical Exfoliation

The rGO is synthesized by chemical exfoliation technique. In this method graphite flakes (1.5gm) were mixed with 25 ml H_2SO_4 in a 250 ml volumetric flask kept in an ice bath. Volumetric flask along with ice bath was kept on a magnetic stirrer for 5 min maintaining the temperature in the range 0–5 °C. Potassium permanganate (3 gm) was added to the solution over a period of 10 min with constant stirring and maintaining the temperature <20 °C. The solution was then stirred on a magnetic stirrer for 3 h at 35 °C. Distilled water (100 ml) was added carefully to the solution with constant stirring and maintaining the temperature <100 °C in the next 10 min. 30 ml of 30% hydrogen peroxide was directly added to the above solution under stirring condition. 100 ml of 10% HCl was added to the solution and then washed and filtered with distilled water. The filtrate was oven dried at 40 °C for 3 h. The dried filtrate was graphene oxide. Colloidal solution of as obtained graphene oxide was thereafter poured into distilled water with constant stirring on a magnetic stirrer for 35 °C for 30 min. Then hydrazine hydrate (10 ml) dispersed in 30 ml of water was added to the colloidal solution of GO with constant stirring for 3 hr. The solution was filtered and the filtrate obtained was reduced graphene oxide (rGO). The filtrate was oven dried at 35 °C overnight and powdered rGO was extracted from filter paper.

2.2 Synthesis of Nitrogen Doped rGO

The as-prepared rGO was thermally treated with three concentrations of ammonia (NH_3) at 100 °C. In two round bottom flasks each containing 0.1 gms of rGO, 5 ml, and 11 ml of 0.04% NH_3 solution, respectively, were added, and the flasks were tightly capped. Thereafter the flasks were put in a constant temperature oven and the temperature of the oven was maintained at 100 °C for 15 h. After this heat treatment, the contents of the flasks were cooled to room temperature.

2.3 Synthesis of Pd Nanoparticles

The Pd nanoparticles were prepared by polyol method. In this method 0.1 gms sodium tetra chloropalladate was reduced at 100 °C by 20 ml ethylene glycol in the presence of 0.033 gms PVP, which acts as the capping agent to avoid agglomerated growth of the nanoparticles. During synthesis the solution was constantly stirred and 175 μl of NaOH (1 M) solution was slowly added to the reaction mixture (@ 25 μl for seven times within a total time period of 90 min) with the help of micropipette. The color of the solution changed from dark brown to black and stirring and temperature conditions were maintained until the solution color was completely black. The solution was then kept in oven for 24 h at 100 °C for the growth of nanoparticles. Thereafter the solution was cooled and centrifuged at 5000 rpm. Black colored Pd nanoparticles were obtained and the yield was repeatedly washed with distilled water.

3 Characterizations Results & Analysis

X-ray diffraction studies were performed using Bruker D2 Phaser diffractometer. The scanning was done in the 2θ range of 10°–60° with step size of 0.0203°. The Raman spectroscopic measurements were performed with LabRAM ARAMIS (Horiba Jobin Yvon) using a low intense (to avoid sample damage) laser source (473 nm) excitation. High resolution transmission electron microscopy images were obtained using Titan ST 80-300 (Thermo-Fisher Scientific).

3.1 X-Ray Diffraction (XRD)

The structural information and interplanar spacing (or d-spacing) values are obtained from XRD. For pure rGO sample three peaks are observed at 2θ positions 12.68°, 26.15°/26.35° and 54.24°, which correspond to (001), (002)/(002) and (004) planes respectively (Fig. 1). Normally the diffraction peak at 26.5° is due to graphite and the corresponding d-spacing value is ~0.336 nm. So the high intensity peaks at 26.15°/26.35° for pure rGO sample are perhaps due to unreacted graphite flakes. The calculated d-spacing values are 0.34/0.337 nm for the 26.15°/26.35° peaks and are close to the d-spacing value of graphite. The peak at 12.68° is due to the reflections from the GO layers and the calculated interplanar spacing value of ~0.696 nm (Table 1). The increase in the value of interplanar spacing indicates the success of the chemical exfoliation and incorporation of functional groups in the layers. The peak corresponding to 54.24° is also due to the graphitic (004) plane. The existence of (002) and (004) graphite reflections probably indicates the use of excess graphite in the starting reaction, which leads to incomplete chemical exfoliation. Further studies are necessary to quantitatively control the reactant amounts. Two low-intensity peaks at 31.15° and 77.5° may be due to higher order GO reflections.

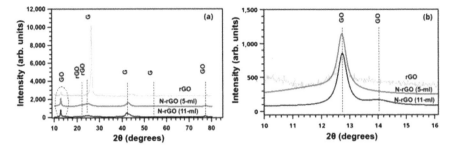

Fig. 1 X-ray Diffraction spectra in 2θ range **a** ~10°–80° **b** −10°–16° (G-Graphite; GO-Graphene oxide; rGO-reduced graphene oxide)

Table 1 Parameter calculated from XRD

Sample	Peak	2θ Position (in degree)	d-spacing (nm)
rGO	(001) GO	12.688	0.696
N-rGO (5 ml)	(001) GO	12.716	0.695
	(002) rGO	21.52	0.412
	(002) rGO	24.62	0.361
N-rGO (11 ml)	(001) GO	12.728	0.694
	(001) GO	14.01	0.631
	(002) rGO	24.5	0.3629

Nitrogen doped rGO (5 ml) shows diffraction peaks at 2θ positions ~12.7°, ~21.5°, ~24.6°, ~42.5, ~54.2 and ~77.2°, in which the ~12.7° and ~77.2° peaks are GO signatures and the diffraction peaks near ~21.5° and ~24.6° can be considered as rGO signatures (Fig. 1). The peak at ~42.5° and ~54.2° correspond to graphitic reflections (JCPDS 75-2078). The unique features of nitrogen doped rGO XRD are (i) Sharp decrease in the intensity of the graphitic (002) peak (ii) shift in the (002) peak position toward lower 2θ values (iii) increase in the peak broadening of the (002) and the existence of multiple peaks (iv) very high (001) GO intensity. All these facts indicate that the chemical exfoliation is also progressing during ammonia treatment. In fact, upon doping with nitrogen there is possibility of three structural configurations, i.e., graphitic N-configuration, pyridinic N-configuration, and pyrollic N-configuration. These configurations arise due to the substitution of some of the carbon atoms by nitrogen atoms and the configuration change is dependent on the ammonia concentration. Also, these structures can tune the attachment of surface functional groups and tune the layer separation due to the difference in electronegativity between oxygen and nitrogen. Such modulation of the surface functional groups probably results in the broad (002) nitrogen doped rGO peak.

With the increase in ammonia content from 5 ml to 11 ml, the diffraction peaks are at 2θ positions ~12.7°, ~14°, ~24.5°, ~41.8, ~43.03°, and ~77.2°. The important difference is additional GO signature at 2θ = 14°. This can be attributed to the formation propensity of different structural configurations arising due to nitrogen incorporation in the carbon rings. Also, the observation of this peak at high 2θ position (14°) probably indicates that the oxygen content in those 2-dimensional patches is relatively less.

It is seen that the d-spacing values change upon ammonia pretreatment. In fact d ~0.35 nm is standard desirable value for few layer graphene. Table 1 shows that the d-spacing reduces (and approaching 0.35 nm) upon the conversion of GO to rGO. Increase in ammonia content affects the oxygen content in layers, which enhances the transformation of GO to rGO and formation of nitrogen induced configurations. Also, in presence of ammonia further layer nucleation at nitrogen substituted positions is likely. However, high ammonia concentration can lead to the squeezing of the layers because the d-spacing of rGO is less than GO. The variation of d-spacing due to the existence of both GO and rGO for a particular sample is revealed from the calculated d-spacing. The existence of graphitic signature for the undoped sample indicates the fact that excess graphite is used in the starting reaction. Although the graphitic signature is reduced in ammonia doped samples, yet it is not eliminated completely.

3.2　High Resolution Transmission Electron Microscopy (HRTEM)

The HRTEM images along with SAED (Selected Area Electron Diffraction) patterns of pure rGO, N-rGO (5 ml) and N-rGO (11 ml) are shown in Figs. 2, 3, and 4, respectively. Veil type morphology with occasional folds of the 2D layers is quite obvious from the HRTEM images (Figs. 2a, 3a, 4a). The magnified images (highlighting the layers and their separation) of the synthesized materials are shown in Figs. 2b, 3b, and 4b. The SAED patterns shown in Figs. 2c, 3c, and 4c are used to calculate the d-spacing. It is clear from the images in Fig. 2 that there is a gradual transformation from graphite to GO and then to rGO. This is evident from the variation in the d-spacing values as highlighted in the Fig. 2b and calculated from Fig. 2c. The values of d-spacing calculated for 1st, 2nd, 3rd, and 6th bright rings (in SAED pattern) are 0.98, 0.552, 0.37, and 0.118 nm, respectively, and they correspond to GO (Fig. 2c).

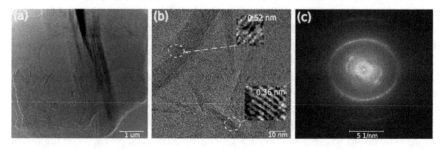

Fig. 2 HRTEM images of pure rGO **a** low magnification **b** high magnification and **c** SAED pattern

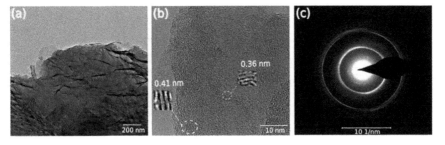

Fig. 3 HRTEM images of N:rGO (5 ml) **a** low magnification **b** high magnification and **c** SAED pattern

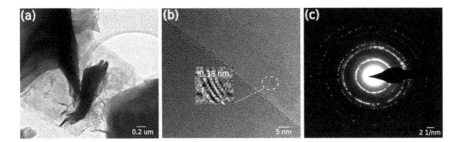

Fig. 4 HRTEM images of N:rGO (11 ml) **a** low magnification **b** high magnification and **c** SAED pattern

The 4th and 5th bright rings having d values 0.33 and 0.205 nm, respectively, correspond to graphite. In fact the value of d-spacing calculated from XRD for pure rGO sample for the first peak is 0.696 nm, which is close to the HRTEM results; also, the graphite peaks near ~26° and ~44° are observed in the XRD spectrum of pure rGO.

The d-spacing calculated from Fig. 3c reveals partial conversion of GO to rGO (maximum d-spacing is ~0.695 nm). Additionally the graphite signature in N:rGO (5 ml) sample is less. The pleated morphology as shown in Fig. 3a probably indicates the formation of high density multilayer graphene. This is in contrast to the pure rGO sample, where the morphology lacks this corrugated edges.

Upon increasing the ammonia amount to 11-ml, the maximum d-spacing is ~0.378 nm (calculated from Fig. 4c), which is a clear indication of the high conversion rate of GO to rGO. This can also indicate the existence of nitrogen-based structural configurations, which probably tunes the interlayer separation due to the electronegativity difference. The graphite signature in N:rGO (11 ml) sample, although less is still present. The rings are not diffuse but very clear and spotty in nature. Hence the multilayer stacking of graphene is appreciably better in comparison to the previous two samples. In fact the 2D graphene flakes are very clearly visible in Fig. 4a.

Apparently it seems from XRD and HRTEM studies that there is a hierarchy in the maximum d-spacing value and it is highest for the pure rGO sample and it is

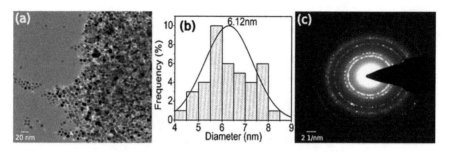

Fig. 5 Pd NPs **a** HRTEM image **b** Particle size histogram and **c** SAED pattern

minimum for the N:rGO (11 ml). This indicates that the conversion efficiency of GO to rGO is high in N:rGO (11 ml) sample, while it is low in pure rGO sample. The N:rGO (5 ml) has an average conversion rate.

The HRTEM images were also obtained for Pd nanoparticles synthesized by polyol method and the size and shape of the nanoparticles were determined (Fig. 5). It is observed that nanoparticles are near spherical in shape (Fig. 5a). The particle size as obtained from the histogram is ~6.12 nm (Fig. 5b). The spotty nature of the SAED reveals excellent crystallinity of the synthesized Pd nanoparticles.

3.3   Raman Spectroscopy

The Raman spectrum of rGO mainly consists of three peaks, viz., D-band, G-band, and 2D-band (Fig. 6). Information regarding the layer stacking and defects are obtained by extracting peak information close to actual positions of D-band, G-band, and 2D-band commonly observed at 1350 cm^{-1}, 1580 cm^{-1}, and 2700 cm^{-1}, respectively, for carbon materials [14].

Fig. 6 Raman spectra of rGO, N-rGO (5 ml), and N-rGO (11 ml)

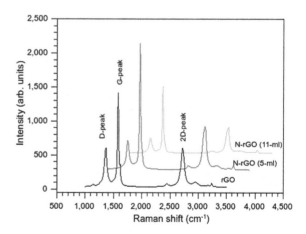

Table 2 Parameters calculated from Raman data

Sample name	Peak position (cm^{-1})			I_D/I_G	I_G/I_{2D}
	D-band	G-band	2D-band		
rGO	1362.1	1579.7	2722.59	0.432	2.318
N-rGO (5 ml)	1356.99	1883.53	2723.44	0.296	2.347
N-rGO (11 ml)	1361.08	1579.7	2738.74	0.245	2.458

The ratio of the intensities of the bands is tabulated in Table 2. The ratio of intensity of D-band to the G-band describes the defects present in graphene structure, while the ratio of the intensity of G-band and 2D-band reveals the layer morphology of graphene. The I_D/I_G value closer to zero indicates less disordered structure of rGO. The high value of I_G/I_{2D} ratio indicates multilayer graphene as the standard value for I_G/I_{2D} is between 0.3 (for monolayer graphene) and 3.1 (for graphite) [15, 16]. For the studied samples the calculated value of I_G/I_{2D} indicates an increase in multilayer graphene stacking upon nitrogen doping. The reduction in the I_D/I_G value upon nitrogen doping probably indicates an improvement in ordered layer stacking with less morphological defects in the layers (Table 2).

4 Device Fabrication and Sensor Study

The synthesized materials (rGO, N-doped rGO, and Pd nanoparticles) were first dispersed in water. The Pd loading was done by mixing rGO and N-doped rGO, with a drop of Pd nanoparticle solution. The thin films (thickness ~ microns) of individual mixtures were then drop casted on clean glass substrates by putting a single drop of the mixed dispersion and subsequent drying on a hot plate at 90-100 °C. Resistive devices were fabricated by manually printing two parallel conducting lines (separation 2 mm) using silver paste and attaching two fine copper wires with the conducting silver paste lines. Thereafter the samples were again dried at 90–100 °C for baking the silver paste.

The hydrogen sensor studies were performed with an indigenously fabricated glass chamber in which gas mixture was introduced using mass flow controllers (Alicat Scientific, USA). The temperature of the chamber was controlled by digital PID temperature controller (make Schneider). The sensor data was recorded by a computer interfaced digital multimeter (Keithley 6487 picoammeter/Voltage source).

4.1 Hydrogen Sensor Study

Films of pure rGO and N-rGO were found to be insensitive to hydrogen in the studied temperature range (RT -125 °C). Maybe the required energy to adsorb and dissociate the hydrogen molecules on rGO and N-rGO surfaces was insufficient. Upon Pd NP decoration of the rGO and N-rGO films, the devices responded to hydrogen. Hence it is apparent that the required activation energies for hydrogen dissociation are available. The nature of response in Pd decorated films was found to depend on the base matrix (rGO or N-rGO) and the operating temperature. Interestingly the response of devices based on only Pd NP to hydrogen was appreciable at RT, and thereafter the performance completely deteriorated. However, the Pd decorated films of rGO and N-rGO showed steady performance in the studied temperature range (RT −125 °C). In addition, on a comparative scale the response (%) of the Pd NP decorated rGO or N-rGO devices are better than devices based on only Pd NP at 30 °C. This implies that the rGO or N-rGO layers are actively participating in the sensing process in Pd NP decorated rGO or N-rGO devices.

Two independent responses (Response-1 and Response-2) are shown when the devices are exposed to hydrogen. These are typical nomenclature used in this discussion to indicate the variation of device resistance during hydrogen exposure (i.e., during the device response). "Response-1" indicates the rise in film resistance (or device resistance) upon hydrogen exposure. The "Response-2" indicates the subsequent decrease in film resistance (or device resistance) after initial increase in resistance while the gas (hydrogen) supply is still ON. Figure 7a shows this typical dual response of Pd NP devices at RT. Interestingly the Pd decorated films of rGO,

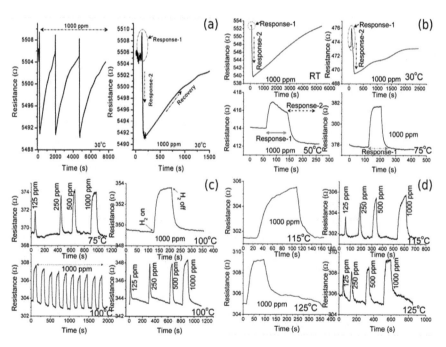

Fig. 7 **a** Response of Pd NP resistive device to hydrogen at 30 °C. **b–d** Response of Pd decorated rGO resistive device to hydrogen

although showing dual response, are also manifesting a transition to single response beyond a certain operating temperature ($t_{transition}$). The response patterns are shown in Fig. 7b–d. Upon nitrogen doping of the rGO matrix, the temperature of transition to single response changes, and this transition temperature is reduced to 30 °C, for doping done with 5-ml ammonia solution (Fig. 8a, b).

The response parameters are calculated from the experimental data and tabulated in Tables 3 and 4. It is apparent from these tables that the transition temperature changes with the change in the base matrix, for instance, $t_{transition} = 100$ °C for Pd NP decorated rGO, $t_{transition} = 30$ °C for Pd NP decorated N-rGO (5-ml), and $t_{transition} = 50$ °C for Pd NP decorated N-rGO (11-ml). Also, the variation of response (%) implies that the double response is favorable for low temperature applications, while the single response is apparent for high temperature applications.

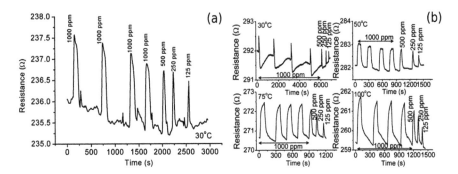

Fig. 8 **a** Response of Pd decorated N-rGO (5-ml) resistive device to hydrogen **b** Response of Pd decorated N-rGO (11-ml) resistive device to hydrogen

Table 3 Response parameters for resistive devices based on Pd NP and Pd NP decorated rGO in 1000 ppm hydrogen

Sample	t (°C)	Response	%Response		Response time (s)		Recovery time (s)
			Response-1	Response-2	Response-1	Response-2	
Pd NP	30 °C	DR	0.085	0.32	12	26.5	1511
Pd NP: rGO	RT	DR	0.01	2.36	3	75	4344
	30 °C	DR	0.47	1.15	13.5	84	2803.5
	50 °C	DR	0.5	0.8	8.5	30	Very slow
	75 °C	DR	0.76	0	23	0	24.5
	100 °C	SR	1.46	–	37	–	47.5
	115 °C	SR	1.37	–	57	–	25
	125 °C	SR	1.42	–	21.5	–	201

DR-Dual Response; SR-Single Response; RT-Room Temperature

Table 4 Response parameters for resistive devices based on Pd NP decorated N-rGO (5-ml) and Pd NP decorated N-rGO (11-ml) in 1000 ppm hydrogen

Sample	t (°C)	Response	%Response		Response time (s)		Recovery time (s)
			Response-1	Response-2	Response-1	Response-2	
Pd NP: N-rGO (5-ml)	30 °C	SR	0.66	–	17	–	89.5
	50 °C	SR	0.77	–	86.5	–	126
	75 °C	SR	2.68	–	79.5	–	283.5
	100 °C	SR	5.68	–	90	–	46.7
	125 °C	SR	7.94	–	86	–	314
Pd NP: N-rGO (11-ml)	30 °C	DR	0.18	0.32	13.5	89.5	806.5
	50 °C	SR	0.34	–	38.5	–	45
	75 °C	SR	0.65	–	44.5	–	74.5
	100 °C	SR	0.84	–	60	–	117
	125 °C	SR	Performance deteriorated				

DR-Dual Response; SR-Single Response

4.2 Proposed Sensing Mechanism

The Pd NPs act as highly active catalytic centers to capture hydrogen molecules and convert them into hydrogen atoms. Since pure Pd NP device responds to hydrogen at 30 °C, it is apparent that splitting of hydrogen molecules happens easily in the studied temperature range (RT-125 °C) for all category of devices; maybe the hydrogen capture and splitting by the Pd NPs is enhanced at elevated temperatures beyond 30 °C. The deterioration of pure Pd NP devices at high temperatures is probably due to the thermal expansion of the nanoparticles and has been elaborately discussed elsewhere [17]. In this study it is observed that materials such as rGO and N-rGO are unable to split hydrogen molecules in the studied temperature range and this maybe attributed to the lack of activation energy to dissociate the molecules. Hence the Pd decoration is necessary. However these layers of rGO (or N-rGO) actively participate in the sensing if hydrogen atoms are available and this is obvious from the fact that the response (%) of Pd decorated rGO is higher than devices made with only Pd NP. Therefore it is apparent that the undoped and doped rGO layers actually control the sensing mechanism.

The rGO or N-rGO normally intercalates the hydrogen atoms in between the layers. In fact, intercalation of H atoms is dependent on the temperature and the interlayer separation. It is reported that thin-film hydrogenation of graphene leads to hydrogen coverage from both sides of the graphene sheet [18]. So crowding of hydrogen atoms is likely if the interlayer separation is less and vice versa. Therefore it is apparent that for a particular gas concentration, the ease of hydrogenation of rGO (or graphane formation) is affected by the layer separation. The hierarchy of interlayer spacing in the samples is studied with the help of material characterization results (XRD and HRTEM) and elaborately discussed in previous sections. Apart from interlayer separation, temperature provides the necessary activation energy that controls the adsorption coverage by the hydrogen atoms. For catalytically active surfaces, the required energy is less, and hydrogenation of rGO or graphane formation proceeds easily at low temperatures; however, in other cases, high temperature may be required.

Therefore based on the above facts, the transition from double response to single response can be thought of as transition from bulk transport to surface transport, respectively. This can be justified on the basis of change in electronic status of reduced graphene oxide. It is reported that upon hydrogenation of graphene, it becomes an acceptor and there is a transition from sp^2 to sp^3 bonding (Fig. 9) [19, 20]. Such a phenomenon is also likely for rGO. So, when the operating temperature is low, the hydrogen adsorption is low, and hence the transformation from donor to acceptor is also low. Hence the rGO matrix is partially compensated due to induction of acceptor characteristics at low operating temperatures. In such a situation there is an increase in resistivity of the bulk as generated charge carriers are neutralized by the in-situ formed graphane (which is an acceptor). With the increase in temperature the hydrogen adsorption and subsequent intercalation increase, and hence the acceptor characteristics of the base matrix. Therefore it is apparent that the bulk is getting completely shielded with respect to electronic conduction with the increase in temperature, and the charge carriers prefer the relatively lower resistive surface path. This is reflected in the response curves because response-2 is completely eliminated at elevated temperatures.

The transition temperature is also dependent on the base matrix. Nitrogen doped graphene provides a relatively better reactive surface for catalysis and sensing applications [18]. Also, the layer separation is small in nitrogen doped samples (as discussed in XRD and HRTEM). Hence, the hydrogenation is high in nitrogen doped graphene, which easily induces acceptor characteristics. So the transition temperature is reduced to 30 °C for 5-ml ammonia treated rGO. However, large amount of

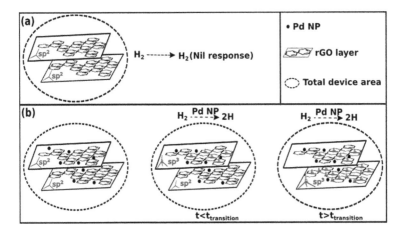

Fig. 9 Hydrogen sensing mechanism **a** without Pd NPs **b** with Pd NPs

ammonia (11-ml) can lead to the formation of pyridinic N-configuration and pyrollic N-configuration that can affect the hydrogenation of the layers. Hence the transition happens at higher temperature (50 °C) compared to 5-ml treated rGO. However, the transition temperatures of ammonia treated samples are still lower than that of pure rGO.

5 Conclusion and Outlook

The sensing results indicate versatility in the application prospectives of rGO (or N-rGO) layers. This is attributed to the dual nature of response that was observed for Pd decorated rGO (or N-rGO) films. In fact, the double response device characteristic is relatively faster with respect to single response. Therefore such materials can be preferred for low temperature gas sensor applications. Also, considering the magnitude and variation of response (%), the single response device characteristics is relatively better than the double response. Hence these materials would be suitable for more efficient gas adsorption applications at relatively lower temperatures. Although the response times and recovery times are large, yet the results indicate a possibility of tuning the time of response and recovery by improving the 2-dimensional material characteristics. The characterization results clearly reveal that the synthesized rGO (or N-rGO) layers have mixed characteristics. For instance, rGO, GO, and graphitic signatures are apparent from the material characterization results. Hence, the sensing performance is likely to improve, if the mixed material (i.e., presence of different material phases) are avoided. In fact, further optimization of the synthesis parameters is expected to yield higher sensing performance.

Acknowledgements The work was taken up with a sponsored research programme (Grant No. EMR/2016/006287) financed by Science and Engineering Research Board (SERB), under the Department of Science and Technology (DST), Government of India. Shikha Sinha gratefully acknowledges the research fellowship from SERB.

References

1. Meyer JC, Geim AK, Katsnelson MI et al (2007) The structure of suspended graphene sheets. Nature 446:60–63. https://doi.org/10.1038/nature05545
2. Lu X, Yu M, Huang H et al (1999) Tailoring graphite with the goal of achieving single sheets. Nanotechnology 10(3):269
3. Stankovich S, Dikin DA, Piner RD et al (2007) Synthesis of graphene-based nanosheets via chemical reduction of exfoliated graphite oxide. Carbon 45(7):1558–1565
4. Stankovich S, Dikin DA, Dommett GHB et al (2006) Graphene-based composite materials. Nature 442:282–286
5. Abideen ZU, Kim HW, Kim SS (2015) An ultra-sensitive hydrogen gas sensor using reduced graphene oxide-loaded ZnO nanofibers. Chem Commun 51(84):15418–15421

6. Ren H, Gu C, Joo SW et al (2018) Effective hydrogen gas sensor based on NiO@ rGO nanocomposite. Sens Actuator B 266:506–513
7. Zou Y, Wang Q, Xiang C et al. (2016) Doping composite of polyaniline and reduced graphene oxide with palladium nanoparticles for room-temperature hydrogen-gas sensing. Intl J Hydrog Energy 5393–5404(11):41
8. Bhati VS, Ranwa S, Rajamani S et al (2018) Improved Sensitivity with Low Limit of Detection of a Hydrogen Gas Sensor Based on rGO-Loaded Ni-Doped ZnO Nanostructures. ACS Appl Mater Interfaces 10(13):11116–11124
9. Wang J, Kwak Y, Lee IY et al (2012) Highly responsive hydrogen gas sensing by partially reduced graphite oxide thin films at room temperature. Carbon 50(11):4061–4067
10. Pandey PA, Wilson NR, Covington JA (2013) Pd-doped reduced graphene oxide sensing films for H_2 detection. Sens Actuator B 183:478–487
11. Wang J, Rathi S, Singh B et al (2015) Alternating current dielectrophoresis optimization of Pt-decorated graphene oxide nanostructures for proficient hydrogen gas sensor. ACS Appl Mater Interfaces 7(25):13768–13775
12. Anand K, Singh O, Singh MP et al (2014) Hydrogen sensor based on graphene/ZnO nanocomposite. Sens Actuator B 195:409–415
13. Parambhath VB, Nagar R, Ramaprabhu S et al (2012) Effect of nitrogen doping on hydrogen storage capacity of palladium decorated graphene. Langmuir 28(20):7826–7833
14. Khan QA, Shaur A, Khan TA et al (2017) Characterization of reduced graphene oxide produced through a modified Hoffman method. Cogent Chem 3(1):1298980
15. Das A, Chakraborty B, Sood AK (2008) Raman spectroscopy of graphene on different substrates and influence of defects. Bull Mater Sci 31(3):579–584
16. Liu W, Li H, Xu C et al (2011) Synthesis of high-quality monolayer and bilayer graphene on copper using chemical vapor deposition. Carbon 49(13):4122–4130
17. Pooja Barman PB, Hazra SK (2018) Role of capping agent in palladium Nanoparticle based hydrogen sensor. J Clust Sci 29:1209–1216
18. Pumera M, Sofer Z (2017) Towards stoichiometric analogues of graphene: graphane, fluoro-graphene, graphol, graphene acid and others. Chem Soc Rev 46:4450–4463. https://doi.org/10.1039/C7CS00215G
19. Pashangpour M, Ghaffari V (2013) Investigation of structural and electronic transport properties of graphene and graphaneusing maximally localized Wannier functions. J Theor Appl Phys 7:9. https://doi.org/10.1186/2251-7235-7-9
20. Pei Q-X, Sha Z-D, Zhang Y-W (2011) A theoretical analysis of the thermal conductivity of hydrogenated graphene. Carbon 49(14):4752–4759. https://doi.org/10.1016/j.carbon.2011.06.083

Carbon-Based Electrodes for Perovskite Photovoltaics

Arun Kumar, Naba Kumar Rana, and Dhriti Sundar Ghosh

Abstract It has long been concluded that solar energy holds the best potential for meeting the planet's long-term energy needs, however, as of now, more than 70% of the global energy demand is still being fulfilled by non-renewable sources. Recently, perovskite solar cells (PSCs) have attracted enormous interest because they can combine the benefits of low cost and high efficiency with the ease of processing. In the last decade, PSCs have seen a remarkable improvement in terms of efficiency, however, there are still some hurdles in its path before they can be commercialized. The major obstacle being their long-term stability in harsh environmental conditions. In addition, the use of vacuum deposited transparent conductive oxides and noble metals as an electrode is also hindering its prospects. Among potential candidates, carbon-based materials provide a good alternative because of their suitable work function, high- carrier mobility, electrical conductivity, stability, and flexibility. The chapter discusses detailed information about different carbon-based materials and their properties which make them a front-runner in future generation PSCs. We will also discuss different advantages like flexibility, photostability, thermal stability, and scalability which will lead to a pathway toward the commercialization of PSCs using carbon-based electrodes.

1 Introduction

Since the last decade, PSCs (perovskite solar cells) have been showing significant growth in power conversion efficiency (PCE). These cells have recorded a remarkable rise in PCE from 3.8% to 25.2% in a very short duration as compared to other technologies [1]. Apart from this, the perovskite materials have attracted keen attention from researchers around the world due to their all-round properties like tunable bandgap, high absorption coefficient, high carrier mobility, long diffusion length, and low exciton binding energy. Also, the fact that perovskite as an active layer material in

A. Kumar · N. K. Rana · D. S. Ghosh (✉)
Department of Physics, Indian Institute of Technology Bhilai, Bhilai, India
e-mail: dhriti.ghosh@iitbhilai.ac.in

© The Author(s), under exclusive license to Springer Nature Singapore Pte Ltd. 2021
A. Hazra and R. Goswami (eds.), *Carbon Nanomaterial Electronics: Devices and Applications*, Advances in Sustainability Science and Technology,
https://doi.org/10.1007/978-981-16-1052-3_16

solar cells is easily processable via well-known techniques, viz., evaporation, spray, and solution processing makes it a perfect versatile material. The inexpensive materials together with scalable low-cost processing make them industrial friendly and a suitable candidate for low-cost fabrication. In addition, compatibility with roll-to-roll printing, doctor blading, and screen printing further makes it capable of large-scale production. Owing to these fascinating properties, PSC technology constantly drags researcher's interest toward it. Chemically, perovskite is a material having a structure of ABX_3, where A can be CH_3NH_3I, $HC(NH_2)_2$, Cs among others; B is usually Pb and Sn; while C is a halogen like Cl, Br, or I. The perovskite material which is most commonly being used in PSCs is $CH_3NH_3PbI_3$. It exhibits a bandgap of 1.57 eV which is quite suitable for light harvesting and thus an efficient solar cell operation.

Among numerous device architectures, the most widely used device architecture for PSC is FTO/TiO_2/perovskite/spiro-OMeTAD/Au. Here, fluorine-doped tin oxide (FTO) acts as a transparent electrode and titanium dioxide plays the role of an electron transporting layer (ETL). Spiro-OMeTAD which stands for 2,2′,7,7′-Tetrakis[N,N-di(4-methoxyphenyl)amino]-9,9′-spirobifluorene is used as a hole transporting layer (HTL). Finally, gold (Au) is the other electrode, usually a thick layer and opaque, which collects the holes. The light enters the device from the FTO side and reaches the perovskite layer where it creates an electron–hole pair. Although the perovskite layer, as mentioned above, is inexpensive and can be deposited via a variety of low-cost methods, there are other layers in the device which are expensive and require vacuum deposition. The organic material Spiro-OMeTAD is very expensive and unstable and requires inert atmosphere handling and fabrication for stability. Also, the use of noble metals like Au and Silver (Ag) as back electrodes thereby raises the cost due to the material itself and their deposition requiring high energy-consuming vacuum deposition techniques. Such material also hinders the scalability of these devices. In addition, the use of brittle indium tin oxide (ITO) and FTO as a transparent electrode in a device makes PSCs not particularly suitable when it comes to flexible cells. Apart from cost, it has also been observed that the use of Au and Ag as electrode degrades the device performance due to migration of halogen ions resulting in the formation of metal halides and also by the migration of metal into perovskite layer through HTL [2, 3]. Honestly speaking, long-term stability is the biggest hurdle that PSCs are facing right now when it comes to commercialization. The stability of PSCs toward light, heat, and moisture is the biggest issue to resolve before it is considered for widespread commercialization.

Many attempts have been made to enhance the stability of PSCs which includes using inorganic materials in place of the organic ones, using stable interfacial layers, using 2D/3D hybrid perovskite active layer, HTL-free device architecture, use of encapsulating layers, and incorporating carbon-based materials. Of all these approaches, the use of carbon-based materials comes out to be more promising in terms of long-term stability improvement. Along with that, the excellent properties like high conductivity, low-temperature processing, low cost, mechanical strength, abundance, high carrier mobility, and chemical stability in harsh environment make them an attractive material to be used in PSCs. Moreover, the suitable work function allows it to be used as an anode in PSCs while the hydrophobic nature of carbon materials helps to enhance the stability by reduction of moisture penetration. Also, carbon not only serve as an electrode but can also act as a hole extractor [4] leading to cost reduction and stability improvement.

As mentioned, one of the ways to include carbon-based materials in PSCs which can in turn also improve long-term stability, reduce cost, and induce scalable production of PSCs is to replace traditional electrodes with carbon-based electrodes. The first PSC with a carbon-based counterelectrode was demonstrated by Ku and co-workers [4] in 2013 showing a PCE of 6.64%. They used an HTL-free architecture with carbon black/graphite as an electrode. To the best of our knowledge, the PCE of carbon electrode-based PSCs has reached 19.2% till now [5]. Over 1-year stability under 1-Sun illumination has already been shown by carbon-electrode-based PSCs [6]. Long-term stability under high thermal stress of 100 °C has also been observed with PSCs using a carbon-based electrode [7]. High stability has also been observed under adverse moisture conditions [8]. Apart from the improvement of long-term stability, carbon electrode-based PSCs are also promising candidates for flexible and scalable photovoltaics [9, 10]. Taking all these parameters into account it will not be wrong to claim that the commercialization of PSCs will be possible in the near future with the help of carbon-based electrodes. In the subsequent sections, we will describe different carbon-based electrode candidates, their properties, fabrication method, and detailed literature survey of work done using the material for incorporating them in PSCs.

2 Suitable Carbon-Based Electrode Materials

2.1 Graphene

Graphene is a two-dimensional allotrope of carbon that forms the basic structure of other allotropes like graphite, carbon nanotubes, etc. It belongs to the polycyclic aromatic hydrocarbon family. The tensile strength of graphene is 130 GPa which means it is a hundred times stronger than steel along with a lower surface mass of 0.763 mg/m^3. The considerable properties that make it compatible with PSCs are its high transparency and high electrical conductivity. If we talk about optical properties, graphene shows transmittance of more than 97% at 550 nm [11]. It has to be mentioned that higher transmittance is an essential criterion for a materials to be used as a transparent electrode. Graphene also has high carrier mobility (both electron and hole), more than 15000 cm^{2}V^{-1}s^{-1} which is a hundred times more than noble metals like gold [12]. Also, it has high conductivity values around 10^8 S/m [13] which helps in fast as well as an efficient collection of charge carriers and also lowers the probability of recombination. Apart from that, its high flexibility makes it a good choice while fabricating flexible cells (Figs. 1 and 2).

In general, graphene is synthesized using CVD technique on a copper foil. Methane as a carbon resource and hydrogen gas as a reducing agent are used. To

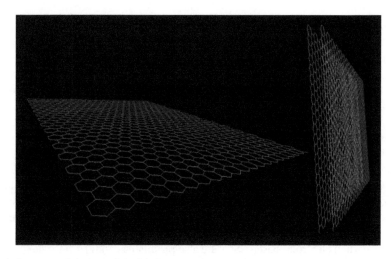

Fig. 1 Structure of single- and double-layer graphene

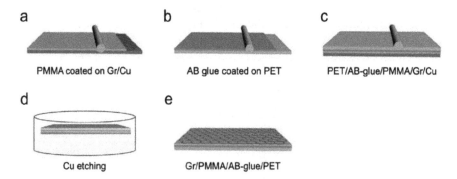

Fig. 2 Transfer process for graphene. Reprinted with permission [11]

transfer the synthesized graphene layer to the desired substrate, a supporting polymer layer of poly (methyl methacrylate) PMMA or poly (3-hexylthiophene) P3HT is spin-coated over graphene on copper foil and annealed for half an hour at a temperature of around 90 °C. In this way, polymer-coated graphene is obtained by etching Cu foil in aqueous ammonium persulfate followed by washing with DI water. This polymer-coated graphene can be transferred to graphene-coated Cu foil followed by etching, and this step can be repeated to increase the number of graphene layers coated. Now, this polymer-coated graphene is transferred to the desired substrate polyethylene terephthalate (PET)/polyethylene naphthalate (PEN) and the polymer is removed by rinsing it with acetone or chlorobenzene for three times [14].

2.1.1 Graphene as Transparent Electrode

ITO and FTO are mostly being used from years as transparent electrodes in photovoltaic devices due to their superior trade-off between good transmittance and low sheet resistance, but due to their higher cost which increases even more due to the presence of rare material availability (indium), high-temperature fabrication (requirement of annealing), and poor flexibility owing to its brittle nature, other materials are being explored which can replace them. Among them, metal nanowires have emerged as a potential candidate to replace ITO/FTO owing to their superior trade-off between transmittance and sheet resistance which is even better than ITO and FTO. Graphene because of its versatile mechanical and opto-electrical properties as explained before can be a good competitor as a transparent electrode. Although the transfer of good quality graphene over a large area is still a challenge, many researches are working on the integration of graphene with PSCs.

In 2016, Liu and co-workers [14] fabricated the first flexible PSC using the graphene electrode as transparent electrode. Instead of glass, a 20 μm PET substrate was used to make the device flexible and light in weight. The device architecture was graphene/P3HT/$CH_3NH_3PbI_3$/PCBM/Ag (Fig. 3). Here, the graphene layer was deposited using CVD technique and transferred using both P3HT and PMMA as a supporting substrate. A better transfer was observed with P3HT and a PCE of 11.5% along with which a high power per unit weight of 5.07 W/g was achieved, which is considered good for wearable electronics. When tested for bending stability, these devices showed only 14% degradation after 500 bending cycles. In the same year, Sung and co-workers [15] fabricated inverted PSC using graphene as a transparent electrode. The device architecture for this was graphene/MoO_3/PEDOT:PSS/$CH_3NH_3PbI_3$/C_{60}/BCP/LiF/Al. Here, single-layer graphene was grown using CVD technique. MoO_3 changes the work function of graphene from 4.23 to 4.71 eV which reduces the energy barrier between HTL and graphene that results in better charge collection. Also, MoO_3 coating

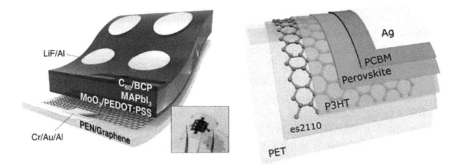

Fig. 3 Device architecture of PSCs using graphene as transparent electrode. Reprinted with permission [14, 19]

reduces the water contact angle from 90.4° to 30.0° which leads to better absorption of the HTL layer. The transmittance and sheet resistance were 90% and 500 $\Omega/\square$, respectively, for the graphene electrode and a PCE of 17.1% was achieved. A similar device using ITO as a transparent electrode showed a PCE of 18.8% which reveals that device with graphene as a transparent electrode can have more than 90% PCE than that of a PSC using conventional transparent conductive oxides (TCOs). Later in 2016, Batmunkh and co-workers [16] fabricated PSC using solution-processed graphene film as a transparent electrode. It was the first demonstration for a PSC with graphene as a transparent electron-collecting electrode, i.e., cathode. The device architecture was graphene/TiO$_2$/CH$_3$NH$_3$PbI$_{3-x}$Cl$_x$/spiro-OMeTAD/Au. Graphene layer was synthesized by oxidation of graphite followed by exfoliation and chemical reduction. Sheet resistance and transmittance of 3.08 K$\Omega/\square$ and 55%, respectively, were obtained and a PCE of 0.62% was obtained. In 2017, Heo and co-workers [17] fabricated an inverted PSC using AuCl$_3$-doped graphene as a transparent electrode. The device architecture was AuCl$_3$-doped graphene/PEDOT:PSS/CH$_3$NH$_3$PbI$_3$/PCBM/Al. Graphene layer in the device was synthesized using CVD and transferred using PMMA as a supporting substrate. For AuCl$_3$ doping, gold chloride powder was dissolved into nitromethane at different concentrations and spin-coated over graphene. Work function was controlled between -4.52 eV and -4.86 eV while the sheet resistance was reduced from 890 $\Omega/\square$ to 70 $\Omega/\square$ by doping. Sheet resistance and transmittance decrease with an increase in the concentration of doping. Hence, 0.75 mM concentration leads to a maximum PCE of 17.4%. Pristine graphene shows a PCE of 11.5%–12.6%. These devices also show good stability under light soaking conditions. In 2017, Jeon and co-workers [18] fabricated an inverted flexible PSC with graphene and single-wall carbon nanotubes (SWCNTs) as a transparent electrode. The device architecture was ITO, graphene, or SWCNTs/PEDOT:PSS/MoO$_3$/CH$_3$NH$_3$PbI$_3$/C$_{60}$/BCP/LiF/Al. Monolayer graphene was synthesized using alcohol catalytic CVD technique. Sheet resistance and transmittance for the graphene layer were about 500 $\Omega/\square$ and 90%, respectively, and that for SWCNTs were about 200 $\Omega/\square$ and 75%, respectively. Devices based on ITO, graphene, and SWCNTs show PCE of 17.8%, 14.2%, and 12.8% respectively. The high PCE of the graphene-based device is due to high transmittance and better morphology. A flexibility test was done to demonstrate the bending stability of these devices. Both the devices using graphene and SWCNTs as a transparent electrode retained about 90% of their initial PCE even after 1000 bending cycles, whereas the PCE of the device with ITO dropped to a value less than 60% of the initial PCE value. In 2017, Yoon and co-workers [19] fabricated highly efficient and flexible PSCs using graphene as a transparent electrode. The device architecture was graphene/MoO$_3$/PEDOT:PSS/CH$_3$NH$_3$PbI$_3$/C$_{60}$/BCP/LiF/Al (Fig. 3). Single-layer graphene was synthesized using CVD method and transferred using PMMA as a supporting substrate. The transferred graphene layer, here, shows transmittance and sheet resistance of about 97% and 550 $\Omega/\square$, respectively. Although sheet resistance is very high in comparison to ITO, high transmittance and better band alignment help to obtain comparable open-circuit voltage (V_{oc}) and short-circuit current (J_{sc}) values. A PCE of 16.8% with negligible hysteresis was obtained. These devices retain 90% of initial PCE after 1000 bending cycle and 85% even after 5000 bending cycles with a bending radius of 2 mm, whereas a similar device with ITO degrades to less than 40% on 1000 bending cycles with 4 mm bending radius. No significant cracks are found in the graphene layer even after bending for 5000 times. It was the first flexible PSC to show a reduction of less than 20% in PCE even after 5000 bending cycles. In 2018, Luo and co-workers [9] fabricated PSC with all carbon-based electrodes. The device architecture was graphene/TiO$_2$/PCBM/CH3NH3PbI3/CSCNTs + spiro-OMeTAD. Double-layer graphene was synthesized using CVD technique. Transmittance and sheet resistance of about 87.3% and 290 $\Omega/\square$, respectively, were obtained which lead to a device with PCE of 11.9% and 8.4% without using spiro-OMeTAD as HTL. These devices with graphene/CSCNTs show quite a good bending stability for bending radii varying between 8 mm and 2.2 mm with no change up to 4 mm and a small decrease to 87% at 2.2 mm. On the other hand, a similar device with ITO/Au reduced to 87% at 4 mm and the device got fully damaged at 2.2 mm. When bent repeatedly, ITO/Au-based devices fell to 13% of initial PCE after 1500 bending cycles at 4 mm, whereas graphene/CSCNTs devices retained 84% PCE even after 2000 bending cycles. In 2018, Jang and co-workers [20] fabricated semi-transparent and flexible PSC using graphene both as cathode and anode. The device architecture was triethylenetetramine (TETA)-graphene/ZnO/CH$_3$NH$_3$PbI$_3$/PTAA/PEDOT:PSS/bis (trifluoromethanesulfonyl)-amide (TFSA)-graphene where graphene was synthe- sized using CVD technique. For doping, TFSA powder was dissolved into nitromethane, and TETA in ethanol and solution was spin-coated over the graphene surface. The work function for graphene was 4.53 eV, for TFSA-graphene was 4.88 eV, and TETA-graphene was 4.42 eV, respectively. The sheet resistance of graphene with TFSA and TETA doping reduced to about 163 and 212 $\Omega/\square$, respec- tively. A PCE of 10.56/10.73 % and 10.96/11.16 % was obtained with and without using Ag reflector where a similar device with ITO and Au as electrode showed 12.87/13.17% PCE. Even after 1000 bending cycles at an 8 mm radius, the PCE of the device with graphene remained about 70% of the initial value.

2.2 CNTs (Carbon Nanotubes)

A carbon nanotube is an allotrope of carbon that exists in the form of hollow cylin- drical tubes with a radius in the order of nanometres. These can be assumed to be made from rolling up of graphene sheets into a cylindrical shape (Fig. 4). The elec- trical and optical properties of carbon nanotubes make it capable of becoming a part of the photovoltaic system. Carbon nanotube-based electrodes can have an optical transmittance of more than 85% [21]. This can lead to better exposure of the active layer of a solar cell to the visible light which ultimately leads to a better performance of the device. If we talk about electrical properties, they have a high conductivity of the order of 10^6 to 10^7 S/m [13]. They can have a sheet resistance of less than 100 $\Omega/\square$ [21]. A carrier mobility of more than 100,000 cm^{2}V^{-1}s^{-1} at room temperature ensures its turning out to be a good material for charge collection and also reduces

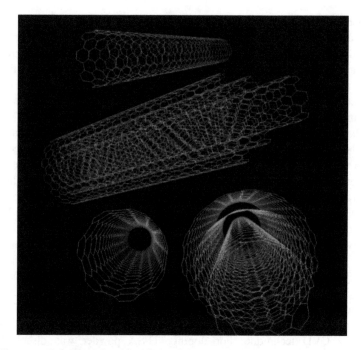

Fig. 4 Structure of single- and multi-walled carbon nanotubes

the chances of recombination [22]. CNTs can have a length-to-diameter ratio of 100000000:1, which means very large CNTs of about a half meter can be produced, which seems interesting from a large-scale fabrication point of view. They have a high tensile strength of about 63 GPa besides holding commendably high flexibility, durability, and stability. All these properties make CNTs a good contender for being an effective replacement of traditional electrodes.

In general, CNT films are synthesized using aerosol CVD method. CVD is carried out in large tubes of the order of 100 mm. Vaporization of catalyst precursor is done by passing CO through ferrocene powder. For stabilization, CO_2 is added along with CO. CNTs are collected by filtration through a silver membrane filter. The ferrocene vapors are then introduced through a water-cooled probe into a high-temperature zone of ceramic tube reactor and additional CO is mixed. Then at a high temperature of about 880 °C, ferrocene vapors are thermally decomposed into the gas phase of the aerosol CVD reactor. Carbon monoxide flowing at 4L/min decomposes on iron nanoparticles which results in the growth of CNTs. These prepared CNTs can be collected by passing through microporous filters. To control transparency and sheet resistance, variations in the collection time can be brought. The collected CNTs can be transferred to various substrates by dry press transfer method [23, 24].

Another approach of cross-stacked superaligned CNTs (CSCNTs) is also being used by scientists. Long CNTs with a length of about 300 μm are synthesized over silicon wafer using acetylene as precursor and iron as a catalyst in a low-pressure

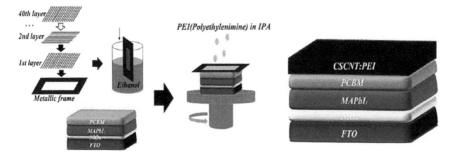

Fig. 5 Transfer process for cross-stacked nanotubes. Reprinted with permission [26]

CVD. Iron thin-film-coated silicon wafer was placed and heated up in the tube furnace in the argon gas flow to a temperature of about 700 °C. To initiate the superaligned CNTs array growth, hydrogen and acetylene are added to the argon gas. In this way, a superaligned array of length of about 300 μm can be obtained on a silicon wafer under 500 and 50 sccm of acetylene and hydrogen, respectively. The formation of a superaligned array can be scaled by varying the silicon wafer size. To obtain CSCNTs, layer-by-layer sequential stacking can be done on a metallic frame keeping one layer perpendicular to the other. To shrink and tighten, these sheets are dipped into ethanol for 1 min (Fig. 5). The number of layers stacked can be changed to alter the conductivity while doping and heat treatment can also be employed for the same [25, 26].

2.2.1 CNTs as Transparent Electrode

In 2014, Qiu and co-workers [27] fabricated a PSC in the form of a flexible fiber. They developed a co-axial structure for PSCs involving lower investments. The device architecture was stainless steel/TiO$_2$/CH$_3$NH$_3$PbI$_3$/spiro-OMeTAD/CNTs. A flexible stainless steel fiber and CNT sheets were used as anode and cathode, respectively. A PCE of 3.3% was obtained. The fiber shows almost the same PCE irrespective of the incident angle of light. Different layers were coated by dipping stainless steel fiber into solutions, and finally CNTs were wrapped to form the transparent electrode. This type of device structure is useful for wearable electronics and future textiles. Meanwhile, Wang and co-workers [28] demonstrated a flexible PSC using CNTs as a transparent electrode. The cell architecture was Ti foil/TiO$_2$ NTs + CH$_3$NH$_3$PbI$_3$/CNTs + spiro-OMeTAD where CVD technique was used for CNTs synthesis. Using this architecture, they obtained a PCE of 8.31%. They demonstrated a flexibility test for this cell with a bending radius of 0.75 cm using a cell with a length of 2.5 cm. These devices retain about 85% of initial PCE even after 100 bending cycles. In the year 2015, Jeon and co-workers [29] fabricated PSC using diluted nitric acid-doped SWCNTs as a transparent electrode. The cell architecture was HNO$_3$—SWCNT/PEDOT:PSS/CH$_3$NH$_3$PbI$_3$/PCBM/Al with (35% v/v)—HNO$_3$-doped SWCNTs are used for top contact. HNO$_3$ doping helps to make SWCNTs hydrophilic so that PEDOT:PSS layer can be easily formed over it. They have obtained a PCE of 6.32% on glass and 5.38% on PET substrate. For this composition (35% v/v), transparency was about 65% and sheet resistance of 25.6 $\Omega/\square$. Without HNO$_3$ doping, a PCE of 4.27% was obtained with a similar configuration. Two years later, Jeon and co-workers [18] fabricated a flexible inverted PSC using CNT as a transparent electrode. The cell architecture was SWCNTs/PEDOT:PSS/MoO$_3$/CH$_3$NH$_3$PbI$_3$/C$_{60}$/BCP/LiF/Al (Fig. 6). SWCNTs were synthesized using an aerosol CVD method in CO atmosphere. Here the CNT films show transparency of about 70% and sheet resistance about 200 $\Omega/\square$. In this way, they have managed to obtain PCEs of 12.8% on glass and 11% on flexible PEN substrate. These solar cells retain 90% of their PCEs after 1000 bending cycles. Again in 2017, Jeon and co-workers [23] fabricated a PSC that used CNTs both as anode and cathode. The device architecture was CNTs/PCBM/CH$_3$NH$_3$PbI$_3$/PEDOT:PSS/CNTs. The CNTs were synthesized using aerosol CVD method as already explained above. CNT films showed sheet resistance of 200 and 20 $\Omega/\square$ with transparency of 90% and 50%, respectively. A PCE of 7.32% was obtained. By opting for such a device architecture, the material cost certainly reduces to approximately 33% of PSC fabricated with conventional materials. These devices show better bending stability up to a bending radius of 1 mm in comparison to devices with conventional electrodes. Again in 2019, Jeon and co-workers [30] demonstrated solution-processed double-walled carbon nanotubes (DWCNTs) as a transparent electrode in PSCs. The cell architecture was DWCNTs/PTAA/MA$_{0.6}$FA$_{0.4}$PbI$_{2.9}$Br$_{0.1}$/C$_{60}$/BCP/Cu. Here DWCNTs were synthesized using high-temperature catalytic CVD method and coated on glass substrates using slot-die coating technique. For HNO$_3$ doping, diluted HNO$_3$ in DI water (30% v/v) was spin-coated on DWCNTs and for TFMS doping, diluted TFMS solution in chlorobenzene (8% v/v) was spin-coated. They obtained a PCE of 15.6% for pristine DWCNTs, 16.7% for HNO$_3$-doped DWCNTs, and 17.2% for TFMS-doped DWCNTs.

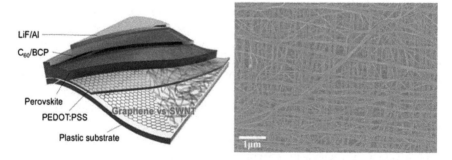

Fig. 6 Device architecture of PSC using CNTs as transparent electrode (left). SEM image of cross-stacked superaligned carbon nanotube film (right). Reprinted with permission [18, 26]

2.2.2 CNTs as Back Electrode

The noble metals like Au, Ag, Cu, and Al are being used as back electrode material in photovoltaic devices from a very long time. A very high conductivity, suitable work function, and high reflectivity make them an amazing choice for the back electrode of PSCs. Under lab-scale and controlled atmosphere, devices with these materials work quite well. But in real-time operation, such devices show performance degradation over time when exposed to temperature and moisture. These materials are very costly and their vacuum deposition processes are quite energy-consuming. So, when it comes to large-scale fabrication, these metal electrodes are not ideal. On the other hand, CNTs are inexpensive, easily processable using solution-based techniques. Ambient atmosphere processing and good mechanical strength also allow upscaling and roll-to-roll printing techniques. Low sheet resistance even less than 10 $\Omega/\square$ has been obtained with nearly opaque CNT layers [21].

In 2014, Li and co-worker [31] fabricated a PSC using CNT films as the back electrode. The device architecture was FTO/TiO$_2$/CH$_3$NH$_3$PbI$_3$/CNT (Fig. 7). A PCE of 9.90 and 6.87% was obtained with and without using Spiro-OMeTAD. A similar device without spiro-OMeTAD using Au as electrode shows a lower PCE of 5.14%. CNTs synthesized using CVD were collected on nickel foil and free-standing CNT thin films were formed and laminated onto perovskite layer with the help of toluene. In 2015, Wei and co-workers [32] fabricated an HTL-free PSC based on multi-walled carbon nanotubes (MWCNTs) back electrode. The cell architecture was FTO/TiO$_2$/CH$_3$NH$_3$PbI$_3$/MWCNTs. They got a PCE of 10.30% with 0.75 fill factor (FF) for one layer of MWCNTs and a PCE of 12.67% with 0.80 FF for two layers of MWCNTs. The J-V characteristics of the devices were observed at different scan rates of 10, 50, and 100 mV/s. During all these observations, devices show a negligible hysteresis with a difference factor of about 1.5%. The proper size of CNTs forms a better contact with the perovskite layer and interpenetrated crack-free 1D chain structure allows high conductivity. In 2016, Aitola and co-workers [33] demonstrated SWCNTs film-based back electrode for PSCs. The cell architecture was FTO/TiO$_2$/(FAPbI$_3$)$_{0.85}$(MAPbBr$_3$)$_{0.15}$/spiro-OMeTAD/SWCNTs or Au. They used this configuration both with and without spiro-OMeTAD. Here they transferred SWCNT film onto the perovskite layer from filter paper by simple press transfer method. Chlorobenzene was drop casted to densify

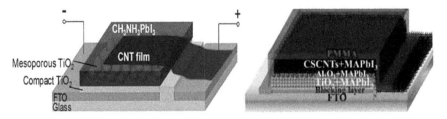

Fig. 7 Device architecture of PSCs using CNTs as back electrode. Reprinted with permission [25, 31]

this CNT layer. The PCEs for devices with HTL using Au and SWCNTs as back electrodes are 17.7% and 13.6%, respectively, and that without using HTL are 5.0% and 9.1%, respectively. In 2016, Luo and co-workers [25] fabricate cross-stacked superaligned carbon nanotubes (CSCNTs)-based HTL-free PSC. The cell architecture was FTO/TiO$_2$/CH$_3$NH$_3$PbI$_3$/CSCNTs (Fig. 7). Superaligned CNTs of around 300 μm length were fabricated using LP-CVD technique on a silicon wafer with Fe thin film, and then a large number of 25, 50, and 75 layers were cross stacked one over another. Doping with iodine (I-CSCNTs) and thermal treatment (T-CSCNTs) were done to improve the performance. A device with 50 stacking layers shows sheet resistance of 43, 30, and 16 $\Omega/\square$ for pristine, thermally treated, and iodine-doped CSCNTs and the corresponding PCEs are 7.57, 8.26, and 9.64%. This is due to better wettability and hence better contact of iodine-doped CSCNTs. Encapsulated devices with PMMA show good stability in the dark as well as under light soaking conditions. Again in 2017, Luo and co-workers [34] fabricated a PSC using SnO$_2$-coated CSCNTs film as a back electrode on a rigid as well as flexible substrate. The cell architecture was FTO/NiO$_2$/Al$_2$O$_3$ + CH$_3$NH$_3$PbI$_3$/SnO$_2$@CSCNTs + CH$_3$NH$_3$PbI$_3$. The sheet resistance of hybrid SnO$_2$@CSCNTs (51 $\pm$ 2.3 $\Omega/\square$) is equivalent to the sheet resistance of CSCNTs (44 $\pm$ 1.7 $\Omega/\square$), which shows that joints in CNTs are good even after SnO$_2$ coating. In this way, they obtained an average PCE of 9 $\pm$ 0.9% for 30 devices. The cells fabricated on flexible substrates with 9% PCE retain nearly 8% PCE after 300 bending cycles. Also, all these cells were tested for humidity, heat, and photodegradation for about 500 h and retained 90% of their PCEs. It has been observed that these cells are more stable than similar cells fabricated using Ag as the back electrode. In 2017, Aitola and co-workers [24] fabricated a triple cation PSC using both Au and SWCNTs separately as back contact. The cell architecture was FTO/TiO-$_2$/FAPbI$_3$:MAPbBr$_3$/spiro-OMeTAD/Au or SWCNTs. SWCNTs were synthesized using a floating catalyst-CVD technique and were transferred to perovskite substrate using a filter paper by applying some pressure. In this work, the maximum PCE of 14.3% was achieved using SWCNTs as back contact. Both types of cells fabricated using Au and SWCNTs as back contact were tested at MPPT under 1-Sun equivalent irradiance at 60 °C for 140 h. Cells fabricated with Au as back contact degrade to 30% of its initial PCE value, whereas cells with SWCNTs as back contact show a negligible drop in PCE. In 2017, Jeon and co-workers [23] fabricated a PSC using CNTs both as cathode and anode. The corresponding device architecture was CNT/PEDOT:PSS/CH$_3$NH$_3$PbI$_3$/PCBM/CNT. A maximum PCE of 10.5% was obtained. They compared the bending stability for three different pairs of electrodes, viz., ITO/CNT, CNT/Al, and CNT/CNT at different bending radii. It is evident from their work that the device with both CNT electrodes can tolerate more bending in comparison to others. In 2018, Zhou and co-workers [26] fabricated an inverted PSC using PEI-modified CSCNTs as the back electrode. The cell architecture was FTO/NiO$_x$/CH$_3$NH$_3$PbI$_3$/PCBM/CSCNT:PEI. The CSCNTs were synthesized using CVD technique on a silicon wafer, densified by immersing into ethanol, and were deposited onto the PCBM layer by solvent-assisted transfer. For a device with CSCNT:PEI (0.5 wt%), they got a PCE of 10.8% with negligible hysteresis. For a similar configuration with Ag back contact, they got 14.9% PCE. The device retains

94% of its initial PCE after 500 h at room temperature and ambient air without any encapsulation, whereas the PCE of the device with Ag back contact falls to 50% of its initial value. At an elevated temperature of 60 °C, the PCE of the device with Ag falls to 40% of the initial value within 200 h and CSCNT devices retain nearly 90%. When devices were tested in high relative humidity of 60% at 60 °C the Ag-based device PCE reduces to 10% of its starting value whereas CSCNTs retain 70% which was further improved to 85% by using PMMA encapsulation. In 2018, Luo and co-workers [9] fabricated highly flexible planar PSC using CSCNTs as the back electrode. The device architecture was graphene/TiO$_2$/PCBM/CH$_3$NH$_3$PbI$_3$/CSCNTs + spiro-OMeTAD. For these devices, a PCE of 11.9% and 8.4% was obtained with and without using HTL. A similar set of devices with ITO/Au as electrodes was also fabricated for comparison. The devices with CSCNTs show great bending stability with a bending radius of up to 2 mm for 2000 bending cycles. Also, the devices with CSCNTs are more stable than those with conventional electrodes under light soaking and thermal stress conditions for more than 1000 h. The improved stability is because the carbon electrode exerts very little effect on the perovskite layer and also CSCNT layer is thick and hydrophobic which provides encapsulation to the perovskite layer. The device architecture was CNTs/PCBM/CH$_3$NH$_3$PbI$_3$/PEDOT:PSS/CNTs. The CNTs were synthesized using aerosol CVD technique. They got a PCE of 7.32%. By opting for such device architecture, the material cost reduces to approximately 33% of the PSC fabricated with conventional materials. These devices show better bending stability up to a bending radius of 1 mm in comparison to devices with conventional electrodes (Fig. 8).

Fig. 8 Structure of graphite (left) and carbon black (right)

2.3 Graphite and Carbon Black

Graphite is one of the most common naturally occurring and most stable allotropes of carbon. It is composed of a hexagonal lattice of carbon atoms and is a soft material. It has a high electrical conductivity of the order of 10^5 S/m. The carrier mobility in graphite is 20100 $cm^2V^{-1}s^{-1}$ for electrons and 1590 $cm^2V^{-1}s^{-1}$ for holes. It is also thermally very stable till 700 °C. Also, the hydrophobic nature of graphite provides additional stability toward moisture and humidity. At the same time, the small size of carbon black or carbon nano-powder provides better contact between active perovskite layer and charge-collecting graphite. In this way, a combination of graphite and carbon black in a suitable ratio provides good contact with active layer and enhanced charge collection. These interesting properties make it a capable candidate for being used as an electrode in solar cells.

In general, a carbon paste consisting of carbon black and graphite flakes or graphite nano-powder in different weight ratios is prepared. In some cases, ZrO_2 is also added to make scratch-resistant films. A finely blended mixture of carbon powder is then dissolved into hydroxypropyl cellulose or its solution with ethyl acetate and polyvinyl acetate. Ethylcellulose and terpineol can also be used to dissolve carbon powder. This mixture is placed in a ball mill at 300–400 rpm for several hours to form a viscous paste. The carbon paste thus obtained is directly coated onto perovskite film using screen printing or doctor blading followed by drying at room temperature. Alternatively, in some cases, mesoporous carbon paste is screen printed directly onto TiO_2 followed by sintering at around 450 °C, and then perovskite solution is dripped into mesoporous structure followed by annealing at 70–90 °C [35–37]. In another approach, a similar carbon paste is doctor bladed over a Teflon film and a carbon layer is formed after solvent evaporation. Then these carbon films can be transferred to the perovskite layer by hot press method at a temperature around 85 °C (Fig. 9) [38].

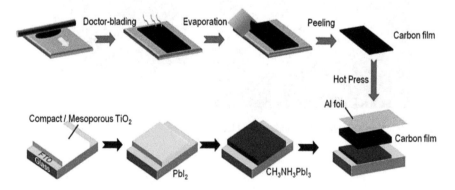

Fig. 9 Schematic of integrating carbon film with PSCs. Reprinted with permission [38]

2.3.1 Graphite and Carbon Black as Back Electrode

In 2013, Ku and co-workers [4] fabricated PSC with a fully printable carbon counterelectrode. The cell architecture was FTO/TiO$_2$/ZrO$_2$/CH$_3$NH$_3$PbI$_3$/carbon. Here two different carbon pastes were prepared, one with flaky graphite + carbon black and other with spheroidal graphite + carbon black. The carbon paste was printed over the ZrO$_2$ layer and sintered at 400 °C. The perovskite layer was drop coated over the carbon electrode. A PCE of 4.08% was obtained using flaky graphite and that of 6.64% was obtained using spheroidal graphite. The performance is improved in the case of spheroidal graphite because of its high conductivity and suitable morphology which provide better pore fillings. The cell with spheroidal graphite was tested for stability for 840 h in dark and it retained 6.5% PCE. In 2014, Xu and co-workers [39] from the same group developed a highly ordered mesoporous carbon counterelectrode for PSC. The cell architecture was FTO/TiO$_2$/ZrO$_2$/CH$_3$NH$_3$PbI$_3$/carbon layer (Fig. 10). Here they have used highly ordered carbon structured films of two different types with graphite and one film with carbon black and graphite. They observed that film with carbon black and graphite shows low PCE of 5.17% and that with highly order carbon was 6.30% and 7.02%. Low charge transfer resistance for ordered carbon structured films, the higher surface area of contact, and better pore filling results into higher PCEs. In 2014, Yang and co-workers [35] use a mesoscopic carbon layer along with a flexible graphite sheet to fabricate HTL-free carbon electrode-based PSC. The cell architecture was FTO/TiO$_2$/CH$_3$NH$_3$PbI$_3$/mesoscopic carbon/graphite sheet. For electrode fabrication, carbon paste was deposited using screen printing and then a graphite sheet was pressed over it. They have tried carbon paste with only graphite and graphite + carbon black and also fabricated one cell

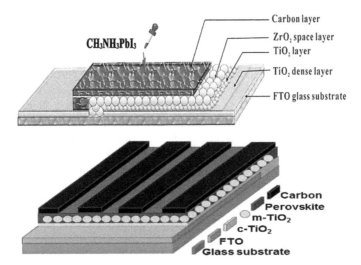

Fig. 10 Device architecture of PSCs using graphite and carbon black as back electrode. Reprinted with permission [39, 44]

with Au as back contact for comparison. They have used differently sized (1, 3, and 20 μm) graphite flakes and observed that smaller size (1 μm) gives better results as it makes better contact at perovskite/electrode interface. They have obtained a PCE of 7.21% and 10.20% for only graphite and carbon black mixed graphite paste, respectively. The PCE for Au back contact is nearly the same (10.73%) as that of graphite + carbon black (10.20%). In 2015, Wei and co-workers [38] developed a thermoplastic carbon film having high conductivity and flexibility to be used as a back electrode in PSCs. The cell architecture was $FTO/TiO_2/CH_3NH_3PbI_3/carbon$ film. Carbon paste was prepared by graphite flakes and carbon black in different weight ratios. The carbon film was deposited on Teflon film and transferred to the perovskite layer by hot pressing at 85 °C for 15 s at different pressures (0.15, 0.25, and 0.40 MPa). An Al foil was used to prevent adhesion of carbon film with a hot plate. Maximum PCE of 13.53% was obtained for PSC with 3:1 graphite-to-carbon black ratio carbon film at 0.25 MPa hot press pressure. Whereas a similar HTL-free PSC with Au as back contact gives PCE of 10.69%. These cells show good reproducibility as 20 separate devices show an average PCE of 12.03%. Also, these cells were tested for stability in the dark under ambient conditions and a reduction of 5% in PCE was observed which in the case of spiro/Au electrode-based cell was 21%. The complete fabrication of the electrode takes place at a temperature below 100 °C. In 2015, Wang and co-workers [37] fabricated a PSC with a carbon black electrode based on pure graphite, pure carbon back, and a mixture of graphite and carbon black. The cell architecture was $FTO/TiO_2/ZrO_2/CH_3NH_3PbI_3/graphite +$ carbon black. They have taken five different compositions of graphite + carbon black to make carbon paste, where carbon black was 0, 10, 20, 30, and 100% by weight. The carbon electrode was deposited using screen printing. The maximum PCE of 7.08% was obtained for the cell with 20% of carbon black. In ambient atmosphere and room temperature, the cell was kept for more than 900 h and it shows a negligible change in PCE. In 2015, Zhang and co-workers [40] fabricated a fully printable PSC with a carbon back electrode. The cell architecture was $FTO/TiO_2/ZrO_2/CH_3NH_3PbI_3/carbon$. The carbon paste was prepared using carbon black and graphite and ZrO_2. This paste was printed and the perovskite layer was dipped into the mesoscopic carbon layer. To make the carbon electrode different thicknesses of carbon film (5, 7, 9, 12, and 15 μm) were used. The cell with an electrode having 9 μm thickness shows a maximum PCE of 11.63%. Again, the experiment was repeated for different sized (500 nm, 3 μm, and 8 μm) graphite flakes. It was observed that 8 μm flakes give the best PCE of 11.65% followed by 500 nm and 3 μm. This happens because of low sheet resistance, charge transfer resistance, and series resistance of 8 μm sample in comparison to others. Also, the bigger pore size in 8 μm electrode allows better pore filling while perovskite dripping. In 2016, Yue and co-workers [41] designed a PSC with a carbon counterelectrode at low temperature. The cell architecture was $FTO/TiO_2/CH_3NH_3PbI_3/carbon$ black + graphite. To optimize the electrode, different weight ratios of carbon black (0, 15, 25, 35, and 100%) were taken with graphite. It has been observed that electrode with 25% carbon black shows the best PCE of 7.29%. While using 100% graphite, there was poor contact between perovskite graphite interface and graphite–graphite interface which was improved with the addition of carbon black but when carbon black was used in excess the PCE decreases because of poor charge transfer due to poor conductivity and porous structure of carbon black. Also, electrochemical impedance spectroscopy (EIS) shows the lowest hole transport resistance and the lowest series resistance for 25% carbon black-based electrode which leads to excellent device performance. In 2016, Liu and co-workers [36] fabricated a low temperature and flexible carbon electrode for PSCs. The cell architecture was $FTO/TiO_2/CH_3NH_3PbI_3/carbon$. For preparing carbon paste, different ratios of carbon black, graphite flakes, and graphite nano-powder were taken and a small amount (8%) of ZrO_2 was added to make film scratch resistant. A 65 μm carbon film was deposited on the perovskite layer using doctor blade technique. Four different configurations with 50% 10 μm graphite flakes, 17% 10 μm graphite flakes with 33% 400 nm graphite powder, 17% 10 μm graphite flakes with 33% 40 nm graphite powder, and 8% 10 μm graphite flakes with 42% 40 nm graphite powder were mixed with 17% by weight with 40 nm carbon black and solvent. Sample with 17% graphite and 33% 40 nm graphite powder shows the highest PCE of 6.88%. In 2017, Li and co-workers [42] fabri-cated an HTL-free PSC using graphite as a back electrode at low temperatures. The cell architecture was $FTO/TiO_2/CH_3NH_3PbI_3/graphite$ paste. In this work, they have made a contact study for different thicknesses and concentrations of the active and mesoporous layer. They got the best efficiency of 10.4% with a V_{OC} of 0.82 V for the optimized device. In 2017, Duan and co-workers [43] fabricated a fully printable HTL-free PSC using ultrathin graphite as back contact. The cell architec-ture was $FTO/TiO_2/ZrO_2/perovskite/carbon$ electrode. For electrode fabrication, two types of carbon pastes were formed by mixing bulk graphite and ultrathin graphite with carbon black. This electrode was fabricated at a low temperature of 50 °C. They have obtained a PCE of 12.63% single bulk graphite and 14.07% using ultrathin graphite. The reason for the increase in efficiency is increasing in specific surface area in the case of ultrathin graphite which leads to better contact and better absorp-tion of perovskite. Photoluminescence (PL) and time-resolved photoluminescence (TRPL) confirm the better extraction of holes in ultrathin graphite than bulk graphite. In 2018, Zhang and co-workers [44] fabricated PSC with low-temperature printable carbon back electrode. The cell architecture was $FTO/TiO_2/CH_3NH_3PbI_3/graphite$ + carbon black (Fig. 10). They have used a simple vibration technique to make better arrangement of graphite and carbon black in the electrode, which resulted in better contact between the perovskite layer and electrode. By optimizing the vibration time in cells, they achieved an improvement of 22% in PCE than those not treated with vibration. A commercial carbon paste (carbon black + graphite) diluted with chlorobenzene was coated over the perovskite layer using doctor blade technique in ambient conditions. Immediately after coating, it was vibrated for different times (0, 5, 10, and 20 min), and then dried for 20 min at 100 °C. The best PCE of 11.49% was obtained for a cell that was vibrated for 10 min. The contact between carbon black and graphite and that is between carbon layer and perovskite got improved due to vibration, which leads to better charge transfer and hence better cell performance. The decrease in efficiency of cell treated for 20 min is due to the degradation of the perovskite layer by the solvent present in the carbon paste. In ambient air, these cells retain about 77% of their PCE without any encapsulation. In a recent work in 2019, Poli and co-workers [45] fabricated a PSC with carbon paste as the back electrode protected with graphite. The cell architecture was $FTO/TiO_2/CsPbBr_3/carbon$. The carbon paste (carbon black + graphite) was coated on the perovskite layer using doctor blade technique. A self-adhesive graphite sheet was stuck onto the carbon layer and finally encapsulated with silicon and epoxy resin. The device was tested for stability in water for about 5 h and it shows no loss in performance.

3   Advantages of Carbon-Based Electrodes

As discussed, poor long-term stability is one of the major issues in the commercialization of PSCs. If we talk about stability, PSCs with carbon-based electrodes are more stable than those with traditional metal electrodes. One of the biggest issues with stability testing is that there are no standard test protocols for PSCs which makes lots of confusion while comparing the stabilities of different PSCs. Although different research groups are testing their cells under different conditions of light, moisture, thermal stress, and bending.

3.1   Photostability

Testing the long-term stability under illumination and maximum power point tracking (MPPT) conditions is more significant than that in a dark or controlled atmosphere, as it will be more realistic considering actual operating conditions of a solar cell. The active perovskite layer shows very poor stability when it comes to continuous light soaking, especially under UV light. So, it becomes important to make such cells that are stable under continuous light soaking for a long period. Many researchers have tested their devices under real operating conditions under 1-Sun illumination [6, 46–52].

In 2014, Mei and co-workers [46] fabricated a mesoporous PSC with carbon as a back electrode. All the mesoporous layers were printed on FTO and the perovskite layer was drop casted. They obtained a PCE of 12.8%. These cells show long-term stability for 1008 h under 1-Sun illumination in ambient air. In 2018, Meng and co-workers [47] developed a full carbon-based PSC with no hysteresis. The cell architecture they used was $FTO/C_{60}/CH_3NH_3PbI_3/carbon$ and obtained a PCE of 15.38%. They demonstrated these cells for stability under 1-Sun illumination at a relative humidity of 40–60% for 180 h with MPPT. The cells retain 95% of its initial PCE without any encapsulation in ambient air.

In 2017, Grancini and co-workers [6] demonstrated the most stable PSC with carbon as a back electrode. The device architecture they used for carbon-based PSC was $FTO/TiO_2/ZrO_2/perovskite/carbon$. A complete module of $10 \times 10 \text{ cm}^2$ was prepared using these cells. For an individual cell with a carbon electrode, a PCE of

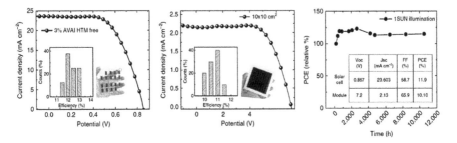

Fig. 11 J-V curves and stability data of PSCs and module with carbon electrode. Reprinted with permission [6]

11.9% was obtained, and that for the complete module was 10.10%. Using HTL and metal electrode, the PCE falls to 60% just in 300 h. On the other hand, HTL-free carbon electrode-based PSC retains its full PCE for more than 10000 h under AM 1.5 1-Sun illumination at 55 °C (Fig. 11).

In 2017, Hashmi and co-workers [48] fabricated PSC with a carbon back electrode having a mesoporous structure. These cells were demonstrated under continuous 1-Sun illumination for 1046 h without any encapsulation, which then retained about 95% of its initial PCE. In 2017, they [49] again fabricated carbon electrode-based PSC with similar structure. They prepared two batches of PSCs, one without any protective encapsulation and other encapsulated with commonly available epoxy. Both the batches were investigated for UV light soaking under 1.5 1-Sun illumination. After 751 h, the first batch with no encapsulation faced a decrease of 25% and 28% in short-circuit current and PCE, respectively. The second batch encapsulated with epoxy retained its full PCE even after 1002 h of illumination (Fig. 12). In 2018, Chu and co-workers [50] fabricated carbon electrode-based planar PSC. The cell architecture was FTO/SnO$_2$ @TiO$_2$/perovskite/P3HT + graphene/carbon. Here, with this architecture, they obtained a PCE of 18.1%. They also demonstrated these cells for stability under dark in the air without any encapsulation for 1680 h and under 1-Sun illumination with encapsulation for 600 h. These cells showed negligible efficiency drop in the dark and retained 89% of initial PCE under illumination (Fig. 13).

3.2 Thermal Stability

When working under real-time operating conditions, the temperature of solar cells increases by 30–40 °C from normal atmospheric temperature. So, it becomes important to analyze thermal stability of PSCs under high thermal stress at about 60–90 °C. It has been observed that at such high temperature, in PSCs, the degradation of HTL interface takes place due to ion migration from perovskite [53]. Metal migration

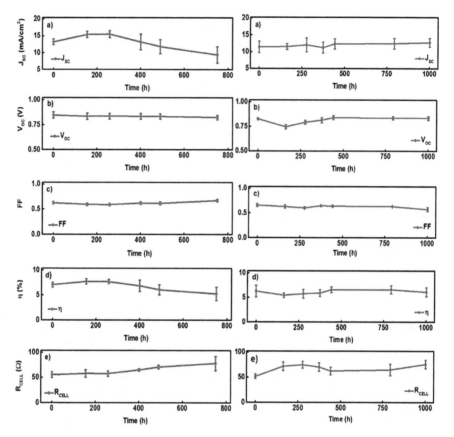

Fig. 12 Variation of different parameters of carbon electrode-based PSCs without encapsulation (red) and with encapsulation (blue). Reprinted with permission [49]

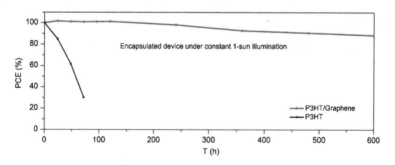

Fig. 13 Stability under 1-Sun illumination. Reprinted with permission [50]

due to high temperature also has been observed [2]. To tackle this problem, a meso-porous carbon electrode-based PSC has appeared as a potential candidate. Many researchers have analyzed stability of carbon electrode-based PSCs at high thermal stress [7, 9, 24, 51, 54–57].

In 2016, Baranwal and co-workers [54] fabricated a carbon electrode-based meso-scopic PSC. Although a low PCE was obtained, these fabricated cells were sealed using UV-curing glue and tested for thermal stability at 100 °C for over 1500 h. The PCE remained unaffected in this duration at such a high temperature. Again in 2019 [7], they demonstrated a similar stability test for a similar PSC with PCE 4.93% for a longer duration of 7000 h at high thermal stress of 100 °C. For the first 4500 h, the PCE falls to 10% of the initial PCE but after that this degradation accelerated, and in 7000 h, PCE fell to 45% of the initial value. In 2017, Aitola and co-workers [24] fabricated PSC with SWCNT as a counterelectrode with spiro as HTL. They fabricated another set with Au as a back electrode to compare the performance. PCEs of 18% and 14.3% were obtained using Au and SWCNTs, respectively. These devices were tested for stability under 1-Sun illumination at 60 °C temperature for 140 h. With Au as back contact PCE falls drastically at 60 °C where devices with SWCNTs as back electrode show a little change in PCE over a duration of 140 h. In 2018, Luo and co-workers [9] fabricated an all-carbon electrode-based flexible PSC. Device architecture was graphene/TiO_2/PCBM/$CH_3NH_3PbI_3$/spiro-OMeTAD/CSCNTs. Another set of devices with ITO and Au/Ag as electrodes was also prepared to compare the performance. Carbon electrode-based PSC obtained PCE of 11.9% and 8.4% with and without spiro-OMeTAD. These devices were exposed to a thermal stress of 60 °C for 1570 h. The devices with carbon electrodes retained 89% of initial PCE which makes it quite more stable in comparison to metal and ITO-based PSCs.

In 2018, Lee and co-workers [55] fabricated HTL-free 2D/3D PSC with carbon as a back electrode. By introducing a 2D perovskite interlayer, the PCE increased from 11.5 to 14.5%. Also, it provided better contact between the perovskite and the carbon layer. To compare the performance and stability, another device with conventional spiro/Au was also prepared. When these devices were tested under high thermal stress of 100 and 150 °C, 3D perovskite/carbon device retained 85% of the initial PCE and 3D/2D perovskite/carbon shows PCE 102% of the initial value. This is because the 2D interlayer prevented the ion migration and PEAI in the carbon electrode suppressed the thermal decomposition. On the other hand, the device with Au as the back electrode showed a drastic fall in PCE due to hygroscopic dopant in HTL and unstable HTL/Au interface (Fig. 14).

In 2019, Wu and co-workers [56] fabricated PSC with carbon electrode. A maximum PCE of 14.5% was achieved. The devices were tested for stability under a double aging condition at 85 °C and 85% relative humidity. After 192 h, these devices retained 77% of the initial PCE. In 2019, Zhou and co-workers [57] fabricated PSC with carbon as a back electrode incorporating MWCNTs into mixed cation perovskite. The whole device structure was prepared at a temperature of less than 150 °C. Device architecture was ITO/SnO_2/perovskite/carbon. A PCE of 15.93% was obtained. The fabricated devices with different concentrations of MWCNTs

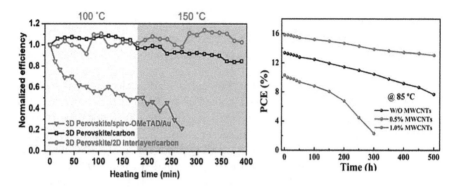

Fig. 14 Stability under high temperature. Reprinted with permission [55, 57]

were tested under high thermal stress of 85 °C continuously for 500 h. The devices with 0.5% MWCNTs retained 82.5% of their initial PCE (Fig. 14).

3.3 Moisture Stability

Humidity/moisture is one of the major problems faced by PSCs. Most of the perovskite materials are found to be highly unstable toward moisture. So, we need an electrode that can protect the active layer from moisture. Commonly used ways for the same are the use of carbon electrodes. The hydrophobic nature of carbon materials makes carbon electrode-based PSC capable of sustaining harsh humid conditions. Many researchers have been successful in showing the capability of carbon electrode-based PSCs against moisture degradation [8, 25, 56, 58–63].

In 2016, Yu and co-workers [8] fabricated PSCs which were highly stable toward moisture and thermal stress. They used a carbon electrode-based HTL-free structure to minimize instability caused by HTL degradation. They fabricated a bilayer back electrode having one layer with carbon/MAI with thickness of 10 μm and another layer of hydrophobic carbon to prevent the device from moisture. To test the moisture stability, the fabricated devices were flushed under running water for 30 s and immersed into water for 15 min. Before any exposure to water, the devices exhibited a PCE of 12.62%. These devices retained 98% and 92% of their initial PCE on flushing and immersing into water, respectively.

In 2016, Mali and co-workers [58] fabricated a waterproof and air-stable PSC with a similar configuration. The device architecture was FTO/TiO$_2$/MAPbI$_{3-x}$Cl$_x$/carbon + MAI/carbon. Using this configuration, they obtained a PCE of 13.87%. For stability comparison, they fabricated similar devices with spiro/Au. A stability test was performed by flushing the devices underwater. The PCE of the device with spiro/Au fell from 15% to 0.33% on the first exposure to water, whereas the PCE of the device with carbon electrode fell from 13.87% to 12.25% and showed a PCE of 5.93% even after seven exposures to water flush (Fig. 15a). In

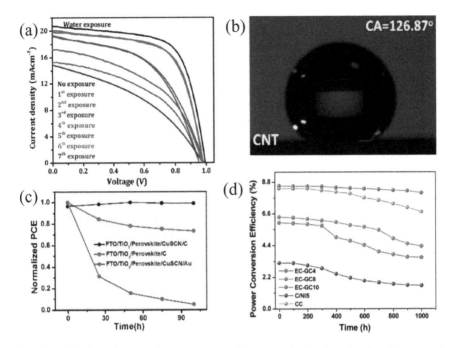

Fig. 15 **a** Stability under several water exposures, **b** water contact angle on CNT surface, and (**c**), **d** stability of PSCs with carbon electrode under high humidity. Reprinted with permission [58–60]

2019, Wu and co-workers [59] fabricated a PSC with carbon as a back electrode and CuSCN/CNT as HTL. They obtained a maximum PCE of 17.58% with good reproducibility. The water contact angle of 126.87^0 of CNTs showed hydrophobic nature of it (Fig. 15b). The fabricated devices with different configurations were tested for stability under 80% relative humidity for 100 h. The device with Au back contact completely degraded in this duration whereas the devices having a carbon electrode with and without HTL retained 95 and 75% of the initial PCE, respectively (Fig. 15c). Recently in 2020, Pitchaiya and co-workers [60] fabricated a porous graphitic carbon HTL/counterelectrode for PSC. They prepared graphitic carbon from Eichhornia Crassipes, an invasive plant species. As graphitization improves at different annealing temperatures (450, 850, and 1000 °C), corresponding to this, three different types of devices were prepared with EC-GC 4/8/10 as the back electrode. The device architecture was $FTO/TiO_2/CH_3NH_3PbI_{3-x}Cl_x/EC-GC10@CH_3NH_3PbI_{3-x}Cl_x/EC-GC4/8/10$. A maximum PCE of 8.52% was thus obtained using EC-GC10. These devices were kept under 70% relative humidity for 1000 h to check their moisture stability. After 1000 h, the device with EC-GC10 retained 94.4% of its initial PCE value (Fig. 15d).

3.4 Flexibility

The flexibility of PSCs can play a key role in the development of foldable, portable, and wearable electronics. The flexibility of photovoltaic devices will also be helpful in building and vehicle-integrated photovoltaics. Most of the flexible solar cells use ITO/PEN as transparent electrodes, but on repeated bending, these devices show drastic performance degradation as ITO is brittle leading to the formation of cracks. Also, flexible solar cells require gold or silver as back contact which is not commercially favorable. A carbon-based electrode, when employed as a flexible electrode, shows lower degradation, which makes these electrodes more preferable while choosing a flexible electrode. Many works have shown the potential of carbon material as a flexible electrode for PSCs [9, 18, 19, 23, 64].

In 2017, Yoon and co-workers [19] demonstrated a flexible PSC with graphene as a transparent electrode. The device architecture was PEN/graphene/MoO_3/PEDOT:PSS/$CH_3NH_3PbI_3$/C_{60}/BCP/LiF/Al. Another device with ITO as a transparent electrode was also prepared to compare the performance. PCEs of 16.8% and 17.3% were obtained for the device with graphene and ITO, respectively. The device with ITO after 1000 bending cycles at radii 6 mm and 4 mm retains 60% and 25% of its initial PCE, respectively. On the other hand, graphene-based device retains 90% of its PCE at 4 mm and 6 mm. When bending cycles increase to 5000, the device with graphene shows only a small change in PCE but that for ITO falls to a very low value just in 100 bending cycles (Fig. 16a, b). Again in 2017, Jeon and co-workers [23] fabricated a complete solution-processable

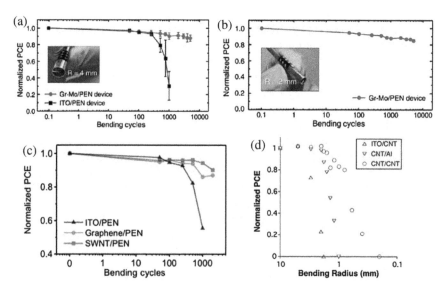

Fig. 16 Stability under different bending radii for large number of bending cycles. Reprinted with permission [18, 19, 23]

PSC. Here, they used CNTs both as anode and cathode and the corresponding device architecture was CNT/PEDOT:PSS/CH$_3$NH$_3$PbI$_3$/PCBM/CNT. A maximum PCE of 10.5% was obtained. They compared the bending stability for three different pairs of electrodes, viz., ITO/CNT, CNT/Al, and CNT/CNT at different bending radii. The device with both CNT electrodes can tolerate more bending in comparison to others (Fig. 16d). Again in 2017, Jeon and co-workers [18] fabricated flexible PSC using three different transparent electrodes, viz. ITO, graphene, and SWCNTs. They obtained PCE of 17.8, 14.2, and 12.8%, respectively, for ITO-, graphene-, and SWCNT-based devices. While performing bending stability tests, they observed that the loss in PCE is negligible for SWCNTs and very small for graphene-based devices but ITO-based devices showed a drastic drop in PCE after 1000 bending cycles (Fig. 16c). The reason for this performance loss was an increase in sheet resistance of transparent electrodes with bending cycles. In 2018, Luo and co-workers [9] fabricated PSC using all-carbon electrodes for moving towards flexible photovoltaics. The device architecture they used was PET/graphene/TiO$_2$/PCBM/CH$_3$NH$_3$PbI$_3$/spiro-OMeTAD/CSCNTs. They obtained PCEs of 11.9% and 8.4% with and without using spiro as HTL. Similar devices with ITO/Au as electrodes were also prepared for comparison. To check the flexibility, these devices were tested at different bending radii and several bending cycles were performed. It was observed that up to 4 mm, device with graphene/CSCNTs did not show any significant change in PCE where ITO/Au device showed a reduced PCE of 87% of the initial value. With further reduction in the radius of curvature, ITO/Au device got damaged; on the other hand, graphene/CSCNTs still showed a PCE of 85% of initial value at the bending radius of 2.2 mm. The PCE of the device with ITO/Au reduced to 13% of its initial value after 1500 bending cycles, where the device with graphene/CSCNTs retained 84% of the initial value even after 2000 bending cycles.

3.5 Scalability

Most of the PSCs fabricated at laboratory scale have an active area less than 1 cm^2, but when it comes to module development, we need large-area cells to collect an appreciable amount of solar energy. A typical silicon solar cell is of an area about 6 inch × 6 inch, so the inactive area between two adjacent cells is quite low. At the same time, if we have 1 cm × 1 cm cells the ratio of inactive to the active area will be very large and hence there will be more land requirement, which is again a big problem. So, to overcome this problem, we have to fabricate large-area cells. Researchers are working on roll-to-roll, spray coating, and other conventional techniques to make large-area cells and trying some methods to integrate them into large modules to harvest more energy (Fig. 18).

In 2016, Priyadarshi and co-workers [65] fabricated a highly stable and highly efficient large-area PSC module. They used two different substrate sizes of 5 × 10 cm^2 and 10 × 10 cm^2 to make a module with 31 cm^2 and 70 cm^2 (Fig. 17b) active area. They used graphite and carbon nanoparticle paste for back electrode

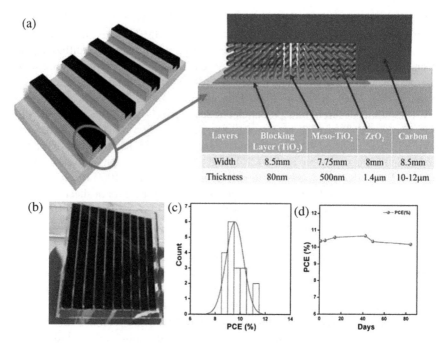

Fig. 17 **a** Large-area module, device architecture, **b** module with area 70 cm², **c** PCE versus number of cells, and **d** variation in PCE over time. Reprinted with permission [65]

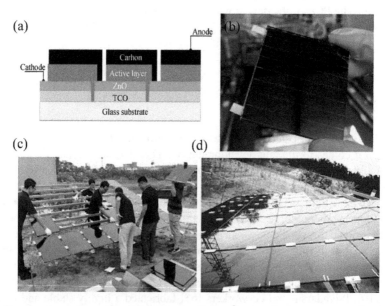

Fig. 18 **a** Device structure, **b** module with area 17.3 cm², **c** installation of panels, and **d** outdoor situated 45 × 65 cm² power station. Reprinted with permission [66]

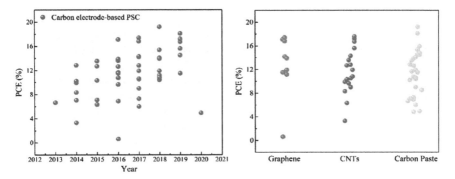

Fig. 19 PCE of carbon electrode-based PSCs over several years (left), PCE of PSCs with different carbon electrode materials (right). The data points on the left for a particular material are the work published before than on the right

fabrication. A simple fabrication process using screen printing was employed which is easily up scalable to roll-to-roll printing techniques for the industrial purpose. The PCEs of 10.46% and 10.74% were obtained, respectively, for modules with 31 and 70 cm^2 area. These devices show over 2000 h stability in an ambient atmosphere with a relative humidity of 65–70% with a small loss of 5% in PCE (Fig. 17d). Also, at MPP under 1-Sun illumination, these devices show extreme stability for 72 h of continuous operation.

In 2017, Cai and co-workers [66] developed large-area PSC module with planar structure. The device architecture was FTO/ZnO/CH$_3$NH$_3$PbI$_3$/carbon (Fig. 19a). On a substrate of 5 × 5 cm^2 they made 17.3 cm^2 active area module consisting of seven cells connected in the series (Fig. 19b). They achieved a PCE of 10.6% with an average PCE of (9.97 ± 0.35) % for 50 modules showing high reproducibility. These modules were sealed with glass and waterproof sealant and placed outdoor. No sign of degradation was observed during 140 days of outdoor testing. After this, they increased the module size to 45 × 65 cm^2 and made a power station with 32 such panels (Fig. 19c, d). In 2017, Hu and co-workers [10] developed a fully printable PSC module with a triple mesoporous layer scaffold. The device architecture was FTO/TiO$_2$/ZrO$_2$/carbon/(5-AVA)$_x$MA$_{1-x}$PbI$_3$. A large-area module of 100 cm^2 with an active area of 49 cm^2 was designed which consisted of 10 subcells connected in series. These modules are highly reproducible as 10 modules show an average PCE of above 9%. These modules were tested under continuous illumination for 1000 h at 54% relative humidity. During the test, surface temperature of devices reached 50 °C and still showed excellent stability. The encapsulated devices were exposed to outdoor conditions in Wuhan, China where average humidity was around 80% for 1 month and showed no sign of degradation. Unsealed devices showed good stability in the dark over 1 year with no degradation. They also fabricated a large module with 7 m^2 area for practical application. Every 1 m^2 contained 96 modules with a different number of subcells.

Although PCEs of PSCs with carbon-based electrodes are not as high as that with traditional electrodes, it is quite evident that the values are continuously improving (Fig. 19) and are approaching the PCEs of PSCs with traditional electrodes. It is just a matter of time that carbon electrode-based PSCs will have PCE values similar to that with traditional electrodes with added advantages of stability and low cost.

4 Summary and Outlook

In this chapter, we have discussed the possibility of replacing traditional electrodes with carbon-based materials for future generation PSCs. The chapter discusses various carbon-based materials, their electro-optical properties, fabrications methods, and advantages for integrating them in PSCs. The work done by various groups, giving up-to-date results, has been included and discussed giving researchers working in the field a comprehensive and complete overview of the work done till now. Different carbon-based electrode materials, viz., graphene, carbon nanotubes, graphite, and carbon black are discussed in the chapter. These materials are inherently stable, inexpensive, and their electro-opticals are such that apart from electrodes some of them can also be used as a charge transport and collection layers in PSCs. We have mainly focussed on the application of these materials as electrodes in PSCs. Graphene and carbon nanotubes do quite well as a transparent electrode because of their better trade-off between electrical and optical properties whereas CNTs and carbon paste (graphite and carbon black) can successfully replace the traditional noble metal back electrodes in PSC device architecture. In many research works reviewed in the chapter, PSCs with carbon electrodes show PCE over 19% exhibiting their suitability. Long-term stability up to 1 year under different conditions of light, heat, and moisture has also been tested for these cells. Mechanical flexibility with very low bending radii has also been verified with promising results compared to traditional electrodes. In terms of scalability to large-area modules, these carbon-based electrodes again present better opportunities in terms of processing ease, time, and cost. It is quite evident from the review of work done in this field that carbon electrodes definitely are a long-term solution and its incorporation can lead to widespread commercialization of third-generation photovoltaics.

References

1. National Renewable Energy Laboratory. http://www.nrel.gov. Accessed 25 May 2020
2. Li J, Dong Q, Li N, Wang L (2017) Direct evidence of ion diffusion for the silver-electrode-induced thermal degradation of inverted perovskite solar cells. Adv Energy Mater 7. https://doi.org/10.1002/aenm.201602922
3. Domanski K, Correa-Baena JP, Mine N et al (2016) Not all that glitters is gold: metal-migration-induced degradation in perovskite solar cells. ACS Nano 10:6306–6314. https://doi.org/10.1021/acsnano.6b02613

4. Ku Z, Rong Y, Xu M, et al (2013) Full printable processed mesoscopic CH3 NH3 PbI$_3$/TiO$_2$ heterojunction solar cells with carbon counter electrode. Sci Rep 3. https://doi.org/10.1038/sre p03132

5. Zhang H, Xiao J, Shi J, et al (2018) Self-adhesive macroporous carbon electrodes for efficient and stable perovskite solar cells. Adv Funct Mater 28. https://doi.org/10.1002/adfm.201802985

6. Grancini G, Roldán-Carmona C, Zimmermann I, et al (2017) One-Year stable perovskite solar cells by 2D/3D interface engineering. Nat Commun 8. https://doi.org/10.1038/ncomms15684

7. Baranwal AK, Kanda H, Shibayama N et al (2019) Thermal degradation analysis of sealed perovskite solar cell with porous carbon electrode at 100 °C for 7000 h. Energy Technol 7:245–252. https://doi.org/10.1002/ente.201800572

8. Yu Z, Chen B, Liu P et al (2016) Stable organic-inorganic perovskite solar cells without hole-conductor layer achieved via cell structure design and contact engineering. Adv Funct Mater 26:4866–4873. https://doi.org/10.1002/adfm.201504564

9. Luo Q, Ma H, Hou Q, et al (2018) all-carbon-electrode-based endurable flexible perovskite solar cells. Adv Funct Mater 28. https://doi.org/10.1002/adfm.201706777

10. Hu Y, Si S, Mei A, et al (2017) Stable large-area ($10 \times 10\,cm^2$) printable mesoscopic perovskite module exceeding 10% efficiency. Sol RRL 1. https://doi.org/10.1002/solr.201600019

11. Cai C, Jia F, Li A et al (2016) Crackless transfer of large-area graphene films for superior-performance transparent electrodes. Carbon N Y 98:457–462. https://doi.org/10.1016/j.carbon.2015.11.041

12. Geim AK, Novoselov KS (2007) The rise of graphene. Nat Mater 6:183–191. https://doi.org/10.1038/nmat1849

13. Wang Y, Weng GJ (2018) Electrical Conductivity of Carbon Nanotube- and Graphene-Based Nanocomposites. In: Micromechanics and nanomechanics of composite solids. Springer International Publishing, pp 123–156

14. Liu Z, You P, Xie C et al (2016) Ultrathin and flexible perovskite solar cells with graphene transparent electrodes. Nano Energy 28:151–157. https://doi.org/10.1016/j.nanoen.2016.08.038

15. Sung H, Ahn N, Jang MS, et al (2016) Transparent conductive oxide-free graphene-based perovskite solar cells with over 17% efficiency. Adv Energy Mater 6. https://doi.org/10.1002/aenm.201501873

16. Batmunkh M, Shearer CJ, Biggs MJ, Shapter JG (2016) Solution processed graphene structures for perovskite solar cells. J Mater Chem A 4:2605–2616. https://doi.org/10.1039/c5ta08996d

17. Heo JH, Shin DH, Kim S et al (2017) Highly efficient CH3NH3PbI3 perovskite solar cells prepared by AuCl3-doped graphene transparent conducting electrodes. Chem Eng J 323:153–159. https://doi.org/10.1016/j.cej.2017.04.097

18. Jeon I, Yoon J, Ahn N et al (2017) Carbon nanotubes versus Graphene as flexible transparent electrodes in inverted perovskite solar cells. J Phys Chem Lett 8:5395–5401. https://doi.org/10.1021/acs.jpclett.7b02229

19. Yoon J, Sung H, Lee G et al (2017) Superflexible, high-efficiency perovskite solar cells utilizing graphene electrodes: Towards future foldable power sources. Energy Environ Sci 10:337–345. https://doi.org/10.1039/c6ee02650h

20. Jang CW, Kim JM, Choi SH (2019) Lamination-produced semi-transparent/flexible perovskite solar cells with doped-graphene anode and cathode. J Alloys Compd 775:905–911. https://doi.org/10.1016/j.jallcom.2018.10.190

21. Kaskela A, Nasibulin AG, Timmermans MY et al (2010) Aerosol-synthesized SWCNT networks with tunable conductivity and transparency by a dry transfer technique. Nano Lett 10:4349–4355. https://doi.org/10.1021/nl101680s

22. Dürkop T, Getty SA, Cobas E, Fuhrer MS (2004) Extraordinary mobility in semiconducting carbon nanotubes. Nano Lett 4:35–39. https://doi.org/10.1021/nl034841q

23. Jeon I, Seo S, Sato Y et al (2017) Perovskite solar cells using carbon nanotubes both as cathode and as anode. J Phys Chem C 121:25743–25749. https://doi.org/10.1021/acs.jpcc.7b10334

24. Aitola K, Domanski K, Correa-Baena JP, et al (2017) High temperature-stable perovskite solar cell based on low-cost carbon nanotube hole contact. Adv Mater 29. https://doi.org/10.1002/adma.201606398

25. Luo Q, Ma H, Zhang Y et al (2016) Cross-stacked superaligned carbon nanotube electrodes for efficient hole conductor-free perovskite solar cells. J Mater Chem A 4:5569–5577. https://doi.org/10.1039/c6ta01715k

26. Zhou Y, Yin X, Luo Q et al (2018) Efficiently improving the stability of inverted perovskite solar cells by employing Polyethylenimine-modified carbon nanotubes as electrodes. ACS Appl Mater Interfaces 10:31384–31393. https://doi.org/10.1021/acsami.8b10253

27. Qiu L, Deng J, Lu X et al (2014) Integrating Perovskite Solar Cells into a Flexible Fiber. Angew Chemie Int Ed 53:10425–10428. https://doi.org/10.1002/anie.201404973

28. Wang X, Li Z, Xu W et al (2015) TiO2 nanotube arrays based flexible perovskite solar cells with transparent carbon nanotube electrode. Nano Energy 11:728–735. https://doi.org/10.1016/j.nanoen.2014.11.042

29. Jeon I, Chiba T, Delacou C et al (2015) Single-walled carbon nanotube film as electrode in indium-free planar heterojunction perovskite solar cells: investigation of electron-blocking layers and dopants. Nano Lett 15:6665–6671. https://doi.org/10.1021/acs.nanolett.5b02490

30. Jeon I, Yoon J, Kim U, et al (2019) High-performance solution-processed double-walled carbon nanotube transparent electrode for perovskite solar cells. Adv Energy Mater 9. https://doi.org/10.1002/aenm.201901204

31. Li Z, Kulkarni SA, Boix PP et al (2014) Laminated carbon nanotube networks for metal electrode-free efficient perovskite solar cells. ACS Nano 8:6797–6804. https://doi.org/10.1021/nn501096h

32. Wei Z, Chen H, Yan K et al (2015) Hysteresis-free multi-walled carbon nanotube-based perovskite solar cells with a high fill factor. J Mater Chem A 3:24226–24231. https://doi.org/10.1039/c5ta07714a

33. Aitola K, Sveinbjörnsson K, Correa-Baena JP et al (2016) Carbon nanotube-based hybrid hole-transporting material and selective contact for high efficiency perovskite solar cells. Energy Environ Sci 9:461–466. https://doi.org/10.1039/c5ee03394b

34. Luo Q, Ma H, Hao F, et al (2017) Carbon nanotube based inverted flexible perovskite solar cells with all-inorganic charge contacts. Adv Funct Mater 27. https://doi.org/10.1002/adfm.201703068

35. Yang Y, Xiao J, Wei H et al (2014) An all-carbon counter electrode for highly efficient hole-conductor-free organo-metal perovskite solar cells. RSC Adv 4:52825–52830. https://doi.org/10.1039/c4ra09519g

36. Liu Z, Shi T, Tang Z et al (2016) Using a low-temperature carbon electrode for preparing hole-conductor-free perovskite heterojunction solar cells under high relative humidity. Nanoscale 8:7017–7023. https://doi.org/10.1039/c5nr07091k

37. Wang H, Hu X, Chen H (2015) The effect of carbon black in carbon counter electrode for CH3NH3PbI3/TiO2 heterojunction solar cells. RSC Adv 5:30192–30196. https://doi.org/10.1039/c5ra02325d

38. Wei H, Xiao J, Yang Y et al (2015) Free-standing flexible carbon electrode for highly efficient hole-conductor-free perovskite solar cells. Carbon N Y 93:861–868. https://doi.org/10.1016/j.carbon.2015.05.042

39. Xu M, Rong Y, Ku Z et al (2014) Highly ordered mesoporous carbon for mesoscopic CH3NH3PbI3/TiO2 heterojunction solar cell. J Mater Chem A 2:8607–8611. https://doi.org/10.1039/c4ta00379a

40. Zhang L, Liu T, Liu L et al (2015) The effect of carbon counter electrodes on fully printable mesoscopic perovskite solar cells. J Mater Chem A 3:9165–9170. https://doi.org/10.1039/c4ta04647a

41. Yue G, Chen D, Wang P et al (2016) Low-temperature prepared carbon electrodes for hole-conductor-free mesoscopic perovskite solar cells. Electrochim Acta 218:84–90. https://doi.org/10.1016/j.electacta.2016.09.112

42. Li J, Yao JX, Liao XY et al (2017) A contact study in hole conductor free perovskite solar cells with low temperature processed carbon electrodes. RSC Adv 7:20732–20737. https://doi.org/10.1039/C7RA00066A

43. Duan M, Rong Y, Mei A et al (2017) Efficient hole-conductor-free, fully printable mesoscopic perovskite solar cells with carbon electrode based on ultrathin graphite. Carbon N Y 120:71–76. https://doi.org/10.1016/j.carbon.2017.05.027

44. Zhang Y, Zhuang X, Zhou K et al (2018) Vibration treated carbon electrode for highly efficient hole-conductor-free perovskite solar cells. Org Electron 52:159–164. https://doi.org/10.1016/j.orgel.2017.10.018

45. Poli I, Hintermair U, Regue M, et al (2019) Graphite-protected CsPbBr3 perovskite photoanodes functionalised with water oxidation catalyst for oxygen evolution in water. Nat Commun 10. https://doi.org/10.1038/s41467-019-10124-0

46. Mei A, Li X, Liu L, et al (2014) A hole-conductor–free, fully printable mesoscopic perovskite solar cell with high stability. Science (80) 345:295. https://doi.org/10.1126/science.1254763

47. Meng X, Zhou J, Hou J, et al (2018) Versatility of carbon enables all carbon based perovskite solar cells to achieve high efficiency and high stability. Adv Mater 30. https://doi.org/10.1002/adma.201706975

48. Hashmi SG, Martineau D, Li X, et al (2017) Air processed inkjet infiltrated carbon based printed perovskite solar cells with high stability and reproducibility. Adv Mater Technol 2. https://doi.org/10.1002/admt.201600183

49. Hashmi SG, Tiihonen A, Martineau D et al (2017) Long term stability of air processed inkjet infiltrated carbon-based printed perovskite solar cells under intense ultra-violet light soaking. J Mater Chem A 5:4797–4802. https://doi.org/10.1039/c6ta10605f

50. Chu QQ, Ding B, Peng J et al (2019) Highly stable carbon-based perovskite solar cell with a record efficiency of over 18% via hole transport engineering. J Mater Sci Technol 35:987–993. https://doi.org/10.1016/j.jmst.2018.12.025

51. Chang X, Li W, Zhu L et al (2016) Carbon-Based CsPbBr 3 perovskite solar cells: all-ambient processes and high thermal stability. ACS Appl Mater Interfaces 8:33649–33655. https://doi.org/10.1021/acsami.6b11393

52. Ito S, Mizuta G, Kanaya S et al (2016) Light stability tests of $CH_3NH_3PbI_3$ perovskite solar cells using porous carbon counter electrodes. Phys Chem Chem Phys 18:27102–27108. https://doi.org/10.1039/c6cp03388a

53. Kim S, Bae S, Lee SW, et al (2017) Relationship between ion migration and interfacial degradation of CH3NH3PbI3 perovskite solar cells under thermal conditions. Sci Rep 7. https://doi.org/10.1038/s41598-017-00866-6

54. Baranwal AK, Kanaya S, Peiris TAN et al (2016) 100 °C thermal stability of printable perovskite solar cells using porous carbon counter electrodes. Chemsuschem 9:2604–2608. https://doi.org/10.1002/cssc.201600933

55. Lee K, Kim J, Yu H et al (2018) A highly stable and efficient carbon electrode-based perovskite solar cell achieved: Via interfacial growth of 2D PEA2PbI4 perovskite. J Mater Chem A 6:24560–24568. https://doi.org/10.1039/c8ta09433k

56. Wu Z, Liu Z, Hu Z, et al (2019) Highly efficient and stable perovskite solar cells via modification of energy levels at the perovskite/carbon electrode interface. Adv Mater 31. https://doi.org/10.1002/adma.201804284

57. Zhou J, Wu J, Li N et al (2019) Efficient all-air processed mixed cation carbon-based perovskite solar cells with ultra-high stability. J Mater Chem A 7:17594–17603. https://doi.org/10.1039/c9ta05744g

58. Mali SS, Kim H, Kim HH et al (2017) Large area, waterproof, air stable and cost effective efficient perovskite solar cells through modified carbon hole extraction layer. Mater Today Chem 4:53–63. https://doi.org/10.1016/j.mtchem.2016.12.003

59. Wu X, Xie L, Lin K et al (2019) Efficient and stable carbon-based perovskite solar cells enabled by the inorganic interface of CuSCN and carbon nanotubes. J Mater Chem A 7:12236–12243. https://doi.org/10.1039/c9ta02014d

60. Pitchaiya S, Eswaramoorthy N, Natarajan M, et al (2020) Perovskite solar cells: a porous graphitic carbon based hole transporter/counter electrode material extracted from an invasive plant species Eichhornia Crassipes. Sci Rep 10. https://doi.org/10.1038/s41598-020-62900-4

61. Wei Z, Zheng X, Chen H et al (2015) A multifunctional C+ epoxy/Ag-paint cathode enables efficient and stable operation of perovskite solar cells in watery environments. J Mater Chem A 3:16430–16434. https://doi.org/10.1039/c5ta03802b

62. Zhou H, Shi Y, Dong Q et al (2014) Hole-conductor-free, metal-electrode-free TiO2/CH3NH3PbI3 heterojunction solar cells based on a low-temperature carbon electrode. J Phys Chem Lett 5:3241–3246. https://doi.org/10.1021/jz5017069

63. Bashir A, Shukla S, Lew JH et al (2018) Spinel Co3O4 nanomaterials for efficient and stable large area carbon-based printed perovskite solar cells. Nanoscale 10:2341–2350. https://doi.org/10.1039/c7nr08289d

64. He S, Qiu L, Son DY et al (2019) Carbon-based electrode engineering boosts the efficiency of all low-temperature-processed perovskite solar cells. ACS Energy Lett 4:2032–2039. https://doi.org/10.1021/acsenergylett.9b01294

65. Priyadarshi A, Haur LJ, Murray P et al (2016) A large area (70 cm2) monolithic perovskite solar module with a high efficiency and stability. Energy Environ Sci 9:3687–3692. https://doi.org/10.1039/c6ee02693a

66. Cai L, Liang L, Wu J, et al (2017) Large area perovskite solar cell module. J Semicond 38. https://doi.org/10.1088/1674-4926/38/1/014006

Emerging Carbon Nanomaterials for Organic and Perovskite-Based Optoelectronics Device Applications

Monojit Bag⊙, Ramesh Kumar, and Jitendra Kumar

Abstract Nanostructured carbon allotropes have gained much attention in the scientific community due to their vast applications in various fields. The carbon nanomaterials can be found in form of various different hybridization states and each of them having unique electronic properties. The reduced dimensionalities with sp^2 bonded graphitic carbon form a highly delocalized electronic state, suggesting their applicability as high mobility electronic materials. Additionally, the tunable optical band gap makes them promising materials for optoelectronic device applications. Recently, carbon nanomaterials such as zero-dimensional fullerenes, one-dimensional carbon nanotubes (CNTs), and two-dimensional graphene and their derivatives have been widely investigated for the latest generation photovoltaic and optoelectronic device applications because of their chemical stability, mechanical stability, and exceptional optoelectronic properties. For example, fullerene derivatives are widely used as acceptor materials for efficient bulk heterojunction solar cells because of their ability to well-diffuse into the semiconducting polymer films and to form an intermixed layer with desired morphology, while these are used as electron transport/injection layer for perovskite solar cells and light-emitting diodes. CNTs and graphene show promise as transparent conducting electrodes for flexible and lightweight organic and perovskite optoelectronic devices. In addition, CNTs are gaining popularity as active layer components of photovoltaic devices owing to their higher aspect ratio and electrical conductivity. Furthermore, graphene oxide has also been used in the active layer of organic solar cells for efficient charge separation and charge transport owing to their large acceptor/donor interface area and continuous pathway. Recently, graphene oxide (GO) is used as a buffer layer in organic and perovskite optoelectronic devices. Specifically, GO and poly(3,4-ethylenedioxythiophene) polystyrene sulfonate (PEDOT:PSS) composite has been used as hole transport layer for improving the stability and efficiency of inverted perovskite solar cells. This is due to the reduced contact barrier between perovskite active layer and charge transport layer, enhanced crystallinity of perovskite morphology, and suppressed leakage current.

M. Bag (✉) · R. Kumar · J. Kumar
Advanced Research in Electrochemical Impedance Spectroscopy, Indian Institute
of Technology Roorkee, Roorkee 247667, India
e-mail: monojit.bag@ph.iitr.ac.in

© The Author(s), under exclusive license to Springer Nature Singapore Pte Ltd. 2021   419
A. Hazra and R. Goswami (eds.), *Carbon Nanomaterial Electronics: Devices
and Applications*, Advances in Sustainability Science and Technology,
https://doi.org/10.1007/978-981-16-1052-3_17

Keywords Carbon · Nanostructured allotrope · Hybridization · Bulk heterojunction · Perovskite · Acceptor · Donor · Morphology · Light-emitting diode

1 Introduction

In the late 1970s, the discovery of electrical conductivity in organic materials has revolutionized state-of-the-art semiconductor physics. Variety of applications such as gas sensors, [9, 16, 66] field-effect transistors (FETs), [48, 54, 56] photodiodes [29, 44, 68], and photodetectors [40, 55, 70] have been realized using carbon nanomaterial in pristine form as well as in the functionalized form. In the last decade, inorganic–organic hybrid perovskite-based solar cells, light-emitting diodes, and detectors have been added to that list. Perovskites, originally discovered by Lev Alekseyevich von Perovski in 1839 are a group of materials which exhibit the chemical formula ABX_3. Here, A and B are mono and divalent cations respectively and X is a monovalent anion. Methylammonium lead iodide ($CH_3NH_3PbI_3$) is an archetypal example of hybrid perovskite, which has been extensively used for the fabrication of perovskite solar cells. Intense research has been carried out over the years to develop these high-performance optoelectronic devices which generally involve an electron transporting or injecting layer (ETL/EIL), an active layer, and a hole transporting or injecting layer (HTL/HIL). Particularly, the performance of these perovskite-based devices heavily depends on each of these charge transport layers. As perovskite materials are unstable and prone to thermal and photoinduced degradation under humid condition, high mechanical strength and chemical stability of carbon nanomaterials can be utilized to provide environmental stability to perovskites-based devices. In this chapter, we will discuss the functionalization of various layers with carbon nanomaterials in order to enhance the device performance of organic and perovskite-based optoelectronic devices. Generally, these nanomaterials are present in various forms such as nanoparticles, nano-diamonds, fibers, cones, scrolls, whiskers, graphite polyhedral crystals, and nanoporous carbon, it's quite interesting to see that the material properties at nanoscale heavily depends on their geometrical shape and size. Although a variety of structures have been discovered, we will limit our discussion to quantum dots, single and multiwall carbon nanotubes, and graphene and its derivatives. Carbon atoms can undergo sp, sp^2, or sp^3 hybridization, diamond with sp^3 and graphene with sp^2 hybridization are two widely known allotropic forms of carbon. However, carbon nanomaterials with sp^2 hybridization are widely used as electronic materials for device application. Here, two-dimensional graphene, one-dimensional carbon nanotubes (CNTs), and zero-dimensional fullerenes have been discussed briefly for their electronic properties.

Graphene is a monolayer of carbon atoms; these atoms are tightly packed into a two-dimensional (2D) honeycomb lattice. Carbon atoms in graphene are joined to each other by σ and π bonds in sp^2 hybridization with an interatomic distance of around 0.142 nm. One of the many great properties of graphene is that the charge carriers in graphene move at extremely high carrier mobility in the range of 15000–20000 $cm^2 V^{-1}S^{-1}$ at room temperature (RT) [30]. Electrons in defect-free graphene sheets can move at an extraordinary speed as there is no effective mass. Graphene is considered as the mother of various other carbon nanomaterials, as it can be wrapped into the form of quantum dots (QDs) rolled as nanotubes or stacked in three dimensions to produce graphite.

CNTs are basically graphene layers rolled in the form of cylinders with their diameters in the range of a few nanometres which is the characteristic feature of the nanotubes. CNTs have high aspect ratios in the order of 10^2–10^7 [30]. These nanotubes can have multiple layers and chirality vectors. Depending on the number of layers, CNTs can be categorized in two basic categories, namely, single-walled carbon nanotube (SWCNT) and multi-walled carbon nanotubes (MWCNT). Depending on their chirality, each of the CNT can again be grouped into two categories: semiconducting carbon nanotubes and metallic carbon nanotubes. Electronic properties of CNTs are mainly determined by their chirality. For example, charge transport properties of SWCNTs depend upon their characteristic (n, m) index, this (n, m) index is also called roll-up vectors. If n-m = 3q where q is any integer/zero, the SWCNTs are metallic. If $n-m \neq 3q$, the SWCNTs behave as a semiconductor with a finite optical bandgap. If n = m, the nanotubes are known as arm-chair. If m = 0, they are known as zig-zag; otherwise, they are known as chiral. The optical bandgap of semiconducting CNT is known to be inversely proportional to the tube diameter, the relation between its bandgap and diameter (d) is given as $E_g \sim 0.9/d$ (d in nm).

The first fullerene molecule was discovered in 1985 by Sir H. W. Kroto and R. F. Smalley. Geometrically a fullerene is a stable cluster of carbon atoms (n > 20) having a spherical surface. Fullerenes are also called buckyballs because of their shape. C_{60} is the most common among material scientists for years as it has served as electron transport or injection material in organic solar cells (OSCs), perovskite solar cells (PSCs), and perovskite light-emitting diodes (PeLEDs). The C_{60} fullerene molecule is composed of 12 pentagons and 20 hexagons of sp^2-hybridized carbon atoms. In buckyballs each carbon atom is attached to three nearest neighbors, three out of four valence electrons of atoms are involved in σ bonding. Therefore, remaining π electrons are delocalized through the structure resulting in the conducting behavior of buckyballs.

2 Carbon Nanomaterials for Organic Optoelectronic Devices

Over the last two decades, optoelectronic devices based on organic semiconductors have rapidly gained attention due to many advantages of organic semiconducting materials, such as solution-processable fabrication onto flexible substrates, low material cost, tunable material properties, great diversity, and higher production volume

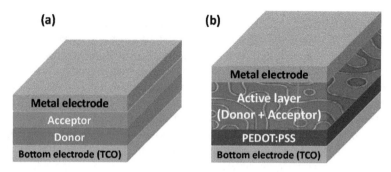

Fig. 1 Device architecture of organic solar cell **a** bilayer **b** bulk heterojunction

as compared to conventional semiconductor technology [18]. These technological advantages are attractive for flexible large-area applications such as organic solar cells (OSCs) and organic light-emitting diodes (OLEDs). However, the main obstacles of this technology are low carrier mobility and susceptibility to environmental instability, which directly impact the device performances (power conversion efficiency, external quantum efficiency, and lifetime) [23]. As a possible way to overcome these obstacles, the scientific community has begun to explore integrating organic materials with inorganic nanoparticles such as quantum dot and carbon nanocomposite material to realize hybrid materials with enhanced carrier mobility and stability while preserving the advantage of facile solution-processable fabrication.

Typical OSCs consist of a substrate normally glass or flexible substrate; transparent bottom electrode typically used transparent conducting oxide (TCO); organic active layer, and top metal electrode as shown in Fig. 1a. An interlayer is normally used between electrodes and the photoactive layer in order to impede surface recombination, which reduces the device performance. In efficient OSCs, the photoactive layer consists of electron donor and acceptor to form bulk heterojunction (BHJ) for efficient charge generation and transport (Fig. 1b). The incorporation of carbon nanomaterials (CNMs), such as fullerenes, CNTs, and graphene, is an effective way to maximize power conversion efficiency (PCE) in OSCs. CNMs (Fullerenes, CNTs, Graphene) show greater carrier mobility and conductivity than organic semiconductors [18]. The fullerene derivatives are electron acceptor materials in organic optoelectronic devices, while graphene and CNTs are used as promising transparent counter electrodes for flexible devices. In fact, poly[(5,6-difluoro-2,1,3-benzothiadiazol-4,7-diyl)-alt-(3,3'''di(2-octyldodecyl)2, 2';5',2'';5'',2'''-quaterthiophen-5,5'''-diyl)] (PffBT4T-2OD) derivatives as donors and C_{71}-fullerene derivatives as acceptors based OSCs have reported above 11% power conversion efficiency [36]. Recently, it has been shown that other electron acceptor materials can also be used to achieve high device performance comparable with fullerene derivatives-based devices. Implementation of CNMs into the device architecture not only enhanced the active layer stability and charge transport properties of the device but its optoelectronic properties are also modified due to the concentration and the shape of the carbon nanostructure. The chemical stability, mechanical flexibility, and elemental abundance of carbon nanomaterials present unique opportunities for optoelectronic devices [18]. Therefore, the carbon nanomaterials can be used to have different roles in the device architectures to improve their performance and their lifetime, as described in subsequent sections.

2.1 Carbon Nanomaterials in Photoactive Layer of Organic Solar Cells

In OSCs, photons are absorbed in the organic active layer and generate Frenkel excitons (typical binding energy ranges over 0.1–1.0 eV) [15]. These excitons diffuse into the photoactive layer and get dissociated at the donor/acceptor interface of OSCs. Therefore, separated electron and hole charge carriers are transported to the charge transport layer or electrodes via the electron donor or acceptor layer, respectively. Integration of CNMs such as CNTs and graphene into the organic photoactive layer provides exciton dissociation sites and charge carriers pathways in the active layer [23].

Conventional architecture of OSCs consists of poly-3-hexylthiophene (P3HT)/Phenyl-C_{61}-butyric acid methyl ester (PCBM) photoactive layer. Typically, PCE of these systems varies between 3 and 5% [45, 59]. Therefore, researchers have started to use dispersed CNTs or graphene as electron acceptors in P3HT and poly-3-octylthiophene (P3OT) polymers. However, the maximum PCEs reported for OSC with CNT/P3OT and RGO/P3HT as photoactive layers are 0.22 and 1.1%, respectively [5, 33]. The PCE of these systems is lower than conventional P3HT/PCBM photoactive layer-based solar cells. Many factors may impede the performance of CNTs and graphene-based OSCs, such as (1) nanophase separation between donor and acceptor within exciton diffusion length is lacking for these systems and (2) lower carrier selectivity of CNTs and graphene which leads to undesired carrier recombination during transport [23]. High fractions of CNT clusters and graphene can cause devices to short circuit easily. Moreover, several nanophase separation strategies for P3HT/PCBM-based solar cells have been reported, including solvent [1] and thermal annealing [49].

Researchers have used several strategies to overcome these difficulties, such as the small incorporation of CNTs into the BHJ OSCs. The incorporation of 3 wt% graphene nanoplate (GNP) into the BHJ OSC photoactive layer resulted in maximum PCE = 3.61%. The addition of CNTs below 1 wt% into the photoactive layer using homogeneously dispersed CNTs in a P3HT/PCBM bulk-heterojunction matrix did not affect nanophase separation [25]. For high dispersion, CNTs are functionalized by alkyl-amide groups and homogeneously mixing them with P3HT/PCBM solution as shown in Fig. 2. The device performance of this system was enhanced up to 40% due to the enhancement of carrier mobility.

Moreover, the integration of doped CNTs and graphene to the photoactive layer of BHJ OSC has also been suggested [24, 35]. The work function of CNMs can be

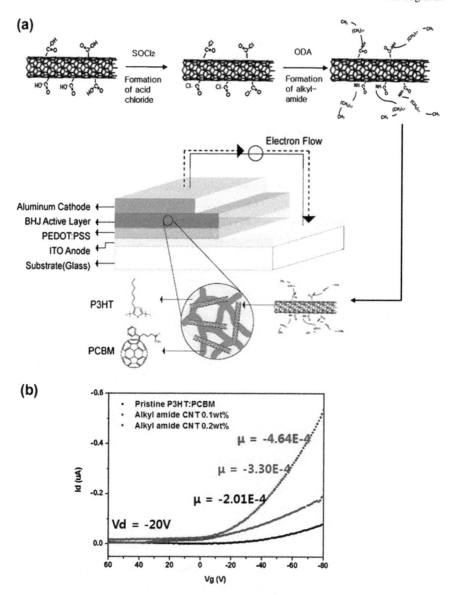

Fig. 2 **a** A schematic representation of the functionalization of CNTs for a homogenous dispersion inorganic solvent and device structure of BHJ system. **b** The transfer characteristics of an organic thin-film transistor. (Reprinted with permission from Ref. [25]. Copyright (2012) Elsevier)

modified by doping the CNMs in OSCs. Kim and co-workers have demonstrated the work function modulation of CNTs by nitrogen (N) and boron (B) doping [35]. N- or B-doped-based carbon nanotubes (CNTs) uniformly dispersed in the photoactive layer of P3HT/PCMB BHJ systems selectively enhanced charge carrier transport due to work function modulation of CNTs. The addition of 1.0 wt% B- and N-doped CNTs into the active layer results in PCEs of OSCs are 4.1% and 3.7%, respectively. These values are higher than undoped CNT-based BHJ OSCs (2.6%) due to balanced electron and hole transport. Moreover, the maximum PCE of OSCs (8.6%) was achieved by integrating of N-CNTs in the PTB7: $PC_{71}BM$ BHJ system [43].

Jun et al. also reported enhanced PCE in BHJ systems by integrating charge selective nitrogen-doped reduced graphene oxide (N-RGO) layer into the organic photoactive layer [24]. For N-RGO process, graphene oxide (GO) was reduced through thermal treatment with ammonia gas (Fig. 3a). N-doped graphene provides electrons charge transport pathways through the modified band structure of composite material as shown in Fig. 3c. The PCE of OSCs with N-RGO shown around 40% enhancement compared to an OSC without N-RGO.

However, Gong et al. have utilized CNTs as an electron donor instead of an acceptor in the photoactive layer of $CNT/PC_{71}BM$ BHJ system [11]. The PCE of this system was reported over 2.5–3.1% by using a mixture of $PC_{71}BM$ and SWNTs.

In OSCs, graphene quantum dots (GQDs) have also emerged as a potential candidate for better performance. Owing to the quantum confinement effect, GQDs show remarkable semiconducting properties, such as (1) tuneable energy band gap by tailoring size, edge, and functional groups, (2) high absorption coefficient, (3) nontoxicity, (4) good solubility, (5) GQDs facilitate the formation of charge pathways in OSC photoactive layers without recombination and (6) effective exciton dissociation due to smaller size in photoactive layers. Gupta et al. have demonstrated that the GQDs incorporated in conjugated polymer show enhanced OPV and OLED

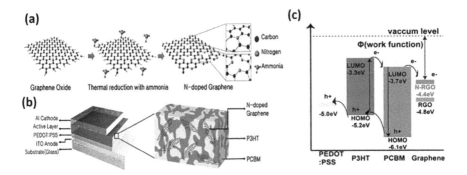

Fig. 3 a Illustration of the nitrogen doping process of RGO **b** BHJ solar cell architecture using the N-doped graphene/P3HT:PCBM active layer and **c** Energy band diagram of a RGO and N-RGO/P3HT:PCBM OSC system. (Reprinted with permission from Ref. [24]. Copyright (2013) Energy Environ. Sci.)

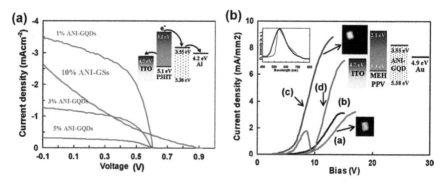

Fig. 4 **a** J–V characteristics of the OSC based on functionalized GQDs with aniline (ANI-GQDs) in AM 1.5G 100 mW illumination. **b** J–V characteristics of GQDs based OLED. (Reprinted with permission from Ref. [13]. Copyright (2011) J. Am. Chem. Soc.)

characteristics as compared to graphene due to improved optical and morphological characteristics as shown in Fig. 4 [13]. Moreover, Hong's group incorporated three different types of GQDs—graphene oxide quantum dots (GOQDs), 5 h reduced GQDs, and 10 h reduced GQDs into BHJ systems, and enhanced optical absorptivity and charge carrier extraction (Fig. 5) [28]. Therefore, the partially reduced GQDs yielded remarkable improved PCE of 7.6% in BHJ solar cells owing to balancing optical absorptivity and electrical conductivity.

2.2 Carbon Nanomaterials in Transport Layer

OSCs consist of either an electron or hole blocking layer, or both, depending on the OSC architecture. In a regular OSC architecture, an electron blocking (hole transport) layer such as poly(3, 4 ethylenedioxythiophene):poly(styrene sulfonate) (PEDOT:PSS) is used to transport holes from the active layer to the transparent anode and match the Fermi level of the highest occupied molecular orbital (HOMO) energy level of photoactive layer to anode. In an inverted OSC geometry, a hole blocking (electron transport) layer such as metal oxide (ZNO, TiO_2) is used for electron transport layer and blocking of holes from the active layer to the transparent cathode. In OSCs architecture, CNMs (CNTs, graphene) can be used as an interlayer between the polymer and the transparent electrode to avoid energy level mismatch. In OSCs, CNTs and graphene have been used as electron blocking (hole transport) layer between polymer active layer and anode. For example, Ozkans group have demonstrated novel processing techniques for the addition of SWCNTs at different locations in the P3HT/PCBM BHJ systems and investigated their photophysical properties and device performance [8]. Incorporation of SWCNTs into PEDOT:PSS layer coated on ITO leads to enhanced PCE with 4.9%. However, SWCNTs between a photoactive layer and the metal electrode resulted in a drastic drop in device performance due

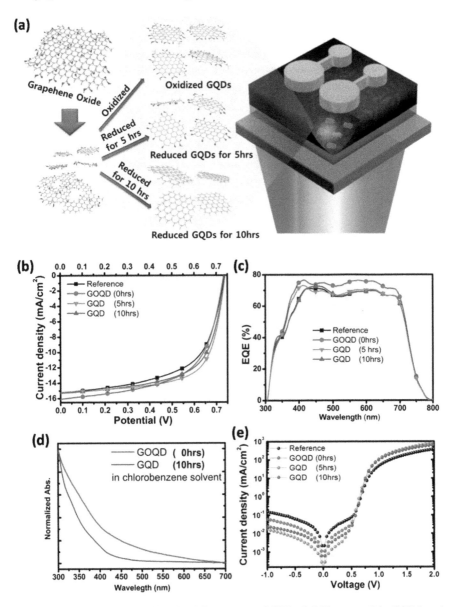

Fig. 5 **a** Schematic of a BHJ OSC with different types of GQDs. **b** J–V curves of the GQD based BHJ devices. **c** Incident photon to charge carrier efficiency (IPCE) of the BHJ systems. **d** UV–visible adsorption spectra of GOQD and GQD. **e** Dark J–V characteristic of the BHJ systems. (Reprinted with permission from Ref. [28]. Copyright (2013) ACS Nano)

to energy level mismatch. Jin et al. reported that CNT/PEDOT:PSS nanocomposites synthesized with non-covalently functionalized CNTs exhibited higher conductivity than that of PEDOT:PSS polymer and similar optical transparency [22]. Owing to higher electrical conductivity, BHJ solar cells with CNT/PEDOT:PSS nanocomposites as a hole transport layer shows about 30% enhanced PCE compared to BHJ OSCs made with PEDOT:PSS polymer (Fig. 6).

Graphene has also been employed as hole transport materials in the BHJ OSCs to enhance chemical stability and electrical conductivity [41, 63]. Li et al. used

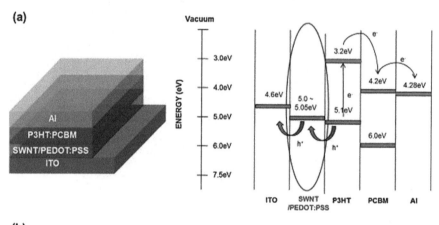

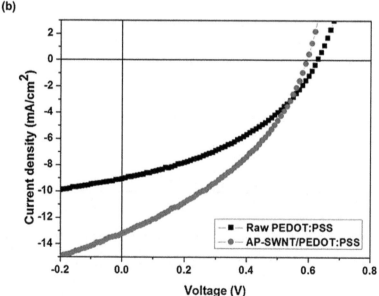

Fig. 6 a Illustration of device structure and energy level diagram of BHJ solar cell with SWNTs/PEDOT:PSS nanocomposites as a hole transport layer **b** J–V characteristics of BHJ OSCs. (Reprinted with permission from Ref. [22]. Copyright (2013) Synth. Met.)

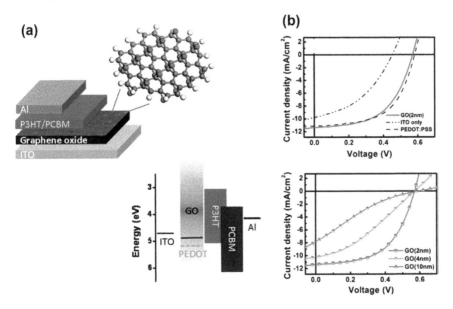

Fig. 7 **a** Schematic of the BHJ solar cell and energy level diagram with GO as hole transport and electron blocking layer. **b** Current density–voltage characteristics of different photovoltaic devices. (Reprinted with permission from Ref. [39]. Copyright (2010) ACS Nano)

graphene oxide (GO) as the hole transport and electron blocking material in OSCs [39]. The addition of GO thin film between the P3HT/PCBM photoactive layer and the transparent anode (ITO) to transport hole and block electrons from photoactive layer to ITO anode as shown in Fig. 7. This BHJ OSC exhibited a PCE of around 3.5%. However, the device performance gradually decreased with increases in the thickness of GO layer due to the insulating properties of GOs (Fig. 7b). These results indicate that GO materials could be alternative to conventional PEDOT:PSS polymer as the effective hole transport layer for OSCs and OLEDs. Several groups have reported graphene/polymer nanocomposites as a hole transport and electron blocking layer (HTL) in the BHJ OSCs [2]. For example, Bae et al. demonstrated a blend of poly(styrene sulfonic acid) grafted with polyaniline (PSSA−g−PANI) and graphene oxide (GO) as a hole transport material for the P3HT/PCBM BHJ system [2]. PSSA−g−PANI/GO nanocomposites show high electrical conductivity and optical transparency. This BHJ OSC exhibits an enhanced PEC of about 23% compared to conventional HTL material, PEDOT:PSS. Moreover, graphene/polymer nanocomposites show higher electrical conductivity and optical transparency than conventional PEDOT:PSS. They can be an effective hole transport layer for high-performance BHJ OSCs. However, Liming Dai group has reversed the charge extraction property of GO through charge neutralization of the –COOH groups of GO with Cs_2CO_3 in BHJ devices [42]. Therefore, GO can be used as an excellent hole and an electron transport layer in the same BHJ system.

3 Applications of Carbon Nanomaterials in Perovskite-Based Devices

Hybrid organic-inorganic perovskites (HOIP) possess some exceptional properties which are generally absent in inorganic or organic semiconductors. HOIPs show exceptional defect tolerance, ambipolar charge transport, and high photoluminescence quantum yield. These materials are solution processible, and the efficiency of the perovskite-based solar cell (PSC) is now comparable to that of the crystalline silicon solar cells. Conventionally a PSC as well as perovskite light-emitting diode (PeLED) is a multilayer device architecture. The PSC is composed of an ETL an active layer and an HTL, these layers are sandwiched between two electrodes namely cathode and anode. Depending on their layer construction PSCs or PeLEDs can be categorized as n-type-intrinsic-p-type (n-i-p) or p-type-intrinsic-n-type (p-i-n) devices. Here, n-type material is usually a high electron affinity material, whereas a p-type material is a low electron affinity material normally used to match the energy band diagram of the device. In a complete device, an n-type material provides low conduction band offset whereas it has to have a high valence band offset so that it blocks holes efficiently, therefore in some cases an electron transport layer is also called a hole blocking layer. Similarly, a hole transport layer has a low valence band offset with the active layer, whereas it has a high conduction band offset, see Fig. 8b. The active layer of PSCs or PeLED is usually composed of a pristine halide perovskite. HOIPs have some excellent properties to be utilized as solar cells or as PeLEDs; these properties include high tolerance factor, [31] large carrier diffusion lengths, [7], and ambipolar charge transport [38, 53]. Although perovskites show high efficiency, it is also prone to degradation when these materials come in contact with oxygen, moisture, or UV irradiation. CNMs can be utilized to further improve their electrical and optical properties by functionalization of these materials with a suitable form of carbon nanostructures. In this section, various studies on the use of

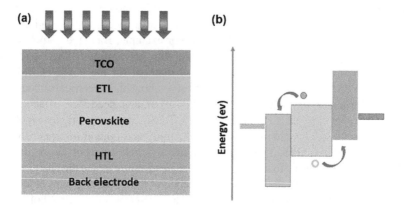

Fig. 8 **a** Schematic diagram of a perovskite solar cell **b** energy band diagram of a typical perovskite solar cell

carbon nanomaterials to boost perovskite-based device performance as well as the stability will be demonstrated.

3.1 Graphene and Its Derivatives in Perovskite Solar Cells

Currently best performing PSCs usually utilize a p-type organic hole conductor, 2,2',7,-7'-tetrakis (N,N-di-p-methoxyphenylamine)-9,9'-spirobifluorene (spiro-OMeTAD) as HTL. The low intrinsic carrier mobility of charge transport layers becomes an issue for improving device performance. Although, doping can be utilized to overcome the problem of low carrier mobility to some extent, the presence of hygroscopic dopants can accelerate the deterioration of the device at high temperatures. Hydrophobic graphene derivatives can not only improve the device stability over moisture degradation but can also improve carrier mobility. Graphene derivatives can be utilized in band alignment not just in hole transporting material but also in inverted perovskite solar cells, these graphene derivatives can be utilized to improve the optoelectronic properties of electron transporting layers. Graphene derivatives can be added into charge transport layers or directly a thin layer of graphene derivatives can be placed between the active layer and the charge transport layer. Graphene was first introduced in perovskite solar cells by Wang and co-workers as an electron transport layer. Wang et al. developed a low temperature processed composite of pristine graphene nanoflakes and anatase TiO_2 nanoparticles and used it as an electron transport layer in PSC [60], as shown in Fig. 9.

The application of graphene nanoflakes with TiO_2 eliminates the need of high temperature (~500 °C) sintering which opens up the possibility of PSCs utilizing TiO_2 as an electron transport material to be processed at temperatures not more than 150 °C. This demonstration opens up a new avenue for low cost and flexible devices. Graphene-TiO_2 composite can improve device performance by reducing series resistance and recombination losses. Apart from pristine graphene it's derivatives such as graphene oxide (GO) and reduced graphene oxide (rGO) can be used as a charge extraction layer. Mahdi, et al. used reduced graphene oxide as the interlayer between perovskite and TiO_2 film. Using time-resolved photoluminescence (TRPL) and electrochemical impedance spectroscopy (EIS) they have illustrated that rGO improves electron extraction from perovskite by reducing charge transfer resistance [58]. Recently it has been found that when sulfonic acid functionalized graphene oxide (SrGO) is used as a hole interfacial layer in a p-i-n structured PSC, device performance improves significantly. Mann and co-workers showed that using double HTL (PEDOT:PSS/SrGO) effectively increases the work function of HTL layer and thus improves charge transport capability that resulted a higher short circuit current (J_{sc}), higher fill factor and improved power conversion efficiency (PCE). Double HTL (PEDOT:PSS/SrGO)-based devices also show improved stability compared to single HTL (PEDOT:PSS only) device, as shown in Fig. 10 [47].

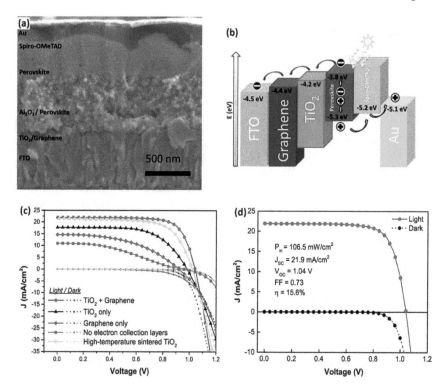

Fig. 9 **a** Cross-sectional FESEM image of a complete perovskite solar cell. **b** Schematic illustration of energy levels of different materials used and shown in (**a**). **c** Current–voltage characteristics of different electron collection layers under simulated AM 1.5, 100 mW/cm² solar irradiation (solid line), and in the dark (dotted line). **d** The best performing ($\eta = 15.6\%$) solar cell based on a graphene-TiO₂ nanocomposite under simulated AM 1.5, 106.5 mW/cm² solar irradiation (solid line), and in the dark (dotted line), which processed at temperatures not exceeding 150 °C. (Reprinted with permission from Ref. [60]. Copyright (2013) Nano Lett.)

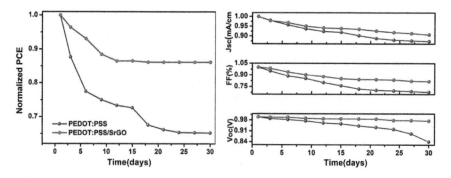

Fig. 10 Normalized photovoltaic parameters as the function of time for the PSCs with different HTLs in air ambient conditions. (Reprinted with permission from Ref. [47]. Copyright (2020) J. Alloys Compd.)

3.2 Graphene and Its Derivatives in Perovskite Light Emitting Diode

Graphene as anode in PeLEDs not only provides alternative to TCO but also opens up a route towards flexible LEDs. Two major issues using ITO electrodes are that they are brittle and therefore unsuitable for flexible electronics and indium ions may diffuse into the transport layer and degrade them. PSC that utilize graphene electrodes are also found to have comparable efficiency to that of the ITO-based devices [64], but the application of graphene electrodes in light emitting diodes is not much reported. High sheet resistance, low work function and surface hydrophobicity are main problem in application of graphene in producing high efficiency PeLEDs, although doping of suitable material can improve graphene properties to become compatible for application in PeLEDs. To overcome the problem of graphene hydrophobicity, Qing et al. used a modified PEDOT:PSS layer on top of graphene, in this case PEDOT:PSS is doped with small-molecule non-ionic surfactant Triton X-100 and Dimethyl sulfoxide (DMSO). Incorporation of Triton X-100 and DMSO improved the surface wettability of graphene, and improved the surface adhesion energy [34]. Suitable doping of Triton X-100 and DMSO reduces the work function of PEDOT:PSS from 4.81 eV to 4.56 eV, thus using graphene electrode along with modified PEDOT:PSS and perovskite quantum dots, Qing et al. were able to fabricate PeLEDs with current conversion efficiency 11.37 cd/A and 2.58% external quantum efficiency (EQE), see Fig. 11. However this efficiency is quite low compared to ITO-based devices but this could be a good step towards flexible device, and these LEDs can be utilized in low intensity applications [67].

In another study, Hong-Kyu et al. reported a very bright electroluminescence (EL) ($L_{max} > 10\ 000$ cd m^{-2}) and conversion efficiency ($CE_{max} = 18.0$ cd A^{-1} EQE 3.8%) based on a graphene anode for the first time. Although at that time device efficiencies were not that high and they showed that the graphene anode-based device in this study outperformed the ITO-based perovskite LED ($CE_{max} = 10.0$ cd A^{-1} EQE 2.2%). In this study, they used a 4 layer chemically doped graphene which was doped using Nitric acid (HNO_3) vapor. The improved performance in this case can be attributed to the chemical stability of graphene. Using time-of-flight secondary ion mass spectrometry (TOF-SIMS) they have shown that in ITO-based devices In and Sn atoms diffuse from ITO to upper layers. These metal ions may form interfacial trap states and reduce the hole injection. Therefore, using graphene anode in place of ITO avoids this ion diffusion problem and improves device performance [51].

3.3 CNTs in Perovskite Solar Cells

Recently a two-layer (two hole transport layers) strategy was employed by Severin et al. where they utilize P3HT wrapping to disperse SWCNT in commonly used solvents, this polymer coating to SWCNT makes electron transfer unfavorable and

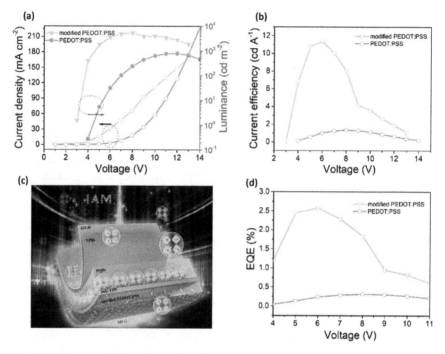

Fig. 11 a Current density and luminance curves, and **b** current efficiency characteristics of graphene-based $FA_{0.8}Cs_{0.2}PbBr_3$ NCs LEDs with PEDOT:PSS and PEDOT:PSS:Triton X-100(0.05 wt%):DMSO (2 vol.%). **c** Schematic Diagram of MAPbBr$_3$ PQD-based LEDs fabricated on graphene. **d** EQE curves of graphene-based $FA_{0.8}Cs_{0.2}PbBr_3$ NCs LEDs. (Reprinted with permission from Ref. [67]. Copyright (2019) Org. Electron.)

leads to selective hole transfer to nanotubes [10]. They have observed that incorporation of SWNT interlayer exhibited close to 1 order of magnitude faster charge carrier extraction. They also observed that incorporation of SWNT does not change the recombination rate significantly compared to undoped hole transport which indicates the enhancement in device performance is mainly because of improved hole extraction throughout the HTL [14]. In another study, Rachelle et al. used nearly monochiral (5, 6) s-SWNT ultra-thin (5 nm) interfacial layer between active perovskite layer and hole transport layer, that resulted an improved charge extraction and slow recombination, therefore overall device efficiency improved from 14.7 to 16.5% [17]. One of the critical issues of PSC is the hysteresis of I-V curves, [17, 19, 27, 50, 58] and its one the major concern among the scientific community, since it is related to PCE and stability of device [26]. Hysteresis factor of a solar cell is defined as

$$\text{Hysteresis factor} = \frac{\text{PCE}(reverse) - \text{PCE}(forword)}{\text{PCE}(reverse)} \quad (1)$$

Huijie et al. used a combination of ultraviolet photoelectron spectroscopy (UPS), space-charge-limited current (SCLC) and EIS measurements to show that a highly efficient and hysteresis free PSC can be realized by using a CNT functionalized tin oxide (SnO_2) as an electron selective layer. The hysteresis factor for the devices with pure SnO_2 and CNT functionalized SnO_2 have shown a hysteresis factor of 14.26% and 2.42%, respectively. They have shown that using CNT functionalized SnO_2 has lower trap density, lower charge transfer resistance, higher conductivity and higher electron mobilities compared to pristine SnO_2 electron transporting layer, therefore using CNT functionalized SnO_2 ultimately enhances the PCE from 17.90 to 20.33% [57]. There are efforts to incorporate CNTs into the perovskites also, though CNTs are hydrophobic and do not dissolve in polar solvents such as Dimethylformamide (DMF) or DMSO which are generally used to dissolve perovskite precursors. Bag et al. used a supramolecular strategy that uses π-π interactions to non-covalently attach the chain end of a polymer to MWCNT to make a stable (stable up to 30 days) solution of MWCNTs [21]. They utilized, poly(methyl methacrylate) (PMMA)-functionalized hexabenzocoronene (HBC) to achieve stable CNT dispersion in DMF. They showed that with the incorporation of MWCNTs in active layers of a PSC they achieved up to 87% reduced charge recombination and 70 meV improved open circuit voltage while keeping J_{sc} and fill factor intact, see Fig. 12 [3].

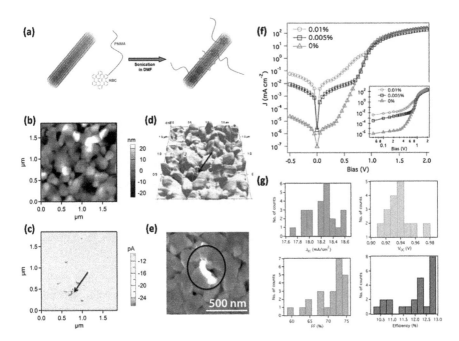

Fig. 12 **a** schematic of HBC-functionalized MWCNTs **b** AFM, **c** and **d** conductive AFM **e** FESEM image of MWCNT in perovskite thin film **f** light current-voltage graph of perovskite solar cells containing MWCNTs **g** histogram of device parameters. (Reprinted with permission from Ref. [3]. Copyright (2016) Chem. Phys. Lett.)

3.4   Carbon Quantum Dots in Perovskite Solar Cells

C_{60} has maintained its best performance in organic and perovskite solar cells for years, C_{60} and its derivatives such as PCBM have resulted some of the best devices. Zhiping, et al. demonstrated a 21.1% PCE and showed that their encapsulated devices maintained 90% of their original PCE after 150 h of operation under 10 suns [61]. It is also seen that the incorporation of PCBM either by mixing it with the bulk or inserting a layer into the device significantly reduces the hysteresis factor. It could be related to enhanced electron transfer [65] or it could be related to passivation of charge trap states in the bulk of perovskite film during thermal annealing process [52]. Yu, et al. used PCBM and its polymerized form (PPCBM) to investigate the influence of PCBM in resolving hysteresis problem. Its well known that ion migration in perovskites is responsible for hysteresis in perovskite-based devices, using in situ wide-field photoluminescence (PL) imaging microscopy and X-ray photoemission spectroscopy (XPS) depth profiling Yu and coworkers showed that the diffusion of PCBM into the bulk perovskite reduces ion migration and helps to suppress hysteresis, they proposed that its possible that diffused PCBM may tie up with migrating iodine ions by direct electron transfer from iodine to PCBM, alternatively PCBM acts as a physical hindrance for ion migration [69]. Apart from this, carbon quantum dots (CQDs) can directly be used in the bulk of perovskite to act as a defect passivating agent. Yuhui, et al. used hydroxyl and carbonyl functional groups containing CQDs into a perovskite precursor solution to passivate the uncoordinated lead ions on grain boundaries. Introducing CQDs in perovskite precursor solution both PCE as well as stability of PSC devices was significantly improved. This improved device performance was attributed to the presence of larger grain size which is basically the result of retarded growth rate. Interaction between Pb^{2+} and carbonyl groups attached to CQDs retarded the growth rate of perovskite grains, (Fig. 13). CQD employed devices retained 73.4% of its initial PCE after being aged for 48 h under 80% humidity at room temperature [46].

3.5   Carbon Quantum Dots in Perovskite LEDs

The structure of a PeLED is exactly same as that of perovskite solar cell; only difference is in their working conditions. An efficient perovskite solar cell requires efficient charge extraction at the perovskite/charge transport layer whereas an efficient perovskite light emitting diode requires balanced charge injection at the perovskite/charge transport layer [32]. Although there are a number of studies incorporating graphene in PSCs there are only fewer reports which focus on the use of graphene-based materials in PeLEDs. Xi et al. found that the incorporation of suitable amount of GQDs into methylammonium lead tribromide ($CH_3NH_3PbBr_3$) films resulted in reduced grain size with much compact and passivated grain boundaries, which gives rise to 2.7-fold increased PL intensity. In this study PLQY was improved

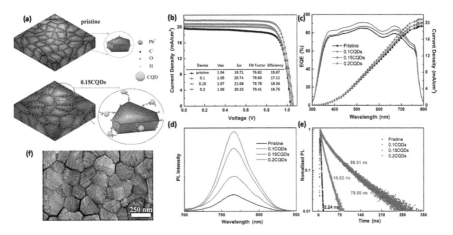

Fig. 13 **a** Schematic diagram of the mechanism of CQD passivation. Pb^{2+}, C, O, H, and CQDs are illustrated in orange, black, red, green, and cyan, respectively. **b** J–V characteristics of champion cells without and with various CQD additive perovskites as the absorbers. **c** EQE and integrated current intensities of all the champion devices. **d** PL spectra and **e** TRPL spectra of devices with different amounts of CQD additive perovskites as the absorbers. **f** SEM image of perovskite films containing CQD. (Reprinted with permission from Ref. [46]. Copyright (2019) ACS Appl. Mater. Interfaces)

from 2.5 to 9.6% which indicates that incorporation of GQDs can help us to prepare perovskite films suitable for light emitting diodes, where a small grain size is desirable to improve the radiative recombination [12]. In a conventional PeLED, PEDOT:PSS is most commonly used HTL. Alternatively, Zhibin et al. used CQDs as HTL and found that CQDs in PeLED can replace the highly acidic PEDOT:PSS and facilitate efficient hole injection. They have fabricated $CsPbBr_3$-based green PeLED exhibiting 13.8% EQE and long term operational stability. Using conventional HTL such as PEDOT:PSS, Polyvinylcarbazole (PVK) or Poly[(9,9-dioctylfluorenyl-2,7-diyl)-co-(4,4′-(N-(4-sec-butylphenyl)diphenylamine)] (TFB) usually require an ultrathin buffer layer to control the surface energy in order to fabricate good quality surface passivated active layer, however precise control over thickness of these ultrathin layers is difficult, on the other hand CQDs could be attached with various functional groups such as -NH$_2$, -COOH or -OH and therefore by controlling the type and amount of these functional groups one can tune the surface energies and also the band energies of CQD-based HTLs [62].

3.6 Carbon Nanomaterials in Photodetectors

Although graphene has a number of excellent properties, but when comes to light absorption properties graphene shows very weak light absorption [4] which is the main reason of low quantum yield for graphene only devices. On the other hand, perovskite has good absorption coefficient but low carrier mobility compared to graphene and CNTs. Therefore, its good to explore hybrid perovskite/graphene or perovskite/CNT photodetectors, there are a number of studies based on perovskite/graphene or perovskite/CNT-based detectors, few of them are discussed in this chapter. Surendran et al. reported $CsPbBr_xI_{3-x}$ nanocrystals (NCs) with higher iodide-to-bromide ratio as the photo absorption layer and graphene as the transport layer, perovskite nanocrystal − graphene-based photodetector exhibits maximum responsivity of 1.13×10^4 A/W and specific detectivity of 1.17×10^{11} Jones under low light intensity ($\sim$80 μW/cm^2). Surendran and co-workers also found that under low intensity of light a common issue of ion separation and segregated halide domains in mixed halide is absent. This indicates that the mixed halide NCs could be potential candidates for photodetection applications, when a low power sensor is required [55]. Zou et al. demonstrated a high photoresponsivity and high photo detectivity by fabricating a perovskite ($CH_3NH_3PbBr_3$)/graphene hybrid vertical photodetector. Zou and co-workers utilized a mono-layer graphene and a perovskite single crystal to realize 1017.1 A/W photoresponsivity and a photo detectivity of 2.02×10^{13} Jones. This high performance of hybrid perovskite/graphene vertical photodetector was attributed to effective charge transfer between perovskite graphene heterojunction [70]. It is well established that the presence of a large number of grain boundaries trigger the degradation in perovskites [6, 37], therefore low-dimensional perovskites such as quasi 2D, 2D or 1D perovskite structure show improved stability compared to conventional 3D perovskites. Shun-Xin et al. fabricated an ultra-stable MAPbBr$_3$ single-crystal microwire arrays (SCMWAs)-based photodetector by utilizing protective hydrophobic Trichloro(1H,1H,2H,2Hperfluorooctyl) silane (FOTS) molecular layer to isolate SCMAs. These SCMAs enjoy the protection of FOTS as well as being low dimensional they do have reduced density of grain boundaries therefore these SCMWAs are less prone to degradation. Shun-Xin et al. demonstrated a high-performance flexible photodetector with responsivity of 20 A/W and detectivity of 4.1×10^{11} Jones which is capable of maintaining 96% of the initial photocurrent after 1 year while exposed to air [40].

3.7   Carbon Nanomaterials as Electrodes

Generally, perovskite photovoltaic devices require metal electrode of suitable work function to collect charge carriers, and these metal electrodes have to be deposited at high vacuum to avoid oxidation of metal. Alternatively, carbon-based materials such as nanotubes can be used as back electrodes to make a fully solution processible photovoltaic devices. Although using CNTs or graphene electrodes in the place of metal electrode is expected to reduce the device performance unless or until these are well engineered, but at the same time they can reduce the cost and effort significantly. The reduction in PCE of solar cells can be understood as there are two main reasons responsible for this (1) finite transparency of back electrodes reduces the number of reabsorbed photons, (2) limited conductivity of CNT hinders charge collection.

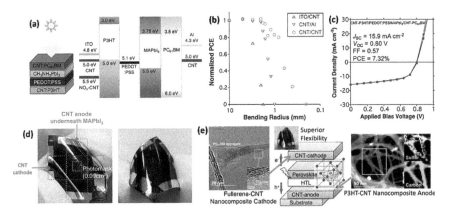

Fig. 14 **a** Schematic illustrations of a both-CNT-electrode PSC and its energy level diagram. **b** Bending radius tests of the CNT cathode devices (ITO/CNT), the CNT anode devices (CNT/Al), and the both CNT-electrode devices. **c** Forward (red) and reverse (blue) bias J–V curves with photovoltaic parameters **d** Photographs of the both-CNT-electrode PSC. **e** SEM image of nanocomposite anode and cathode materials with device structure. (Reprinted with permission from Ref. [20]. Copyright (2017) J. Phys. Chem. C)

However, Xin, et al. used multi walled carbon nanotubes as back contact in PSCs, as a result they have been able to achieve PCE over 17% with improved stability. Optimized device in their study maintains 80% of its initial power conversion efficiency up to 1000 h of continuous illumination at 1 sun [64]. Similarly, Jeon et al. used double walled carbon nanotubes (DWCNT) layer as the front electrode in PSCs and achieved a hysteresis free and 17.2% PCE. They also demonstrated the effect of doping in CNTs, the controlled device in their study has shown a PCE of 15.6% which can be improved to 16.7 and 17.2% upon HNO_3 and trifluoromethanesulfonic acid (TFMS) doping respectively; the improvement in PCE was attributed to deeper fermi level of doped DWCNT compared to the fermi level of pristine CNT films [46]. Jean et al. also demonstrated a comparable device performance by using CNTs both as anode and cathode, which enables us to fabricate PSCs using entirely solution processible layers with a minor loss of PCE, see Fig. 14. A cost analysis show that using both anode and cathode made of CNTs, this method reduces the fabrication cost to 33% of that of conventional devices [20].

4 Conclusion

Carbon nanomaterials are widely used in optoelectronic devices including organic and hybrid perovskite-based solar cells, light-emitting diodes, and detectors. There has been a plethora of research to improve device performance and stability upon utilizing carbon nanomaterials either in the active layer as nanocomposite or at the semiconductor/electrode interfaces. Fullerene and its derivatives are widely used as acceptor type materials for high-efficiency organic bulk heterojunction solar cells. They are also very good electron transport layers for perovskite solar cells and light-emitting diodes. However, the major problem to control the bulk heterojunction morphology using Fullerene derivatives at a large scale remains elusive in organic solar cells. Carbon nanostructures especially single wall and multiwall carbon nanotubes and reduced graphene oxides are widely used as conducting transparent electrode materials for flexible electronics. This can overcome the shortcoming of transparent conducting oxide-based substrates having limited conductivity as well as mechanical and environmental instability. Carbon nanomaterials-based conducting transparent electrodes can improve the environmental stability by increasing the chemical stability, decreasing oxygen and moisture diffusion through electrode as well as improving flexibility of the transparent substrates. Very recently, all carbon-based optoelectronic devices are being fabricated using various solution processing techniques which can reduce not only the fabrication cost but also improve environmental safety and stability. Carbon nanomaterials can open up a new era of technology in optoelectronic industries to complement existing Si-technology. The major challenges still remaining in organic and perovskite-based optoelectronic devices are the overall devices' stability and reproducibility compared to the state-of-the-art Si-technology. Therefore, much more research and understanding are still needed to address the instability and irreproducibility in this new technology.

References

1. Al-Ibrahim M, Ambacher O, Sensfuss S, Gobsch G (2005) Effects of solvent and annealing on the improved performance of solar cells based on poly(3-hexylthiophene): Fullerene. Appl Phys Lett 86:1–3. https://doi.org/10.1063/1.1929875
2. Bae S, Lee JU, Park HS et al (2014) Enhanced performance of polymer solar cells with PSSA-g-PANI/graphene oxide composite as hole transport layer. Sol Energy Mater Sol Cells 130:599–604. https://doi.org/10.1016/j.solmat.2014.08.006
3. Bag M, Renna LA, Jeong SP et al (2016) Evidence for reduced charge recombination in carbon nanotube/perovskite-based active layers. Chem Phys Lett 662:35–41. https://doi.org/10.1016/j.cplett.2016.09.004
4. Bera KP, Haider G, Huang YT et al (2019) Graphene sandwich stable perovskite quantum-dot light-emissive ultrasensitive and ultrafast broadband vertical phototransistors. ACS Nano 13:12540–12552. https://doi.org/10.1021/acsnano.9b03165
5. Berson S, De Bettignies R, Bailly S et al (2007) Elaboration of P3HT/CNT/PCBM composites for organic photovoltaic cells. Adv Funct Mater 17:3363–3370. https://doi.org/10.1002/adfm.200700438
6. Boyd CC, Cheacharoen R, Leijtens T, McGehee MD (2019) Understanding degradation mechanisms and improving stability of perovskite photovoltaics. Chem Rev 119:3418–3451. https://doi.org/10.1021/acs.chemrev.8b00336
7. Brenes R, Guo D, Osherov A et al (2017) Metal halide perovskite polycrystalline films exhibiting properties of single crystals. Joule 1:155–167. https://doi.org/10.1016/j.joule.2017.08.006
8. Chaudhary S, Lu H, Müller AM et al (2007) Hierarchical placement and associated optoelectronic impact of carbon nanotubes in polymer-fullerene solar cells. Nano Lett 7:1973–1979. https://doi.org/10.1021/nl0707171

9. Suehiro J, Zhou G, Hara M (2003) Fabrication of a carbon nanotube-based gas sensor using dielectrophoresis and its application for ammonia detection by impedance spectroscopy. J Phys D Appl Phys 36:L109–L114. https://doi.org/10.1088/0022-3727/36/21/L01

10. Dissanayake NM, Zhong Z (2011) Unexpected hole transfer leads to high efficiency single-walled carbon nanotube hybrid photovoltaic. Nano Lett 11:286–290. https://doi.org/10.1021/nl103879b

11. Gong M, Shastry TA, Xie Y et al (2014) Polychiral semiconducting carbon nanotube-fullerene solar cells. Nano Lett 14:5308–5314. https://doi.org/10.1021/nl5027452

12. Guo X, Han B, Gao Y, et al (2020) Enhanced emission from $CH^3 NH_3 PbBr_3$ perovskite films by graphene quantum dot modification. Mater Res Express 7:016415. https://doi.org/10.1088/2053-1591/ab61a6

13. Gupta V, Chaudhary N, Srivastava R et al (2011) Luminscent graphene quantum dots for organic photovoltaic devices. J Am Chem Soc 133:9960–9963. https://doi.org/10.1021/ja2036749

14. Habisreutinger SN, Leijtens T, Eperon GE et al (2014) Enhanced hole extraction in perovskite solar cells through carbon nanotubes. J Phys Chem Lett 5:4207–4212. https://doi.org/10.1021/jz5021795

15. Halls JJM, Pichler K, Friend RH et al (1996) Exciton diffusion and dissociation in a poly(p-phenylenevinylene)/C 60 heterojunction photovoltaic cell articles you may be interested in. Appl Phys Lett 68:3120. https://doi.org/10.1063/1.115797

16. Huang CS, Huang BR, Jang YH et al (2005) Three-terminal CNTs gas sensor for N2 detection. Diam Relat Mater 14:1872–1875. https://doi.org/10.1016/j.diamond.2005.09.006

17. Ihly R, Dowgiallo AM, Yang M et al (2016) Efficient charge extraction and slow recombination in organic-inorganic perovskites capped with semiconducting single-walled carbon nanotubes. Energy Environ Sci 9:1439–1449. https://doi.org/10.1039/c5ee03806e

18. Jariwala D, Sangwan VK, Lauhon LJ et al (2013) Carbon nanomaterials for electronics, optoelectronics, photovoltaics, and sensing. Chem Soc Rev 42:2824–2860

19. Jeon NJ, Noh JH, Kim YC et al (2014) Solvent engineering for high-performance inorganic-organic hybrid perovskite solar cells. Nat Mater 13:897–903. https://doi.org/10.1038/nmat4014

20. Jeon I, Seo S, Sato Y et al (2017) Perovskite solar cells using carbon nanotubes both as cathode and as anode. J Phys Chem C 121:25743–25749. https://doi.org/10.1021/acs.jpcc.7b10334

21. Jeong SP, Boyle CJ, Venkataraman D (2016) Poly(methyl methacrylate) end-functionalized with hexabenzocoronene as an effective dispersant for multi-walled carbon nanotubes. RSC Adv 6:6107–6110. https://doi.org/10.1039/c5ra19883f

22. Jin SH, Il Cha S, Jun GH et al (2013) Non-covalently functionalized single walled carbon nanotube/poly(3, 4ethylenedioxythiophene):poly(styrenesulfonate) nanocomposites for organic photovoltaic cell. Synth Met 181:92–97. https://doi.org/10.1016/j.synthmet.2013.08.018

23. Jin S, Jun GH, Jeon S, Hong SH (2016) Design and application of carbon nanomaterials for photoactive and charge transport layers in organic solar cells. Nano Converg 3:8. https://doi.org/10.1186/s40580-016-0068-8

24. Jun GH, Jin SH, Lee B et al (2013) Enhanced conduction and charge-selectivity by N-doped graphene flakes in the active layer of bulk-heterojunction organic solar cells. Energy Environ Sci 6:3000–3006. https://doi.org/10.1039/c3ee40963e

25. Jun GH, Jin SH, Park SH, et al (2012) Highly dispersed carbon nanotubes in organic media for polymer: fullerene photovoltaic devices. In: Carbon, pp 40–46

26. Kang DH, Park NG (2019) On the current-voltage hysteresis in perovskite solar cells: dependence on perovskite composition and methods to remove hysteresis. Adv Mater 31:1–23. https://doi.org/10.1002/adma.201805214

27. Kim HS, Jang IH, Ahn N et al (2015) Control of I-V hysteresis in CH3NH3PbI3 perovskite solar cell. J Phys Chem Lett 6:4633–4639. https://doi.org/10.1021/acs.jpclett.5b02273

28. Kim JK, Park MJ, Kim SJ et al (2013) Balancing light absorptivity and carrier conductivity of graphene quantum dots for high-efficiency bulk heterojunction solar cells. ACS Nano 7:7207–7212. https://doi.org/10.1021/nn402606v

29. Kirezli B, Gucuyener I, Kara A et al (2019) Electrical and optical properties of photodiode structures formed by surface polymerization of P(Egdma-Vpca)-Swcnt films on n-si. J Mol Struct 1198: https://doi.org/10.1016/j.molstruc.2019.126879
30. Kour R, Arya S, Young S-J et al (2020) Review—recent advances in carbon nanomaterials as electrochemical biosensors. J Electrochem Soc 167: https://doi.org/10.1149/1945-7111/ab6bc4
31. Kovalenko MV, Protesescu L, Bodnarchuk MI (2017) Properties and potential optoelectronic applications of lead halide perovskite nanocrystals. Science (80-) 358:745–750. https://doi.org/10.1126/science.aam7093
32. Kumar R, Kumar J, Srivastava P et al (2020) Unveiling the morphology effect on the negative capacitance and large ideality factor in perovskite light-emitting diodes. ACS Appl Mater Interfaces 12:34265–34273. https://doi.org/10.1021/acsami.0c04489
33. Kymakis E, Amaratunga GAJ (2002) Single-wall carbon nanotube/conjugated polymer photovoltaic devices. Appl Phys Lett 80:112–114. https://doi.org/10.1063/1.1428416
34. Lee I, Kim GW, Yang M, Kim TS (2016) Simultaneously enhancing the cohesion and electrical conductivity of PEDOT:PSS conductive polymer films using DMSO additives. ACS Appl Mater Interfaces 8:302–310. https://doi.org/10.1021/acsami.5b08753
35. Lee JM, Park JS, Lee SH et al (2011) Selective electron- or hole-transport enhancement in bulk-heterojunction organic solar cells with N- or B-doped carbon nanotubes. Adv Mater 23:629–633. https://doi.org/10.1002/adma.201003296
36. Li W, Cai J, Cai F et al (2018) Achieving over 11% power conversion efficiency in PffBT4T-2OD-based ternary polymer solar cells with enhanced open-circuit-voltage and suppressed charge recombination. Nano Energy 44:155–163. https://doi.org/10.1016/j.nanoen.2017.12.005
37. Li X, Ibrahim Dar M, Yi C et al (2015) Improved performance and stability of perovskite solar cells by crystal crosslinking with alkylphosphonic acid ω-ammonium chlorides. Nat Chem 7:703–711. https://doi.org/10.1038/nchem.2324
38. Li F, Ma C, Wang H et al (2015) Ambipolar solution-processed hybrid perovskite phototransistors. Nat Commun 6:1–8. https://doi.org/10.1038/ncomms9238
39. Li SS, Tu KH, Lin CC et al (2010) Solution-processable graphene oxide as an efficient hole transport layer in polymer solar cells. ACS Nano 4:3169–3174. https://doi.org/10.1021/nn100551j
40. Li SX, Xu YS, Li CL et al (2020) Perovskite single-crystal microwire-array photodetectors with performance stability beyond 1 year. Adv Mater 2001998:1–10. https://doi.org/10.1002/adma.202001998
41. Liu Y, Summers MA, Edder C et al (2005) Using resonance energy transfer to improve exciton harvesting in organic-inorganic hybrid photovoltaic cells. Adv Mater 17:2960–2964. https://doi.org/10.1002/adma.200501307
42. Liu J, Xue Y, Gao Y et al (2012) Hole and electron extraction layers based on graphene oxide derivatives for high-performance bulk heterojunction solar cells. Adv Mater 24:2228–2233. https://doi.org/10.1002/adma.201104945
43. Lu L, Xu T, Chen W et al (2013) The role of N-doped multiwall carbon nanotubes in achieving highly efficient polymer bulk heterojunction solar cells. Nano Lett 13:2365–2369. https://doi.org/10.1021/nl304533j
44. Ma Z, Han J, Yao S et al (2019) Improving the performance and uniformity of carbon-nanotube-network-based photodiodes via yttrium oxide coating and decoating. ACS Appl Mater Interfaces 11:11736–11742. https://doi.org/10.1021/acsami.8b21325
45. Ma W, Yang C, Gong X et al (2005) Thermally stable, efficient polymer solar cells with nanoscale control of the interpenetrating network morphology. Adv Funct Mater 15:1617–1622. https://doi.org/10.1002/adfm.200500211
46. Ma Y, Zhang H, Zhang Y et al (2019) Enhancing the performance of inverted perovskite solar cells via grain boundary passivation with carbon quantum dots. ACS Appl Mater Interfaces 11:3044–3052. https://doi.org/10.1021/acsami.8b18867
47. Mann DS, Seo YH, Kwon SN, Na SI (2020) Efficient and stable planar perovskite solar cells with a PEDOT:PSS/SrGO hole interfacial layer. J Alloys Compd 812: https://doi.org/10.1016/j.jallcom.2019.152091

48. Novoselov KS, Geim AK, Morozov SV, Jiang D, Zhang Y, Dubonos SV, Grigorieva IV, Firsov AA (2016) Electric field effect in atomically thin carbon films. 306:666–669

49. Padinger F, Rittberger RS, Sariciftci NS (2003) Effects of postproduction treatment on plastic solar cells. Adv Funct Mater 13:85–88. https://doi.org/10.1002/adfm.200390011

50. Prochowicz D, Tavakoli MM, Solanki A et al (2018) Understanding the effect of chlorobenzene and isopropanol anti-solvent treatments on the recombination and interfacial charge accumulation in efficient planar perovskite solar cells. J Mater Chem A 6:14307–14314. https://doi.org/10.1039/c8ta03782e

51. Seo HK, Kim H, Lee J, et al (2017) Efficient flexible organic/inorganic hybrid perovskite light-emitting diodes based on graphene anode. Adv Mater 29. https://doi.org/10.1002/adma.201605587

52. Shao Y, Xiao Z, Bi C et al (2014) Origin and elimination of photocurrent hysteresis by fullerene passivation in CH3NH3PbI3 planar heterojunction solar cells. Nat Commun 5:1–7. https://doi.org/10.1038/ncomms6784

53. Srivastava P, Bag M (2020) Elucidating tuneable ambipolar charge transport and field induced bleaching at CH3NH3PbI3/electrolyte interface. Phys Chem Chem Phys 22:11062. https://doi.org/10.1039/d0cp00682c

54. Star A, Han T-R, Joshi V, Stetter JR (2004) Sensing with Nafion coated carbon nanotube field-effect transistors. Electroanalysis 16:108–112. https://doi.org/10.1002/elan.200302925

55. Surendran A, Yu X, Begum R et al (2019) All inorganic mixed halide perovskite nanocrystal-graphene hybrid photodetector: from ultrahigh gain to photostability. ACS Appl Mater Interfaces 11:27064–27072. https://doi.org/10.1021/acsami.9b06416

56. Szafranek BN, Fiori G, Schall D et al (2012) Current saturation and voltage gain in bilayer graphene field effect transistors. Nano Lett 12:1324–1328. https://doi.org/10.1021/nl2038634

57. Tang H, Cao Q, He Z et al (2020) SnO2–carbon nanotubes hybrid electron transport layer for efficient and hysteresis-free planar perovskite solar cells. Sol RRL 4:1–8. https://doi.org/10.1002/solr.201900415

58. Tavakoli MM, Tavakoli R, Hasanzadeh S, Mirfasih MH (2016) Interface engineering of perovskite solar cell using a reduced-graphene scaffold. J Phys Chem C 120:19531–19536. https://doi.org/10.1021/acs.jpcc.6b05667

59. Troshin PA, Hoppe H, Renz J et al (2009) Material solubility-photovoltaic performance relationship in the design of novel fullerene derivatives for bulk heterojunction solar cells. Adv Funct Mater 19:779–788. https://doi.org/10.1002/adfm.200801189

60. Wang JTW, Ball JM, Barea EM et al (2014) Low-temperature processed electron collection layers of graphene/TiO 2 nanocomposites in thin film perovskite solar cells. Nano Lett 14:724–730. https://doi.org/10.1021/nl403997a

61. Wang Z, Lin Q, Wenger B et al (2018) High irradiance performance of metal halide perovskites for concentrator photovoltaics. Nat Energy 3:855–861. https://doi.org/10.1038/s41560-018-0220-2

62. Wang Z, Yuan F, Sun W et al (2019) Multifunctional p-type carbon quantum dots: a novel hole injection layer for high-performance perovskite light-emitting diodes with significantly enhanced stability. Adv Opt Mater 7:1–9. https://doi.org/10.1002/adom.201901299

63. White MS, Olson DC, Shaheen SE et al (2006) Inverted bulk-heterojunction organic photovoltaic device using a solution-derived ZnO underlayer. Appl Phys Lett 89: https://doi.org/10.1063/1.2359579

64. Wu X, Xie L, Lin K et al (2019) Efficient and stable carbon-based perovskite solar cells enabled by the inorganic interface of CuSCN and carbon nanotubes. J Mater Chem A 7:12236–12243. https://doi.org/10.1039/c9ta02014d

65. Xing G, Wu B, Chen S et al (2015) Interfacial electron transfer barrier at compact TiO2/CH3NH3PbI3 heterojunction. Small 11:3606–3613. https://doi.org/10.1002/smll.201403719

66. Yeow JTW, Wang Y (2009) A review of carbon nanotubes-based gas sensors. J Sens 2009. https://doi.org/10.1155/2009/493904

67. Zhang Q, Lu Y, Liu Z et al (2019) Highly efficient organic-inorganic hybrid perovskite quantum dot/nanocrystal light-emitting diodes using graphene electrode and modified PEDOT:PSS. Org Electron 72:30–38. https://doi.org/10.1016/j.orgel.2019.05.046
68. Zhang J, Xi N, Lai KWC, et al (2007) Single carbon nanotube based photodiodes for infrared detection. In: 2007 7th IEEE International Conference Nanotechnol-IEEE-NANO 2007, Proceedings, vol 1, pp 1156–1160. https://doi.org/10.1109/NANO.2007.4601388
69. Zhong Y, Hufnagel M, Thelakkat M et al (2020) Role of PCBM in the suppression of hysteresis in perovskite solar cells. Adv Funct Mater. https://doi.org/10.1002/adfm.201908920
70. Zou Y, Zou T, Zhao C et al (2020) A highly sensitive single crystal perovskite-graphene hybrid vertical photodetector. Small 2000733:1–9. https://doi.org/10.1002/smll.202000733

Lightning Source UK Ltd.
Milton Keynes UK
UKHW020809260522
403559UK00002B/22